IFIP Advances in Information and Communication Technology

470

Editor-in-Chief

Kai Rannenberg, Goethe University Frankfurt, Germany

IFIP – The International Federation for Information Processing

IFIP was founded in 1960 under the auspices of UNESCO, following the first World Computer Congress held in Paris the previous year. A federation for societies working in information processing, IFIP's aim is two-fold: to support information processing in the countries of its members and to encourage technology transfer to developing nations. As its mission statement clearly states:

> *IFIP is the global non-profit federation of societies of ICT professionals that aims at achieving a worldwide professional and socially responsible development and application of information and communication technologies.*

IFIP is a non-profit-making organization, run almost solely by 2500 volunteers. It operates through a number of technical committees and working groups, which organize events and publications. IFIP's events range from large international open conferences to working conferences and local seminars.

The flagship event is the IFIP World Computer Congress, at which both invited and contributed papers are presented. Contributed papers are rigorously refereed and the rejection rate is high.

As with the Congress, participation in the open conferences is open to all and papers may be invited or submitted. Again, submitted papers are stringently refereed.

The working conferences are structured differently. They are usually run by a working group and attendance is generally smaller and occasionally by invitation only. Their purpose is to create an atmosphere conducive to innovation and development. Refereeing is also rigorous and papers are subjected to extensive group discussion.

Publications arising from IFIP events vary. The papers presented at the IFIP World Computer Congress and at open conferences are published as conference proceedings, while the results of the working conferences are often published as collections of selected and edited papers.

IFIP distinguishes three types of institutional membership: Country Representative Members, Members at Large, and Associate Members. The type of organization that can apply for membership is a wide variety and includes national or international societies of individual computer scientists/ICT professionals, associations or federations of such societies, government institutions/government related organizations, national or international research institutes or consortia, universities, academies of sciences, companies, national or international associations or federations of companies.

More information about this series at http://www.springer.com/series/6102

Luis M. Camarinha-Matos · António J. Falcão
Nazanin Vafaei · Shirin Najdi (Eds.)

Technological Innovation for Cyber-Physical Systems

7th IFIP WG 5.5/SOCOLNET Advanced Doctoral
Conference on Computing,
Electrical and Industrial Systems, DoCEIS 2016
Costa de Caparica, Portugal, April 11–13, 2016
Proceedings

 Springer

Editors
Luis M. Camarinha-Matos
NOVA University of Lisbon
Monte da Caparica
Portugal

Nazanin Vafaei
NOVA University of Lisbon
Monte da Caparica
Portugal

António J. Falcão
NOVA University of Lisbon
Monte da Caparica
Portugal

Shirin Najdi
NOVA University of Lisbon
Monte da Caparica
Portugal

ISSN 1868-4238   ISSN 1868-422X (electronic)
IFIP Advances in Information and Communication Technology
ISBN 978-3-319-80979-3   ISBN 978-3-319-31165-4 (eBook)
DOI 10.1007/978-3-319-31165-4

Printed on acid-free paper

This Springer imprint is published by Springer Nature
The registered company is Springer International Publishing AG Switzerland

Preface

This 2016 edition of the DoCEIS proceedings book presents a series of selected articles produced in the context of engineering doctoral programs. The theme is on "Technological Innovation for Cyber-Physical Systems" and contributions reflect the growing interests in research, development, and application of cyber-physical systems. Fast progress on embedded intelligence and interconnection of systems, enabled by advances in pervasive computing, sensing technologies, and computer networks, including developments in the Internet of Things and cloud computing, have led to new architectural approaches to systems engineering. By exploring the synergies between computational and physical components, these systems leverage the emerging "network effect" and induce new advanced applications.

Potential benefits can be found in all engineering fields and at all levels, e.g., supporting systems-of-systems, facilitating the industrial Internet and sensing enterprise, enabling effective smart energy grids, creating the basis for smart environments, etc. This approach can change the way engineering systems are designed while leading to exciting challenges for researchers and industrial practitioners.

DoCEIS is an international forum providing a platform for the presentation of research results from PhD work, and a space for the discussion of post-graduation studies, PhD thesis plans, and practical aspects of PhD work in these inter-related areas of engineering, while promoting a strong multi-disciplinary dialog. As such, participants were challenged to look beyond their specific research question and relate their work to the selected theme of the conference, namely, to identify in which ways their research topics can benefit from, or contribute to, cyber-physical-based solutions.

A basis for innovation nowadays is to embrace the application of multi-disciplinary and interdisciplinary approaches in the context of research. In fact, more and more funding agencies are including this element as a key requirement in their calls for proposals. As such, the challenge put forward by DoCEIS to its authors can be seen as a contribution to the process of acquiring such skills, which are mandatory in the profession of a PhD.

This seventh edition of DoCEIS, which is sponsored by SOCOLNET, IFIP, and IEEE IES, attracted a considerable number of paper submissions from a large number of PhD students and their supervisors from 24 countries. This book comprises the works selected by the international Program Committee for inclusion in the main program and covers a wide spectrum of application domains. Research results and on-going work are presented, illustrated, and discussed in areas such as:

- Enterprise collaborative networks
- Ontologies
- Petri nets
- Manufacturing systems
- Biomedical applications
- Intelligent environments

- Control and fault tolerance
- Optimization and decision support
- Wireless technologies
- Energy: smart grids, renewables, management, and optimization
- Bio-energy
- Electronics

As anticipated, and confirmed by the submissions, it is shown that virtually any research topic in this broad engineering area can either benefit from a cyber-physical systems perspective, or be a direct contributor with models, approaches, and technologies for further development of such systems.

We expect that this book will provide readers with an inspiring set of promising ideas and new challenges, presented in a multi-disciplinary context, and that by their diversity these results can trigger and motivate richer research and development directions.

We would like to thank all the authors for their contributions. We also appreciate the efforts and dedication of the DoCEIS international Program Committee members, who both helped with the selection of articles and contributed with valuable comments to improve their quality.

February 2016   Luis M. Camarinha-Matos
António J. Falcão
Nazanin Vafaei
Shirin Najdi

Organization

7th IFIP/SOCOLNET Advanced Doctoral Conference on COMPUTING, ELECTRICAL AND INDUSTRIAL SYSTEMS

Costa de Caparica, Portugal, April 11–13, 2016

Conference and Program Chair

Luis M. Camarinha-Matos, Portugal

Organizing Committee Co-chairs

Luis Gomes, Portugal
João Goes, Portugal
Pedro Pereira, Portugal

International Program Committee

Alan Jovic, Croatia
Alok Choudhary, UK
Américo Azevedo, Portugal
Andrea Bottino, Italy
Angel Ortiz, Spain
Antoni Grau, Spain
Antonio Maña, Spain
Antonios Tsourdos, UK
Armando Pires, Portugal
Barbora Buhnova, Czech Republic
Carlos Eduardo Pereira, Brazil
David Hutchison, UK
Diego Gachet, Spain
Dimitris Mourtzis, Greece
Enrico Vicario, Italy
Enrique Romero, Spain
Erik Bruun, Denmark
Ezio Bartocci, Austria
Fausto P. Garcia, Spain

Florin G. Filip – Romania
Ghazanfar Safdar, UK
Giuseppe Buja, Italy
Gordana Ostojic, Republic of Serbia
Hans-Jörg Kreowski, Germany
Horacio Neto, Portugal
Ip-Shing Fan, UK
João Goes, Portugal
João Martins, Portugal
João Paulo Pimentão, Portugal
Jorge Dias, Portugal
Jose de la Rosa, Spain
José Igreja, Portugal
José M. Fonseca, Portugal
Juan Jose Rodriguez Andina, Spain
Klaus-Dieter Thoben, Germany
Kleanthis Thramboulidis, Greece
Laura Carnevali, Italy
Luigi Piegari, Italy

Luis Bernardo, Portugal
Luis Gomes, Portugal
Luis M. Camarinha-Matos, Portugal
Luis Oliveira, Portugal
Manuela Vieira, Portugal
Marcin Paprzycki, Poland
Maria Helena Fino, Portugal
Marko Beko, Portugal
Martin Törngren, Sweden
Michael Huebner, Germany
Nik Bessis, UK
Noelia Correia, Portugal
Nuno Paulino, Portugal
Olga Battaia, France
Paulo Miyagi, Brazil
Paulo Pinto, Portugal
Pavel Vrba, Czech Republic
Pedro Pereira, Portugal
Peter Marwedel, Germany
Peter Palensky, Austria

Pierluigi Siano, Italy
Ratko Magjarevic, Croatia
Ricardo Jardim-Gonçalves, Portugal
Ricardo Rabelo, Brazil
Rita Ribeiro, Portugal
Roberto Canonico, Italy
Rolf Drechsler, Germany
Rui Aguiar, Portugal
Rui Melicio, Portugal
Simon Pietro, Italy
Stefano Di Carlo, Italy
Sven-Volker Rehm, Germany
Thilo Sauter, Austria
Thomas Strasser, Austria
Vasos Vassiliou, Cyprus
Vedran Bilas, Croatia
Willy Picard, Poland
Wojciech Cellary, Poland
Zoran Bosnic, Slovenia
Zoran Hadzi Velkov, Macedonia

Organizing Committee (PhD Students)

Ana Paula Correia, Portugal
André Lourenço, Portugal
António Falcão, Portugal
Artem Nazarenko, Ukraine
Bruno Augusti Mozzaquatro, Brazil
Esmaeil Kondori, Iran
Francisco Xavier Fonseca, Portugal

Miguel Fernandes, Portugal
Nazanin Vafei, Iran
Paulo Figueiras, Portugal
Pedro Oliveira, Portugal
Raquel Melo, Portugal
Shirin Najdi, Iran
Vagner Schaefer, Brazil

Technical Sponsors

  Society of Collaborative Networks

IFIP WG 5.5 COVE
Co-Operation infrastructure for Virtual Enterprises
and electronic business

   IEEE−Industrial Electronics Society

Organizational Sponsors

Organized by: PhD Program on Electrical and Computer Engineering FCT-UNL.

Contents

Manufacturing Systems

Biomedical Applications

Intelligent Environments

Energy Management

Optimization in Energy Management

Bio-energy

Flexible and Transparent Oxide Electronics

Collaborative Networks
and Marketplaces

A Decision-Support Tool to Deal with the Strategies Alignment Process in Collaborative Networks

Beatriz Andres[1(✉)], Raul Poler[1], Joao Rosas[2,3], and Luis Camarinha-Matos[2,3]

[1] Research Centre on Production Management and Engineering (CIGIP),
Universitat Politècnica de València (UPV), Calle Alarcón, 03801 Alcoy, Spain
{bandres,rpoler}@cigip.upv.es
[2] Faculty of Sciences and Technology, New University of Lisbon, Lisbon, Portugal
{jrosas,cam}@uninova.pt
[3] Centre of Technology and Systems, Uninova Institute, Caparica, Portugal

Abstract. The alignment of strategies among the enterprises that belong to collaborative networks is of increasing importance due to the influence on the networks operation success in the long term. This paper proposes a Decision-Support Tool for Strategies Alignment (DST-SA) to support SMEs in the selection of strategies that allow higher levels of alignment amongst all the strategies formulated by each partner. The DST-SA includes a mathematical model, a simulation software, and a programmed application, to address the strategies alignment process from a collaborative perspective. The result of the DST-SA is the identification of the strategies that are aligned and the proper time to activate these strategies with the main aim of obtaining higher levels of network performance.

Keywords: Collaborative networks · Strategies alignment · Decision-support tool

1 Formulation of the Research Question and Its Motivation

Small and Medium Enterprises (SMEs) are currently more conscious about the benefits of participating in Collaborative Networks (CN), due to higher levels of competitiveness, agility, responsiveness, and adaptability that are acquired through collaboration. These characteristics allow them to rapidly face the market evolutions and the dynamicity induced by the globalization process. On the other hand, the participation of SMEs in CN has a number of associated challenges, which derive from the lack of resources and capabilities, and the limitations associated to cultural barriers. Generally speaking, European SMEs do not currently have access to advanced collaborative decision-support tools due to their limited resources.

In general, a CN consists of heterogeneous and autonomous partners [1], each one defining its own objectives and formulating its own business strategies. In this context, a wide variety of strategies can be formulated by each of the enterprises that belong to the CN, with the aim of reaching their defined objectives. Therefore, contradictions between the strategies formulated by one enterprise and the objectives defined by another

© IFIP International Federation for Information Processing 2016
Published by Springer International Publishing Switzerland 2016. All Rights Reserved
L. Camarinha-Matos et al. (Eds.): DoCEIS 2016, IFIP AICT 470, pp. 3–10, 2016.
DOI: 10.1007/978-3-319-31165-4_1

enterprise of the network could appear, resulting in potential conflicts and selfish behaviours. Lets describe an intuitive example considering two enterprises that acquire the role of manufacturer and distributor. Each one defines two objectives and two strategies. The manufacturer, Obj_{1Mnf}: Reduce the production cost by 10 %, Obj_{2Mnf}: Reduce fluctuations in production, Str_{1Mnf}: Use lower quality packaging, Str_{2Mnf}: Establish a collaborative production planning; and the Distributor: Obj_{1Dis}: Increase the net demand by 10 % in an exclusive market, Obj_{2Dis}: Sell all the stock next to expire, Str_{1Dis}: Promote the image of an exclusive product, Str_{2Dis}: Acquire a decision support system in the forecast demand process. Using a logical reasoning, a misalignment is observed when activating Str_{1Mnf} due to it has negative influences with the Obj_{1Dis}, whilst Str_{2Mnf} positively influences all the objectives defined by the distributor. This is an intuitive example but when a large amount of enterprises face the decision-making of selecting business strategies the problem increases in difficulty to be solved, specially when considering the CN context.

It is therefore important to study the strategies alignment process, and provide collaborative enterprises with support to proper selection of strategies in order to avoid the activation of misaligned strategies among the CN partners. The main aim of such process is to reduce the lack of coherence and concordance in the pool of strategies activated by each of the enterprises belonging to the CN. The activation of a proper combination of well-aligned strategies reduces the emergence of conflicts within the CN, ensuring its sustainability and convenient operation in the long term.

Despite the importance of the concept of alignment in the CN operation, there is a gap in the literature of adequate methods to formally represent and solve the strategies alignment process. More specifically, there is a lack of approaches that consider (i) the whole set of strategies formulated by the network partners, and (ii) the influence that each of these strategies exert on the wide diversity of objectives defined by each enterprise. Besides this, and to the best of our knowledge, it does not exist any decision support tool that guides the selection of strategies based on the alignment characteristic. Motivated by this situation, the following research questions are addressed:

1. *What would be an adequate decision support tool to guide enterprises in the selection of strategies?*
2. *Amongst all strategies formulated by each enterprise belonging to a CN, which strategies should be selected for achieving high levels of alignment?*

The strategies alignment process involves modelling the impacts that such strategies, once activated, have on the objectives [2]. The heterogeneity and autonomy that characterises the CN enterprises and the multiplicity of the information exchanges related with the formulated strategies and the defined objectives, requires new approaches to cope with the complexity of the process. Consequently, SMEs need a decision support tool to help them achieving a global view of the strategies and objectives with the aim to respond faster and more efficiently to potential contradictions and misalignments. As such, a Decision-Support Tool is proposed to deal with the Strategies Alignment process in CN (DST-SA). DST-SA provides an affordable tool to SMEs with the aim of helping them to reach high levels of alignment among the activated strategies, thus contributing to increase their competitiveness in the global economy.

The remaining of this paper includes identifying in which ways the research in business strategies alignment can benefit from cyber-physical systems (Sect. 2). In Sect. 3, a brief state of the art on the strategies alignment process is provided. The research contribution and innovation, regarding the proposed DST-SA tool, is described in Sect. 4. Finally, in Sect. 5 the conclusions of the work, some discussion, and further research lines are introduced.

2 Cyber Physical Systems

The integration of computing, communication, and control technologies has led to developments in intelligent sensors and a great level of integration between the software and the physical world [3]. The term Cyber Physical Systems (CPS) was devised by Lee [4] motivated by the relevance of the interactions between computation and physical worlds, encouraging real-time information extraction, transmission and analysis for intelligent monitoring, decision-making and control [3]. In this context, CPS, and more specifically the Internet of Things (IoT), are opening a new dimension for innovation, higher levels of integration in enterprise networks, and consequently challenging their business strategies. With CPS, enterprises have much wider and fast access to real-world information, contributing to the notion of "sensing enterprise". This allows enterprises to anticipate future decisions by using multidimensional data, captured through physical and virtual objects, and generating added value information [5]. Enriching enterprises' awareness through intelligent, interconnected, and interoperable smart components and devices, empowers enterprises systems, making them responsive in real time to events related with business strategies.

In the context of decision making for the selection of aligned business strategies, the access to relevant information, among the collaborative partners, is a decisive issue. The use of devices embedded in the enterprises' business environment provides access to larger amounts of information in the decision-making process. Furthermore, access to real time information allows enterprises to be more agile when they need to make adjustments in business strategies, in case a misalignment appears.

The consideration of CPS requires new approaches to strategies alignment. On one hand they need to be developed taking into account this "new referential"; and on the other hand additional requirements and complexity related with the decision making for the selection of strategies, considering a collaborative perspective, increase the need to make the right decisions (high quality decisions) for the whole of the CN partners. As such, there is a need for (i) proper modelling of the relevant variables contained in the strategies alignment process, as well as their interdependencies and influences, and (ii) a decision support tool to help in identifying and assessing strategies alignment process.

3 State of the Art

Different models, guidelines, methods, and tools have been proposed in the literature to address the business strategies alignment process. Nevertheless, the conducted analysis revealed that the contributions, provided so far, refer to the alignment of particular network strategies, such as supply strategies [6, 7], marketing strategies [8], or product design strategies [9]. Despite the importance of the alignment process, namely in terms of avoiding partnership conflicts, to the best of our knowledge, there is a lack of approaches modelling the alignment of the different strategies formulated by individual and heterogeneous enterprises belonging to the CN. Therefore, a holistic approach that allows modelling the influences that all strategies formulated by all partners have on the wide diversity of defined objectives is needed. In order to fill this gap, an integrated approach to holistically model the strategies alignment process has been previously proposed [2, 10, 11]. The strategies alignment model (SAM) [10] deals with the strategies and objectives regardless of their nature and type, considering a CN context. Taking this baseline, a novel decision-support tool is proposed in this paper, which aims at supporting enterprises in the automatic construction of the SAM and allow SMEs to succeed in the establishment of sustainable long-term collaborative relationships using the business strategies alignment mechanism.

4   Research Contribution and Innovation in DST-SA

The research contribution of this paper extends previous work of Andres and Poler [2, 10–12] in which a model to represent the influences between the defined business objectives and the strategies formulated by each of the networked partners, as well as an associated method, the System Dynamics (SD) [13], are proposed to solve the strategies alignment process. The main aim is to identify the set of strategies to activate, while guaranteeing their alignment. Hereafter, and in order to facilitate the understanding of DST-SA, these works are briefly described. The Strategies Alignment Model (SAM) proposes a mathematical representation of the influences and relations between either the improvement or worsening of KPIs when a set of strategies are activated. Potential misalignments among the strategies are identified by the SAM implementation. Thus, its application leads to obtain the set of aligned strategies that positively influence the majority of the objectives defined by the networked partners. The optimization function of SAM focuses on the network level, so that the identified strategies maximise the positive influences (or minimize the negative influences) on the defined KPIs. The decision variables used in SAM for the maximisation function are (i) the number of units of the strategy (u_str_{is}) to be activated, and (ii) the time when the strategies have to be activated (ti_str_{is}). The SD [13] is the method selected to solve the SAM, as it allows representing the causal influences between the strategies and the objectives achievement (through KPIs), within the complex system formed by the enterprises of a CN. The SD method classifies the parameters of SAM using the stock variables, flow variables, and auxiliary variables [11].

The Decision-Making Tool for the Strategies Alignment (DST-SA) is proposed for solving SAM through a SD simulation approach. The DST-SA is based on a performance measurement schema that allows estimating, from a quantitative perspective, the value influences that an activated strategy has on the KPIs used to measure each objective defined by each CN enterprise. A simulation tool is thus included in DST-SA to automatically solve the proposed model, assessing and supporting the strategies alignment process. *AnyLogic* simulation software [14] was selected as a support component of DST-SA, since it allows representing, in SD, the elements of SAM. The optimisation package included in the simulation software allows to automatically solve SAM and obtain the solution of the decision variables $(u\_str_{is}, ti\_str_{is})$ that maximize the network's performance [11]. Moreover, DST-SA includes a component to automatically generate the model introduced in [2, 12] for the strategies alignment process under a SD approach.

The manual modelling of the SAM [2, 10], under SD simulation [11] could be feasible with a small number of enterprises. However, when the modeller faces a network with a high number of enterprises, each one defining several KPIs and formulating a high number of strategies, the amount of parameters, auxiliary variables, flow variables, and stock variables defined in SAM tend to increase exponentially. This results in the increase of the size of the problem to be modelled, which becomes difficult to handle manually. In order to avoid such tedious task, DST-SA includes a component that automatically generates the SAM for the simulation software: the Strategies Alignment model GENerator (SAGEN). DST-SA also includes a database with all data required for building the SAM in SAGEN and feed the SAM to the SD simulation software.

For the design of SAGEN component, the authors assume that models built in AnyLogic SD simulation software have the property of being read in XML language (Extensible Markup Language) with a specific schema. SAGEN allows building an XML file, containing the structured schema of SAM in SD notation, representing the enterprises, objectives, strategies and the influences between them. In order to reproduce the SAM specific schema in SD, the structure of an XML file created by *AnyLogic* is analysed. An example of the XML notation used to represent the Flow and Stock variables in SAM is depicted in Table 1.

Table 1. XML Schemas to define Flow and Stock Variables

```
<Variables>                                          <Variables>
  <Variable Class="Flow">                              <Variable Class="StockVariable">
    <Id>1390479682183</Id>                               <Id>1390479604249</Id>
    <Name>                                               <Name>
      <![CDATA[Name_of_the_Flow_Variable]]>               <![CDATA[Name_of_the_StockVariable]]>
    </Name>                                              </Name>
    <X>120</X><Y>60</Y>                                  <X>110</X><Y>60</Y>
    <Label><X>-45</X><Y>-20</Y></Label>                  <Label><X>0</X><Y>-20</Y></Label>
    <PublicFlag>false</PublicFlag>                       <PublicFlag>false</PublicFlag>
    <PresentationFlag>true</PresentationFlag>            <PresentationFlag>true</PresentationFlag>
    <ShowLabel>true</ShowLabel>                          <ShowLabel>true</ShowLabel>
    <Properties External="false" Constant="false"       <Properties Array="false">
    Array="false">                                        <EquationStyle>classic</EquationStyle>
      <Formula>                                           <Width>20</Width>
        <![CDATA[Formula_of_the_Flow_Variable]]>          <Height>20</Height>
      </Formula>                                          <InitialValue>
      <Color/>                                              <![CDATA[Initial_Value_of_StockVariable]]>
      <ValveIndex>1</ValveIndex>                          </InitialValue>
      <Points>                                            <Color/>
        <Point><X>0</X><Y>0</Y></Point>                 </Properties>
        <Point><X>100</X><Y>0</Y></Point>             </Variable>
        <Point><X>200</X><Y>0</Y></Point>           </Variables>
      </Points>
    </Properties>
  </Variable>
</Variables>
```

From a technical point of view, Lazarus [15] has been used as an Integrated Development Environment (IDE) for the SAGEN development. Lazarus is an open source and free alternative to Delphi, developed as open source project from Free Pascal. Therefore, Pascal programing language was used to build SAGEN, due to its simplicity and easiness to expand the SAGEN component for its use in other domains. The obtained XML file contains all variables and data required to feed the SAM, which is readable by *AnyLogic* simulation software (see Fig. 1). Besides generating XML code, SAGEN offers a user-friendly interface that allows the enterprises to enter the data required to feed the SAM (Fig. 2). The procedure followed in SAGEN is: (i) introduce all the required data to feed the SAM through the interface, (ii) the data is stored in a *Microsoft Access Database* using a *OCDBConnection*, (iii) the procedures described in Lazarus IDE allow to create XML file with all the elements to build both the flow diagram of the SAM (simulation experiment) in SD, and the optimization experiment to be opened in the *AnyLogic* simulation software. SAGEN creates a structured positioning of all objects that form the SAM, increasing the readability and understanding of the model. Figure 3 includes an example of two enterprises, each one formulating two strategies and defining two KPIs.

Summarising, the proposed DST-SA consists of the SAM, the SD-based simulation software, and the SAGEN component to automatically generate the SAM in a format accepted by the *Anylogic* simulation software. With DST-SA, enterprises will be able to build, solve, and assess the strategies alignment process from a collaborative perspective. The network enterprises collaboratively make the decision of identifying the aligned strategies to be activated, and the time frame when to activate them so that the performance of the network is maximised.

Fig. 1. XML file: strategies alignment simulation model in XML language

Fig. 2. SAGEN component interface

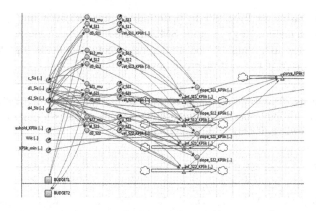

Fig. 3. SAM simulation experiment built with SAGEN

The DST-SA supports the decision-making regarding the selection of those strategies that exert positive influences on the majority of the defined objectives (being the negative influences minimized), and deals with misalignments, reducing potential conflicts.

5   Conclusions and Further Work

In this paper we have presented a decision support tool called DST-SA to help in the selection of aligned business strategies in a CN context. The tool rests on three pillars: the strategies alignment model (SAM) based on SD, the simulation software (*AnyLogic*), and application component (SAGEN) that automatically generates the SD model for the simulation software. DST-SA allows considering from a holistic perspective all the strategies formulated by each of the network partners and the influences that their activation would have on each of the performance objectives defined at the enterprises level. The SAGEN module significantly facilitates the collaborative process of strategies alignment, so that the construction of the model in the SD simulation software is automated. The SAM embedded in the DST-SA aims at maximising the network performance. Thus, not all CN partners necessarily experience an improvement in their performance level, being possible that part of the partners improve their performance at the expense of a loss of other partners. However, in the global context of the CN, the performance will be improved, contributing to its sustainability, and thus, also improving the long-term relationships among the collaborative partners.

The utilization of the DST-SA will serve (i) to show SMEs how they are currently making decisions when selecting the strategies to activate, (ii) to show SMEs how to collaboratively make decisions with the main aim of selecting aligned strategies, (iii) to train companies in the decision-making process so that they collaboratively perform the selection of aligned strategies. DST-SA allows decision makers to get a new vision of the problem of selecting strategies, from a global perspective within the CN. Hence, decision-makers do not only consider the achievement of the objectives of their company, but also take into account the influences that strategies have on the objectives of other network partners. Future research includes the assessment of DST-SA in real CNs, considering various network sizes and industrial sectors and performing a sensitivity analysis to estimate the robustness of the tool to perturbations in objectives and strategies. The development of a methodology to guide SMEs in the strategies alignment process must be performed; for assessing of the strategies selection considering decentralised scenarios, and fostering negotiation processes.

Acknowledgments. This work was supported in part by *Programa Val i + d para investigadores en formación* (ACIF) and Uninova-CTS through FCT (Fundação para a Ciência e Tecnologia) - PEST program UID/EEA/00066/2013 (Impactor project).

References

1. Camarinha-Matos, L.M., Afsarmanesh, H.: Collaborative networks: a new scientific discipline. J. Intell. Manuf. **16**(4–5), 439–452 (2005)
2. Andres, B., Poler, R.: Dealing with the alignment of strategies within the collaborative networked partners. In: Camarinha-Matos, L.M., Baldissera, T.A., Di Orio, G., Marques, F. (eds.) DoCEIS 2015. IFIP AICT, vol. 450, pp. 13–21. Springer, Heidelberg (2015)
3. Yue, X., Cai, H., Yan, H., Zou, C., Zhou, K.: Cloud-assisted industrial cyber-physical systems: an insight. Microprocess. Microsyst. **39**(8), 1262–1270 (2015)
4. Lee, E.A.: Cyber-physical systems—are computing foundations adequate? In: NSF Workshop on Cyber-Physical Systems: Research Motivation, Techniques and Roadmap, pp. 1–9 (2006)
5. Camarinha-Matos, L.M., Goes, J., Gomes, L., Martins, J.: Contributing to the internet of things. In: Camarinha-Matos, L.M., Tomic, S., Graça, P. (eds.) DoCEIS 2013. IFIP AICT, vol. 394, pp. 3–12. Springer, Heidelberg (2013)
6. Cousins, P.D.: The alignment of appropriate firm and supply strategies for competitive advantage. Int. J. Oper. Prod. Manag. **25**(5), 403–428 (2005)
7. Ashayeri, J., Selen, W.: Global sourcing strategy alignment using business intelligence: a conceptual framework. Int. J. Procure. Manag. **1**(3), 342–358 (2008)
8. Green, K.W., Whitten, D., Inman, R.A.: Aligning marketing strategies throughout the supply chain to enhance performance. Ind. Mark. Manag. **41**(6), 1008–1018 (2012)
9. Dell'Era, C., Verganti, R.: Collaborative strategies in design-intensive industries: knowledge diversity and innovation. Long Range Plann. **43**(1), 123–141 (2010)
10. Andres, B., Poler, R.: Improving the collaborative network performance through the activation of compatible strategies. Int. J. Eng. Manag. Econ. **5**(1/2), 35–47 (2015)
11. Andres, B., Poler, R., Sanchis, R.: Collaborative strategies alignment to enhance the collaborative network agility and resilience. In: Camarinha-Matos, L.M., Benaben, F., Picard, W. (eds.) PRO-VE 2015. IFIP AICT, vol. 463, pp. 88–99. Springer, Heidelberg (2015). doi: 10.1007/978-3-319-24141-8_8
12. Andres, B., Poler, R.: Modelling the strategies alignment process in the collaborative network context. In: ICIEOM-CIO-IIIE International Conference on Engineering Systems and Networks: The Way Ahead for Industrial Engineering and Operations Management (2015)
13. Forrester, J.W.: Industrial Dynamics. MIT Press, Cambridge (1961)
14. AnyLogic: AnyLogic. http://www.anylogic.com/. (Accessed from February 2015)
15. Lazarus Free Pascal: Lazarus. http://www.lazarus-ide.org/. (Accessed from February 2015)

XaaS Multi-Cloud Marketplace Architecture Enacting the Industry 4.0 Concepts

Adrián Juan-Verdejo[1,2] and Bholanathsingh Surajbali[3(✉)]

[1] Smart Development Centre, CAS Software AG, Karlsruhe, Germany
[2] Information Systems Chair, Stuttgart University, Stuttgart, Germany
[3] School of Communication and Computing, Lancaster University, Lancaster, UK
b.surajbali@comp.lancs.ac.uk

Abstract. Cloud computing in conjunction with recent advances in Cyber-Physical Systems (CPSs) unravels new opportunities for the European manufacturing industry for high value-added products that can quickly reach the market for sale. The Internet of Things joins the Internet of Services to enact the fourth industrial revolution that digitalises the manufacturing techniques and logistics while pushing forward the development of improved factories with machine-to-machine communication delivering massively customised products tailored to the individualised needs of the customer. Interconnected CPSs using internal and cross-organizational services cooperate in real time increase the business agility and flexibility of manufacturing companies. Using CPS with cloud computing architectures leverage data and services stored and run in cloud environment from different vendors that usually offer different service interfaces to share their services and data. However, when using such architectures data silos appear and different vendors having different service interfaces can easily result in vendor lock-in issues. This paper proposes a multi-cloud marketplace architecture leveraging the existing myriad of different cloud environments at different abstraction levels including the Infrastructure-, Platform-, and Software-as-a-Service cloud models—that is, Everything as a Service or XaaS—delivering services and with different properties that can control the computation happening in multiple cloud environments sharing resources with each other.

Keywords: Cloud computing · Cyber-Physical Systems · Migration · XaaS · Decision making · Industry 4.0 · IoT

1 Introduction

Rapid advancements in the embedded intelligence that mobile devices provide through the use of their wireless sensors in a connected world thanks to modern mobile communications have led to the development of CPSs. Interconnected systems and systems of systems deliver a myriad of applications in the environmental, aerospace, civilian, military, industry, and government sectors. Pervasive computing, sensing technologies, smart environments (such as smart homes or smart grids), and computer networks unveil new opportunities in the Cyber-Physical research and industrial worlds [1]. But the limited ability

© IFIP International Federation for Information Processing 2016
Published by Springer International Publishing Switzerland 2016. All Rights Reserved
L. Camarinha-Matos et al. (Eds.): DoCEIS 2016, IFIP AICT 470, pp. 11–23, 2016.
DOI: 10.1007/978-3-319-31165-4_2

of mobile devices and pervasive systems to run computation-intensive jobs arise challenges because of their limited computing power and battery life. The cloud computing model deliver advanced CPSs to offer solutions to this specific challenges and pave the way for implementing the Industry 4.0 concepts [2]. Industry 4.0 includes the so-called Smart Factory with CPSs monitoring physical processes creating a virtual copy of the physical world to make effective and accurate decentralised decisions [3]. In the Internet of Things (IoT), interconnected CPSs cooperate in real time through internal and cross-organizational services to deliver value-added complex products and services [3]. Disruptive advances in software and hardware together with the cross-pollination of concepts and the integration of information and communication in industrial automation and control systems provide new architectures. These architectures incorporate current advances in industrial automation aiming at increasing the business agility and flexibility not to mention mass customisation [4].

The **motivation** to this research work is to facilitate the delivery of the cloud-based Cyber Physical Systems of tomorrow whereby the upcoming Industry 4.0 and the IoT can build on top of cloud computing systems with many resources storing, processing, and sharing analysed information in real time from anywhere. Regardless the huge potential the interplay of all these technologies offer, at the same time new questions related to the selection of which cloud environment to use and the appearance of the data silo problem due to the vendor lock-in. A repository of fixed data under the control of one organisation's entity in isolation from the rest of the organisation or data silo resembles the grain stored in a farm silo removed from outside elements [1]. For organisations, the decision of which cloud service model to choose for their applications or decision on how to change from one cloud service model to another are becoming increasingly complex due to the cloud offerings heterogeneity and affects the decision making and how to adapt the application so that organisations can migrate it.

This paper answers the **research question** of how to deliver a multi-cloud architecture across multiple layers of the cloud stack that on the one hand avoids the portability and vendor lock-in issues while unravelling the whole potential behind modern computing architectures for IoT deployments that enable enacting the Industry 4.0 concepts. The rest of paper is organised as follows. Section 2 provides the research motivation for CPS, followed by Sect. 3, provides a taxonomy of cloud offerings, followed by Sect. 4 our proposed XaaS Multi-Cloud Marketplace Architecture to address CPS challenges applied to the Industry 4.0. Next, in Sect. 5 presents a discussion of our contribution and the related literature. Finally, we offer our conclusion and future work in Sect. 6.

2   Motivation for Cloud-Enabled Cyber-Physical Systems

Very different domains—from agriculture and meteorology to biomedicine and computer vision including emergency response [1], automation, and aerospace but many others too—use technologies as varied as sensor-actuator networks, machine-to-machine communication, supervisory control, data acquisition. In addition to the multitude of applications domains, the conjunction of the IoT, services that cloud computing architectures deliver, and Cyber Physical Systems have triggered the emergence of the so-called fourth industrial revolution described. The Industry 4.0 collective term considers recent studies that state that in 2020 around twenty-six "Things" communicating through broad-band technologies. The digitalisation of the manufacturing techniques and logistics are pushing forward the development of Smart Factories with machine-to-machine communication delivering massively customised products tailored to the individualised needs of the customer. CPSs such as intelligent machines, storage resources, and utilities cater for a smarter production in real time respecting the stated requirements while delivering a competitive price. The EU therefore funds research aiming at bringing back to Europe production plants now outsourced to cheap-labour countries. This trend will strengthen the European manufacturing industry to deliver value-added products using machinery and robots with enhanced capabilities to achieve higher levels of quality, occupancy rate, and faster processing times. Due to the use of software deployed into systems with big data analysis capabilities and real-time intelligent monitoring and decision-making processes.

New challenges related to Business Intelligence (BI) systems appear with the introduction of the Industry 4.0 [5]. Fast and accurate decision-making at strategic and operative levels paves the way to offering higher productivity, quality, and flexibility. Massive amounts of data coming from different sources—inside a company and across its boundaries—in different formats available in real-time must be analysed in real time to reach efficient solutions to problems and to improve the existing processes. This prompts the need for expertise from different areas including the manufacturing process, procurement, logistics, and product requirements. Given the large amount of data at hand and the time constraints; the analysis is best done in an automated manner based on the models of the existing knowledge domains and the data analysis processes to reach effective decisions. The integration of the

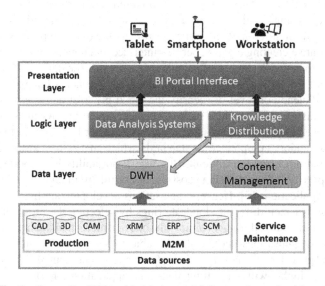

Fig. 1. Generalised BI-enabled Industry 4.0 System (non-cloud-based)

Industry 4.0 and CPSs in a cloud-enabled environment delivers the needed connection through sensors and networks; computation, memory, and data on demand; advance modelling; content and context analysis and correlation; sharing and collaboration; and customisation. Cloud computing architectures facilitate the development of products using (micro) services delivered to consumers so that they configure products and services while reconfiguring manufacturing systems. Industry 4.0, cloud-based design and manufacturing, and cloud manufacturing adapt the cloud computing paradigm to the realm of computer-aided product development and the IoT. Figure 1 exemplifies the complexity of a BI-enabled Industry 4.0 systems whereby this paper shows the difficulty in selecting a multiple cloud offerings and adapting the system to use them in typical BI three-tier architectures. The data layer stores the structured and unstructured data coming from operational systems; the logic layer performs different analysis on those data; and the access layer let BI users access the high-level data they need [1]. Increasing the portability of such a system in a multi-cloud environment works under different interdependent constraints and success criteria specific to each organisation porting their system. Different dimensions that are not always equally important to all organisations and they can trade one for another one.

3 Research Contribution to the Taxonomy of Current Cloud Offerings

Nowadays organisations are increasingly moving the computing infrastructure supporting their manufacturing and enacting the entire supply chain from the traditional on-premises approach to current cloud computing environments. That way, ubiquitous networking service management and application opens the way to the reuse of services, cost reduction, flexibility, and increasing collaboration opportunities. Migrating to cloud deployments and using cloud-based services generate additional value due to the amount of data coming from different data sources. However, the data silo problem and vendor lock-in issues arise creating a gap of service interface and data model interoperability. The data of customers risk staying in individual data silos. At the same time organisations want to benefit from the characteristics of cloud environments but worry about the deployment efficiency or level of security the cloud provider achieves. Additionally, organisations may face the vendor lock-in problem when they want to migrate their system from one cloud provider to another one. The vendor lock-in risk stems from the lack of cloud vendors compliance to standards which makes the portability of data, application, and infrastructure difficult. But even if portability is supported in theory, the high complexity and the switching costs incurred still discourage organisations from porting their applications [6]. Hence, organisations developing applications should be able to choose between different cloud offerings by selecting the most reliable, well-reputed, or cost-efficient offering; or simply the one that meets their technical or business requirements. Organisations should also be able to switch easily and transparently between cloud providers whenever needed—for example, due to an SLA breach or when the cost is too high—without setting data and applications at risk—for example, loss of data or inability to run the application on a different platform. Moreover, they should be able to compare cloud offerings with different characteristics—such as resource, pricing, or quality of service—and to choose the one that best matches the computing needs of their services and applications [7]. A number of different cloud providers exist today making it difficult to select the most adequate cloud environment for an organisation. Organisations choose whether to move their applications to a different cloud service model or provider according to their current and future needs and how to do it. The cloud offering taxonomy the authors propose permeates across the whole architecture of the XaaS Multi-Cloud Marketplace and the Multi-Cloud Portability Tier and Broker. Figure 2 shows the Cloud Offering Decision Making Module that selects a cloud offering configuration to use for a particular application according to the dimensions an organisation selects as important of each of the cloud deployment models from IaaS up to SaaS.

IaaS Offerings: IaaS resides at the bottom of the cloud service model and the rest of levels in the stack build on top of IaaS. Providers offer their resources, both physical and virtual: load balancers, virtual machines, servers, storage, and network. Given the particularities of the IaaS level, literature study, and the BI use case presented; this paper describes six dimensions to analyse different IaaS offerings as shown in Fig. 2. The IaaS Service Characteristics Dimension: Each VM instance type offers different price, number of virtual cores, available memory, and bandwidth for input and output traffic. The IaaS service characteristics include descriptions related to network access and service configuration. Additionally, IaaS let their consumers deploy their systems according to the private, community, public, or hybrid cloud deployment model. The Hardware Infrastructure Dimension: IaaS providers define their hardware infrastructure in terms of processing and memory. They define the types of server, processor (CPU), and memory, as well as the instance capacity. The Network Infrastructure Dimension: Systems deployed on IaaS environments share the provider's network resources with different network connection bandwidth and network infrastructure. The Repository Infrastructure Dimension: IaaS providers supply different database storage and might offer virtual private cloud possibilities. The tendency in regard with messaging services is to commoditise this service. An example is the Amazon Simple Queue Service (SQS) which liberates their consumers from maintaining their own service built by using a messaging implementations such as the Java Message Service (JMS) and the Microsoft Message Queuing. The Infrastructure SLA Dimension: The contractual agreement on the level of service to be provided by the IaaS service provider to the IaaS cloud consumer specifies numerous service performance metrics and their corresponding service-level objectives. Infrastructure SLA includes agreements related to the resource provisioning; and the service availability, reliability, security, geographical location, and elasticity. The SLA also specifies the service and auditing, disaster management, energy consumption, and the cost. The infrastructure cost includes defining the price level, as well as the billing metrics. The IaaS Portability Standardisation Dimension: Standardisation minimises the cost of porting software already developed for an IaaS environment to a new platform. Respecting portability standards and following portability methodologies increases portability and avoids vendor lock-in at the IaaS level.

IaaS cloud consumers need to port the software runtime environments including configurations and APIs. Typically, this involves VMs. The Distributed Management Task
Force (DMTF) proposed the Open Virtualization Format (OVF) which is often used in
IaaS. OVF provides a format for open packaging and distribution which outlines how
to deploy, manage, and run a virtual appliance on a VM.

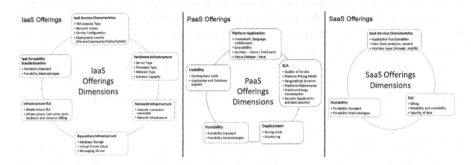

Fig. 2. IaaS, PaaS, and SaaS offerings dimensions

PaaS offerings: Figure 2 shows the six PaaS offerings dimensions that come out from
the analysis of the use case scenario described in Sect. 2 and differentiate PaaS offerings
from one another. The Application Dimension consists of four sub-dimensions tightly
related to the application whereas the SLA Dimension specifies the service contract in
terms of quality of service, pricing model, geographical coverage, and availability. PaaS
providers usually charge differently according to resources usage with billing that is
fixed, metered, or hybrid [8]. Geographical coverage presents legal concerns such as
those posed to EU-based companies that are not allowed to transfer any of their
customer-related data outside of the European Union [9]. The Deployment Dimension
describes how provider tools to *validate* and *test* the deployment of application and data
components, from model-driven approaches [10] to organisations requirements validation tools [1, 7], and software development life-cycle tooling facilities. The data layer
represents the asset of the BI user organisation interested on porting the BI system to
PaaS environments. The portability of this layer to PaaS environments poses important
security concerns for the organisation related to the trustworthiness of the PaaS provider
in charge of storing the data layer and relates to the portability dimension. The Portability
Dimension considers PaaS offerings providing information on the different standards
they comply to in order to ensure application and data portability. The portability process
itself has to keep data integrity and let users access them effectively. For example some
users work with the data from different geographical locations. The authors consider
three BI portability models for the private, public and hybrid PaaS cloud models. The
private PaaS model whereby the organisation wants to ensure total data ownership
through self-data security enforcement. The public PaaS model whereby the organisation trust cloud provider offerings. The hybrid PaaS model whereby the system can use
more computational resources on demand while keeping sensitive data in the organisation premises. The Usability Offering Dimension considers the set of rapid PaaS application development tools that PaaS providers offer with different levels of support.

Furthermore, PaaS offerings application and database usability varies as well as the training and documentation they offer. The usability dimension determines the application and database knowledge needed to maintain and extend an application deployed to a PaaS offering.

SaaS Offerings: Finally, the most abstract of the three fundamental cloud service models provides less degrees of freedom for the SaaS cloud consumer. As a result, fewer dimensions drive the migration decision to SaaS. The most important factor is the data migration as the solutions on SaaS are at the level of services: Customer Relationship Management (CRM), e-mail, enterprise resource planning (ERP), virtual desktop, management information systems (MIS), communication, games, computer-aided design (CAD) software. This paper proposes three dimensions within the SaaS level as shown in Fig. 2. The SaaS Service Characteristics Dimension: At this higher level of abstraction within the cloud computing service stack, the service characteristic dimension only involves the application functionalities, data store, and interface layer. The latter defines how the user of SaaS applications will access the service, such as browser or mobile interfaces. The SLA Dimension: SaaS providers formally define their services as a part of the service contract. The contractual agreement on the SaaS service level defines billing, reliability and availability, and security of the data. The Portability Dimension to enable applications to achieve SaaS portability by respecting *portability standards and methodologies*. SaaS portability can help them deal with cost increases, business stability issues, or SLA breaks by switching to a different SaaS provider. Data portability enables the use of data components across different applications.

4   Proposed XaaS Multi-Cloud Marketplace Architecture

The XaaS Multi-Cloud Marketplace Architecture in Fig. 3—as an architecture evolving from previous research works the authors did [7]—consists of three basic components including the XaaS Multi-Cloud Marketplace Broker, the XaaS Multi-Cloud Decision Support Tier and the XaaS Multi-Cloud Portability Tier. The XaaS Multi-Cloud Marketplace Broker allows end users to track IaaS, PaaS, and SaaS offerings, services, interdependencies, and offers through a marketplace. The main goal of the decision support broker is to make it easier, safer, and more productive for cloud adopters to navigate, integrate, consume, extend, and maintain cloud services through a marketplace. In this respect, the broker offers a range of value-added services to both cloud vendors and consumers, allowing organisations to lookup cloud offerings matching their requirements. Deploy the business logic of their application on one cloud offering and the underlying data on another without using additional software, changing application settings, or putting the application and data at risk. Efficiently select an alternative cloud offering and port the application to the new cloud offering with little development and decision effort. Transparently exchange data and application components with other cloud providers.

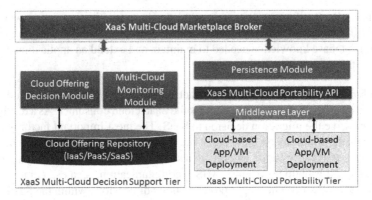

Fig. 3. Proposed XaaS Multi-Cloud Marketplace Architecture

The XaaS Multi-Cloud Decision Support Tier: this tier is responsible of the cloud offering selection and assessments taking into account the cloud offerings taxonomy from Sect. 3 above. The Cloud Decision Support contains the modules, which maintain the Cloud Offering Repository consistent and synchronised with the current offerings available. The Cloud Offering Decision Module supports seamless interaction between the cloud consumer and the cloud offering. Cloud consumers find available cloud offerings, which comply with their application's requirements and their criteria for porting their application. Additionally, the broker supports the application porting to cloud environments. This module implements the core functionalities such as cloud offering discovery, cloud offering analysis, cloud offering recommendation, and the InCLOUDer cloud offering selection and migration support from our previous work [1]. InCLOUDer addresses the multi-dimensional decision problem with the presented structured hierarchy of migration criteria. The hierarchy of criteria makes it easier for organisations to weight different criterion to one another and also complies with the natural nature of the migration criteria [7]. The cloud offering decision module facilitates the rating of a particular cloud offering by a SME's Software Engineer. Each engineer can leave a comment and rate a particular cloud provider, thus offering expressing their satisfaction or dissatisfaction with regards to quality, usability, reliability, and user-friendliness of the offering. The Multi-Cloud Monitoring Module monitors the deployed application and the cloud environment and manages the application's life-cycle. The Cloud Offering Repository includes the Unified Cloud API and the Cloud Offering Repository interacting with heterogeneous cloud offerings. It holds information on the capabilities, requirements, services, and restrictions of the cloud services across the three cloud service models. Once the cloud decision support tier recommends a cloud offering, the XaaS Multi-Cloud Portability Tier helps with the application deployment and porting from one cloud vendor to another one. Several models cater for deployment and porting process support. The Persistence Module handles the application dependencies that affect the processes to deploy, undeploy, start, stop, and port the target application to the target cloud provider. The dependencies range from operating system to executable formats, including compilers and libraries. The persistence module includes all these dependencies in the migration and allows developers to add more themselves.

The XaaS Multi-Cloud Portability API supports the management of the deployed applications independently of the specific API of the underlying cloud offering. The Middleware is an adaptor between the cloud portability module and the native API of the cloud offering. More specifically, the middleware module translates the functions of the cloud Portability API to the native API of the cloud offering, and vice versa. As a result, the middleware module, allows the seamless communication among cloud offerings addressing the vendor lock-in concern.

The proposed XaaS Multi-Cloud Decision Marketplace Architecture addresses several challenges that Industry 4.0 practices. The architecture provides the necessary infrastructure catering for the end-to-end digital integration along the entire value chain. The cloud deployment provides advanced communication mechanisms within a virtualised infrastructure that makes smart factories oblivious to which factory executes which processes. The proposed architecture aims at allowing for the vertical integration that paves the way to achieving the high flexibility needed to support the zero (or near zero) lot size. The resulting cloud infrastructure allows for integrating assets such as sensors, actuators, mechanical parts, documents, or Intellectual Property with the interfaces between the "real world" and its IT representation and it continues with its communication towards the information layer to finally move to the functional layer that is the gate to the achievement of business goals. The architecture uses standards to allow worldwide value chains exchanging information for transfer of data and utilisation facilitating the horizontal integration from products to the connected world and the needed Collaborative Networks. Network partners can access competences of one another and respond to market needs in a sustainable manner. The proposed broker allows the access to collaboration spaces whereby companies have well defined access and writing when collaborating and where not so as to increase the levels of trust by offering control over the information companies exchange with other companies with which they compete. The portable cloud-based deployments allow for real-time monitoring of large amounts of sensor data, including RFID or similar technologies, about product data, the status and position of goods, and the status of the different components of a CPS that the architecture can seamlessly port to different environments to overcome the vendor lock-in and data silo problems. The resulting deployment connects the value chain across the entire product life cycle and allows for the agility needed for massive customisation.

5 Discussion and Related Literature

The current evolution of general computation for individuals transforming into increasingly advanced pervasive devices has been accompanied with computing infrastructures moving to cloud computing environments. In conjunction with the trend towards miniaturisation, ubiquitous computing is a reality that the modern industry puts into play to improve and automate manufacturing, transport, and logistics. The interplay of technologies in the industrial domain ushers in the convergence of the physical and virtual world in the CPS allowing for the delivery of the infrastructure and ubiquitous computing fundamental to the fourth industrial revolution. Smart objects networked with the relatively new IPv6 addresses aggregate the information into modern cloud environments hosting advanced analytic (micro)services enhancing the decision making and control. As an example, manufacturers can timely react to changes such as in the supplier chain suddenly during production. Industry 4.0 opens the way for smoother changes in this network with manufacturers anticipating distortions in the supply chain by simulating, evaluating and selecting the best supply alternative. Organisations adopting the concepts behind the Industry 4.0 concepts have very special requirements tailored to their highly complex systems and can profit from the selection of particular cloud offerings at different levels of the cloud deployment stack. Additionally, organisations can minimise the data silo problem and portability issues with the use of multi cloud environments. In that respect, this paper proposes a multi cloud architecture with the use of a decision broker to schedule and deploy components to the appropriate XaaS cloud environment as well as services and (micro)services offered across the whole cloud service stack. Another contribution of our research work concerns the decision process involved in migrating to a cloud-enabled deployment by providing the basic mechanisms to help organisations find the best cloud provider offerings through a marketplace and mechanisms for cloud environment selection using the taxonomy of cloud offerings.

XaaS encompasses cloud-related services—from the Infrastructure- to the Software-as-a-Service cloud model including the Platform-as-a-Service model [11]—running different cloud service models according to the famous NIST definition [12]. Concepts coming from cloud computing such as broad network access and resource pooling and from Service-Oriented Infrastructures (SOI) such as on-demand business-driven infrastructure and service-orientation [13] come together in XaaS. The selection of the best cloud model matching the current and future organisations' needs are becoming increasingly complex due to the number of cloud offerings to consider [1]. The proposed decision broker builds on top of the Analytic Hierarchy Process allowing formalising the cloud migration problem by defining the cloud migration criteria, the cloud service offerings, and the architecture and properties of the application. It is a multi-dimensional decision problem affected by a lot of factors, sub-factors, and parameters [14, 15] and our research work automate the process differing from some works [16, 17] improving the usage for real-case scenarios. The decision broker weights these alternatives by taking into account criterion metrics [1] and their importance relative to other criterion. When no metric is provided the weighting is done manually by the organisation based on their requirement needs. The migration of existing applications to cloud environments demands from organisations to select a suitable target cloud service [17, 18] and adapt their applications to them according to their needs and the specifics of their application [15]. Organisations have to evaluate many conflicting criteria before making any decision and the problem becomes a multiple criteria decision making problem [19]. Cloud4SOA differs from the proposed approach in that it requires the annotation of each provider offerings [20], and the annotation process of each offering is only feasible if there are established benchmarks of PaaS offerings so that all PaaS vendor offerings can be taken into consideration. The mOSAIC project [21] focuses on application portability at both IaaS and PaaS using a Cloud ontology to analyse the offerings of the cloud and using a brokering mechanism to search offerings requirements. Numerous Cloud API standardisation try to unify incompatible APIs coming from cloud providers with respect to cloud standards such as OCCI [22], TOSCA [23] and CIMI [24]; to PaaS standards such as the OASIS CAMP [25] which help avoiding vendor lock-in. However, the adherence of a cloud standard at one cloud model for example at IaaS using OCCI with another provider adherence to another cloud at the PaaS using CAMP results in a standard incompatibility. To address standard incompatibility in portability across cloud models, we envisage extending our middleware layer in the cloud broker architecture in Fig. 3 to support cross-standard to interoperate with each other inspired by the research work from [26].

6 Conclusion and Future Work

Organisations adopting the concepts behind the Industry 4.0 concepts have very special requirements tailored to their highly complex systems and can profit from the selection of particular cloud offerings at different levels of the cloud deployment stack. We have presented a generic architecture-based approach to provide the necessary end-to-end digital integration along the entire value chain of cloud infrastructures. The information exchange for data transfer and utilisation facilitates the horizontal integration via standards to connect everything from products to the connected world in current worldwide value chains. In Collaborative Networks, network partners can access competences of one another and sustainably react to the market needs. Our broker is the door to collaboration spaces allowing organisations collaborate but keeping high levels of trust. This trust stems from the secured and controlled access of the information organisations exchange with one another including competing entities. The cloud-enabled advanced communication mechanisms connects smart factories executing different processes. In addition to horizontal integration, the architecture facilitates the vertical integration needed to provide high flexibility to adjust the production and lot size. The proposed cloud-based infrastructure integrates assets with their interfaces with its IT representation allows the communication and moving knowledge up to the information layer to finally from the functional layer achieve the organisations business goals. Using our approach and our XaaS Multi-Cloud Marketplace Broker organisations can minimise the data silo problem and portability issues with the use of multi-cloud environments and thereby addressing the vendor lock-in issue. Additionally, our proposed architecture eases the decision process using the author InCLOUDer framework to migrate cloud-enabled deployment from one cloud environment to another and provides the basic mechanisms to help organisations find the best cloud provider offerings through a marketplace. This is achieved using the cloud offering taxonomy that facilitates the selection of cloud offerings (IaaS, PaaS and SaaS) based on a number of offerings classification criteria with minimal effort.

As future work the authors will firstly face the validation phase with the quantitative and qualitative analysis of the ported cloud-enabled CPS infrastructure respecting the Industry 4.0 practices in terms of the cloud offerings dimensions relevant to the Smart Factory organisation and the end-to-end integration. Additionally we will confront the security and trust dimensions within the appropriate hierarchy in the cloud deployment model with the levels of trust and security at the level of Collaborative Networks.

In a further phase, we will address the challenges with respect to standards incompatibility by building an interoperability middleware framework across cloud models where applications differences are seen in terms of incompatible interface signatures and data content and communication protocols across standardised layers of the cloud. The work will support the marketplace presented here with additional portable services tailored to the Industry 4.0. Services that will pool the monitoring information related to complex products and services to optimise the smart factory with regard to the logistics and manufacturing automation so as to seamlessly integrate with other business management applications. Organisations making use of the market place will have access to re-usable (micro)services supporting the addictive and subtractive manufacturing and to a myriad of off-the-self 3D designs ready to use.

Acknowledgements. The European Commission supported this work through the OSMOSE project FP7 610905 and the German Ministry of Education and research, BMBF, through the SysPlace Joint Project 01IS14018D.

References

1. Al Ali, R., Gerostathopoulos, I., Gonzalez-Herrera, I., Juan-Verdejo, A., Kit, M., Surajbali, B.: Towards a framework for compute-intensive pervasive systems in dynamic environments. In: 2nd International Workshop on Hot Topics in Cloud service Scalability. ACM (2014)
2. Kagermann, H., et al.: Recommendations for Implementing the Strategic Initiative INDUSTRIE 4.0: Securing the Future of German Manufacturing Industry; Final Report of the Industrie 4.0 Working Group. Forschungsunion (2013)
3. Hermann, M., Pentek, T., Otto, B.: Design principles for Industrie 4.0 scenarios: a literature review (2015)
4. Brettel, M., et al.: How virtualization, decentralization and network building change the manufacturing landscape: an Industry 4.0 Perspective. Int. J. Sci Eng. Technol. **8**(1), 37–44 (2014)
5. Givehchi, O., Trsek, H., Jasperneite, J.: Cloud computing for industrial automation systems —a comprehensive overview. In: 18th Conference on Emerging Technologies & Factory Automation (ETFA). IEEE (2013)
6. McKendrick, J.: Does Platform as a Service have interoperability issues. In: Cloud Commons (2010)
7. Juan-Verdejo, A., et al.: InCLOUDer: a formalised decision support modelling approach to migrate applications to cloud environments. In: 2014 40th EUROMICRO Conference on SEAA, Verona, Italy (2014)
8. Kolb, S, Wirtz, G.: Towards application portability in platform as a service. In: Proceedings of the 8th IEEE International Symposium on Service-Oriented System Engineering (SOSE) (2014)
9. Strobl, J., Cave, E., Walley, T.: Data protection legislation: interpretation and barriers to research. BMJ J. **321**, 1031–1032 (2000)
10. Fleurey, F., Steel, J., Baudry, B.: Validation in model-driven engineering: testing model transformations. In: Proceedings of the 2004 First International Workshop on Model, Design and Validation. IEEE (2004)
11. Duan, Y., et al.: Everything as a Service (XaaS) on the Cloud: Origins, Current and Future Trends. In: Cloud 2015. IEEE (2015)

12. Mell, P., Grance, T.: The NIST definition of cloud computing (2011)
13. The-Open-Group SOCCI Framework Technical Standard: Service-Orientation and Cloud Synergies (2014)
14. Armbrust, M., et al.: A view of cloud computing. Commun. ACM **53**(4), 50–58 (2010)
15. Andrikopoulos, V., et al.: How to adapt applications for the Cloud environment. Computing **95**(6), 493–535 (2013)
16. CSMIC: Introducing the service measurement index. In: Cloud Commons (2011)
17. Garg, S.K., Versteeg, S., Buyya, R.: Smicloud: a framework for comparing and ranking cloud services. In: 2011 Fourth IEEE International Conference on Utility and Cloud Computing (UCC). IEEE (2011)
18. Tran, V.X., Tsuji, H., Masuda, R.: A new QoS ontology and its QoS-based ranking algorithm for Web services. Simul. Model. Pract. Theor. **17**(8), 1378–1398 (2009)
19. Zeleny, M., Cochrane, J.L.: Multiple Criteria Decision Making, vol. 25. McGraw-Hill, New York (1982)
20. D'Andria, F., et al.: Cloud4soa: multi-cloud application management across PaaS offerings. In: 2012 14th International Symposium on Symbolic and Numeric Algorithms for Scientific Computing (SYNASC). IEEE (2012)
21. Petcu, D., et al.: Experiences in building a mOSAIC of clouds. J. Cloud Comput. **2**(1), 1–22 (2013)
22. Edmonds, A., Metsch, T.: Open Cloud Computing Interface–Infrastructure. In: Standards Track. The Open Grid Forum Document Series, no. GFD-R. Open Cloud Computing Interface (OCCI) Working Group, Muncie (2010)
23. Binz, T., Breitenbücher, U., Haupt, F., Kopp, O., Leymann, F., Nowak, A., Wagner, S.: OpenTOSCA – a runtime for TOSCA-based cloud applications. In: Basu, S., Pautasso, C., Zhang, L., Fu, X. (eds.) ICSOC 2013. LNCS, vol. 8274, pp. 692–695. Springer, Heidelberg (2013)
24. CIM: Cloud Infrastructure Management Interface (2014)
25. OASIS Cloud Application Management for Platforms Version 1.1 – Draft 03 (2014)
26. Bromberg, Y.-D., Grace, P., Réveillère, L.: Starlink: runtime interoperability between heterogeneous middleware protocols. In: 2011 31st International Conference on Distributed Computing Systems (ICDCS). IEEE (2011)

Towards a Collaborative Business Ecosystem
for Elderly Care

Thais Andrea Baldissera[✉] and Luis M. Camarinha-Matos

Faculty of Science and Technology, Uninova-CTS,
NOVA University of Lisbon, Campus de Caparica, Caparica, Portugal
{tab,cam}@uninova.pt

Abstract. In fast changing environments, companies and organizations have to continuously adapt their operating principles in response to new business or collaboration opportunities in order to remain competitive. The growing demand for high-quality services is taking organizations to format their operations to offer more personalized service packages and seek collaboration with other stakeholders. Moreover, provision of personalized services depends on frequent contextual information analysis. In the case of elderly care, the use of assistive technology is expected to have a positive contribution to the diversity of required services that support aging well. Demographic trends show that the percentage of elderly population is increasing, while ageing entails several limitations, calling for assistance services adapted to the specific needs of each person. These needs can evolve along the ageing process, requiring an evolution of the care services. In this paper, we make an overview of related concepts and propose a personalized and evolutionary care services model supported on collaborative networking, context-awareness and Internet of Things.

Keywords: Collaborative business services · ICT and Ageing · Collaborative networks · Context awareness

1 Introduction

In the last decades, one of the most important demographic changes in the history of humanity has been taking place: the aging of the global population. In fact, current trends [1, 2] clearly suggest that elderly population will surpass young people in many regions of the world. This change is already a consolidated fact in countries such as Portugal, where the population over 60 years is 10 % higher than people under 15 years old, a situation that was inversely proportional 30 years ago. By 2050, it is expected that the population over 80 years exceeds the young [2]. This situation seem to concern mainly the developed world, but it no exclusive theirs.

Current estimates indicate that 60 % of elderly people live by themselves or in the company of another elderly person, and strive to guarantee their autonomy and perform their own duties. With continuous advances in medicine and wellness areas, people aspire to an active and enjoyable aging [1], although aging might entail several

L. Camarinha-Matos et al. (Eds.): DoCEIS 2016, IFIP AICT 470, pp. 24–34, 2016.
DOI: 10.1007/978-3-319-31165-4_3

limitations due to physical and cognitive decay. In extreme cases, regular daily activities such as cooking, personal hygiene, housework, etc., might be affected [1, 2].

Traditional approaches to deal with this problem require intense care from family. Relatives need to actively participate in the aging process and support elderly. As such, many people live with their family, especially when they begin to loose capabilities. Those who remain living alone either care for themselves or require caregivers to assist them on daily activities. Alternatively, the elderly may stay in nursing homes. All these changes can directly affect the senior life style [3].

The use of technology in assisting elderly is expected to have a positive contribution to overcome identified challenges by supporting a diversity of services that promote independent living [1, 3–5]. However, most elderly people still face several barriers in an increasingly technological society. This fact puts older adults at a disadvantage, requiring that researchers and practitioners deliver more suitable products for this community [3, 4, 6]. For each elderly individual, specific care and assistance services might be necessary, according to his/her life context. A care service may be adequate for one elderly and completely useless for another one. As a single care organization can hardly meet all needs of each individual senior with the best services available in the market, collaboration between companies may be a promising way for delivery of integrated and personalized services.

In this context the main research question addressed by this work is: *How to provide personalized and evolutionary collaborative care services for elderly in an effective and reliable way?*

The pursued hypothesis is: *Effective and reliable evolutionary services for elderly care can be provided if a suitable set of multi-provider business services are composed and integrated in the context of a collaborative network environment and supported by context awareness methods, mechanisms and systems.*

In this paper, we propose a Collaborative Business Ecosystem Model for elderly care based on collaborative networking and relevant context-aware technologies that can be applied to personalization and evolution of care and assistance services. The Collaborative Networks paradigm (Sect. 3) supports provision of integrated services and helps services providers to acquire agility and better survive in market. In particular, we present collaborative business services in the elderly care domain and context-awareness technologies for personalization and evolution of care and assistance services. A Collaborative Business Ecosystem (CBE) Model (Sect. 4) gives the adequate organizational context for collaboration. Finally, we conclude (Sect. 5) with a discussion on the main advantages and concerns on emergent technologies for elderly care and assistance.

2 Benefits from Cyber-Physical Systems

It is expectable that with the increased intelligence in the new systems, interaction with technology becomes part of daily life for seniors, family members, caregivers and integrated service providers, while respecting the individual life context. Some examples of technological solutions with impact in aging well include smart devices, like wearable devices, sensors and associated software services that provide real-time data about an individual and the environment he/she lives in; a bracelet that uses a GPS sensor to set up a "safe zone" for people with cognitive disorders; an at-home patient-tracking device with a panic button for emergencies; cameras and sensors that capture movements, calculate body weight and make the measurement of heart-bit rate; and intelligent home appliances [9].

With advances in sensor technology, a growing number of sensors that communicate with each other with minimum human intervention will be attached to objects around us, generating a massive amount of data. Context-aware computing allows us to extract contextual information linked to sensor data so the interpretation can be done easily and more meaningfully [7, 8]. The adoption of context-aware analysis techniques, combined with smart devices and sensors, are likely to enable a great number of new care and assistance services. These technologies can be a strong ally in helping on the identification and analysis of the user's context in order to provide relevant inputs for the evolution of assistance services.

In the elderly care domain, context aware technologies supported by CPS can: (i) provide inputs to personalize or evolve the service based on multi-source contextual information, (ii) monitor individuals to provide best options and recommendation of services, (iii) enable businesses to redefine key aspects of their customer relationships, and (iv) cater direct interaction between senior and his/her living environment (devices and people) and services provided.

3 Related Areas

Collaborative Networks. In highly dynamic domains, such as health and personal services, companies are challenged to be able to efficiently interplay with multiple organizations to compose personalized offers without losing competitiveness and quality in their services. In such context, new collaborative strategies can facilitate the engagement and interaction of distinct stakeholders in any effort towards common or compatible goals. Advances in ICT, namely Internet and pervasive computing, have enabled or induced the emergence of new collaboration paradigms. However, the rapid formation of a collaborative network (CN) to respond to a business opportunity faces a number of challenges, whereas two are most relevant: (i) the large heterogeneity of the autonomous participants involved in the process, in terms of their technological infrastructures, business practices, culture, etc., and (ii) the time needed to build trust [10]. To face these challenges, the concepts of Virtual Organization Breeding Environment and Collaborative Business Ecosystem were proposed [10].

Participation in collaborative networks has the potential of bringing benefits to the involved entities such as: survivability in a context of turbulence, reaching higher levels of agility, acquisition of a larger apparent dimension, access to new/wider markets and new knowledge, sharing risks and resources, joining of complementary skills, and better achieving common goals [10–12].

In the elderly care domain, CNs can: (i) provide organizational structures and governance models for groups of care and assistance stakeholders; (ii) support provision of integrated services; (iii) enable offering services tailored to each senior and his/her life style;

(iv) facilitate composition of new services in line with new demands (current situation, specific requirements, new technologies); (v) facilitate lobbying & market influence; (vi) provide a framework for organizing collaborative processes among stakeholders.

Collaborative Business Services. Research and development in elderly care services traditionally focused on the development of isolated services, often considering a single service provider and showing an excessive techno-centric flavor. However, to satisfy current demanding market challenges, organizations must collaborate to overcome their weaknesses and strengthen their expertise, so that they can offer better integrated and user-centric services, and gain competitive advantage [13]. Furthermore, in the elderly care context there is a need for personalized services that respect the individuality of each senior and the evolution of needs as senior's life context changes [8]. In order to support this perspective of service provision, the term Collaborative Business Ecosystem (CBE) can be adopted [14]. A CBE can assist in the integration of different services from distinct providers. For the elderly care domain the term collaborative business service can be named as Collaborative Care and Assistance Service. Such services can be seen as the result of collaboration among various stakeholders, including local communities, governmental institutions, care professionals, family and caregivers, and thus require a supporting collaboration environment.

Context Aware Technologies. Context awareness is about capturing a broad range of contextual attributes (such as the user's current position, activities, and surrounding environment) to better understand what the customer is trying to accomplish, and what services might be of interested [15]. According to Costa et al. [16] context-awareness seeks to exploit human-computer interactions by providing computing devices with knowledge of the users' environment, i.e., with context. Context awareness computing is a recent and development [9] which appears quite promising for the elderly care domain. Many recent technologies contribute to the development of context-aware solutions. Scoble and Israel [9] highlight five forces that have shaped and cooperated with the advances on context-aware technologies: *Mobile, Social Media, Data, Sensors, and Location-based services*. These forces can directly influence decisions through a context analysis and supported by a wide range of sensors, smart devices, and new evolutionary services. These devices in a communicating–actuating collaborative network are part of the IoT, where sensors and actuators blend seamlessly with the environment around us.

4 Collaborative Business Ecosystem Model

The proposed CBE model for elderly care includes:

Service Composition: A care service can be either a single (atomic) or integrated business service. Therefore, service provision can be done by a single entity or by several entities. Some entities participating in a composite service may already be a composite service provider. This leads to a hierarchical categorization of service providers. For instance, in Fig. 1 two providers deliver one collaborative service CS1, and three other service providers offer atomic services SS2, SS3, and SS4. CS1 is seen by the CBE like a simple service, which combined with other services, compose the personalized service CS2. This collaborative service (CS2) is provided to the customer as one composite service, delivered by a collective provider (i.e. a collaborative network).

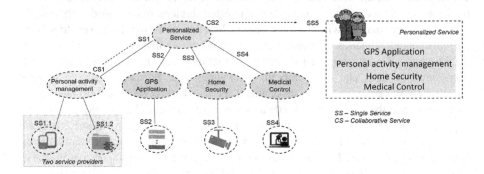

Fig. 1. Example of creation of composite services from providers of distinct types

Service Personalization and Evolution for Elderly Care: In the elderly care domain, personalization involves the analysis of the senior´s life context. For instance, let us consider a senior, 68 years old that lives alone and still works in a company. The organization responsible for delivering care and assistance services identifies specific requirements for this context and builds a service package for this customer, including, for instance, adjusted working space and home security monitoring services. At a later stage in life additional needs are identified as he got (in this scenario) some chronic diseases, diabetes, and less vision. As such, new services need to be delivered, e.g. personal management notices for application of insulin, control of medication, and continuous health monitoring. One year later, he recovered from some disease and is now retired: his life context changed again. So the care and assistance provider organization identifies the new situation and adjusts the provided service. The customer does not need an adjusted working space in the company anymore and this service will be removed from the package, but a social program for meeting with friends and services allowing the participation in virtual professional senior communities is added to his package. In other words, the service package evolves to cope with the new life stage.

For each new context change, the care and assistance service provider organization shall analyze the situation (in collaboration will all relevant stakeholders), and evolve the service to fit that context. This perspective is illustrated in Fig. 2.

Under this perspective, the notion of evolutionary service [7, 13, 17, 18] means that the provided service adapts to the senior's needs, to the environment and to any changes that affect the senior´s life context. In the previous example, the care provider organization adjusts the service due to new senior's context requirements (e.g. when he retired). But this service evolution could also be done due to technological reasons, as sensors and devices are improving and becoming cheaper, or new technologies emerge better meeting the user needs. Additionally, market factors (companies may disappear or merge with others) or new governmental regulations may also lead to service evolution.

CBE Structure: The proposed CBE is based on the ARCON reference model [19] and the Ambient Assisted Living (AAL) Ecosystem concept [5], focusing on a care and assistance community of interacting entities, comprising both organizations and individuals participating in elderly care. It is aimed that the ecosystem community is supported by common operating principles and cooperation agreements, and also includes the basic profiles of all stakeholders [13].

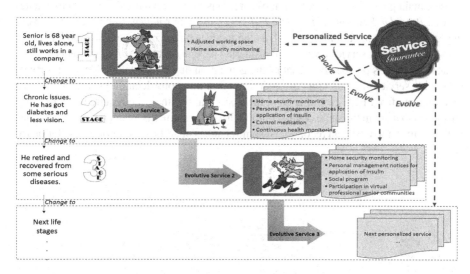

Fig. 2. Example of Personalized and Evolutionary Service Stages

In such ecosystem, as illustrated in Fig. 3, the main intervenient actors are divided into two groups:

Elderly Living Environment (ELE), representing the elderly costumers and their ambient and context, including informal caregivers such as family and friends, and all applications and devices with which they interact such as context aware devices;

Services Providers Environment (SPE), representing the care provision stakeholders with the goal to provide care and assistance services, facilitating support, companionship, all based on community participation with established trust. The SPE may include private companies and/or governmental institutions;

To operate the CBE, three main modules (Table 1) are considered:

Table 1. CBE Modules

Module 1 - Service Design	Module 2 - Service Evolution	Module 3 - Service Runtime
Responsible for receiving service requests from seniors and family and searching the providers and best services available for the senior and his/her living environment (*ELE costumer*)	Responsible for analyzing sensorial information and proposing a personalized and evolutionary care and assistance service for ELE costumer	Responsible for notifying the *SPE* and delivering to *ELE* the care and assistance services according to the corresponding agreed terms

Besides atomic business services, the CBE, with the characteristics of a Virtual Organizations Breeding Environment, also facilitates the rapid composition of business services into integrated business services, from one or several stakeholders, being thus one of the requirements that services are prepared to collaborate with each other [10]. Additionally, the planned CBE Structure considers that service descriptions are kept in a *Service Catalogue*, to be consulted whenever needed.

The main goal of these functionalities (Fig. 3) is to support a middle layer between the *Service Provider Environment* and the *Elderly Living Environment,* supporting both parts in service provision. At a higher level, as the focus is on environments, the model is also intended to support service provision considering several customers' requirements under a context awareness perspective. That is, the generation of a multi-stakeholders consortium to provide services to a certain customer also takes into consideration historical data and requirements from ELE customers and their personal routines, the requirements of other customers that are in the same context (although each customer is treated respecting his/her individuality), the CBE can recommend some changes or additional services (evolutionary service).

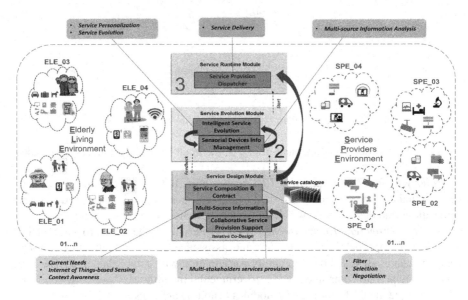

Fig. 3. Collaborative Business Ecosystem (CBE Structure)

Whenever a customer wants to make a request or subscribe a certain service, Module 1, through a *Collaborative Service Provision Support* activity, collects information about ELE costumer´s care and assistance service requirements (through questionnaires, similarities with other customers, etc.) and verifies if a suitable service description is already defined in the service catalogue. If it is an undefined service, this module is responsible for conducting a new care and assistance service design involving the relevant stakeholders [13].

In the first interaction, after performing the data analysis, the CBE suggests a service to the senior and offers an experimental period. During this period, the *Multi-source Information* activity collects sensorial data from multiple devices, such as: home sensorial information (motion detectors, pressure sensors, etc.); mobile *sensorial* information (mobile phones, wearable/ubiquitous devices/computing, etc.); and medical devices (mobile electrocardiogram, blood glucose monitoring systems, etc.) and helps the *Collaborative Service Provision Support* activity to store the new description in the service catalogue (iterative co-design). This service represents the first evolution of the service based on context awareness analysis. Furthermore, Module 1 initiates a negotiation round with suitable service providers in order to create a consortium capable of fulfilling the request in the best optimized manner. Due to the characteristic of the ELE customers, the mentioned optimization can also include sensorial and contextual information of other ELE customers, so that SPE can fulfill the request considering other activities that are already being carried out in the similar contexts.

Figure 4 illustrates an adapted i* Rationale Strategic Model for the Service Design Module (illustrative example), including the main dependencies of this layer with other actors and systems.

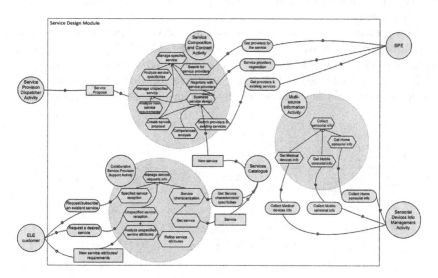

Fig. 4. Adapted i* Rationale Strategic Model for the Service Design Module

Module 2, *Service Evolution*, is responsible for proposing a personalized and evolutionary care and assistance service for the ELE costumer through data analysis (current needs, IoT-based sensing and, context awareness). The *Sensorial Devices Information Management System* gets information from the *Multi-Source Information* activity and analyzes these data. The result is sent to *Intelligent Service Evolution* activity which, by using relevant information, is able to recommend and/or suggest changes in evolutionary services for each ELE customer. Types of input information include: (i) sensorial information, essentially for context-awareness; (ii) relevant information from the ELE customer, to verify if additional services are necessary; (iii) analysis of the ELE

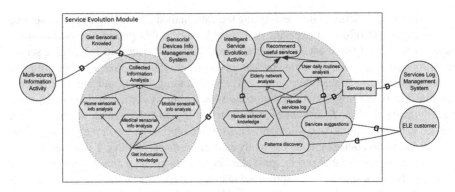

Fig. 5. Adapted i* Rationale Strategic Model for the Service Evolution Module

customer's daily habits to recommend or suggest a care and assistance services adapted to the corresponding routines; and (iv) analysis of ELE to discover new patterns in ELE customer's behavior. Figure 5 shows a preliminary adapted i* Rationale Strategic Model for Service Evolution Module, with the main dependencies of this module and the other actors and systems.

The suggested evolutionary care and assistance service implies a SPE provider (or several) for service provision for the ELE customer. Module 3 delivers the care and assistance service. Also, this module is responsible for service report to SPE.

Figure 6 shows an adapted i* Rationale Strategic Model for Service Runtime Module, with the main dependencies of this layer and the other actors and systems.

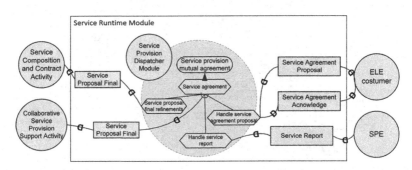

Fig. 6. Adapted i* Rationale Strategic Model for the Service Runtime Module

5 Conclusions and Future Work

In the era of IoT, many services will promote the extension of capabilities of elderly people for active aging. In this paper we highlight the design of a CBE model for elderly care based on collaborative networking and context-awareness to support active and productive aging. The adoption of context-aware analysis techniques, combined with smart devices and sensors, will enable a great number of seamless and, evolutionary and personalized services. As the intelligence in new systems grows, the technology becomes more interactive and gradually will be part of daily life of seniors, family members, caregivers and integrated service providers. While envisioned as a revolution on services, it is important to recognize that the economic growth alone will not improve older people's well-being and specific policies are required to address the implications of aging. Many computer tasks are still characterized by having high cognitive demands and, for this reason, it is important to carefully handle age-related changes on cognitive skills, when designing systems that target older adults. Because of the importance of improving the management and delivery of health care and social services, ICT plays a key role on developing quality of life of older people. Nevertheless, such initiatives are expected to assist on the mitigation of situations of isolation and loneliness, and improve family and community relationships. This paper represents the first phase of PhD and future work will rely on the progress of the CBE development and definition of mechanisms for personalization and evolution of services, as well as system's validation through tool prototype, publications, simulation and formulation of case studies (scenarios), comparison with other proposals and discussion with/assessment by stakeholders in elderly care.

Acknowledgments. This work has been funded in part by the Center of Technology and Systems and the Portuguese FCT-PEST program UID/EEA/00066/2013 (Impactor project), and by the Ciência Sem Fronteiras and Erasmus Mundi project (Brazil and European Commission).

References

1. Kearney, A.T.: Understanding the Needs and Consequences of the Ageing Consumer. The Consumer Goods Forum (2013). https://www.atkearney.com/documents/10192/682603/ Understanding+the+Needs+and+Consequences+of+the+Aging+Consumer.pdf/ 6c25ffa3-0999-4b5c-8ff1-afdca0744fdc. Accessed 5 October 2015
2. HelpAgeInternational: Global AgeWatch Index (2014). http://www.helpage.org/global-agewatch/. Accessed 3 October 2015
3. Czaja, S.J., Lee, C.C.: Information technology and older adults. In: Sears, A., Jacko, J.A. (eds.) Human-Computer Interaction: Designing for Diverse Users and Domains, pp. 18–30. CRC Press, Boca Raton (2009)
4. Alwan, M., Wiley, D., Nobel, J.: A program of the American Association of Homes and Services for the Aging (AAHSA). B. S. o. C. Foundation Edition, Washington, D.C. (2007)
5. Camarinha-Matos, L.M., Ferrada, F., Oliveira, A.I., Rosas, J., Monteiro, J.: Integrated care services in ambient assisted living. In: IEEE 15th International Conference on e-Health Networking, Applications and Services (Healthcom 2013), Lisbon, pp. 197–201 (2013)
6. Cresci, M.K., Yarandi, H.N., Morrell, R.W.: Pro-nets versus no-nets: differences in urban older adults' predilections for internet use. Educ. Gerontol. **36**(6), 500–520 (2010)
7. Hong, J., Suh, E.-H., Kim, J., Kim, S.: Context-aware system for proactive personalized service based on context history. Expert Syst. Appl. **36**(4), 7448–7457 (2009). doi:10.1016/ j.eswa.2008.09.002
8. O'Grady, M.J., Muldoon, C., Dragone, M., Tynan, R., O'Hare, G.M.: Towards evolutionary ambient assisted living systems. J. Ambient Intell. Humaniz. Comput. **1**(1), 15–29 (2010)

9. Scoble, R., Israel, S.: Age of Context: Mobile, Sensors, Data and the Future of Privacy. Patrick Brewster Press, Lexington (2014)

10. Camarinha-Matos, L.M., Afsarmanesh, H., Boucher, X.: The role of collaborative networks in sustainability. In: Camarinha-Matos, L.M., Boucher, X., Afsarmanesh, H. (eds.) PRO-VE 2010. IFIP AICT, vol. 336, pp. 1–16. Springer, Heidelberg (2010)

11. Crispim, J.A., de Sousa, J.P.: Partner selection in virtual enterprises. Int. J. Prod. Res. **48**(3), 683–707 (2010)

12. Bititci, U., Garengo, P., Dörfler, V., Nudurupati, S.: Performance measurement: challenges for tomorrow. Int. J. Manag. Rev. **14**(3), 305–327 (2012)

13. Camarinha-Matos, L.M., Rosas, J., Oliveira, A.I., Ferrada, F.: Care services ecosystem for ambient assisted living. Enterp. Inf. Syst. **9**(5–6), 607–633 (2015)

14. Graça, P., Camarinha-Matos, L.M.: The need of performance indicators for Collaborative Business Ecosystems. In: Camarinha-Matos, L.M., Baldissera, T.A., Di Orio, G., Marques, F. (eds.) DoCEIS 2015. IFIP AICT, vol. 450, pp. 22–30. Springer, Heidelberg (2015)

15. Lee, W.-P.: Deploying personalized mobile services in an agent-based environment. Expert Syst. Appl. **32**(4), 1194–1207 (2007)

16. Costa, P., Pires, L.F., van Sinderen, M., Rios, D.: Services platforms for context-aware applications. In: Markopoulos, P., Eggen, B., Aarts, E., Crowley, J.L. (eds.) EUSAI 2004. LNCS, vol. 3295, pp. 363–366. Springer, Heidelberg (2004)

17. Xu, X., Wang, Z.: State of the art: business service and its impacts on manufacturing. J. Intell. Manuf. **22**(5), 653–662 (2011)

18. LeadingAge: A Look into the Future: Evaluating Business Models for Technology-Enabled Long-Term Services and Supports (2011). www.leadingage.org/uploadedFiles/Content/About/CAST/CAST_Scenario_Planning.pdf. Accessed 4 November 2015

19. Camarinha-Matos, L.M., Afsarmanesh, H.: A comprehensive modeling framework for collaborative networked organizations. J. Intell. Manuf. **18**(5), 527–615 (2007)

Ontologies and CPS

Automatic Generation of Cyber-Physical Software Applications Based on Physical to Cyber Transformation Using Ontologies

Chen-Wei Yang[1](✉), Valeriy Vyatkin[1,2], and Victor Dubinin[3]

[1] Department of Computer Science, Electrical and Space Engineering,
Luleå University of Technology, Luleå, Sweden
chen-wei.yang@ltu.se
[2] Department of Electrical Engineering and Automation, Aalto University,
Helsinki, Finland
vyatkin@ieee.org
[3] University of Penza, Penza, Russia
victor_n_dubinin@yahoo.com

Abstract. In this paper, the aim of automatically generating a cyber-physical control system (more precisely, an IEC61499 control system) is discussed. The method is enabled by ontology models, specifically the source plant ontology model and the target control model for the CPS system implemented in the preferred programming language. The transforming of ontologies is enabled by an extension of SWRL (called eSWRL) and it is introduced here. There interpreter of eSWRL is developed using the Prolog language. A case study Baggage Handling System is used to demonstrate how the ontology models are transformed and the corresponding transforming rules that are developed.

Keywords: CPS · BHS · eSWRL · IEC61499 · Ontology transformation

1 Introduction

The complexity of cyber-physical systems design implies the use of Model Driven Engineering (*MDE*) methods [1]. The aim of MDE is to create a destination model from a source model after undergoing several transformation steps, creating intermediate models in the process. MDE are widely used and an example is the work in [2] which suggests Model-Integrated Computing (*MIC*) method, expanding *MDA* to the field of domain-specific modelling languages. Physical system's architecture can often determine the architecture of CPS control hardware and software. However it needs to be considered in conjunction with functional and non-functional requirements. In search of a proper means to represent all this information, the Semantic Web technologies [3] appear as an appropriate candidate. The cornerstone of these is the concept of ontology and the most widespread ontological language is *OWL* language based on description logic [4]. There are two components to an OWL ontology, the *T-Box* and *A-Box*.

© IFIP International Federation for Information Processing 2016
Published by Springer International Publishing Switzerland 2016. All Rights Reserved
L. Camarinha-Matos et al. (Eds.): DoCEIS 2016, IFIP AICT 470, pp. 37–45, 2016.
DOI: 10.1007/978-3-319-31165-4_4

The *T-Box* introduces the terminology of the domain while the *A-Box* captures the asserted relationships between the instances of the T-Box terms.

In this paper, a method to the automatic generation of CPS control systems utilizing the modular component-based IEC61499 CPS control system. The content of the paper includes the discussion on the development of the concept of transforming ontology, developing *eSWRL* [5] and interpretation engine in the *Prolog* language [6] to enable the transformation and lastly, demonstrated on a case study BHS system showing the developed transformation rules.

2   Relationship to Cyber-Physical Systems

Automation systems today are becoming more and more software intensive [7]. Automation systems were primarily designed using the design languages of *IEC* 61131-3 standard, targeting Programmable Logical Controllers (PLC) as hardware platform. The Internet of Things revolution raises interest in distributed systems design, which has been addressed in the automation area by such technologies as the *IEC61499* [8] standard which acts as the reference for designing de-centralized (or distributed) control systems based on the artefacts of FBs introduced in the standard. *IEC61499* uses a top-down approach which decomposes a system (or an application) down to smaller intelligent artefacts represented as FBs. *IEC61499* has already been adopted as the control system in CPS systems such as a Smart Grid [9].

BHS is a good example of a CPS where there exists an integration of collaborative computation between networks of computational elements (control intelligence) within the physical processes (conveyor section, sensors, and actuators). The traditional design paradigm in developing BHS systems is to decouple the physical system from the cyber software design. This means that the physical systems are designed first followed by the software design where there is very little connection between the two design steps. However, this practice is becoming less and less suitable as complexity in BHS systems increases with the integration of mechatronic components, computer hardware and software where it is beneficial for the physical and the cyber components to be designed concurrently. In addition, due to the complexity of BHS systems, it is more suitable to decompose the BHS system down to smaller parts in thus, moving away from a centralized control system to a more distributed control system where there is a need for cyber intelligence to interact with one-another within a cyber network. One of the tenants of MDE design is the automatic generation of software code. This work contributes to this front by using an arbitrary physical layout of a BHS system to automatically generate a CPS control system in IEC61499 using based on transforming ontologies.

3   State of the Art

There a several works [10–14] which are similar to the work presented here. The work in [10] utilizes the *Prolog* language as a mean to model and verify *IEC61499* applications. Ontology or model transformation are not the focus in this work. In [11] automatic generation of formal IEC61499 application model is proposed using graphical transformation methods. In the work, the source model is the IEC61499 application while the destination model is the formal Net Condition/Event System (*NCES*) model and the transformation is performed using the *AGG* tool [15]. In [12], a method using multi-layered ontological knowledge representation and rule-based inference engine is proposed for the purpose of semantically analyse IEC61499 based projects. In [13], an approach based on utilizing Semantic Web Technologies is proposed for the purpose of migrating IEC61131-3 PLC to IEC61499 FBs. The migration is implemented as a transformation of ontological representation of IEC61499 function block system to ontological representation of IEC 61131-3-based system. The migration is based on transforming ontologies from IEC61131-3 to IEC61499. The drawback of this approach is that the rules are developed in complicit of rule coding using XML.

The main difference between this work and the works listed above is that UML is not the core models in our approach. Both the source and target models in this work are in ontological representations and the transformation is performed directly on the ontology models. In addition, the transformation is performed on the level of the class instances and properties rather than on the level of RDF triples. The transformation rule interpreter is also developed as a self-modifying *Prolog* program implemented in *SWI Prolog* using the *OWL Thea* library [16].

4   Transforming Ontology and Development in Prolog

The concept of transformation based on ontology is represented in Fig. 1. In the proposed approach, it is possible to transform not only the *A-Box*, but the *T-Box* as well. The initial models are:

(1) T-Box and A-Box ontology of the physical plant;
(2) T-box ontology of the IEC61499 control system [16] supplemented by the A-Box with
      common and standard FB Types serving as the building blocks of the control system.

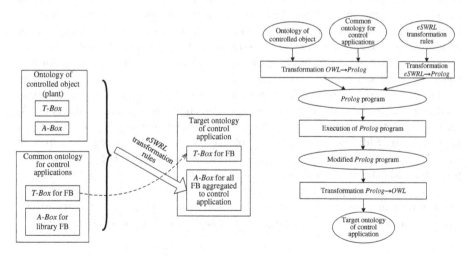

Fig. 1. Concept of proposed ontology transformation (Left) and implementation steps for the ontology transformation using Prolog (Right)

The result of the transformation is the ontology of the control application comprising of an the IEC61499 T-Box and an A-Box derived from the source model. The transformation rules play the central role in transforming the ontology. To represent the rules, the *eSWRL* language [5] is suggested. eSWRL is an extension of *SWRL* where the monotonicity property is waived. As a result, eSWRL has capabilities of ontology self-modification.

The right diagram in Fig. 1 shows the steps taken to implement the ontology transformation using the *Prolog* language. *Prolog* is used as the implementation mechanism of eSWRL rules [5].

Once the ontology model of the source plant model and the target IEC61499 T-Box are developed, the next step is to convert both ontologies to a set of facts and rules in the form of *Prolog*. At the same time, the *Prolog* equivalent of the *eSWRL* rules are also created. In a *Prolog* program, the *Prolog* facts and the rules are executed concurrently modifying the existing *Prolog* database be it creation, deletion or modification of facts. The resulting *Prolog* database (Or from the perspective of ontology, the A-Box) is the resultant target IEC61499 control system. This database is then converted back to ontology and merged with the IEC61499 T-Box to create the ontology model of the final control system. The scope of this work is to demonstrate how ontology can be used for the purpose of transforming models, overcoming the monotonicity restrictions for the purpose of model transformation. The development of the ontology for the source and destination models are out of the scope of this paper.

5   Case Study: Transforming BHS Description to IEC61499

For illustrative purpose, a simplified ontology of an airport BHS is considered for this case study. The foundation of the BHS ontology used for the case study was developed in the works [17, 18]. BHS consists of a set of conveyors. The conveyors can be connected: (1) sequentially when the baggage reached the end of one conveyor must be moved to the next conveyor and (2) in a branching way, when the baggage from the middle of one conveyor can be moved by a diverter to the beginning of another conveyor. It can be assumed that each conveyor has no more than one ejector. At the end of each conveyor a photocell is installed to discover bags and to signal to the control system the need to start the next conveyor. If a conveyor has a diverter, then it is equipped with an additional photocell, which allows push bags at the right time in the given direction. As seen in Fig. 2, there are three classes in the BHS ontology: *the convey, the diverter, the photo_eye*. In addition, there are four object properties which are: *has_diverter, has_photoeye, has_divert_connection, has_straight_connection*. The first two object properties indicates whether the conveyor section has a diverter or a photoeye respectively. Two last object properties shows how the conveyors are connected to each other. Their domain and rank are class *convey*. Moreover, two data properties: *has_id* ("to have a unique numeric identifier") and *has_name* ("to have a symbolic name") are introduced.

Fig. 2. BHS ontology consisting of Classes (left), Object Properties (Center) and Data Properties (Right) shown in Protégé.

A simple example of a step-by-step description of the sequence of transformations of one domain (BHS layout) to another domain (FB applications) is considered. The source BHS, which consists of four conveyers and one diverter, is represented in the left diagram in Fig. 3. Using *T-Box* of BHS ontology, an ontological description of BHS has been developed in *Protégé* tools [19] by developing and adding the *A-Box* description to the *T-Box* description. The graphical representation of the source ontological description of BHS in *Protégé* is represented in the right diagram in Fig. 3. The yellow colour box in Fig. 3 shows the classes of BHS ontology and the purpled coloured boxes shows the instances of classes *convey*, *diverter* and *photo_eye*, respectively.

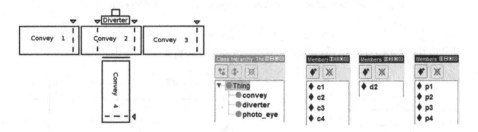

Fig. 3. A fragment of an airport baggage handling system (left) and its ontological representation (Right).

The basic building blocks for the construction of control systems for BHS in this case study are the *IEC61499* FBs. A set of Type of FBs has been developed by means of which a conceptual BHS control system for an arbitrary layout of BHS conveyors can be built. This set of FB includes three Type of FBs: *fb_control_block* is a conveyor control (Fig. 4a); *fb_photo_eye* is a photocells driver (Fig. 4b); *fb_diverter* is a diverter control (Fig. 4c).

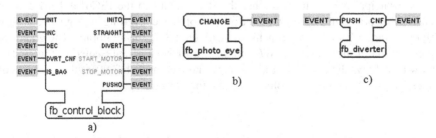

Fig. 4. Type of FBs for building the BHS control system

There are several rules which are required for transforming a full BHS system from an arbitrary layout. In total, there are 5 rules which are used to completely transform the case study BHS system. Due to paper constraints, only three main rules are presented here to illustrate the usage of the rules.

Rule 1: For each conveyor, an FB instance fb_control_block is created

This means that for each conveyor section that exists in the BHS layout, a corresponding control FB fb_control_block in Fig. 4a is created.

Rule 2a: If a conveyor has a sequential connection, a logical connection between Straight and inc (straight->inc) is made as shown in Fig. 5.

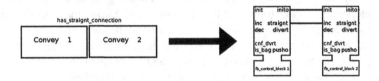

Fig. 5. Sequential conveyor connection and its IEC61499 equivalent

Rule 3: For each fb_control_block instance which is associated with another FB of the same type, an FB instance of photo_eye type is created as shown in Fig. 6.

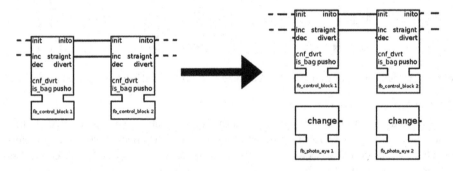

Fig. 6. Creating fb_photo_eye between 2 fb_control_block instance

The principal part of transforming the BHS layout to the IEC61499 FB control system is performed in *SWI Prolog*. The process of then converting Prolog fact database [20] to the target FB ontology is achieved with the *OWL Thea* 2 library. The graphical representation of the resulting *OWL* ontology in *Protégé* is represented in Fig. 7. The yellow box shows the classes of the ontology and the purple boxes are the instances of *fb_control_block* class and the blue boxes are the object properties.

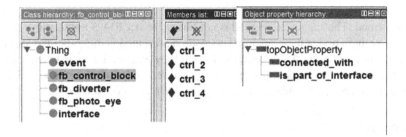

Fig. 7. Resultant IEC61499 ontology in Protégé.

The resulting ontological description of the corresponding IEC61499 FB control system is represented in Fig. 8. This FB system is an IEC61499 application intended to control the BHS from Fig. 3. The final phase in the implementation of BHS control system would be to allocate the generated control application to the resources and devices [8].

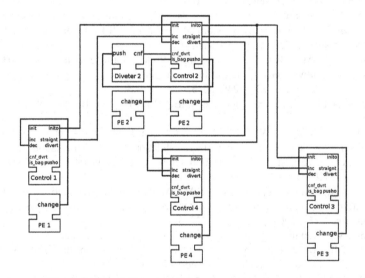

Fig. 8. IEC61499 application to control BHS from Fig. 3.

6   Conclusion and Future Work

This paper presents a method to automate the generation of control software for cyber-physical systems by transforming ontologies. This is enabled by the Prolog language and results in an IEC61499 application. The presented method follows the framework of ontology-driven engineering (ODE) and can be further extended to including refactoring, generation of formal models for analysis, implementation using programming or modelling languages. The method is applied to a BHS case study where a small fragment of the airport BHS is used as the source model and the resultant IEC61499 control system is transformed. The suggested method is not limited to IEC61499 applications and can be used in other application domains. The method presented in this paper transform the ontology directly on the A-box level of the ontology and the T-Box is unused for this work. In addition, further development of a formal semantics of the language *eSWRL* is necessary. Moreover, it is important to verify *eSWRL* transformations.

References

1. Beydeda, S., Book, M., Gruhn, V.: Model-Driven Software Development. Springer, Heidelberg (2005)
2. Ledeczi, A., Bakay, A., Maroti, M., Volgyesi, P., Nordstrom, G., Sprinkle, J., et al.: Composing domain-specific design environments. Computer **34**(11), 44–51 (2001)
3. W3C: Ontology Driven Architectures and Potential Uses of the Semantic Web in Systems and Software Engineering (2006)
4. Grau, B.C., Horrocks, I., Motik, B., Parsia, B., Patel-Schneider, P., Sattler, U.: OWL 2: the next step for OWL. Web Semant. **6**(4), 309–322 (2008)
5. Dubinin, V., Vyatkin, V., Yang, C.-W., Pang, C.: Automatic generation of automation applications based on ontology transformations. In: 2014 IEEE Emerging Technology and Factory Automation (ETFA), pp. 1–4 (2014)
6. Clocksin, W.F., Mellish, C.S.: Programming in Prolog. Springer, Heidelberg (2003)
7. Vyatkin, V.: Software engineering in industrial automation: state-of-the-art review. IEEE Trans. Industr. Inf. **9**(3), 1234–1249 (2013)
8. International Electrotechnical Commission: IEC 61499 Function Blocks, vol. IEC 61499, ed. (2005)
9. Zhabelova, G., Yang, C.-W., Patil, S., Pang, C., Yan, J., Shalyto, A., et al.: Cyber-physical components for heterogeneous modelling, validation and implementation of smart grid intelligence. In: 12th IEEE International Conference on Industrial Informatics (INDIN 2014), pp. 411–417 (2014)
10. Dubinin, V., Vyatkin, V., Hanisch, H.M.: Modelling and verification of IEC 61499 applications using prolog. In: IEEE Conference on Emerging Technologies and Factory Automation, ETFA 2006, pp. 774–781 (2006)
11. Dubinin, V., Vyatkin, V.: Graph transformation-based approach to the synthesis of formal models of IEC 61499 function blocks systems. In: Proceedings of the Institutes of Higher Education, Volga Region, Technical Sciences (2008)
12. Dai, W., Dubinin, V., Vyatkin, V.: Automatically generated layered ontological models for semantic analysis of component-based control systems. IEEE Trans. Industr. Inf. **9**(4), 2124–2136 (2013)
13. Dai, W., Dubinin, V.N., Vyatkin, V.: Migration from PLC to IEC 61499 using semantic web technologies. IEEE Trans. Syst. Man Cybern. **44**(3), 277–291 (2013)
14. Almendros-Jiménez, J., Iribarne, L.: ODM-based UML model transformations using prolog. In: Abelló, A., Bellatreche, L., Benatallah, B., (eds.) Model-Driven Engineering, Logic and Optimization: Friends or Foes? (2011)
15. AGG. http://tfs.cs.tu-berlin.de/agg
16. Thea: A Prolog library for OWL2. http://www.semanticweb.gr/thea/index.html
17. Black, G., Vyatkin, V.: Intelligent component-based automation of baggage handling systems with IEC 61499. IEEE Trans. Autom. Sci. Eng. **7**(2), 337–351 (2010)

18. Yan, J., Vyatkin, V.V.: Distributed execution and cyber-physical design of Baggage Handling automation with IEC 61499. In: 9th IEEE International Conference on Industrial Informatics (INDIN 2011), pp. 573–578 (2011)
19. Protégé. http://protege.stanford.edu
20. Grosof, B.N., Horrocks, I., Volz, R., Decker, S.: Description logic programs: combining logic programs with description logic. Presented at the Proceedings of the 12th International Conference on World Wide Web, Budapest, Hungary (2003)

Semantic BMS: Ontology for Analysis of Building Automation Systems Data

Adam Kučera[✉] and Tomáš Pitner

Laboratory of Software Architectures and Information Systems,
Faculty of Informatics, Masaryk University, Botanická 68a, 602 00 Brno, Czech Republic
{akucera,tomp}@mail.muni.cz

Abstract. Building construction has gone through significant change with the emerging spread of ICT during last decades. "Intelligent buildings" are equipped with building automation systems (BAS) that can be remotely controlled and programmed and that are able to communicate and collaborate. However, BAS aim to facilitate operation of the building and do not provide sufficient support for strategic level decision support. This article presents adaptation of Semantic Sensor Network ontology for use in the field of building operation analysis. The Semantic BMS ontology enriches the SSN with model of building automation datapoints that gather operation data and describe the interconnections between BAS devices, algorithms and influenced or monitored properties of a building. Proposed ontology allows facility managers to query BAS systems in a way that is convenient for tactical and strategic level planning and that is unavailable in current state of the art systems.

Keywords: Computer-aided facility management · Building management systems · Intelligent buildings · Building automation systems · Semantic web · Ontology · Semantic sensor network ontology · Data integration

1 Introduction

Facility management is defined by the International Facility Management Association (IFMA) in following words: "Facility management is a profession that encompasses multiple disciplines to ensure functionality of the built environment by integrating people, place, process and technology."[1]

We can distinguish several systems and/or data sources that can be utilized in order to support and simplify tasks of facility management staff, namely Building Information Model (BIM), Computer Aided Facility Management (CAFM) and Building Automation/Management Systems (BAS/BMS).

---

[1] Available from http://ifma.org/about/whatis-facility-management.

© IFIP International Federation for Information Processing 2016
Published by Springer International Publishing Switzerland 2016. All Rights Reserved
L. Camarinha-Matos et al. (Eds.): DoCEIS 2016, IFIP AICT 470, pp. 46–53, 2016.
DOI: 10.1007/978-3-319-31165-4_5

On the strategic level, CAFM systems are used for benchmarking of building oper-ation efficiency and performance. Benchmarking methods in facility management are covered in EN 15221-7 standard. Requirements placed on Key Performance Indicators (KPIs) are summarized in [1]. Among others, the authors mention flexibility, the quan-titative nature of the KPIs and simplicity of use. BMS data satisfy the first three require-ments, the simplicity of use is a downside of current BMS/BAS solutions. The paper proposes ontology description of the BAS data that largely facilitates use of the BAS data on a strategic level.

The paper is organized as follows: Sect. 2 describes contribution of the research to the field of cyber-physical systems. Section 3 presents state of the art in the field of semantic technologies for building automation. Section 4 describes the basic concept of Semantic BMS ontology. Section 5 demonstrates the capabilities of the proposed ontology on sample SPARQL queries. The last section discusses the advantages of the presented work and proposes future research topics.

2 Relationship to Cyber-Physical Systems

Modern ("intelligent") buildings are equipped with variety of sensors and controllable devices (e.g. HVAC, security systems). Devices are integrated into the Building Auto-mation System (BAS), also referred as Building Management System (BMS). Devices incorporated in BAS can be remotely controlled, monitored and queried. Actions performed in an user interface of BMS have direct impact on the physical world (e.g. opening a valve) and changes in the physical world are reflected by automation algo-rithms (e.g. air conditioning units adjusts operating parameters in response to changes of outdoor air temperature). BAS systems thus can be considered cyber-physical systems.

The BMS contains large amount of a precise, up-to-date and detailed data that are valuable for a building operation analysis and cannot be obtained any other way. Indi-vidual information objects (such as current temperature in particular room measured by a sensor) accessible in the BAS network are referred to as "datapoints" further in the paper.

Absence of structured semantic information prevents efficient querying of the data points for analytical purposes, as it is not possible to select and filter the data based on criteria such as type of source device, location of a measurement or measured quantity kind. If the data from particular data points are required (e.g. electricity consumption for the last month for each of the buildings on the site for comparison), the operators of the system have to manually gather the data point addresses by inspecting the building plans or user interface of the BAS.

The above mentioned problem clearly emerges when operating large BAS system. Masaryk University utilizes BAS network consisting of approximately 1500 devices communicating using BACnet automation protocol with hundreds of thousands data points available. The network covers 35 buildings with overall area of 120 000 m^2 at the site of University Campus Bohunice in Brno, Czech Republic and several more over the whole Brno city. The requirements of effective operation of large-scale installations of automation technologies are discussed in [8].

This paper presents Semantic BMS ontology that aims to provide novel semantic description of the building automation systems. This allows developers of business intelligence applications to effectively query the model and easily gather required building operation data based on parameters unavailable in current semantic description of building automation systems. Such parameters describe relation of measured data with the physical world and objects described by different information systems.

3   Related Work

Standardized building automation protocols such as BACnet (ISO 16484-5), LONWorks (ISO 14908-1), KNX (ISO 14543), or ZigBee generally cover operation of building automation devices, providing specifications for physical communication layer, data link and networking layer and application layer on the highest level. A specific setup of automation network is however unique to each installation site. Automation protocols focus on communication interfaces and do not provide tools for complex and structured description of datapoint semantics.

Advanced ontological representations of building automation systems pursue various goals. Overview of different approaches and aims in the field of building automation ontologies follows.

The concept of "Semantic Agents" ensuring different aspects of building operation (Energy management, Safety, Security, Comfort) is proposed in [2]. Semantic agents are complex applications facilitating semantically described automation data.

In [6], authors define several basic "ontology modules", addressing different aspects of automation systems' semantic description. In [10], the W3C Semantic Sensor Network ontology (see Sect. 4 for further details) is extended by model of physical processes occurring in the building (e.g. adjacent room exchanging energy) in order to provide tool for building operation diagnosis and anomaly detection.

Concept of Building as a Service (BaaS) is introduced in [4], aiming to simplify development and maintenance of building automation installations. In [3], discovery services for smart building are proposed using enriched SSN ontology, however they lack structures needed for complex querying (e.g. hierarchy of locations). Instead, they aim to facilitate development of self-adapting control algorithms. Integration of different automation systems facilitated by ontological generic application model is proposed in [11]. A generic application model enables deployment of platform-independent system configuration to physical devices implementing different automation protocols. In [5], ontology for integration of different BAS is proposed. The ontology describes platform-independent "parameters" (data points) which can be observed or controlled.

The Linked Open Data approach for building automation data streams is facilitated by EDWH Ontology proposed in [9]. The aim of the EDWH ontology is to provide bridge between the SSN ontology and the W3C RDF Data Cube vocabulary, as the data are meant to be analyzed by OLAP data cube techniques. In [7], the authors use SSN-based ontology for energy management based on sensor data. The semantically described BAS data help to establish situation awareness at the strategy level, allowing multi-level evaluation of energy consumption (from organization level to the level of individual appliances).

In general, existing ontologies either aims to describe aspects of BAS different from data analysis, use proprietary/ad-hoc structures for storing semantic data, ignore integration with BIM systems or do not provide domain-specific mapping of BAS systems to the SSN ontology.

4 Semantic BMS Ontology

Goal of the presented research is to provide a BAS-protocol-independent model of intelligent building systems. The proposed ontology aims to represent information (data) available for operation analysis. The Building Automation System can be viewed as a sensor and actuator network for the purposes of data analysis. Semantic description of sensor networks is a subject of extensive research, resulting in frameworks and tools such as SensorML language, Observations & Measurements (O&M) model or Semantic Sensor Network ontology (SSN). However, for the use in the domain of building automation, particular differences have to be taken into account, and SSN was thus extended into the Semantic BMS ontology (SBMS).

A semantic description of the BMS is not required to contain some information that would be duplicate (copy) of data available in the BAS, BIM or CAFM systems. The aim of the presented research is to enrich the BAS/BMS with semantic links to entities present in other systems (BIM, CAFM) and add new layer of semantic metadata that are not available elsewhere. However, in some cases replication of the information is convenient for effective querying.

The SBMS represents spatial relations within the built environment as a tree hierarchy ("site-building-floor-room"). Devices and device types taking part in building automation are modeled as well. This simplified model has to be provided by the BIM system. The ontology repository duplicates BIM data and is updated whenever source data change. This approach was chosen because it allows for efficient retrieval of frequent queries ("get all temperature sensors from the first floor of the building B2") and keeps the semantic model and BIM/OWL translation relatively simple.

The key concepts for accurate semantic annotation of sensor/actuator data are "Observed property" (OP) and "Feature of Interest" (FoI). The FoI represents an object of measurement. The OP represents specific information that we observe. In the domain of building automation, we can demonstrate the concepts on examples such as energy consumption (OP) of a specific building (FoI) or speed (OP) of a specific fan (FoI). The Semantic BMS further specializes the concepts. Namely, it restrict class and property definitions so as valid FoI can be either a device (e.g. valve, pump, engine, or PLC) or a location (site, building, floor, or room). The FoI has to be present in the BIM system.

The building operation data are generally not meant to be publicly available on the Internet. Therefore, there is no need for semantic annotation of individual measurements. For that reason, the SSN Observation concept represents only semantic connection of datapoint, sensor, FoI and OP and do not contains any actual measured data. Annotated data point values can then be directly obtained from the BAS.

The Semantic BMS Ontology introduces simplified model of processes occurring in the BAS system as an association of input and output – room temperature serves as an input for regulation algorithm of a respective AC unit, thus influencing values of various datapoints that the algorithm controls (e.g. fan speed). Furthermore, the ontology describes "indirect" influence observed as a result of physical processes occurring in the built environment. The indirect influence usually occurs between output of one algorithm and input of another. As an example, we can consider a datapoint A representing openness of a valve that mixes cold and hot water. The state of the valve is controlled by a datapoint acting as an output of an algorithm. Different control algorithm uses as an input a datapoint B representing water temperature past the valve. Thus, the datapoint A value indirectly influences the datapoint B value – the measured temperature depends on the state of the valve.

The ontology further specifies domain-specific sensing methods, provides categorization of observed physical qualities and provides model of generic data point types.

5   Results

At the moment, the Semantic BMS ontology is implemented using OWL language and populated with sample data imitating relationships in real facilities.[2] The Semantic BMS ontology defines 107 classes (79 of them however represent device types and are adapted from Industry Foundation Classes 4 standard) and 34 relations (properties). In most of the cases for both classes and relations, the added elements are specializations of existing SSN elements. For testing purposes, the ontology is populated with 175 individuals (e.g. buildings, rooms, devices, datapoints) and their respective relations.

The key classes are named Site, Building, Floor, Room, and Device, representing BIM elements, Address (representing various types of datapoints) and several classes extending original definitions from the SSN ontology: Observation, Scope (representing Feature of Interest), Observed property and Sensing.

As an ontology repository, Apache Jena TDB is used. The repository can be queried using SPARQL query language and allows to select and filter data points according to their relation to physical world and to entities described by a BIM system. Following SPARQL query extracts available information about particular data point (PREFIX declarations are omitted):

SPARQL Query: Description of a data point

---

[2] Available from http://is.muni.cz/www/255658/sbms/v1_0/?lang=en.

```
SELECT * WHERE { values ?bmsId { "2309.AI5" }
    ?dataPoint sbms:hasBMSId ?bmsId.
    ?dataPoint a ?dataPointClass.
    ?dataPointClass rdfs:subClassOf sbms:DataPoint.
    ?dataPoint sbms:expressesObservation ?observation.
    ?observation sbms:observedBy ?source.
    ?source sbim:hasBIMId ?sourceBIMId.
    ?source a ?sourceClass.
    ?sourceClass rdfs:subClassOf* sbim:Device.
    FILTER
        (not exists {?subtype rdfs:subClassOf ?sourceClass.
         FILTER (?subtype != ?sourceClass) }
      && ?sourceClass != sbms:Source).
        ?observation sbms:featureOfInterest ?scope.
        ?scope sbim:hasBIMId ?scopeBIMId.
        ?scope a ?scopeClass.
        ?scopeClass rdfs:subClassOf* dul:PhysicalObject.
        FILTER
            (not exists {?subtype rdfs:subClassOf ?scopeClass.
             FILTER (?subtype != ?scopeClass) }
          && ?scopeClass != sbms:Scope).
        ?observation sbms:sensingMethodUsed ?sensing.
        ?sensing a ?sensingClass.
        ?sensingClass rdfs:subClassOf* sbms:Sensing.
        FILTER
            not exists {?subtype rdfs:subClassOf ?sensingClass.
            FILTER (?subtype != ?sensingClass) }.
        OPTIONAL { ?sensing sbms:hasAggregationTimeWindow
            ?timeWindow }
        ?observation sbms:observedProperty ?property.
        ?property sbms:hasPhysicalQuality ?quality.
        ?property sbms:hasPropertyDomain ?propDomain.
    }
```

The presented query provides following information about the respective datapoint with BACnet address of "2309.AI5":

- Datapoint type – Specifies a role of the datapoint. Possible types can be Input, Output, User defined value, Auxiliary, Algorithm and Historical data.
- Data source – Specifies a device (further described in the BIM database) that provides the BAS with the data.
- Source device class – Describes type of the source device (e.g. Energy meter). The list of device types is derived from the IFC 4 specification.
- Scope (Feature of Interest) – A scope specifies an object the datapoint value is related to (e.g. fan in the case of an data point representing fan speed, room when measuring room temperature or building when measuring energy consumption).
- Scope type – Specifies if the scope is a location or a device.

- Physical quality (Observed Property) – Specifies (usually) physical quality represented by the datapoint (e.g. temperature, humidity, energy, or output power). The list of available qualities is based on the Unified Code of Units of Measure (UCUM) and can be extended to meet specific requirements.
- Sensing method and time window – Different data point implement different sensing methods. The simplest case is direct sensing (e.g. room temperature). Other (indirect) sensing methods employ some kind of aggregation over time period (e.g. energy consumption total for the last month).

Described parameters allow facility managers to query the BAS data storage in a way unavailable in the BAS systems themselves. The SBMS facilitates queries based on an origin of the data and their role in the building operation. Additionally, a query can inspect relations between data points and influenced operational parameters of the building using simplified model of processes.

However, facility managers should not be forced to learn SPARQL language nor to construct complex queries by themselves. The queries are meant to be constructed by an API available as a part of a middleware layer described in the following section. The API then will be used by applications for end users equipped with convenient user interface for data selection and visualization.

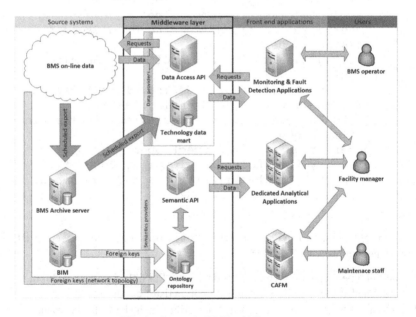

Fig. 1. Semantic BMS middleware.

6   Conclusions and Future Work

The proposed ontology introduces novel approach to querying building automation cyber-physical systems. Querying is based on relations of datapoints with the physical world, contrary to common approach based on network topology based querying. The proposed method allows flexible querying for strategic level decision support in the field of facility management.

Future work includes development of middleware layer providing RESTful APIs providing convenient methods both for semantic querying and data access (see Fig. 1). The APIs will be designed and developed with respect to the typical building performance analysis queries, mainly based on benchmark indicators proposed in "EN 15221-7 Benchmarking in Facility Management" standard.

Other challenges include effective methods to populate the ontology with semantic data and on seamless integration of the Semantic BMS ontology with BIM data sources by providing explicit mapping between Semantic BMS ontology concepts and objects described by Industry Foundation Classes standard used for data exchange in BIM systems.

References

1. Alwaer, H., Clements-Croome, D.J.: Key performance indicators (KPIs) and priority setting in using the multi-attribute approach for assessing sustainable intelligent buildings. Build. Environ. **45**(4), 799–807 (2010)
2. Andrushevich, A., Staub, M., Kistler, R., Klapproth, A.: Towards semantic buildings: goal-driven approach for building automation service allocation and control. In: 2010 IEEE Conference on Emerging Technologies and Factory Automation (ETFA), pp. 1–6, September 2010
3. Bovet, G., Ridi, A., Hennebert, J.: Toward web enhanced building automation systems. In: Bessis, N., Dobre, C. (eds.) Big Data and Internet of Things: A Roadmap for Smart Environments, vol. 546, pp. 259–283. Springer, Heidelberg (2014)
4. Butzin, B., Golatowski, F., Niedermeier, C., Vicari, N., Wuchner, E.: A model based development approach for building automation systems. In: 2014 IEEE Emerging Technology and Factory Automation (ETFA), pp. 1–6, September 2014
5. Caffarel, J., Jie, S., Olloqui, J., Martnez, R., Santamara, A.: Implementation of a building automation system based on semantic modeling. J. Univ. Comput. Sci. **19**(17), 2543–2558 (2013)
6. Legat, C., Seitz, C., Lamparter S., Feldmann, S.: Semantics to the shop floor: towards ontology modularization and reuse in the automation domain. In: Preprints of the 19th IFAC World Congress, pp. 3444–3449 (2014)
7. Curry, E., Hasan, S., O'Riain, S.: Enterprise energy management using a linked dataspace for energy intelligence. In: Sustainable Internet and ICT for Sustainability (SustainIT 2012), pp. 1–6 (2012)
8. Kučera, A., Pitner, T.: Intelligent facility management for sustainability and risk management. In: Hřebíček, J., Schimak, G., Kubásek, M., Rizzoli, A.E. (eds.) ISESS 2013. IFIP AICT, vol. 413, pp. 608–617. Springer, Heidelberg (2013)
9. Mehdi, M., Sahay, R., Derguech, W., Curry, E.: On-the-fly generation of multidimensional data cubes for web of things. In: Proceedings of the 17th International Database Engineering and Applications Symposium, IDEAS 2013, pp. 28–37. ACM, New York (2013)
10. Ploennigs, J., Schumann, A., Lecue, F.: Extending semantic sensor networks for automatically tackling smart building problems. In: 2014 European Conference on Artificial Intelligence (ECAI), pp. 1211–1214 (2014)
11. Reinisch, C., Granzer, W., Praus, F., Kastner, W.: Integration of heterogeneous building automation systems using ontologies. In: 34th Annual Conference of IEEE Industrial Electronics, IECON 2008, pp. 2736–2741, November 2008

Ontological Interaction Using JENA and SPARQL Applied to Onto-AmazonTimber Ontology

Márcio José Moutinho da Ponte[1,2(✉)], Paulo Alves Figueiras[2],
Ricardo Jardim-Gonçalves[2], and Celson Pantoja Lima[1]

[1] Federal University of Western Pará - UFOPA, Santarém, Brazil
{marcio.ponte,celson.lima}@ufopa.edu.br
[2] Institute Development of New Technologies – UNINOVA,
NOVA University of Lisbon, Campus de Caparica, Caparica, Portugal
{paf,gr}@uninova.pt

Abstract. Knowledge representation and use are fundamental processes in many areas. The use of a semantic referential (i.e, a domain ontology and a set of related tools to exploit it) to represent knowledge has allowed the development of new mechanisms of semantic search, inferences, and analysis of complex content, but the development of a semantic referential is still a complex task, time-consuming and fundamentally performed by knowledge holders. Taking that into account this work discusses the development of a semantic referential applied to botanical identification process in the Brazilian Amazon area, mainly focused on the mechanisms of interaction and access to a domain ontology (named Onto-Amazon-Timber) based on JENA API and SPARQL queries. The main aspects of the development of this work are presented and discussed here. Current challenges and open points are also addressed.

Keywords: Ontology · Interaction · JENA · SPARQL

1 Introduction

The Knowledge Management technological platforms frequently include a Semantic Referential (SR) in order to support the knowledge life cycle (i.e, creation, formalisation, sharing, dissemination, acquisition). Essentially a SR includes a controlled vocabulary (e.g. ontology, taxonomy, and thesaurus), semantic vectors, and additional services/tools allowing the proper use of the SR [1].

The use of SR to represent knowledge on the web and in other fields has allowed the development of new inference mechanisms [2]. However, the development of a SR is a complex and lengthy process, since it requires the development of software tools, construction or adaptation of controlled vocabularies involving knowledge experts [3].

The formal representation of knowledge and its complex relationships is the goal of a SR [4]. In our work, the SR is enabled by a domain ontology providing formal and explicit specifications of shared concepts from botanical domain, aiming to capture and

© IFIP International Federation for Information Processing 2016
Published by Springer International Publishing Switzerland 2016. All Rights Reserved
L. Camarinha-Matos et al. (Eds.): DoCEIS 2016, IFIP AICT 470, pp. 54–61, 2016.
DOI: 10.1007/978-3-319-31165-4_6

to explain the vocabulary used by experts from the domain. By doing that, the SR aims to ensure a communication free of ambiguities (as much as possible) [5].

Structurally speaking, an ontology contains Concepts, Individuals, Axioms, Properties, and Relations, interrelated by semantic liaisons supporting both management and use of the knowledge hold by the ontology. The semantic liaisons pave the way to: (i) make inferences about acquired knowledge; (ii) create reasoning mechanisms to explore the richness of the existing knowledge; (iii) help detecting structural problems such as inconsistent relations, absence of concepts, individuals, or properties, due to mismanagement of the ontology.

This work aims to develop software artefacts to handle the semantic liaisons previously described based on services from JENA API (Application Programming Interface) and queries SPARQL (Protocol and RDF Query Language). The software artefacts created here are to be assessed using the Onto-AmazonTimber ontology, which holds botanical knowledge and, in our scenarios, specifically targeting the botanical identification process, bringing a collection of features and characteristics of a vast amount of forest species of the Amazon.

The paper is structured as follows. Section 2 highlights the link between the this work and Cyber-Physical Systems (CPS). Section 3 shows the related works relevant to this one. Section 4 gives a short explanation of the case studies selected to assess this work. Finally, Sect. 5 draws some conclusions and discusses the future research.

2   Contribution to Cyber-Physical Systems

Cyber-Physical Systems (CPSs) integrate the dynamics of the physical processes with those of the software and communication, providing abstractions and modelling, design, and analysis techniques for the integrated whole [6]. The interaction of computers, networking, and physical systems happens in multiple ways that require fundamentally new design technologies. The technology depends on the multi-disciplines such as embedded systems. Additionally, it is worth recalling that the software is now embedded into devices with other purposes than computation (e.g. cars, medical devices, scientific instruments, and intelligent transportation systems) [7].

Computers are integrated into the physical world in a transparent way through Embedded Systems, Real-Time Systems, and mobile computing applications. The latter is deeply characterised by highly visual processing software embedded into smartphone platforms, such as apple iOS and Google's Android. Mobile applications, commonly referred as *apps*, allow users from remote areas to have access to the whole world of ubiquous services, ranging from internet banking to health services such as image analysis and diagnosys.

Part of the work developed here is to be assessed in a smartphone-enabled scenario and, as such, both local and remote services will be available to test and validate our SR, which can be considered as a kind of Cyber-Physical System. Having said that, if on the one hand th semantic-based applications are technologically demanding, in terms of memory and performance, on the other hand mobile phones have grown from simple cell phones to highly powerful and sophisticated devices potentially able to host any sort of application in a near future, including semantic-based.

3   Related Work

The identification of botanic species is an integral part of any forest inventory, essential for forest management plan and, therefore, mandatory for commercialisation of wood. However, the usual process of botanical identification has been traditionally based on empirical knowledge coming from native experts of the forest area (Bushmen), who adopt popular names to identify and classify the species. However, such terminology normally does not match to the scientific names cataloged by taxonomists [8].

In this context, the Onto-Amazon Timber ontology (Fig. 1) is a domain ontology essentially focused on the botanic identification process of Amazonian species, exploiting pattern recognition concepts and technologies, which allow increased accuracy of the botanical identification process, thus reducing the differences of knowledge representation between Bushmen and taxonomists.

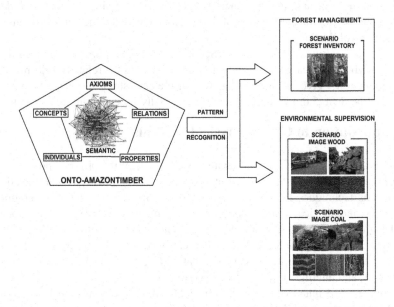

Fig. 1.   Scenarios of assessment of Onto-Amazon Timber ontology

The application scenarios (Fig. 1) are used for recognition of patterns and features extracted from the images (stored in the Onto-Amazon Timber). Those patterns and features are obtained through the use of axioms and image processing on the stored images. Such application scenarios provide inputs deemed relevant to support the decision-making process in botanic identification process, in the cases depicted in Fig. 1, namely forest inventory, inspection of transportation of legal wood, and species recognition using coal-based images.

4 Case Study Onto-AmazonTimber: Semantic Structure and Ontology Interaction

It is worth recalling that the main focus of this work is to create a common ground, semantically speaking, allowing bushmen and taxonomists to exchange knowledge without ambiguities. In other words, the Onto-Amazon Timber aimis at offering an equivalence between popular and scientific names in the botanical identification process of the amazonian species, in order to increase the accuracy of results.

The ontological Relations are formed by regular expressions that are organised in Objects and Data properties. The former links the related ontological entities, providing the required degree of semantics to build knowledge. The latter has the function to connect typed data to ontological entities in order to characterise them.

The Onto-AmazonTimber offers the following Objects Properties: *Classified by, Composed by, Formed by, Included in, It has, It has popular name, It has scope, It has synonymy.* The following expressions represent the set of Data properties in Onto-AmazonTimber: *Has image, It has function, It has measured.* The axioms are represented by semantic relationships consist of expressions, entities, literal data.

The axioms define ontological entities and the appropriate restrictions on their interpretation, allowing to infer new knowledge from the existing knowledge. The development of axioms requires the participation of experts with good knowledge from the domain of work, in order to help creating the sets of logic sentences to be inferred when handling the ontological entities. These axioms are necessary and sufficient to express these issues and characterize their solutions. Moreover, any solution to an issue of competence should be described by the axioms of ontology and should be consistent.

For illustrative purposes only, the axiom with high semantic value in the ontology, which describes the characteristics of each botanical species (Fig. 2). The entity *Specie* contains several individuals and, among them, the *Dipteryx odorata* which contains various semantic relations restrictions grouped in *Objects properties* defined by (*Dipteryx odorata included in list most marketed, Dipteryx odorata it has popular name Cumaru*) and *Data properties* defined by (*Dipteryx odorata has image "c:\\imagemontologia/..."*). The latter is the link to access the image of that species, allowing the ontology to offer image processing capabilities. These axioms make it possible to infer what is the botanical species holding the characteristics expressed by the restrictions of the axioms.

Fig. 2. The semantic structure of Onto-AmazonTimber ontology.

4.1   Ontology Interaction with JENA

JENA consists of a Java framework allowing to work on programming environment with dynamic handling of Resource Description Framework (RDF) models, represented by resources, properties and literals, forming tuples (*predicate, [subject] [object]*) that originate the objects created by Java. It presents a set of features to support application development in the context of ontologies. In addition to the features for manipulating the OWL language and use the Simple Protocol and RDF Query Language (SPARQL) [9].

The API JENA offers a set of methods giving access to the elements of a given ontology (classes, properties and individuals). Examples of methods are *listClasses (),* *listIndividuals (), or listSubClasses (),* as shown in Table 1. Those methods call the *toList ()* method in order to get the elements through an instance of the *java.util.List* class. Additionally to that, JENA offer two basic methods allowing to identify which class or instance is being manipulated in the iterations, namely *getURI ()* and *getLocal-Name ()*. Whilst the former returns the full name or URI (prefix + name) of the object, the latter returns only the name of the given object.

Table 1. Few methods from JENA API

Methods	Source
listClasses	OntClass concept : newM.listClasses().toList()) concept.getLocalName()
listIndividuals	Individual intance : newM.listIndividuals().toList()) intance.getLocalName()
listSubclasses	OntClass concept = (OntClass) newM.getOntClass(uri); OntClass subConcept : concept.listSubClasses().toList()) subConcept.getLocalName()

JENA offer a larger set of methods to access the ontological structure. For instance, the *getObjectsFromObjectTriple* method gets the list objects from a concept A that are related through a specific property with another object of concept B. The excerpt of Java code below illustrates the invocation of that method.

OntologyInteraction listCaracteristics = **new** OntologyInteraction() ;

ArrayList<String> objects = listCaracteristics.
getObjectsFromObjectTriple("Dipteryx_odorata", "ClassifiedBy");

In the given example, the invoked method lists the botanical characteristics of *Dipteryx_odorata* botanical species that are interconnected by *Classified By* property. Such semantic relationship may be observed in Fig. 3, where the concept Species has a series of objects including the *Dipteryx_odorata*, which presents some property of objects that create relationships with other objects, such as *Heartwood_Distinct_Color* instance of *Heartwood_Color* class.

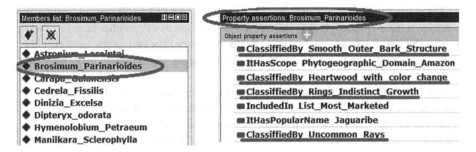

Fig. 3. Semantic relationships obtained by the *getObjectFromObjectTriple* method.

4.2 Ontology Interaction with SPARQL

Several knowledge management software applications require the integration of data from distributed, autonomous data sources. Until recently it was rather difficult to access and query data in such a setting because there was no standard query language or interface. With SPARQL [10], a W3C recommendation for an RDF query language and protocol, this situation has changed. It is now possible to make RDF data available through a standard interface and query it using a standard query language.

The RDF created from the semantic modeling presented in ONTO-AMAZON-TIMBER ontology stands as environment for SPARQL queries of this work, for illustrative purposes only, one relevant SPARQL query is showed in the source below, which aims to identify a botanical species using as criteria some features supplied by the user.

```
PREFIX rdf: http://www.w3.org/1999/02/22-rdf-syntax-ns#

PREFIX mm:<http://www.semanticweb.org/2016/ontology8#>

SELECT DISTINCT ?Especie WHERE {

        ?Species mm:IncludedIn mm:List_Most_Marketed.

        ?Species mm:itHasSynonymy mm:Coumarouna_odorata.

        ?Species mm:itHasPopularName mm:Cumaru.

        ?Species mm:itHasScope mm:Amazonia.

        ?Species mm:classifiedBy mm:Cener_Blue.}
```

Most forms of SPARQL query contain a set of triple patterns called a *basic graph pattern*. Triple patterns are like RDF triples except that subject, predicate, or object may be a variable. A basic graph pattern *matches* a subgraph of the RDF data when RDF terms from that subgraph may be replaced by the variables and the result is the RDF graph equivalent to the subgraph [11].

The consultation is a two-step process, as follows: the SELECT clause identifies the variables selected in the query represented by a question mark, another function shown is the DISTINCT whose function is to exclude clauses repeated in the answer. Other function is the WHERE acting as restrictive filter using the semantic relations of the ontology.

The basic pattern graphic in the case study is composed of several patterns of three tuples with a single variable (*?Species*) as the object. Tuples make such references as the characteristics (*List_Most_Marketed; Coumarouna_odorata; Cumaru; Amazonia; Heartwood_Bluish*) selected criterion for selection of Species, integrated by their property objects (*Included In; It Has Synonym, It Has Popular Name, It has Scope, Classified By*).

The semantic relationships referred to in this SPARQL query are depicted in Fig. 3, observing the peculiarities and distinctions of access methods, which use the botanical characteristics as variable for selection in *getObjectsFromObjectTriple* method or as selection criteria for the SPARQL method.

5   Conclusions and Future Work

This paper presented a set of software mechanisms allowing interaction and access of the domain ontology named Onto-Amazon Timbe ontology, using the JENA API integrated with SPARQL queries.

The Onto-Amazon Timber ontology has fundamentally been conceived to support the botanical identification process of Amazonian species. Therefore, a short illustrative example of semantic and conceptual structure in the domain of botany was given. The work presented here is part of the assessment of the Onto-Amazon Timber ontology which belongs to a broader context, represented by a semantic framework supported by a solid conceptual model and application scenarios, using pattern recognition and images for identification of botanical species in the Amazon.

Additionally, this work (among others) help reinforcing the potential of JENA API as a powerful tool to handle ontologies. This API offers numerous features that, combined with the Java language, accelerate and facilitate the implementation on new software tools due to the fact that Java already offer packages containing various classes and implemented interfaces and flexibility to interact with other languages such as the queries SPARQL used in this research.

Finally, it is worth emphasising the relevant role of ontologies in the current development of the semantic web. Therefore, the evolution of technologies, mechanisms, and tools to handle semantic resources is a must in this quest, such as the combination of the JENA API and SPARQL queries, presented in this paper.

Future work covers the assessment and validation of the SR here presented, including the development of new application scenarios, as well as integration with mobile devices and technologies of artificial intelligence in order to optimize the botanical identification algorithms to increase the accuracy of the results produced.

References

1. Lima, C., Zarli, A., Storer, G.: Controlled vocabularies in the European construction sector: evolution, current developments, and future trends. In: Loureiro, G., Curran, R. (eds.) Complex Systems Concurrent Engineering, pp. 565–574. Springer, London (2007)
2. Bittencourt, I.I., Isotani, S., Costa, E. Mizoguchi, R.: Research directions on semantic web and education. J Scientia **19**(1), 59–66 (2008)
3. Guarino, N.: Formal ontology and information systems. In: Proceedings of the 1st International Conference on Formal Ontologies in Information Systems. IOS Press, Trento (1998)
4. Legg, C.: Ontologies on the semantic web. Ann. Rev. Inf. Sci. Technol. **41**, 407–451 (2007)
5. Breitman, K.K.: Web semântica: a internet do futuro. LTC, Rio de Janeiro (2005). CNPq. Sala de Imprensa, 19 September 2008. http://www.cnpq.br/saladeimprensa/noticias/2008/0919e.htm. Accessed April 2009
6. http://chess.eecs.berkeley.edu/cps/
7. Zhang, F.M., Szwaykowska, K., Wolf, W., Mooney, V.: Task scheduling for control oriented requirements for Cyber-Physical Systems. In: Proceedings of the 2008 Real-Time Systems Symposium, pp. 47–56 (2005)
8. Procópio, L.C., Secco, R.S.: A importância da identificação botânica nos inventários florestais: o exemplo do—tauari (*Couratari* spp. e *Cariniana* spp. - Lecythidaceae) em duas áreas manejadas no estado do Pará, pp. 31–44. Acta Amazonica, Manaus (2008)
9. Apache.org. Apache JENA (2015). http://jena.apache.org/. Accessed November 2015
10. Prud'hommeaux, E., Seaborne, A.: SPARQL Query Language for RDF. W3C Recommendation, January 2008. http://www.w3.org/TR/rdf-sparql-query/
11. W3C – World Wide WEB Consortium. eXtensible Markup Language (XML) (2014). http://www.w3.org/XML/

Petri Nets

Combining Data-Flows and Petri Nets for Cyber-Physical Systems Specification

Fernando Pereira[1,2,3(✉)] and Luis Gomes[1,3]

[1] Faculdade de Ciências e Tecnologia, Universidade Nova de Lisboa, Lisbon, Portugal
fjp@deea.isel.ipl.pt, lugo@fct.unl.pt
[2] ISEL, Instituto Superior de Engenharia de Lisboa, Lisbon, Portugal
[3] CTS, UNINOVA, Caparica, Portugal

Abstract. This paper proposes a new modeling formalism for the specification of cyber-physical systems, combining the functionality offered by Petri nets and synchronous data flows. Petri nets have been traditionally used to model the behavior of reactive systems, whose state evolves depending on the interaction with external events. On the opposite, data-flow formalisms have been used predominantly to describe data-driven systems that produce output data through mathematical transformations applied to input signals. The proposed formalism covers both kinds of problems, offering support for the design of mixed systems containing linear control and signal processing operations along with event driven elements. Model composition using multiple components communicating through input and output signals and events, enable the implementation of distributed cyber-physical systems. The new formalism and the respective execution semantics are presented, with special attention to the bidirectional interaction between Petri net elements and data-flow nodes.

Keywords: Cyber-physical systems · Embedded systems · Petri nets · Data-flow

1 Introduction

Cyber-physical systems assume a growing importance in all fields of the modern world, with many applications that include industrial machines, home appliances, entertainment systems and gadgets. The fast dissemination of the Internet and the wide availability of inexpensive networking technology, brought Internet connectivity to the recent generations of embedded devices, contributing to the birth of the Internet of Things. This evolution enabled the development of new applications and services, including access to automatic payment systems, connection to social media platforms and solutions based on distributed networks of remote devices, like smart grids, city traffic control systems, in-vehicle systems and wireless sensor networks.

These advances opened a gap for novel development solutions adapted to the new design challenges. Model based development formalisms, from which Petri nets can be highlighted, promise to answer these questions: offering high level design concepts that

© IFIP International Federation for Information Processing 2016
Published by Springer International Publishing Switzerland 2016. All Rights Reserved
L. Camarinha-Matos et al. (Eds.): DoCEIS 2016, IFIP AICT 470, pp. 65–76, 2016.
DOI: 10.1007/978-3-319-31165-4_7

hide the low level platform details, contribute to accelerate development time, minimize the probability of coding errors and reduce time-to-market.

From the existing Petri net [1] based tools, the IOPT tools framework [2] and the underlying Petri net class [3] have been designed for embedded system controller development. However, the IOPT tools currently do not offer component based model composition and lack support for complex data manipulation operations. The formalism proposed in this paper was specified to address both problems: provide support for model composition and introduce a complementary formalism to deal with data driven problems. These extensions simplify the modeling of mixed linear/event-driven controllers and systems that make extensive use of mathematical data operations. Model composition bring advantages to the modeling of complex systems, support component re-use and allow the creation of component libraries.

2   Relationship to Cyber-Physical Systems

This paper proposes a new development formalism designed to support the development of embedded systems and cyber-physical systems. The proposed formalism supports model composition using components. The external interface of the components is defined by input and output signals and events.

The formalism supports both centralized and distributed implementations, where each component can be located on remote Internet locations. A complete Cyber-physical system may be specified as a single model composed by multiple components and the components may be executed at different locations. A communication protocol for remote control, monitoring and debug of embedded controllers [12, 13] satisfies the requirements to support the component interactions.

3   Related Work

The formalism proposed in this paper is the result of previous work around the IOPT tools framework [2] and the IOPT Petri net class [3]. Other Petri net based tools that support the modeling of embedded systems have been proposed by different authors, from which the NCES [4], SNS [5], SIPN [6] and CPN [7] should be mentioned.

However, the traditional Petri net based formalisms have primarily focused on the reactive part of the controllers and do not offer good support to solve data-driven problems. Petri net classes have usually relied on text based mathematical expressions that are enabled when specific places are marked or transitions fire. In some classes these expressions may call procedures written using standard programming languages, as Java and Standard-ML. In contrast, this paper proposes a hybrid formalism that employs Petri nets and data-flows. Petri nets provide good modeling capabilities to design reactive controllers. Data-flows offer advantages to solve data driven problems, including digital signal processing, linear control of systems and support systems that require many data transformation operations.

Some Petri net based dialects, as the NCES [4] and SNS [5], support model composition based on components communicating through signals and events. One of the main applications of this capability is the study of the interaction between controllers and controlled systems (called plants). However, the plant models often require intensive data processing and are not adequately modeled using Petri nets. The new formalism offers both model composition and data processing capabilities and is well adapted to model the plants and the controllers.

Other formalisms, as the synchronous data flows [8, 9], IEC61499 [10] and Matlab Simulink [11] offer support for data driven problems, but are not centered on Petri nets and do not benefit from all the available Petri net model-checking tools.

4 Graphical Representation

A DSPnet model (Data-flow, Signals and Petri nets), is a directed graph composed by five types of nodes: Petri net places, Petri net transitions, input/output signals, inputs/output events and data-flow operations. The nodes may connected using two types of arcs: normal-arcs and read-arcs. Normal arcs correspond to the traditional Petri net arcs. Read arcs are used to transmit data between graph nodes and may be used to connect signals, data-flow operations and Petri net nodes.

Figure 1 presents an example model. Petri net places as drawn as yellow circles and transitions as cyan rectangles. Input signals are presented as green circles and input events as green diamonds. In the same way, output signals and events are drawn as red circles and red diamonds. The data-flow nodes that perform mathematical operations, are represented by gray trapezoids, with green anchors to attach input arcs and red anchors to attach output arcs. Normal Petri net arcs are drawn as solid black and read arcs are drawn as dashed blue.

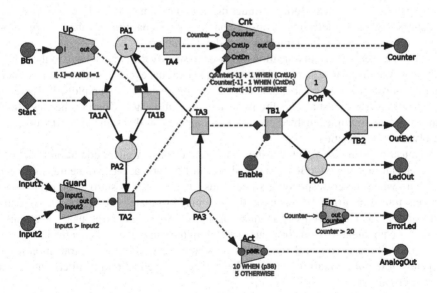

Fig. 1. Example model (Color figure online)

Data-flow nodes, called operations, employ mathematical expressions to calculate new values. Each operation has a set of input and output anchors that may be connected to other nodes using read arcs. Each output anchor holds one expression, used to compute the corresponding value. Input anchors are used as operands in the mathematical expressions. Each anchor has an associated name and data-type. Computation time is considered instantaneous, with no propagation delays, implementing a synchronous data-flow behavior. As a consequence, signal loops, where the output of an operation is directly or indirectly feedback to an input, are forbidden. When a signal loop is desired, expressions may employ a delay operator «[−n]», to refer the value of the feedback input from previous execution steps. In the example on Fig. 1, the *Cnt* operation employs the delay operator to read the previous value of the *Counter* output signal («Counter[−1]»).

Input and output signals and events are used to define the external interface of the model and establish the communication with the external world. Input signals may be connected to sensors or user interface items like buttons and switches. Output signals may be connected to LEDs, power-electronics, relays or mechanical actuators. When used as components in distributed cyber-physical systems, the input and output signals and events may be used to communicate with other sub-systems. Each signal has an associated data-type. Available data-types are Boolean, integer ranges and fixed-point ranges.

Events represent instantaneous actions that may (or not) happen on any execution step. Input events are triggered by external sources, but internal or output events may be triggered by the firing of transitions or as the result of data-flow computations. For example, a data-flow operation may detect the crossing of a predefined threshold on an input signal and produce an event. In this case, the data-type associated with the operation output must be defined as an «event». Events may also be used as input for data-flow operations, treated as Boolean values inside expressions. In Fig. 1, the *Up* operation produces an event by detecting an up edge on the input signal *Btn,* and the *Cnt* operation interprets two events to implement and up-down counter.

The Petri net places and transitions are used to specify the control logic of the cyber-physical systems, whose state evolves reacting to external events and changes in signals and data-flow elements. A maximal step execution semantics is employed, where all enabled transitions must always fire in the next execution step. Conflicts between transitions, when multiple enabled transitions compete for the same place tokens, are resolved using priorities.

Transition firing is inhibited using guards and events. A guard condition is defined by a read arc starting on a Boolean signal or data-flow operation, and the transition can only fire when the corresponding value holds true. Read arcs starting on events also prevent transition firing. In this case, the arcs may originate on external events, data-flow nodes producing events or another transition. When the arc originates on a transition, it is equivalent to a synchronous channel. In the same way, when a read arc starts on a place, it creates a guard condition that is equivalent to a test arc: transition firing depends of the place marking but no tokens are consumed and no conflicts with other transitions are raised.

Figure 1 presents several examples of guard conditions and transition input events. Arcs terminating on a transition, with a solid dot near the end, denote a guard condition and arcs ending with a diamond denote an event. Transitions *TA1*, *TA2* and *TB1* are conditioned by input events, while transitions *TA2*, *TA4* and *TB1* have guard conditions. In particular, transition *TA4* has a test arc from place *PA1* and transitions *TA3* and *TB1* are connected using a synchronous channel.

In the opposite direction, place marking and events generated by transition firing, may be used to influence the data-flow calculations. Read arcs starting on places read the number of place tokens. Read arcs starting on a transitions transmit events. In Fig. 1 the value of *AnalogOut* is calculated from place PA3, with value 10 when the place is marked and 5 otherwise. The value of the *Counter* output is defined by the *Cnt* operation. This operation implements an up/down counter controlled by events triggered by the TA2 and TA4 transitions.

Arcs may be visualized using two formats: in addition to the usual graphical representation, a symbolic format contributes to minimize clutter and improve readability. When this format is chosen, instead of drawing an arrow, the identifier of the source node is presented near the target node. Figure 1 includes two symbolic arcs. Both arcs start at the *Counter* output, ending at the *Cnt* and *Err* operations.

Input and output signals and events may be connected directly to Petri net nodes. In Fig. 1, transition *TB2* generates an output event and the marking on place *POn* defines the value of output *LedOut*.

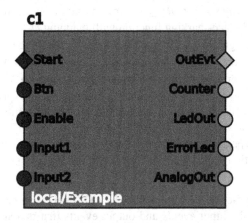

Fig. 2. External interface (Color figure online)

A typical approach to the design of complex systems, is the assembly of multiple components that implement individual sub-systems. This way, the same component sub-systems may be re-used on multiple projects. Existing models might be used as components to create higher level solutions. Figure 2 displays the external interface of the model presented on Fig. 1. The interface has a set of anchors corresponding to the input and output signals and events defined in the original model, which may be directly used in other models using read arcs. Distributed implementations may place the components on different hardware platforms, or even on remote network locations. Centralized implementations may start with the assembly of a flat model containing the contents of all components.

5   Formal Definition

A DSPnet is a directed graph composed by three parts: an external interface composed by input and output signals and events, a state control part consisting of a low-level Petri net and a data-processing part composed by data-flow operations and internal signals.

Definition 1: A system specified by a DSPnet is described as a tuple $DSPnet = (P, T, S, E, O, A, R, m_0, s_0, w, pt, ex, st, ot)$ satisfying the following requirements:

(1) P is a finite set of places
(2) T is a finite set of transitions
(3) S is a finite set of signals
(4) E is a finite set of events
(5) O is a finite set of data-flow nodes, called operations
(6) $P \cup T \cup S \cup E \cup O = \phi$
(7) A is a finite set of normal Petri net arcs with $A \subseteq (P \times T) \cup (T \times P)$
(8) R is finite set of read arcs with
      $R \subseteq (S \times S) \cup (S \times O) \cup (S \times T) \cup (O \times S) \cup (O \times O) \cup (O \times T) \cup$
      $(P \times T) \cup (O \times E) \cup (E \times O) \cup (E \times E) \cup (E \times T) \cup (T \times T)$
(9) $\forall s \in S$, $\#\{(x \times s)|(x \times s) \in R\} \leq 1$ (signals have no more than one input arc)
(10) $\forall e \in E$, $\#\{(x \times e)|(x \times e) \in R\} \leq 1$ (events have no more than one input arc)
(11) m_0 is the initial place-marking function with mapping $m_0: P \rightarrow N_0$
(12) s_0 is the initial signal values partial function with mapping $s_0: S \rightarrow \{N_0, -\}$
(13) w is the normal-arcs weight function with mapping $w: A \rightarrow N_0$
(14) pt is the transition priority function with mapping $pt: T \rightarrow N_0$
(15) ex is a function applying operations to mathematical expressions (where all expression non-literal operands are the source of the operation input arcs)
(16) $ex : O \rightarrow exp$, where $\forall nlop \in exp(O)$, $nlop \in \{x|(x, O) \in R\}$
(17) st is a signal type function with mapping $st: S \rightarrow t$, $t \in \{Boolean, Range\}$
(18) ot is an operation result type function with mapping $ot: O \rightarrow t$, $t \in \{Boolean, Range, Event\}$.

The external interface of a system defined by a DSPnet is composed by a set of input signals, output signals, input events and output events that is a subset of the system signals and events.

Definition 2: The external interface of system specified by a DSPnet is a tuple $EIF = (IE, IS, OE, OS)$ satisfying the following requirement:

(1) $IE \subseteq E$
(2) $IS \subseteq S$
(3) $OE \subseteq E$
(4) $OS \subseteq S$
(5) $IE \cap IS \cap OE \cap OS = \phi$
(6) $\forall s \in IS$, $\{(x \times s)|(x \times s) \in R\} = \phi$ (input signals have no input driver arcs)
(7) $\forall e \in IE$, $\{(x \times e)|(x \times e) \in R\} = \phi$ (input signals have no input driver arcs).

A system defined by a DSPnet can be decomposed in two parts, the state control logic and the data-processing part:

Definition 3: The state control logic part of a *DSPnet* is a low level Petri net defined by a tuple $PN = (P, T, A, m_0, w, tp, R^-)$ where:

(1) P is the DSPnet set of places
(2) T is the DSPnet set of transitions
(3) A is the DSPnet set of normal arcs
(4) m_0 is the DSPnet initial marking mapping $m_0\colon P \to N_0$
(5) w is the DSPnet arc weight mapping $w\colon A \to N_0$
(6) tp is the DSPnet transition priority mapping: $T \to N_0$
(7) R_P is a subset of the DSPNet set of read arcs such as $R_P \subseteq R \wedge R^- \subseteq (P \times T) \cup (T \times T)$ (test arcs and synchronous channels).

Definition 4: The data processing part of a *DSPnet* is a synchronous data-flow $SDF = (O, S, E, R^+, s_0, ex, st, ot)$ where:

(1) O is the DSPnet set of data-flow operation nodes
(2) S is the DSPnet set of signals
(3) E is the DSPnet set of events
(4) R_D is a subset of the DSPnet read arcs $R_D = R - R_P$
(5) s_0 is the DSPnet initial signal values partial function $s_0\colon S \to \{N_0, -\}$
(6) *ex* is the DSPnet operation expressions function
(7) st is the DSPnet signal types function
(8) *ot* is the DSPnet operation results type function.

The *ex* mathematical expressions described in Definition 1, produce integer range values, Boolean values, or events and can include the following items:

- Literal operands: decimal values or hexadecimal values starting with the «0x» prefix
- Variable operands corresponding to the graph nodes directly connected through input arcs
- The arithmetic operators +, −, *, / and *MOD*, plus the unary operator −
- The comparison operators =, <>, <, <=, > and >=
- The logical operators AND, OR, XOR and the unary operator NOT
- The « bit » operators and (&), or (|) and not (!) in addition to shift left («) and shift right (»), - Sub-expressions inside parentheses (and)
- The delay operator ([−n]) in association with variable operands, to refer past values from previous execution steps
- The array index operator ([+ i]) associated with tables of constant values stored in operation nodes, to implement mathematical functions based on tables of values
- The conditional operators *WHEN* and *OTHERWISE* to build if/case constructs.

In Fig. 1, the *Cnt* up-down counter was implemented using the expression:

```
Counter =      Counter[-1]        +1        WHEN       CntUp,
               Counter[-1]        -1        WHEN       CntDn,
               Counter[-1] OTHERWISE
```

Meaning that the new value of the *Counter* output signal is calculated as:

(a) The previous value of *Counter* plus 1 if the *CntUp* event happens
(b) The previous value of *Counter* minus 1 if *CntDn* happens
(c) The value remains unchanged when none of these events occur.

6   Execution Semantics

The execution semantics of the proposed modeling formalism inherits principles from both synchronous data flows [9] and low level Petri nets [1], in particular from the IOPT Petri net class [3].

The evolution of the system state is performed in quantum steps, called execution steps that typically occur at a certain frequency, with variable of fixed time intervals. Each execution step is considered instantaneous, but may be divided in a finite number of micro-steps.

As presented in Definition 3, the state control logic part of a model consists on a low level Petri net. As a consequence, the main aspect defining the evolution of a Petri net are the firing rules associated with transitions: in order to fire, a transition must be simultaneously enabled and ready.

Definition 5: A transition is enabled when all input places, connected through normal Petri net arcs, hold a number of tokens that is equal or more than the respective arc weights. For a transition t: $\forall (p \in P \mid (p,t) \in A), m(p) > w(p,t)$.

Definition 6: A transition guard condition is defined by a read arc ending in the transition, originating on a node containing a Boolean value or a range value, evaluated as true when different from zero.

Definition 7: A transition input event is defined by a read arc ending in the transition, originating on an event, a transition or an operation producing a result of type event.

Definition 8: A transition is ready when all guard conditions and input events are true

Definition 9: Maximal step execution semantics - all transitions simultaneously enabled and ready are forced to fire on the next execution step.

Definition 10: Conflict – Two or more transitions are in conflict when all of them are simultaneously enabled, but the number of tokens on the shared input places is not enough to fire all of them.

Definition 11: Conflict resolution – Conflicts between transitions are solved using priorities. Priority criteria is composed by: (1) execution micro-step, (2) transition priority and (3) transition unique identifier.

A read arc starting on a transition propagates events produced when the transition fires, called transition output events. These events can be forwarded to other transitions or used by data-flow operations. For example, a read arc directly connecting two transitions creates a synchronous channel, where the source transition can be viewed as a master and the target as a slave. Synchronous channels are typically used to synchronize transitions located inside different components. Systems with multiple components may contain large chains of master-slave transitions.

However, as both the master and slave transitions must fire on the same execution step (when enabled and ready), the execution semantics rules must ensure that the firing of the master transitions is evaluated before evaluating the slave transitions. Any data-flow operations that depend on events produced by the masters must also be calculated before evaluating the slave transitions. This problem lead to the concept of micro-steps, that define a precise sequence of evaluation. All DSP-net nodes are associated with a micro-step number, including transitions and data-flow operations.

Definition 12: Micro-step assignment:

(1) Nodes with no input read arcs are assigned to micro-step 1.
(2) Nodes with input read arcs are assigned a micro-step number corresponding to the maximum micro-step associated with these input read arcs, according to the following rules:
   (a) Read arcs used in inside mathematical expressions in association with the delay operator «[−n]», are assigned to micro-step 1, as the expression uses values stored from previous executions steps.
   (b) Read arcs starting on a transition, propagating transition output events, are assigned a micro-step number equal to the transition micro-step plus 1.
   (c) Read arcs starting on non-transition nodes, are assigned the same micro-step number as the source node.

In the same way as transition dependencies lead to the concept of micro-steps, the existence of dependencies between data-flow operations evaluated in the same micro-step, require the definition of another sequencing mechanism to fine-tune the order of operations evaluation, called nano-steps.

Definition 13: Nano-step assignment:

(1) Nodes with no input read arcs are assigned to nano-step 1.
(2) Nodes with input read arcs are assigned a nano-step number corresponding to 1 the maximum nano-step associated with these input read arcs, according to the following rules:
   (a) Read arcs used in inside mathematical expressions in association with the delay operator «[−n]», are assigned to nano-step 1.
   (b) Read arcs starting on nodes from past micro-steps, including all places and transitions, are assigned nano-step 1.

(c) Read arcs starting on nodes from the same micro-step, are assigned 1 plus the source node nano-step number.

Lemma 1: Any data-flow operation nodes sharing the same micro-step and nano-step numbers can be evaluated by any execution order, or executed in parallel.

It is important to notice that micro-step and nano-step numbers are just used to sort the operation/transition nodes and define a deterministic sequence of evaluation, but does not imply any clocking scheme for hardware or software implementations.

In order to be syntactically correct, the number of micro-steps and nano-steps must be finite and a DSPnet cannot exhibit loops where nodes with higher micro/nano step numbers are used by nodes with lower numbers. This would contradict the rules on definitions 12 and 13, even if the loops occur across nodes located on different components. To avoid loops, the external interface of the components must include information about the dependencies between input and output signals and events. When models of the individual components are available, a flat model containing all the nodes from all components should be constructed and analyzed.

In conclusion, the following algorithm may be used to implement an execution step of a system described by a DSPnet:

```
read input-signals, input-events
for-each                            place                        do
    avail-marking[place] = marking[place]
    add_marking[place] = 0
done
for micro-step = 1 to n-micro-steps do
    for nano-step = 1 to n-nano-steps[micro-step] do
        execute data-flow-operations[micro-step][nano-step]
    done
    for-each transition[micro-step] (sort by priority,identifier)
    do
      if transition-is-enabled and transition-is-ready
      then
          for-each input-place[transition] do
              avail-marking[place] = marking[place] - arc-weight
          done
          for-each output-place[transition] do
              add-marking[place] = add-marking[place] + arc-weight
          done
      end if
done
for-each                            place                        do
    marking[place] = avail-marking[place] + add_marking[place]
done
for-each signal do
    if use-delay-operators(signal)     then shift-registered-
values(signal)
done
write output-signals, output-events
```

7   Conclusion and Future Work

This paper proposes a new modeling formalism for the design and specification of cyber-physical systems, addressing both reactive, data-driven and mixed systems, employing a hybrid language that joins Petri nets and data-flows. Model composition based on components that communicate using input and output signals and events, simplify the design of distributed solutions based on networks of remote components located on the internet.

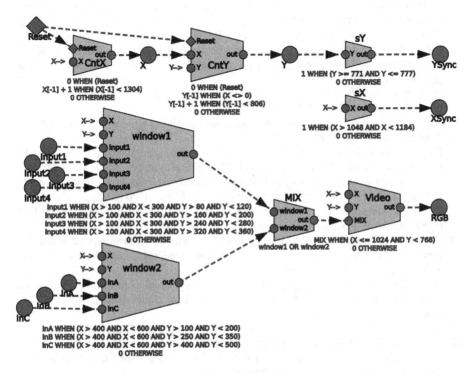

Fig. 3. Video image generator model (Color figure online)

Relative to previous solutions, the new formalism offers many advantages to model data-driven systems that do not employ traditional state-machines, with advantages over purely Petri net centered based languages. As an example, Fig. 3 presents a simple data flow model of a video image generator model. This model may be easily translated to VHDL and implemented on a FPGA device, to generate a video image that presents two windows with rectangular buttons that lighten whenever several input signals are active. A Web based graphical editor for the proposed formalism is currently under development, and all the examples presented in this paper were designed using this tool. The editor implements an algorithm to calculate the micro-step and nano-step number according to the definitions presented in this paper and displays the results. Future work includes the development of an entire tool framework, including a Web based simulator, model-checking tools, automatic code generation tools for software (C) and hardware

(VHDL) based platforms. An existing protocol to support distributed implementations and remote debug and monitoring over the internet [12, 13] will be ported to the new formalism.

References

1. Reisig, W.: Petri Nets: An Introduction. Springer, New York (1985)
2. Pereira, F., Moutinho, F., Gomes, L.: IOPT-Tools - towards cloud design automation of digital controllers with Petri nets. In: ICMC 2014 International Conference on Mechatronics and Control, Jinzhou, China, 3–5 July 2014
3. Gomes, L., Moutinho, F., Pereira, F., Ribeiro, J., Costa, A., Barros, J.-P.: Extending input-output place-transition Petri nets for distributed controller systems development. In: ICMC 2014 - International Conference on Mechatronics and Control, Jinzhou, China, 3–5 July 2014, pp. 1099–1104 (2014)
4. Hanisch, H.M., Lüder, A.: A signal extension for Petri nets and its use in controller design. Fundamenta Informaticae **41**(4), 415–431 (2000)
5. Starke, P., Roch, S.: Analysing Signal-Net Systems. Humboldt-Universiät at zu Berlin, Institut fur Informatik (2002)
6. Frey, G.: Hierarchical design of logic controllers using signal interpreted Petri nets. In: Proceedings of the IFAC AHDS 2003, Saint-Malo, France, vol. 12, pp. 401–406 (2003)
7. Jensen, K., Kristensen, L.M.: Colored Petri nets: a graphical language for formal modeling and validation of concurrent systems. Commun. ACM **58**(6), 61–70 (2015)
8. Lee, E.A., Messerschmitt, D.G.: Synchronous data flow. Proc. IEEE **75**(9), 1235–1245 (1987). doi:10.1109/PROC.1987.13876
9. Colaço, J.L., Pagano, B., Pouzet, M.: A conservative extension of synchronous data-flow with state machines. In: Proceedings of the 5th ACM International Conference on Embedded Software, pp. 173–182. ACM (2005)
10. Vyatkin, V.: Instrument Society of America. IEC 61499 function blocks for embedded and distributed control systems design (p. o3neida). ISA-Instrumentation, Systems, and Automation Society (2007)
11. Dabney, J.B., Harman, T.L.: Mastering Simulink. Pearson, Upper Saddle River (2004)
12. Pereira, F., Gomes, L.: Minimalist architecture to generate embedded system web user interfaces. In: Camarinha-Matos, L.M., Tomic, S., Graça, P. (eds.) DoCEIS 2013. IFIP AICT, vol. 394, pp. 239–249. Springer, Heidelberg (2013)
13. Pereira, F.; Melo, A.; Gomes, L.: Remote operation of embedded controllers designed using IOPT Petri-nets. In: 13th IEEE International Conference on Industrial Informatics; 22–24 July 2015, Cambridge, UK (2015)

Systematization of Performance Evaluation Process for Industrial Productive Systems Considering Sustainability Indicators

Edson H. Watanabe[1,3(✉)], Robson M. da Silva[2], Fabrício Junqueira[1],
Diolino J. Santos Filho[1], and Paulo E. Miyagi[1]

[1] University of São Paulo, São Paulo, Brazil
{edsonh.watanabe,fabri,diolinos,pemiyagi}@usp.br
[2] State University of Santa Cruz, Ilhéus, BA, Brazil
rmsilva@uesc.br
[3] Instituto Federal de Santa Catarina, Florianópolis, Brazil
edsonh@ifsc.edu.br

Abstract. Available industrial standards do not explicitly consider how to treat sustainability indicators in PS design and its control system. Therefore, this paper proposes a framework to systematize the performance evaluation process for industrial PS considering indicators that qualify and quantify its sustainability. The framework adopts Petri net technique and extensions of the standard ANSI/ISA95. Simulation-based analysis, decision making techniques and a PS´s classification based on product green seal are also considered. Furthermore, the framework considers the processing information, storing and accessing each component using a Cyber Physical Technology due to the trend of PSs to be, in fact, a network for companies that are, in general geographically dispersed.

Keywords: Productive system · Petri Net · Sustainability indicator

1 Introduction

Over the years, industrial productive systems (PSs) have been modified to include innovation [1], such as serialization, standardization and reconfiguration capabilities, however, without worrying about the waste of natural resources. Since the mid-80 s, due to the scarcity of raw materials, non-governmental organizations, such as Roman Club, have been warned about the need to include sustainability in the PS design [1–3]. Thereby the governmental initiatives arose through the United Nations such as World Commission on Environment and Development [4], and events as Rio 92, Kyoto 97 and more recently Doha 2015. Currently, the PSs performance must be concerned to sustainability indicators, such as: reduction of negative impacts in conservation of energy and natural resources, management practices for safety assurance of the employees, communities, consumers and best practices for business feasibility and profitability). However, available industrial standards like ANSI/ISA95 do

L. Camarinha-Matos et al. (Eds.): DoCEIS 2016, IFIP AICT 470, pp. 77–85, 2016.
DOI: 10.1007/978-3-319-31165-4_8

not explicitly consider how to treat sustainability indicators into the PS design and its control system [5]. Therefore, this paper proposes a framework to systemize the PS sustainability performance evaluation. Productive systems (PSs) concept used in this work is all industrial automated process, developed to execute activities to produce specific product, defined by stages such as: material preparation, assembly, validation test, and expedition. The framework considers enhancement of PS design requirements related to sustainability, adoption of Production Flow Schema (PFS) and Petri Net (PN) modeling techniques, extensions in the ANSI/ISA95 standard in order to include sustainability indicators. To evaluate the sustainability of PSs, a set of indicators must be measured to quantify and qualify the PS performance related to them. In turn, these indicators also must be used to guarantee certain grade of sustainability for PSs, positive impact on the environment, satisfaction of the employees, proper use of technology and profitable manufactured products. Thereby the framework also supports product classification based on sustainability seal. The performance and sustainability evaluation is based on computational environment of Cyber Physical Systems (CPS), which can be explored in PSs to create an infrastructure for data processing and acquisition, connecting elements to monitor variables that compose the sustainability indicators. According to a cloud computing vision, PS must explore CPS to assure a collaborative environment to (re)configure online productive processes that are executed in disperse PSs independently of its geographical localization.

The text is structured in sections: Sect. 2 describes the importance of CPS to monitor sustainability indicators. Section 3 presents the considerations about industrial standard ANSI/ISA95, sustainability indicators and green seal. Section 4 shows the framework considered for the systematization of the performance evaluation process. Section 5 describes an example of the simulation and the analysis procedure. Section 6 reports the conclusions and further works.

2   Cyber Physical System and Sustainability

The deployment of cyber-physical systems (CPS) in for PSs is fundamental to use resources efficiently providing time economy, waste and cost reduction. PSs must be designed according to sustainable and service-oriented business practices, optimizing production processes to attend customer demand considering product features, deadline, costs, security, reliability, logistic and sustainability, also to achieve resource efficient production.

Then, CPS must be explored to create an smart infrastructure for data processing and acquisition, connecting elements to monitor variables that will compose the sustainability indicators. According to a cloud computing vision, PS must explore CPS to assure a collaborative environment to (re)configure online productive processes that are executed in disperse structure, independently of its geographical localization. These systems form the basis of emerging and future intelligent services, and improve the quality of life in many areas [6–8], providing the foundation of this proposal, including its infrastructure.

3   Standards and Indicators

Available industrial standards do not explicitly consider how to treat sustainability indicators in PS design and its control system. Therefore, a new approach must be considered, i.e., in this paper the ANSI/ISA95 standard is reviewed to meet the sustainability requirements.

According to organizational structure of an industrial company established in ANSI/ISA95 [5], the production information is processed at level 3 - Manufacturing Execution System (MES). The results from the production performance analysis executed at MES are sent to the business level (upper level) that assists managers to make decisions (Fig. 1, on the left side). The proposed review of this work is: the level 3 is re-interpreted to include a Sustainability Management System (SMS) module to treat sustainability, which is responsible for processing the data collected from the lower levels to calculate the sustainability indicators through information from PS infrastructure. In case of any indicators discrepancy, the SMS acts close to the existing MES modules in order to indicate and notify the higher level and to send commands for lower levels in accordance to directive established by the business level.

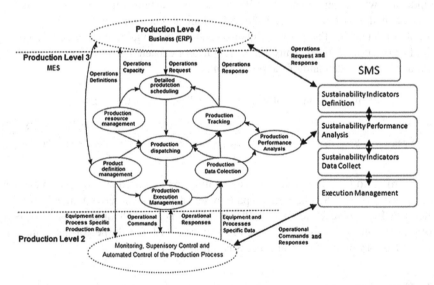

Fig. 1. ANSI/ISA-95 norm and the proposed SMS module.

The SMS is composed by sub-modules:

- "sustainability indicators data collection", that stores the PS indicators data;
- "sustainability performance analysis", that calculates the performance of the PS based on data received from both the MES and PS sustainability indicators;
- "sustainability indicators definition", that deliveries the interface of a performance and a evaluation of sustainability indicators from other PSs that composes the disperse system to ensure compatibility among them;

- "execution management", that coordinates interactions among the SMS sub-modules and equipment located on inferior levels.

3.1 Metrics and Indicators

According to [9], sustainability indicators have three main objectives: enhance awareness and understanding, inform decision-making, and measure progress toward established goals. These indicators are qualitative or quantitative values used to evaluate the sustainability aspects of a system [10]. However, according to [11], the measure of the sustainability is more than just a set of indicators and there are different approaches to be considered, such as the definition of the actions set, in which the indicators must be verified. For e.g., [12] states that the result of this measurement should support the identification of specific areas to apply enhancements related to sustainability in PS activities. Analyzing the data achieved and its interpretation is other fundamental phase, since the difficulties are due to the complexity related to definition of several indicators [13, 14]. Inter-relationships may bring the conclusions about the level of sustainability and decisions of future improvements.

Based on [15] and other references, due to limited space in this work, Table 1 lists an indicator which can be used to a PS considering sustainability in four dimensions: environmental, economical, social and technological. For example: through the table is possible to determine the "Energy intensity index" by dividing energy consumed and unit of product (kWh/unit). Thus, for each indicator chosen it is necessary to determine the index based on demand of the production.

Table 1. Example of sustainability indicators (adapted from [15])

Dimension	Sub-category	Indicator	Quantification method (yearly)
Environmental	Resource consumption	Energy intensity (kWh/unit)	Energy consumed/unit of product

According to [16], the grade of sustainability may be used as a metric to evaluate the performance of PS. There are a pattern set of processes performance indicators, called Key Performance Indicator (KPI), which are measured to quantify and to qualify process performance evaluation. In the ISO standards [17, 18], the performance measure is treated as part of an industrial process creation value.

3.2 Green Seal

Based on similar initiative to encourage industries to produce in accordance to sustainability factors, such as economical, social, environmental and technological, it is also suggested a "green seal". This seal is also a register that the framework is in working order. All customers that buy products with green seal have the guarantee that they are helping to keep a better world. This way, the industry can show that it is doing something for the welfare of people and nature, and it also produces an extra motivation for its employees and local community.

4 Framework

Based on previous works [19, 20], there is necessity to systematize the performance evaluation process for industrial PS, and a way do this, is through of a framework. Thus, the framework defines a procedure to evaluate the performance of PS considering indicators that qualify and quantify sustainability in PSs.

PSs can be approached as a Discrete Event System (DES) [21, 22] and based on this, Petri Net (PN) technique [23] can be adopted as a tool to systematize the modeling procedure, analysis and control specification (Fig. 3). In fact, there are other techniques to model DES, but when the implementation of control solutions in industrial process is relevant; the models based on PN are considered the most effective an easy way to program industrial controllers [24]. Even though, the introduction of the sustainability concept does not change the nature of PS, its consideration at the system design stage is not trivial. Therefore, it is presented in Fig. 2 the framework for the performance evaluation of sustainable PS [19, 20]. This Figure also shows a simplified flow of information.

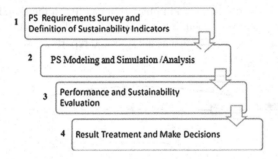

Fig. 2. Framework for the analysis procedure of performance in sustainable PSs.

The proposed framework to perform evaluation of sustainable and dispersed PSs considers:

1. Specifications of physical machine operations and the types of technologies involved in the processes. Based on these specifications, data are extracted. The environmental resources data, materials and processes that composes the environmental information and are previously defined.
2. The processes in the PS are described by using a top-down procedure that generates PN models [21]. The PN models are structurally and functionally analyzed including simulation techniques for quantitative/qualitative evaluation for different scenarios. The process modeling describes a practical and systematized way of assessing the performance of a sustainable PS by monitoring the indicators defined according to four dimensions of sustainability: environmental, economical, social and technological.
3. The expected KPIs related to sustainability, which are obtained from the PN models are stored at SMS database and used to compare the information collected from the productive plants on-line. It is supposed that the information about the current status of the

productive plants is available at the MES database; however, there are cases where direct communication from SMS to the supervisory level (lower level) is necessary.

4. The evaluation of the differences among the expected values of KPIs and measured values are reported to superior level (level 4). Although there are cases where some activation commands are previously established (derived from decision of upper level). In this case, a message must be send to MES to update the tasks to be carried out in the PS.

5   Simulation and Analysis

The proposed systematization procedure specifies how to execute the evaluation process of sustainability indicators and it could be applied in any type of PSs considering its particularity and complexity. The framework associated with the proposed systematization supports the specification for data acquisition system of equipment, sensors and data network of all information into the industrial infrastructure. The data acquisition systems work continually while the production is operating, this way at any time the responsible staff of the production administration can log in the system and evaluates performance in any network. Then, it can be used Petri editor/Simulator as an analysis tool or any other discrete event tool. The analysis and reports are sent to the upper level system to make a decision.

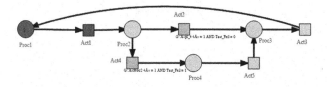

Fig. 3.   Production representation using Petri Net in IOPT environment.

Figure 3 represents an example of production line where a flow of material goes through PN places indicating three different productive processes: Proc1-> Proc2 -> Proc3, at approved situation. In each place that an operation is executed it demands energy consumption. If a fail occur in Proc2, the material flow is switched to Proc4 to repair the material and then follows to Proc3. The PN arc between PN transition Act3 and PN place Proc1 assures the sequence of activities. In parallel to production line, there is a measurement system working represented in Fig. 4. The elements Sen1_C, Act_R, Sen1_O and Act_W represent a sensor operation that acts when Proc2 in Fig. 3 starts operation, thus the sensor reads the energy consumption and the value read is compared with a pattern value stored in a database. There are three possibilities for the index: normal, regular and high. In all cases after checking the result it is stored in a database. This is executed by tasks (Fig. 4): ProcSen1_2, ProcSen1_3 and ProcSen1_3. These results will be used to make decisions at upper level system (level 4).

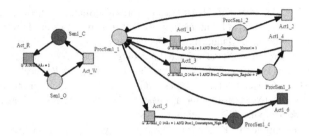

Fig. 4. Measure consumption in Proc2 representation

This measurement structure is applied where the sustainability indexes values are relevant to evaluate the production line.

6   Conclusions and Further Work

To evaluate the performance of a PS considering sustainability the ANSI/ISA95 standard structure is reviewed. This paper also defines a set of indicators which can be used to qualify and quantify sustainability in PSs and presents a framework that considers these indicators. The framework adopts the PN technique to consider the sustainable PS design, simulation-based analysis, decision making and classification techniques based on green seal of products. The seal is a register that the framework is in working order. In the adopted approach, the process modeling describes a practical and systematized way of assessing the performance of a sustainable PS by monitoring the indicators defined according to four dimensions of sustainability: environmental, economic, social, and technological. The proposed systematization specifies how to execute the evaluation process of sustainability indicators and it could be applied in any type of PSs considering its particularity and complexity. But the framework needs to be to know its limitations mainly in large systems. Due to limited space in this work, the case shown in Sect. 4 is a small sample of simulation and analysis features that they will be detailed in further works.

Acknowledgements. The authors would like to thank CNPq, CAPES, FAPESP for the financial support.

References

1. Senge, P.M., Carstedt, G., Porter, P.L.: Innovating our way to the next industrial revolution. MIT Sloan Manag. Rev. **42**(2), 24 (2001). Winter 2001, ABI/INFORM Global
2. Goldstone, J.A.: Efflorescences and economic growth in world history: rethinking the "rise of the west" and the industrial revolution. J. World Hist. **13**(2), 323–389 (2002). Fall 2002
3. McDonough, W., Braungart, M.: The NEXT industrial revolution. The Atlantic Monthly **282**(4), 82–92 (1998)
4. WCED - World Comission on Environment and Development: Our Commom Future. Oxford University Press, Oxford and New York (1987)

5. ANSI/ISA-95.00.03.2005: ANSI-American National Standart Institute and ISA–The Instrumentation Systems and Automation Society, Enterprise-Control System Integration Part3: Activity Models of Manufacturing Operations Management (2005)
6. NIST- National Institute of Standards and Technology: Cloud Computing and Sustainability: The Environmental Benefits of Moving to the Cloud NIST, Cyber-Physical Systems: Situation Analysis of Current Trends, Technologies, and Challenges, National Institute of Standards and Technology (NIST). Columbia, Maryland (2012). http://events.energetics.com/NISTCPSWorkshop/pdfs/CPS_Situation_Analysis.pdfi
7. Sundmaeker, H., Guillemin, P., Fries, P., Woelffle, S.: Vision and challenges for realising the internet of things. In: CERP-IoT Cluster of European Research Projects on the Internet of Things (2010)
8. Colombo, A.W., Karnouskos, S., Bangemann, T.: A system of systems view on collaborative industrial automation. In: IEEE International Conference on Industrial Technology (ICIT 2013), 25–28 Feb 2013, Cape Town, South Africa (2013)
9. Veleva, V., Hart, M., Greiner, T., Crumbley, C.: Indicators of sustainable production. J. Cleaner Prod. **9**, 447–452 (2001)
10. O'Brien, C.: Sustainable production - a new paradigm for a new millennium. Int. J. Prod. Econ. **60**, 1–7 (1999)
11. Amrina, E., Yusof, S.M.: KeyPerformance indicators for sustainable manufacturing evaluation in automotive companies. In: 2011 IEEE International Conference on Industrial Engineering and Engineering Management (IEEM), 6–9 Dec 2011
12. Joung, C.B., Carrell, J., Sarkar, P., Feng, S.C.: Categorization of indicators for sustainable manufacturing. Ecol. Ind. **24**, 148–157 (2013)
13. OECD - Organization for Economic Co-operation and Development: Sustainable Development: Critical Issues. OECD Publishing (2001)
14. OECD - Organization for Economic Co-operation and Development: Sustainable manufacturing toolkit - Seven steps to environmental excellence, START-UP GUIDE. OECD Publishing (2011)
15. Tan, H.X., Yeoa, Z., Nga, R., Tjandraa, T.B., Song, B.: A sustainability indicator framework for Singapore small and medium-sized manufacturing enterprises. In: The 22nd CIRP Conference on Life Cycle Engineering, Procedia CIRP, vol. 29, pp. 132–137 (2015)
16. US Department of Commerce: Sustainable manufacturing initiative. In: Proceedings of the 2nd Annual Sustainable Manufacturing Summit, Chicago, USA (2009)
17. ISO Std.22400-1:2010, Automation systems and integration—key performance indicators (KPIs) for manufacturing operations management–Part 1: overview, concepts and terminology (2010)
18. ISO Std.22400-2:2014, Automation systems and integration—key performance indicators (KPIs) for manufacturing operations management–Part 2: definitions and descriptions of key performance indicators (2014)
19. Watanabe, E.H., Blos, M.F., da Silva, R.M., Junqueira, F., Santos Filho, D.J., Miyagi, P.E.: A framework to evaluate performance of disperse productive system through the sustainability performance indicators. In: 15th IFAC/IEEE/IFIP/IFORS Symposium Information Control Problems in Manufacturing, Ottawa, Canada (2015)
20. Watanabe, E.H., Blos, M.F., da Silva, R.M., Junqueira, F., Santos Filho, D.J., Miyagi, P.E.: Key performance indicators of disperse productive system to evaluate performance and sustainability. In: 23rd ABCM International Congress of Mechanical Engineering, Rio de Janeiro, RJ, Brazil (2015)
21. Miyagi, P.E.: Controle Programável - Fundamentos do Controle de Sistemas a Eventos Discretos. (In Portuguese). Editora Edgard Blücher, São Paulo, Brasil (2001)

22. Villani, E., Miyagi, P.E., Valette, R.: Modelling and analysis of hybrid supervisory systems: a Petri net approach. Springer, London (2007)
23. Silva, M.: Half a century after Carl Adam Petri´s PhD thesis: a perspective on the field. Annu. Rev. Control **37**(2), 191–219 (2013)
24. da Silva, R.M., Watanabe, E.H., Blos, M.F., Junqueira, F., Santos Filho, D.J., Miyagi, P.E.: Modeling of mechanisms for reconfigurable and distributed manufacturing control system. In: Camarinha-Matos, L.M., Baldissera, T.A., Di Orio, G., Marques, F. (eds.) DoCEIS 2015. IFIP AICT, vol. 450, pp. 93–100. Springer, Heidelberg (2015)

Extending IOPT Nets with a Module Construct

José Ribeiro[1,2,3(✉)], Fernando Melício[1], and Luis Gomes[2,3]

[1] Instituto Superior de Engenharia de Lisboa, Instituto Politécnico de Lisboa,
1959-007 Lisbon, Portugal
{jribeiro,fmelicio}@deea.isel.ipl.pt
[2] Faculdade de Ciências e Tecnologia, Universidade Nova de Lisboa,
2829-516 Caparica, Portugal
lugo@fct.unl.pt
[3] UNINOVA - CTS, Monte de Caparica, 2829-516 Caparica, Portugal

Abstract. Input-output place-transition nets (IOPT nets) is a Petri net based formalism targeted for the development of embedded systems controllers. It is an extension to common place-transition Petri nets, introducing constructs to model the communication between the controller and the environment and using an execution semantics assuring a deterministic behavior. However, IOPT nets and the supporting tools framework - the IOPT-Tools - do not have a mechanism to support model structuring. Since models are flat, all the graphical components and annotations are visualized in the same page. Systems with several dozens of nodes become very difficult to manage. In this paper a modular construct for IOPT nets is presented, helping to manage large-scale systems, and the reuse of model components across projects. The algebraic specification of the model is provided and an example illustrating the concept is presented.

Keywords: Modularity · Composition · Low-level Petri nets · IOPT nets

1 Introduction

The development of embedded systems is a challenging task due to their strict requirements of safety, correctness and real time constraints, among others restrictions [1]. Embedded systems are dedicated computational devices, which controls a physical system through a communication interface. These systems are characterized by states, changing from one state to another by the occurrence of discrete events.

Among several formal methods to model discrete event systems, Petri nets have the advantage of allowing to model the structure and the dynamics of the system and are provided with powerful analysis methods [2]. Aspects as concurrent execution and synchronization of actions are naturally modeled with Petri nets. There are also a vast number of extensions that allow the modeling of distributed components or the reactive behavior of systems [3, 4]. The use of formal methods allows the analysis and validation of the system, which is crucial, in particular when the compliance with the requirements has to be assured previously to their implementation [5]. Formal methods, particularly those with a graphical representation, as Petri nets, also enable the implementation

© IFIP International Federation for Information Processing 2016
Published by Springer International Publishing Switzerland 2016. All Rights Reserved
L. Camarinha-Matos et al. (Eds.): DoCEIS 2016, IFIP AICT 470, pp. 86–95, 2016.
DOI: 10.1007/978-3-319-31165-4_9

of development approaches, based on models, that supports the entire development flow and the automation of some tasks.

The IOPT nets are a Petri net based modeling language, extending the place/transition Petri net class, addressed to the development of embedded systems controllers. Its main features are the constructs to explicitly model the communication between the net and the environment and a deterministic execution semantics. One aspect in which the Petri nets exhibit a major limitation is in the structuring of models. The representation of large-scale systems with flat models is highly inadequate. Modular mechanisms are essential for the expressivity and compactness of the models and to raise the efficiency of the modeling development process [6].

Concerning the IOPT nets, the lack of a structuring mechanism makes difficult the construction and management of large-scale and complex models. All the models must be built from scratch, with basic primitives, becoming the design task inefficient and preventing the reuse of model components into other projects.

Considering this, we state the following research question: *What structuring mechanisms should be defined to build generic subnets, reusable across different projects, allowing to manage complex and large models, while keeping all analysis capabilities?*

Hypothesis: *Modules with dynamic interfaces and configurable parameters, at instance level, allow the design of more compact models and its use in different situations. Converting a structured model into an equivalent flat model assures the use of existing analysis tools to validate the model.*

The modular mechanism presented here is an extension for the IOPT nets [7, 8] and is being implemented in the IOPT-Tools development framework [9] (available online at http://gres.uninova.pt/IOPT-Tools), the supporting tool for the IOPT nets. Figure 1 shows a diagram with the IOPT nets extensions and the IOPT-Tools features.

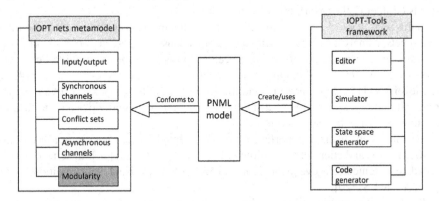

Fig. 1. IOPT Net and IOPT-Tools main features.

Concerning the proposing modular extension, the interface of the modules are nodes from the encapsulated net. Modules may be defined with more than one interface for being used in different situations. Several instances of the same module may be used in a containing net with different parameters.

The following Section of this paper cover the relationship of this work with the Cyber Physical Systems (CPS). In Sect. 3 it is reviewed some of the proposals of modular constructs that have been made in the context of Petri nets. The description of the modules for IOPT nets and its mathematical model is presented in Sect. 4. An example of a net with modules is presented in Sect. 5, followed by a discussion and conclusions.

2   Relationship to Cyber-Physical Systems

This work is about a modular extension for a modeling language (IOPT nets) addressed to the development of embedded systems controllers. An embedded system (ES) is a dedicated computational system embedded in a physical system which controls it. In [10] Cyber-Physical Systems (CPS) are defined as "the tight conjoining of and coordination between computational and physical resources". The main difference between ES and CPS is in the focus that is given to the physical system, being the CPS a more complete approach. While in embedded systems the main emphasis is on the computational part, from the point of view of a CPS an additional attention is given to the physical system, resulting in a greater integration of computational and physical systems. Although IOPT nets are suited to the development of ES controllers, the modular structuring presented here allows the modeling of the controller and the controlled physical system. Thus one can use the whole model for simulation and analysis purposes and the controller part of the model for the synthesis of the controller using a CPS development approach.

3   Related Literature

There are multiple languages for the development of embedded systems controllers [11], namely those where the communication with the controlled system has a fundamental role, as Grafcet [12] or Mark Flow Graphs [13].

In this section we overview some of the structuring mechanisms proposed in the field of Petri nets, namely those related with the modular construct presented in this work.

Many structuring mechanisms have been proposed, in order to provide Petri nets (PNs) with abstraction and composition capabilities, either for low-level PNs [14] or high-level PNs [15, 16]. Concerning composition, most of the proposals uses a module, or a similar entity, which models a small part of a system and provides an interface to communicate with the container net.

In some mechanisms the interface is composed by places or transitions, as Modular PNML [17, 18] and Hierarchical Coloured Petri nets (HCP nets) [16]. In some others it is used new elements, as events or signals in Signal Nets [19] and Net Condition Event Systems (NCES) [20].

The kind of communication can be synchronous, usually made through transition fusion, as in HCP [16] nets and modular PNML [17] or asynchronous, usually made by place fusion or by message sending, as in Object Petri nets (OP nets) [21].

Concerning instantiation it can be static or dynamic. Static instantiation has a substitution semantics, where an instance, or a macro, is substituted by the module. In this kind of instantiation the structure of the net remains unchanged during their execution and can be transformed into an equivalent unstructured net. Transition substitution in

HCP nets and modular PNML are examples of static instantiation. Dynamic instantiation consists in the creation of a module instance, during the execution of the net, and its subsequent destruction. The creation and destruction of the instance is governed by an event of the net, as the firing of a transition or a marking in a place. This change in net structure makes it very difficult to analyze. Dynamic instantiation is used in classes inspired in object-oriented paradigm as OP nets and Reference Nets [22]. The invocation transition mechanism of HCP nets also uses dynamic instantiation although it was never implemented in the CPN Tools, the supporting tool of the HCP nets.

4   Research Contribution and Innovation

This section contains the definition of the modules for IOPT nets. This mechanism uses nodes as interface and static instantiation. At first it is presented the definition of unstructured IOPT nets, their main features and semantics. Then the modular extension is presented and formalized.

4.1   IOPT Nets

As stated in the introduction, the IOPT nets have explicit constructs to model the communication with external devices. This communication is done through input and output signals and events. An IOPT net models a controller for an embedded system, thus must have a deterministic behavior. To accomplish this behavior it was adopted an execution semantics with maximal step combined with a single server semantics. Maximal step semantics means that all the transitions ready to fire, will fire in a given step. With single server semantics a transition fires only once in one step, even if it remains ready. The firing of transitions is synchronized with an external clock. A step is the occurrence of all the transitions ready to fire. A transition t, is enabled if for all its input places, $\bullet t$, $M(\bullet p) \geq w(p, t)$ with $(p, t) \in A$. An enabled transition is ready to fire if the transition guard is true and the event associated with the transition occurs.

Output events can be generated by the firing of a transition, and output signals can be updated by expressions evaluated by the firing of a transition, or by a specific marking of a place.

4.2   IOPT Net Mathematical Model

An IOPT net is defined as a tuple: $IOPT = (NG, IO, AN, Const, M0)$, where,

- NG is the net graph, $NG = (P, T, F)$ where,
  - $P \cup T$ are the set of nodes, satisfying the condition $P \cap T = \emptyset$,
  - F is the set of arcs which defines the flow relation. Each arc has a type from the set $AT = \{normal,\ test\}$, such that $F_{normal} \subseteq (P \times T) \cup (T \times P)$ and $F_{normal} \subseteq (P \times T)$.
- $IO = S_{in} \cup S_{out} \cup E_{in} \cup E_{out}$ is a finite set of input/output signals and input/output events.
- $AN = (A, TG, TIE, TOE, POS)$ is a finite set of annotations, where,

- $A : F \to \mathbb{N}$, is the weight function, assigning each arc with a positive integer.
- $TG : T \nrightarrow Exp_{(bool)}$ is a partial function that annotates transitions with boolean guard expressions.
- $TP : T \to \mathbb{N}$, is a function that annotates each transition with a priority value.
- $TIE : T \nrightarrow E_{in}$, is a partial function that annotates transitions with input events.
- $TOE : T \nrightarrow E_{out}$, is a partial function that annotates transitions with output events.
- $POS : P \nrightarrow (S_{out}, Exp_{(bool)})$ is a partial function that annotates places with output signals. The signal is updated when the place is marked and the boolean expression is true.
- *Const* is a set of symbolic constants whose values are naturals. A constant may be used as the initial marking of a place.
- $M_0 : P \to \mathbb{N}_0$, is the initial marking of the net, assigning a nonnegative integer to each place.

The transition *guard* and place *output expression* are built with operands from signal values, constants and place marking values (the number of dots), and operators from the set $O = \{O_{arithmetic}, O_{comparison}, O_{logic}\}$, with $O_{arithmetic} = \{+, -, *, /, \%\}$, being the usual arithmetic operators, $O_{comparison} = \{<, >, =, <=, >=, =\}$, the set of comparison operators and $O_{logic} = \{and, or, xor, not\}$ the set of logic operators.

4.3 Modules

An IOPT net uses the element page as the container of the net elements: graph elements, annotations and input/output declarations through which is done the communication with the environment. A module is also a container, with an encapsulated net, with one or more interfaces. The interface of a module is a subset of the module set of nodes.

A module is used in a net by their instance. An instance of a module is a copy of the module, represented in the net by a rectangle with the interface nodes. Each instance uses one interface. The interface nodes of the instance are copies of its counterpart in the module net.

An interface node may be of type input, output or, by default, input/output. Input nodes can only have incoming arcs from the container net. Output nodes can only be linked with outgoing arcs.

An IOPT net composed with modules has an equivalent at net, where the instances are substituted by their referenced modules. The interface nodes, from the instance, are fused with the respective nodes of the module. In the following definition of module it is considered a simple module, that is, a module without instances of other modules in its composition. A module is a structure *Module = (IOPT, ITF, IT)* where,

- IOPT is a IOPT net as defined in Sect. 4.2.
- $ITF = itf_1, itf_2, \ldots$ is a set of interfaces, with, $itf_i : P \cup T \nrightarrow \{in, out, in/out\}$, each interface is a partial function that assigns nodes with an attribute *in, out, in/out*, for input, output or input/output nodes.
- $IT: S_{in} \cup E_{in} \cup Const \to \{module, instance\}$ is a function that assigns an attribute to each input signal, input event and symbolic constant declared in the module.

For signals and events, if the attribute is *module* the signal/event is unique and shared by all the instances. If the attribute is *instance*, each instance has its own input signal/event. Constants with the attribute *instance* may be overridden and have a different value in each instance. A constant with an attribute *module* cannot be overridden at instance level.

4.4   Composition of Modules

A module is used through instances. An instance is graphically represented by a rectangle, a round rectangle or an oval. The interface nodes are drawn over the instance object. The nodes in the instance are copies of the respective interface nodes of the module. The encapsulated net of a module may be composed with instance of other modules. However it is not allowed to define a module with instances of itself.

When more than one instance of the same module exists in the same net, input signals, input events and symbolic constants may be shared by all the instances (attribute *module*), or exist as separate entities in each instance (attribute *instance*). Output signals and output events are unique for each instance of the same module.

The equivalent plain net is obtained by substituting the instances by the net of the respective module. The interface nodes of the instance are fused with the nodes of the module and the element names are prefixed with the name of the instance.

5   Example

The following example illustrates the definition of modules and a net built with instances of that modules. The equivalent flat net, obtained from the conversion of the net with module instances into an equivalent net without module instances, shows the semantics of the modules. The example is adapted from [23]. It is used a parking lot with an entrance and an exit as shown in the Fig. 2. When a car is detected near the entrance gate, a ticket is printed. When the driver gets the ticket, the gate opens and he can enter

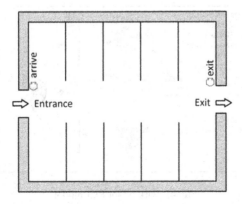

Fig. 2. Parking lot scheme.

the park. To leave the park, the car must be detected near the exit gate and after paying the ticket, the gate opens. The sensors *entrance* and *exit* detect the arrival of a car at the entrance and exit, respectively. The signal of these sensors have a positive edge trigger when a car is detected and a negative edge trigger when the detection ceases.

The models of the park entrance and park exit are implemented, respectively, in the module *Entrance* (Fig. 3), and in the module *Exit* (Fig. 4). Each of these modules have a declaration of two interfaces.

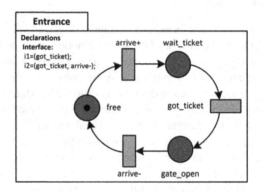

Fig. 3. Parking lot Entrance model.

A module may have multiple interfaces declared, supporting its use in different situations. For each instance of a module the designer chooses the most suited interface to communicate with the surrounding elements of the net.

In this example the interface i_1 of the module *Entrance* (composed by the node *got_ticket*) is used to decrease de number of available places and increase the occupied ones. The interface i_2 (nodes *got_ticket* and *arrive-*) could be used to, additionally, control a traffic light, using the transition *got_ticket* to switch to green and *arrive-* to switch to red.

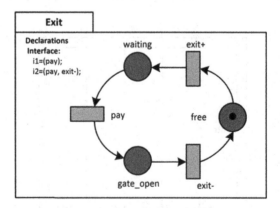

Fig. 4. Parking lot Exit model.

Figure 5 is the net composed by the instances of the modules. The places *occupied* and *free* models the number of places occupied/free in the parking lot. The equivalent plain net is shown in Fig. 6.

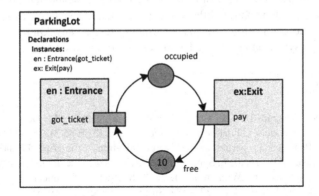

Fig. 5. Parking lot model with modules.

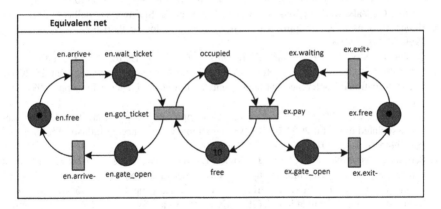

Fig. 6. Parking lot equivalent net.

6   Discussion and Conclusions

It was presented a modular extension for IOPT nets, a Petri net based language targeted for the development of embedded systems. The aim of the work was to create a structuring mechanism to represent models in a more compact way, while ensuring a greater distinction among different components of a system, whether they are physical or logical.

Comparing with similar proposals, as the substitution transition in HCP nets [16] and modular PNML [18], the mechanism presented here distinguishes by supporting multiple interfaces composed by concrete nodes (interface nodes in [18] are reference nodes). Defining multiple interfaces for the same module, supports different ways to connect the module within the container net and therefore a more flexible usage.

The parameterization of some module elements as instance elements (events, signals and constants) also increase the flexibility, enabling that different instances of the same module have its one parameters.

Although it is used the fusion of nodes as the composition mechanism between the surrounding net and an instance of the module, this is done in a transparent manner for the designer. The connections are made through arcs, or any kind of channels, as synchronous or asynchronous, which facilitates the composition.

References

1. Edwards, S., Lavagno, L., Lee, E.A., Sangiovanni-Vincentelli, A.: Design of embedded systems: formal models, validation, and synthesis. Proc. IEEE **85**(3), 366–389 (1997)
2. Murata, T.: Petri nets: properties, analysis and applications. Proc. IEEE **77**, 541–580 (1989)
3. Moalla, M., Pulou, J., Sifakis, J.: Synchronized Petri nets: a model for the description of non-autonomous sytems. In: Winkowski, J. (ed.) Internet – Technical Development and Applications. LNCS, vol. 64, pp. 374–384. Springer, Heidelberg (1978)
4. Moutinho, F., Gomes, L.: Asynchronous-Channels within Petri net-based GALS distributed embedded systems modeling. IEEE Trans. Ind. Informatics **10**(4), 2024–2033 (2014)
5. Girault, C., Valk, R.: Petri Nets for Systems Engineering. Springer, Heidelberg (2001)
6. Huber, P., Jensen, K., Shapiro, R.M.: Hierarchies in coloured Petri nets. Adv. Petri Nets **1990**, 313–341 (1989)
7. Gomes, L., Moutinho, F., Pereira, F., Ribeiro, J., Costa, A., Barros, J.P.: Extending input-output place-transition Petri nets for distributed controller systems development. In: ICMC 2014 International Conference on Mechatronics and Control, Jinzhou, China, pp. 1099–1104, July 2014
8. Gomes, L., Barros, J.P., Costa, A., Nunes, R.: The input-output place-transition Petri net class and associated tools. In: 2007 5th IEEE International Conference on Industrial Informatics, pp. 509–514, July 2007
9. Pereira, F., Moutinho, F., Ribeiro, J., Gomes, L.: Web based IOPT Petri net editor with an extensible plugin architecture to support generic net operations. In: IECON 2012 - 38th Annual Conference on IEEE Industrial Electronics Society, pp. 6151–6156, October 2012
10. National Science Foundation (NSF), Cyber-Physical System (CPS) (2011). http://www.nsf.gov/pubs/2011/nsf11516/nsf11516.htm
11. Gomes, L., Barros, J.P.: Models of computation for embedded systems. In: Zurawski, R. (Editor-in-Chief) The Industrial Information Technology Handbook, (Section VI – Real Time Embedded Systems; Chapter 83), pp. 83:1–83:17. CRC Press, Boca Raton (2005)
12. Thomas, B.H., McLean, C.: Using Grafcet to design generic controllers. In: International Conference on Computer Integrated Manufacturing, 1988, pp. 110–119, 23–25 May 1988
13. Hasegawa, K., Takahashi, K., Masuda, R., Ohno, H.: Proposal of mark flow graph for discrete system control. Trans. Soc. Instrum. Control Eng. **20**(2), 122–129 (1984)
14. Zuberek, W.M., Bluemke, I.: Hierarchies of place/transition refinements in Petri nets. In: 1996 IEEE Conference on Emerging Technologies and Factory Automation, 1996, EFTA 1996, Proceedings, vol. 1, pp. 355–360 (1996)
15. He., X.: A formal definition of hierarchical predicate transition nets. In: Proceedings of the 17th International Conference on Application and Theory of Petri Nets, pp. 212–229 (1996)
16. Huber, P., Jensen, K., Shapiro, R.M.: Hierarchies in coloured Petri nets. Adv. Petri Nets **1990**, 313–341 (1989)

17. Kindler, E., Petrucci, L.: Towards a standard for modular Petri nets: a formalisation. In: Franceschinis, G., Wolf, K. (eds.) PETRI NETS 2009. LNCS, vol. 5606, pp. 43–62. Springer, Heidelberg (2009)

18. Kindler, E., Weber, M.: A universal module concept for Petri nets-an implementation-oriented approach. Informatik-Bericht 150, Humboldt-Universität zu Berlin, Institut für Informatik (2001)

19. Juhás, G., Lorenz, R., Neumair, C.: Modelling and control with modules of signal nets. In: Desel, J., Reisig, W., Rozenberg, G. (eds.) Lectures on Concurrency and Petri Nets. LNCS, vol. 3098, pp. 585–625. Springer, Heidelberg (2004)

20. Rausch, M., Hanisch, H.M.: Net condition/event systems with multiple condition outputs, vol. 1, pp. 592–600 (1995)

21. Lakos, C.: Object oriented modelling with object Petri nets. In: Agha, G.A., De Cindio, F., Rozenberg, G. (eds.) Concurrent OOP and PN. LNCS, vol. 2001, pp. 1–37. Springer, Heidelberg (2001)

22. Kummer, O.: Referenznetze. Logos-Verlag, Berlin (2002)

23. Gomes, L., Barros, J.P.: Structuring and composability issues in Petri nets modeling. IEEE Trans. Ind. Inform. 1(2), 112–123 (2005)

Manufacturing Systems

Agent-Based Data Analysis Towards the Dynamic Adaptation of Industrial Automation Processes

Jonas Queiroz[1,2(✉)] and Paulo Leitão[2,3]

[1] Faculty of Engineering, University of Porto, 4200-465 Porto, Portugal
jonas.queiroz@fe.up.pt
[2] LIACC - Artificial Intelligence and Computer Science Laboratory,
University of Porto, 4169-007 Porto, Portugal
[3] Polytechnic Institute of Bragança, Campus Sta Apolónia, 5300-253 Bragança, Portugal
pleitao@ipb.pt

Abstract. Industrial complex systems demand the dynamic adaptation and optimization of their operation to cope with operational and business changes. In order to address such requirements and challenges, cyber-physical systems promotes the development of intelligent production units and products. The realization of such concepts requires, amongst others, advanced data analysis approaches, capable to take advantage of increased availability of data, in order to overcome the inherent dynamics of industrial environments, by providing more modular, adaptable and responsiveness systems. In this context, this work introduces an agent-based data analysis approach to support the supervisory and control levels of industrial processes. It proposes to endow agents with data analysis capabilities and cooperation strategies, enabling them to perform distributed data analysis and dynamically improve their analysis capabilities, based on the aggregation of shared knowledge. Some experiments have been performed in the context of an electric micro grid to validate this approach.

Keywords: Multi-agent system · Distributed data analysis · Adaptive system

1 Introduction

Companies are subject to the market dynamics and competitiveness, demanding highly customized and quality products and services with reduced prices. Additionally, they operate in complex environments, characterized by distributed and heterogeneous systems, which require the dynamic adaptation and optimization of their processes to cope with operational changes caused by technical problems (e.g., equipment damage and resource availability) or changes in business rules (e.g., new product demands or design). In order to address such requirements, Industrie 4.0 vision [1] and Cyber-Physical Systems (CPS) principles [2] promote the use of smart machines, systems and products. To support the realization of such concepts, the use of advanced data analysis approaches should be considered, taking advantage of the increased availability of great amounts of data produced in such environments. The continuous data analysis enables

© IFIP International Federation for Information Processing 2016
Published by Springer International Publishing Switzerland 2016. All Rights Reserved
L. Camarinha-Matos et al. (Eds.): DoCEIS 2016, IFIP AICT 470, pp. 99–106, 2016.
DOI: 10.1007/978-3-319-31165-4_10

to identify and predict the system operational conditions, providing valuable information to support process supervision and control.

The existing approaches handle these issues without worrying about the inherent dynamics and complexity of these environments (e.g., in face of condition changes, be capable to adapt its data analysis capabilities), which require features of modularity, adaptability and responsiveness. In industrial environments, approaches also need to support different data analysis scopes: (1) at operational level, applying distributed streaming data analysis for rapid response, and (2) at supervisory level, applying centralized and more robust data analysis for decision-making.

Multi-agent systems (MAS) [3, 4] have being pointed as a suitable approach to support the design and development of distributed, flexible and dynamic systems. Thus, the goal of the ongoing work is to design an advanced distributed data analysis approach, based on MAS, to support intelligent and adaptive supervisory control applications, towards the dynamic adaptation of industrial automation processes. This approach proposes to endow agents with data analysis capabilities and cooperation strategies, enabling them to perform distributed data analysis, continuously improve and dynamically adapt their local capabilities, based on the aggregation of knowledge. Some experiments, in the context of an electrical micro grid, have being performed to consolidate and validate the proposed approach. The preliminary results shown that agents are able to perform distributed predictive data analysis of energy production.

Although this approach presents a great potential to address the recent challenges faced by industry, there are some open questions regarding to how properly endow agents with data analysis capabilities and how the extracted information could enhance agents' behaviors. The conceptual and technical answers for such questions will enable the design and development of innovative and powerful approaches.

This paper is organized as follows. Section 2 describes the contributions of the work to the realization of CPS. Section 3 presents the literature review and Sect. 4 presents the proposed approach. Section 5 overviews the critical analysis of this proposal and discusses the preliminary results. Finally, Sect. 6 wraps up the paper with the conclusions and states the research directions.

2   Contribution to Cyber-Physical Systems

CPS promotes the integration of physical and virtual worlds, the first is characterized by a large network of interacting heterogeneous hardware devices, while the second provides robust computing infrastructures, replete of software platforms, applications and information technologies. Such integration aims a more effective management of the physical environment and their processes, by embedding computational elements in physical entities and connecting such entities in a cloud-based infrastructure.

CPS have been deployed in several fields, related to smart production, grids and buildings, where large amount of devices should be efficiently sensed and controlled in a reliable, secure, real-time and distributed manner. Additionally, such devices produce large volume of data, requiring advanced data analysis approaches in order to enable the capabilities and features envisioned by CPS, namely self-adaptation, fault tolerance, automated diagnosis and proactive maintenance [5, 6]. While most of existing works focus in the design of control approaches for CPS, this work intends to contribute with the issues and challenges related to supervisory aspects, considering the Big Data features. The main objective is to provide algorithms and mechanisms to support more intelligent and adaptive monitoring and supervisory CPS.

3   Literature Review

Industrial environments are characterized by a large network of heterogeneous devices (endowed with sensors and actuators), which monitor and control related processes [7]. The industrial management systems need to integrate and coordinate such devices, automating the overall process in order to optimize and ensure the quality of outcomes, and keep the plant availability. MAS have been proposed to address the issues of industrial systems, which need to be flexible and adaptive to cope with inherent complexity required to manage dynamic, heterogeneous and distributed components [3, 4]. In MAS, several autonomous, collaborative and self-organizing decision-making entities, called agents, interact and exchange knowledge to achieve their goals [3]. The application of agent-based technology in the industrial domain, to solve problems related to production automation and control, supervision and diagnosis, production planning, and supply chain and logistics, is surveyed in [4, 8, 9], and have been covered by the Industrial Agents [8] research field.

Technological advances in sensor devices have contributed to leverage their use in industrial environments and consequently the amount of collected data [10] further increases the complexity of such environments. In many cases, the produced data is underused, mainly because it is necessary a great expertise and specialized knowledge for its integration and analysis. However, the recent popularization of the Big Data concept and its potential, cached the attention of industry. In this context, data analysis has been widely applied in industrial domain, e.g., at operational level for process monitoring, diagnosis, optimization and control, and at business level for customer relationship management, supply chain, sales and others [11–14]. However, to effectively use data analysis and extract its full potential, several challenges found in industrial scenarios need to be overcome, such as, mechanisms to integrate distributed, heterogeneous, dynamic and streaming data sources [15].

In general, MAS and data analysis have been used successfully, but separately, to address several issues in industrial domain. In particular, MAS is used to develop adaptive and intelligent control systems, while data analysis to provide effectively data-driven decision-making algorithms. In this sense, several works leverage and discuss the potentials and how the integration of these technologies can provide better solutions in various domains [16–18].

4   Research Contribution and Innovation

Considering the assumptions discussed in the previous section, this work intends to design and develop an agent-based data analysis approach towards a flexible and adaptive industrial supervisory control system, capable to cope with the dynamics and high amount of distributed and heterogeneous industrial devices. This approach is more concerned with the supervisory and monitoring aspects than with the control of processes. Therefore, the general objective of this project encompasses mechanisms and algorithms to derive information and knowledge from data of different industrial levels, and then properly provide them for decision-making and process management.

4.1   Agent-Based Data Analysis Features and Requirements

The design of the proposed approach requires the consideration of some essential requirements and features, as illustrated in Fig. 1. They are directly related to ongoing and upcoming industry challenges and issues that are raised by the Industrie 4.0 vision. As already discussed in the previous section, *MAS and Data Analysis* are the basis technologies that will support this approach. The first provides the base infrastructure to achieve the required flexibility and adaptability, while the second, provides the proper tools required to take advantage of increased data availability.

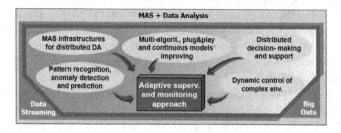

Fig. 1. Essential requirements and features of the proposed approach.

On the other hand, to cover different industry automation levels, such as, the monitoring of the operational process and the supervision of the whole plant, the proposed approach requires to support different data analysis scopes: (1) at operational level, distributed data streaming analysis for rapid response; and, (2) at supervisory level, more robust big data analysis for decision-making. *Big Data* considers the volume, variety and velocity of data, requiring dedicated and usually distributed computing infrastructures to extract valuable information from raw data, while *Data Streaming* considers the analysis of data at real or near-real time, providing simpler information that address rapid response requirements. In the literature some works already discuss approaches to address these two kinds of data analysis scopes [18].

Other requirements consider (Fig. 1): (1) MAS infrastructures for distributed DA (Data Analysis), and, (2) Multi-algorithm, plug&play and continuous models' improving. The first focuses in providing a modular and scalable data analysis infrastructure, by taking advantage of the MAS approach to support and enhance the various data analysis phases. For example, agents can be employed for data retrieve, preprocess, integrate and analyze, in a distributed and cooperative way. The second comprises three related features where the focus is the utilization of MAS to provide a dynamic and adaptive infrastructure to perform data analysis. Multi-algorithms comprises the deployment of different data analysis algorithms and models, e.g., one per agent, which can perform the same task over the data and at the end the results could be combined to obtain more accurate information. Plug&play comprises the use of MAS to provide an open and dynamic infrastructure that enables the seamless addition of new algorithms and data sources to the system. Continuous models' improving comprises mechanisms and algorithms that enable data analysis models to be updated to fit the environments' dynamics. In this case, specialized agents could be in charge to analyze the performance of current data analysis models, updating them to enhance their current accuracy.

While the previous features are more related to infrastructural aspects, there are also others related to industrial supervisory and control aspects. In the Fig. 1, the Distributed decision-making and support element comprises coordination and negotiation mechanisms for agents monitoring and diagnosing the system's conditions. The Pattern recognition, anomaly detection and prediction element represents the common application of data analysis to solve industrial problems, while the Dynamic control of complex environments element comprises the support of dynamic adaptation and optimization of operations and processes in face of changes in the environment or operating process conditions.

4.2   Agent-Based Model

Considering the analysed features and requirements, an agent-based model is proposed (Fig. 2), comprising two layers of agents. In the left side of Fig. 2, at the lower layer, agents are in charge of streaming data analysis, providing simple information about the processes (e.g., operation status, triggers and events), but attending rapid response constraints. In this layer, each agent is responsible to retrieve and analyze the data from process devices, in order to support control actions. These agents could be embedded into devices to perform distributed data analysis and intelligent monitoring, cooperating to identify problems or provide information about the system. At the upper layer, agents are responsible to process and analyze great amounts of historical and incoming data from plant operations, business and also external data, in order to provide information

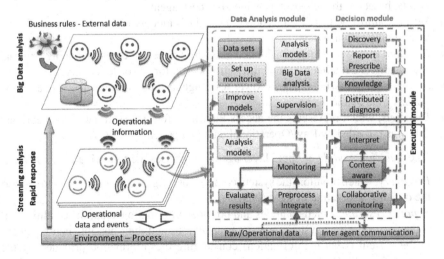

Fig. 2. Agent-based data analysis approach for adaptive industrial supervisory control systems.

for high level decision-making (e.g., performance, quality or degradation indicators, event diagnosis, tends and forecasts). These agents could be deployed in a cloud-based computing environment, taking advantage of such kind of infrastructure and other tools to perform their tasks and also to manage the lower level agents.

In this approach, the agents of each layer comprise three modules (illustrated in the right side of Fig. 2) that group a set of specific components, which define the agent behaviors and capabilities. The *Data Analysis module* defines the components that perform data analysis tasks, the *Decision module* defines the components that process, organize and consolidate the analysis result, and the *Execution module* defines the components that use the consolidated information to act in the environment. Agents from both layers have two common components, the *Raw/Operational data* and the *Inter agent communication*, responsible to retrieve external data from the environment and manage the agent interaction, respectively.

The components of lower layer agents *Data Analysis* module comprise:

- *Preprocess Integrate*, which prepares the raw data to be analyzed;
- *Monitoring*, which performs several types of data analysis;
- *Analysis models*, which comprises all the data analysis models used by the agent.
- *Evaluate results*, which assess the analysis model accuracy (e.g., by comparing its output with a system feedback);

The *Decision module* comprises the *Interpret*, which contextualizes and makes assumptions over the analysis result, the *Collaborative monitoring*, which realizes if the agent need any kind of information that could be provided by other agents, and, the *Context aware*, which provides a local knowledge used by the other components.

The upper layer agents are also defined by several components:

- *Supervision*, which receives monitoring information from lower layer agents and uses it to obtain the status of production stations, plants or the whole process;
- *Improve models*, which retrains or rebuilds data analysis models used by lower layer agents, based on the feedback provided by these agents;
- *Set up monitoring*, which builds new data analysis models, sets up and deploys lower layer monitoring agents;
- *Big Data analysis*, which considers data from different sources, including external and historical data, in order to extract information for a broader context;
- *Analysis models*, which, like in lower level agents, comprises all the data analysis models used by the agent;
- *Data sets*, which represents the access interfaces for historical data, since external data was provided by Raw/Operational data component.

The *Decision module* of upper layer agents comprises the following components:

- *Discovery*, which monitors the system components (e.g., agents or devices) to support the dynamic adaptation of the system;
- *Report Prescribe*, which compiles and provides information about the conditions of some parts or whole system, and suggests actions and their possible consequences in the system (what-if information) considering the information provided by the *Supervision, Big Data analysis*, or *Knowledge* components;

- *Knowledge*, which is related to operational and technical characteristics and constraints associated to some parts or the whole system;
- *Distributed diagnose*, which interact with other upper layer agents to collaborative identify and diagnose the whole system conditions.

5   Discussion of Results and Critical View

The described approach is being designed and verified based on a case study in the context of an electric micro grid comprising some wind turbines and photovoltaic panels. The preliminary results showed that producer agents (PA) are capable to perform distributed analysis of the energy production and weather data from sensors installed in the energy production units. PAs were able to monitor the operational conditions of production units, in order to identify abnormalities in energy production, by performing short-term prediction of energy production using different data analysis models, which were built based on historical data. Through a mid-term prediction of energy production, performed by integrating external weather forecasting data, PAs were able to provide information about the amount of energy expected to be produced in a near future, which could be used by engineers, grid operators and other systems to enhance and optimize the energy distribution and balance. During energy predictions, agents were able to continuously evaluate and improve their analysis models [19].

The experiments performed so far only covered some of the lower agents' aspects of the proposed approach. The preliminary experiments showed promising results, but there are still many features and requirements that need to be addressed in order to verify the expected potentials and benefits. Moreover, these features need to be evaluated by its performance, robustness and scalability regarding the data analysis aspects. Other aspects that should be explored in this case study, comprise the development of predictive capabilities for consumer and storage agents in order to manage energy consumption and power storage of micro grids nodes.

6   Conclusions and Future Work

In industrial domain, MAS have been used as a suitable approach to design and develop flexible and adaptable industrial control systems, while data analysis is being used to provide effective algorithms to support data-driven decision-making. In this context, the proposed approach intends to combine the features of these two technologies to contribute for the realization of CPS principles, in order to attend the requirements imposed by the Industrie 4.0 vision. This work describes an agent-based data analysis approach for intelligent and adaptive industrial supervisory control systems. Moreover, the approach covers the requirements of process monitoring and supervision automation levels. Although the promising perspectives, it is clear that to achieve the desired objectives, some aspects and issues need to be addressed, namely the dynamic, openness and rapid response requirements of industrial environments, and mechanisms for distributed, cooperative and self-improving data analysis.

Future works encompass the detailed specification and the definition of the required mechanisms and strategies to cover the more advanced aspects and features of the proposed approach. Thereafter, the current case study should be further explored, extending the preliminary experiments in order to validate and assess other aspects. Moreover, it is intended to explore another case study scenario in the manufacturing domain.

References

1. Drath, R., Horch, A.: Industrie 4.0: Hit or Hype? [Industry Forum]. IEEE Ind. Electron. Mag. **8**(2), 56–58 (2014)
2. Lee, E.: Cyber physical systems: design challenges. In: 11th IEEE International Symposium on Object Oriented Real-Time Distributed Computing, pp. 363–369 (2008)
3. Wooldridge, M.: An Introduction to Multiagent Systems. Wiley, New York (2009)
4. Leitão, P.: Agent-based distributed manufacturing control: a state-of-the-art survey. Eng. Appl. Artif. Intell. **22**(7), 979–991 (2009)
5. Lee, J., Lapira, E., Bagheri, B., Kao, H.A.: Recent advances and trends in predictive manufacturing systems in big data environments. Manuf. Lett. **1**(1), 38–41 (2013)
6. Sharma, A.B., Ivančić, F., Niculescu-Mizil, A., Chen, H., Jiang, G.: Modeling and analytics for cyber-physical systems in the age of big data. ACM Sigmetrics Perform. Eval. Rev. **41**(4), 74–77 (2014)
7. Chryssolouris, G.: Manufacturing Systems: Theory and Practice. Springer Science & Business Media, New York (2013)
8. Leitão, P., Karnouskos, S.: Industrial Agents: Emerging Applications of Software Agents in Industry. Morgan Kaufmann, San Diego (2015)
9. Metzger, M., Polakow, G.: A survey on applications of agent technology in industrial process control. IEEE Trans. Ind. Inform. **7**(4), 570–581 (2011)
10. Aggarwal, C.C., Ashish, N., Sheth, A.P.: The Internet of Things: A Survey from the Data-Centric Perspective. In: Aggarwal, C.C. (ed.) Managing and Mining Sensor Data, pp. 383–428. Springer, New York (2013)
11. Qin, S.J.: Survey on data-driven industrial process monitoring and diagnosis. Annu. Rev. Control **36**(2), 220–234 (2012)
12. Gröger, C., Niedermann, F., Mitschang, B.: Data mining-driven manufacturing process optimization. Proc. World Congr. Eng. **3**, 1475–1481 (2012)
13. Harding, J.A., Shahbaz, M., Kusiak, A.: Data mining in manufacturing: a review. J. Manuf. Sci. Eng. **128**(4), 969–976 (2006)
14. Choudhary, A.K., Harding, J.A., Tiwari, M.K.: Data mining in manufacturing: a review based on the kind of knowledge. J. Intell. Manuf. **20**(5), 501–521 (2009)
15. Obitko, M., Jirkovský, V., Bezdíček, J.: Big data challenges in industrial automation. In: Mařík, V., Lastra, J.L., Skobelev, P. (eds.) HoloMAS 2013. LNCS, vol. 8062, pp. 305–316. Springer, Heidelberg (2013)
16. Cao, L., Gorodetsky, V., Mitkas, P.: Agent mining: the synergy of agents and data mining. IEEE Intell. Syst. **24**(3), 64–72 (2009)
17. Albashiri, K.A., Coenen, F., Leng, P.: EMADS: an extendible multi-agent data miner. Knowl.-Based Syst. **22**(7), 523–528 (2009)
18. Twardowski, B., Ryzko, D.: Multi-agent architecture for real-time big data processing. In: International Joint Conference on Web Intelligence and Intelligent Agent Technologies, pp. 333–337 (2014)
19. Queiroz, J., Dias, A., Leitão, P.: Predictive data analytics for agent-based management of electrical micro grids. In: Proceedings of the IEEE IECON 2015, pp. 4684–4689 (2015)

Context Awareness for Flexible Manufacturing Systems Using Cyber Physical Approaches

Sebastian Scholze[1(✉)] and Jose Barata[2]

[1] Institut für angewandte Systemtechnik Bremen GmbH, 28359 Bremen, Germany
scholze@atb-bremen.de
[2] Faculdade de Ciencias e Tecnologica, Dep. Eng. Electrotecnica,
Universidade Nova de Lisboa, 2829-516 Caparica, Portugal
jab@uninova.pt

Abstract. The work presented in this paper demonstrates how flexible manufacturing systems (FMS) combined with context awareness can be used to allow for an improved decision support in manufacturing industry. Thereby manufacturing companies shall be supported in a continuous process of increasing efficiency and availability of their production machines. Such optimization has to be embedded in the processes allowing for run time adaptation of the process to various dynamically changing external conditions. Context awareness, based on the information obtained from cyber physical systems, is a promising approach to allow for efficient building of such embedded optimization solutions. The objective of the research presented is to explore how context awareness, using the information from cyber physical systems integrated in the processes, can be applied to build a solution for self-optimization of discrete, flexible manufacturing processes.

Keywords: Context awareness · Cyber physical systems · Flexible manufacturing system · Process optimization · SOA

1 Introduction

Manufacturing industry is nowadays facing several challenges, such as the up-coming customized production, that forces them to move a new concept of FMS. Advanced FMS need to include cyber-physical features, which to assure highest efficiency and availability. This is achieved by optimization capabilities embedded in the cyber physical features. Such capabilities are needed in wide range of applications: e.g. customized production, maintenance activities, etc.

The work presented in this paper demonstrates how context awareness can be used for optimization of various systems in manufacturing industry, especially flexible manufacturing systems. Key objective thereby is to enable the manufacturing companies to move into a continuous process of increasing efficiency and availability of their production machines [1]. Thereby, the assumption is that cyber-physical features

L. Camarinha-Matos et al. (Eds.): DoCEIS 2016, IFIP AICT 470, pp. 107–115, 2016.
DOI: 10.1007/978-3-319-31165-4_11

and ICT based context awareness services are suitable technologies that allow for solving the above mentioned problems.

For realizing an ICT solution for run-time optimization of FMS several pre-requisites for a context aware infrastructure have to be fulfilled. First, the solutions need to be capable of handling high amount of data. Secondly it needs to be capable of handling complex models and algorithms. Thereby, the key challenge is to find a common approach for addressing the following problems: (a) Information on changes need to be gathered in run-time and used for (self) optimization control processes. (b) Advanced data processing is needed to allow for enhancement of the gathered information in order to facilitate generation of knowledge about the changes relevant for operation of manufacturing systems.

There is a need for a solution which can be easily applied to various processes and for optimization of different parameters in the processes, instead to build scattered solutions for each process and parameter. This raises the following research question:

What could be a suitable set of methods and tools to allow the realization of reliable adaptable production systems for dynamic manufacturing processes that can be adapted during run-time assuring high availability of such adaptable production systems?

The above stated research question can be addressed by the following hypothesis:

Reliable and highly available production systems for dynamic manufacturing processes can be achieved if a context aware approach is used to identify the current context of manufacturing processes as a basis for adaptation of process parameters and for sharing of knowledge on manufacturing processes.

2   Context Awareness and Cyber-Physical Systems

In this paper it is investigated how context awareness can be used to achieve a solution for run-time optimization of various FMS, optimizing high variety of parameters. The context awareness approach allows for observation of changes under which a FMS is operated. The identified context in turn allows for a dynamic adaptation of the FMS to these varying conditions.

Cyber Physical Systems (CPS), integrated in the manufacturing processes offer new opportunities to provide information, which are needed for an effective identification of dynamically changing context under which the observed manufacturing system is operating, during run-time. The assumption is that building and adjustments of such generic context aware solution for various specific optimizations and processes is much more time/costs effective than building of classical optimizations solutions. In order to investigate applicability of a new generic context aware solution to wide scope of complex FMS, in this paper the applications for the two different optimization of two processes is experimentally investigated: process energy consumption optimization (as typical example of so-called secondary manufacturing processes) and process control optimization aiming at higher process efficiency and availability (prime processes).

CPS based on a service oriented architecture (SOA) approach offers completely new possibilities for self-adaptive and context aware solutions. Therefore, it is likely that

CPS and SOA based approaches are the most appropriate for the realization of self-adaptive and context aware solutions. By making use of various information sources (sensors, controllers, etc.) through embedded ICT services, such an approach is promising specifically for FMS.

3 Related Work

Context Awareness and Context Modelling: Context Awareness is a concept propagated in the domains of ambient intelligence and ubiquitous computing. Existing research on context can be classified in two categories: context-based, proactive delivery of knowledge, and the capture & utilization of contextual knowledge. In the case of embedded services, the notion of context refers to process preferences of product and process skills of devices, physical capabilities of the equipment and environment conditions. The idea is that computers can be both sensitive and reactive, based on their environment. Using context information is an active area of research, with various context capture methods and context languages defined. Although research on context already started back in 1999 [2–4], the current research on context is primarily oriented towards capturing and utilization of contextual data for actionable knowledge [5]. Furthermore research shows, that knowledge context could be used to classify and organize knowledge in a networked enterprise [6]. However, provided in [7] shows a lack in provision of knowledge context. Several systems to handle context were proposed by the research community [8–12].

By context modeling, the problem of how to represent the context information can be solved. However, how to extract context from the knowledge process and how to manipulate the information to meet the requirement of knowledge enrichment remains to be solved. In the research presented in this paper it is planned to model context with ontologies, and, therefore, context extraction mainly is an issue of context reasoning and context provisioning: how to infer high-level context information from low-level raw context data [13] or monitored sensorial data. Based on the formal description of context information, context can be processed with contextual reasoning mechanisms [14, 15]. The modelling of context in this case presents an additional challenge, especially in highly dynamic and distributed environments [16].

Cyber Physical Systems: In [17] cyber-physical systems (CPS) are defined as '[...] integrations of computation and physical processes [...]'. Research on CPS can be understood as a compound of various disciplines, such as computer science, software engineering systems with mechanical and electrical embedded systems. A CPS represents the idea of combining real-world objects and processes with information processing objects and processes linked together through open, partly global and anytime interconnected information networks like the internet or networking with each other. One example of a typical CPS is an intelligent manufacturing line, where the work of a machine is supported by the communication with its depending components.

Through the coupling of virtual and physical systems, several advantages against conventional computational or embedded systems arise. All accessible data, information and services can be deployed and utilized at any time anywhere in the system.

In this way, a series of novel functionalities, properties and services become possible. Thus, cyber-physical systems' services are independent from location, adapted to current systems requirements, partly autonomously, multifunctional and multimodal, networked and distributed along their application area [18, 19]. Therefore, it is expected that CPS will play an important role in future systems, especially also in manufacturing systems [20].

4   Concept for Context Awareness

The proposed approach to achieve context awareness is summarized in Fig. 1. The objective of the Context Monitoring/Extraction services is to use monitored "raw data" provided by sensors and CPS at the FMS to derive the FMS current contextual situation. Based on the identified current contextual situation, knowledge can be generated that is necessary to provide the basis for operational decisions. This generated knowledge in turn forms the basis for decisions about optimizations of specific manufacturing processes. Decisions regarding the optimization of manufacturing process can be (a) short-term (specific task of a manufacturing process) and (b) long-term (overall manufacturing process). The proposed approach provides context awareness about situations and based on this provides an appropriate decision support.

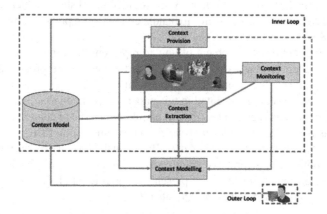

Fig. 1. Context awareness concept

Using CPS and other information sources, data on context of manufacturing processes are collected. This information is used to identify the current context of the processes (current approaches often only use location/user for identifying context). The identification is starts through context monitoring services, which are, e.g. services for monitoring of processes or of a user interacting with a system for changing conditions. The monitored "raw data" is transformed into an "standardized" data format by the monitoring services in order to allow further processing by the context extraction services. The context extraction services identifies current context by instantiating monitored data into the context model. Furthermore, reasoning techniques are used to support context identification. For reasoning previously identified context and the context model is used, which is stored in the context repository. After the current context is identified, it is send to the system adapter services, which are responsible for the system adaptation. In addition, the outer loop supports updating the context. Currently the update is foreseen as manual task, but in future this might be realized as an automatic task.

Context Model: The Context Model forms the foundation for the context extraction. The approach selected to model context is ontology based. Ontologies provide flexibility, expressiveness and extensibility and therefore can be considered as a suitable candidate for representing the context model. They ensure that different entities that use the context data have a common semantic understanding of that data. Furthermore, the possibility to easily use reasoning techniques is an advantage of using ontologies. By using such reasoning techniques it is possible to identify inferred knowledge out of the implicitly stated situations. In order to be able to use a context model for context extraction it is necessary to first define a 'holistic' and dynamic context model. Thereby, it is important to take into account the context of FMS, machines and processes for which context shall be extracted. The ontology based context modelling is promising to be applicable to wide scope of FMS, asking for minimal adjustments developments.

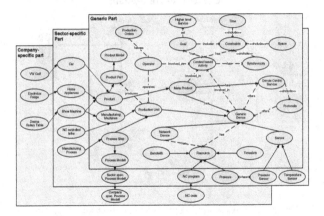

Fig. 2. Proposed context model

The context model defines a layered ontology: Generic Context Model, domain-specific model and an application-specific model. The context model includes identification of the set of features that determine the context and, consequently, identification of the set of parameters to be monitored depending on the services to be made. The Context Model is contrary to many current approaches not intended to model "everything", but only the necessary information required for context extraction. Figure 2 shows an excerpt of an example of the above mentioned layered ontology approach.

5   Implementation

A generic solution for a context aware solution is proposed. The proposed solution can be applied to various FMS as an add-on to the existing system. In order to create a common architecture, several application cases from different industrial sectors have been analyzed. Furthermore, results from former research projects (e.g. Self-Learning, U-Qasar, ProSEco) have been re-used. The proposed architecture is shown in Fig. 3 and follows SOA principles. The architecture includes, among others, the following key components:

- The System Monitor receives raw sensor data and provides aggregated data. To achieve this, it allows monitoring of legacy systems in enterprises via different interfaces. It is therefore able to correlate data from distinct systems, which later serves as a basis for extraction of contexts.
- The Context Extractor uses the monitored "raw data" provided by the System Monitor to identify the current context of the monitored machine/process. The context is extracted by using the monitored data instantiated in the context model.
- The Context Sensitive System Adapter uses the identified context to update the system behavior. Each adaptation is based on the context knowledge, which defines the rules/parameters to be updated for each identified context.

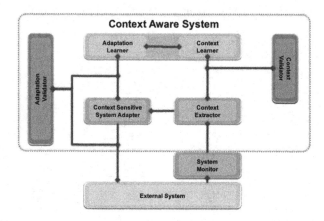

Fig. 3. Proposed architecture for a context-aware system

6   Early Results

This experiment involved control systems, machines and automation systems for shoe industry. The production and manufacturing of shoes, especially the production of shoe soles involves various different materials and a large number of automatic as well as manual operations. Such FMS comprise a set of complex operations that depended on the human operator and therefore depend on the operators skills. The need for automatic recognition of current situations and continuous optimisation of processes has been identified by the machine vendors for the shoe industry.

The proposed solution has been integrated into a real installation of a machine vendor for the shoe industry. The company is designing and delivering complex automated machines/systems for the shoe production worldwide. These automation systems, implying rotary table and injection machines, robots etc. The solution is used to identify the operative context within the production process. Based on the identified current context the solution reacts to changing situations associated with variations in different parameter sets in order to improve error-prone processes and reduce maintenance problems. The considered experiment addresses an improved synchronization of valves of an injection machine. The "mixing head" of the injection machine is injecting different materials into a mould through different non-mechanical connected valves. However, after a number of production cycles the valves might get asynchronous due to a variety of influences such as inconstant air pressure and/or valve abrasion. Due to asynchronous operation of the valves the quality of the final product is affected negatively. By applying the proposed context aware solution in the injection machine, the valves opening times are automatically adjusted during run-time in order to assure initially planned operation. The proposed solution continuously monitors the process parameters and tries to identify changes in this operative context. Whenever such a change is identified by the proposed solution, the identified context is sent to the system adapter. The system adapter in turn starts an adaptation process, which leads to a set of parameters, that need to be updated in order to adjust the opening times of the valves.

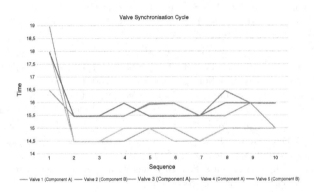

Fig. 4. Results of a Valve Synchronisation test run (including adjustment times) (Color figure online)

The results of automatic valve synchronisation are shown in Fig. 4. It shows that the opening times of five valves that are injecting two types of materials (three valves for material A, two valves for material B) are continuously adapted to assure an optimum working range, which is achieved after an initial training phase.

7   Conclusions and Future Work

Prototypes of the proposed solution have been implemented and are integrated in real industrial test environment. Currently the approach is running through the testing and evaluation phase. The approach is evaluated in two scenarios and several test cases in different business domains in order to validate the results under different conditions. The innovation of the proposed research are: (a) methods on how to achieve context awareness for FMS (this includes guidelines for modelling context) and (b) solution to enable the use of context awareness in FMS, e.g. to enable self-adaptation of FMS (this includes services for monitoring and extracting context).

The results of the first executed results in real industrial environments allow for an optimistic expectation regarding future system applicability.

Future research is planned to focus on using additional reasoning capabilities for improving context extraction results. A second topic for further research will be put on methods on how to update the context model(s). Especially for dynamically changing production environments updates of the context model are important. In optimal case these updates should be executed (semi-) automatically. Focus is hereby on solving the problem of how to feed updates of the specific context models (domain and application specific) back to the generic context model.

References

1. Scholze, S., Barata, J., Kotte, O.: Context awareness for self-adaptive and highly available production systems. In: Camarinha-Matos, L.M., Tomic, S., Graça, P. (eds.) DoCEIS 2013. IFIP AICT, vol. 394, pp. 210–217. Springer, Heidelberg (2013)
2. Weiser, M., Gold, R., Brown, J.S.: The origins of ubiquitous computing research at PARC in the late 1980s. IBM Syst. J. **38**(4), 693–696 (1999)
3. Bouquet, P., Giunchiglia, F., van Harmelen, F., Serafini, L., Stuckenschmidt, H.: C-OWL: contextualizing ontologies. In: Fensel, D., Sycara, K., Mylopoulos, J. (eds.) ISWC 2003. LNCS, vol. 2870, pp. 164–179. Springer, Heidelberg (2003)
4. Voida, S., Mynatt, E.D., MacIntyre, B., Corso, G.M.: Integrating virtual and physical context to support knowledge workers. IEEE Pervasive Comput. Arch. **1**(3), 73–79 (2002)
5. Mladenić, D.: Active Project Website. https://tinyurl.com/jkc23mn. Accessed 10 Jan 2012
6. Tao, Y., Tianyuan, X., Linxuan, Z.: Context-centered design knowledge management. Comput. Integr. Manuf. Syst. **10**, 1541–1545 (2004)
7. Ahn, H.J., Lee, H.J., Cho, K., Park, S.J.: Utilizing knowledge context in virtual collaborative work. Decis. Support Syst. **39**, 563–582 (2005)
8. Bellavista, P., Corradi, A., Montanari, R., Toninelli, A.: Context-aware semantic discovery for next generation mobile systems. IEEE Commun. Mag. **44**, 62–71 (2006)
9. Toninelli, A., Corradi, A., Montanari, A.: Semantic-based discovery to support mobile context-aware service access. Comput. Commun. **31**, 935–949 (2008)
10. Gu, T., Pung, H.K., Zhang, D.Q.: A service-oriented middleware for building context-aware services. J. Netw. Comput. Appl. **28**, 1–18 (2005)
11. Kim, S., Suh, E., Yoo, K.: A study of context inference for Web based information systems. Electron. Commer. Res. Appl. **6**, 146–158 (2007)
12. Chang, J.-W., Kim, Y.-K.: Design and implementation of middleware and context server for context-awareness. In: Gerndt, M., Kranzlmüller, D. (eds.) HPCC 2006. LNCS, vol. 4208, pp. 487–494. Springer, Heidelberg (2006)
13. Scholze, S., Stokic, D., Barata, J., Decker, C.: Context extraction for self-learning production systems. In: INDIN 2012, Beijing (2012)
14. Luther, M., et al.: Situational reasoning – a practical OWL use case, Chengdu, China (2005)
15. Forstadius, J., Lassila, O., Seppänen, T.: RDF-based model for context-aware reasoning in rich service environment. In: Proceedings of the Third IEEE International Conference on Pervasive Computing and Communications Workshops. IEEE Computer Society (2005)

16. Bettini, C., Brdiczka, O., Henricksen, K., Indulska, J., Nicklas, D., Ranganathan, A., Riboni, D.: A survey of context modelling and reasoning techniques. Pervasive Mob. Comput. **6**(2), 161–180 (2010)
17. Lee, E.A.: Cyber physical systems: design challenges, Berkeley, UC (2008)
18. Geisberger, E., Broy, M.: agendaCPS: Integrierte Forschungsagenda Cyber Physical Systems. National Academy of Science and Engineering (2012)
19. Bettenhausen, K.D., Kowalewski, S.: Cyber Physical Systems: Chancen und Nutzen aus Sicht der Automation. In: Thesen und Handlungsfelder (2013)
20. Khaitan, S., McCalley, J.: Design techniques and applications of cyberphysical systems: a survey. Syst. J. **9**(2), 350–365 (2015)

An Approach for Implementing ISA 95-Compliant Big Data Observation, Analysis and Diagnosis Features in Industry 4.0 Vision Following Manufacturing Systems

Kevin Nagorny[1(✉)], Sebastian Scholze[1], José Barata[2], and Armando Walter Colombo[3]

[1] ATB - Institut für angewandte Systemtechnik Bremen, 28359 Bremen, Germany
{nagorny,scholze}@atb-bremen.de
[2] NOVA University of Lisbon, 1099-085 Lisbon, Portugal
jab@uninova.pt
[3] University of Applied Sciences Emden/Leer, 26723 Emden, Germany
awcolombo@technik-emden.de

Abstract. Current trends are showing a technological evolution to an unified Industrial Internet of Things network where smart manufacturing devices are loosely coupled over a cloud to realize comprehensive collaboration and analysis possibilities, and to increase the dynamic and volatile of manufacturing environments. This rising complexity generates also higher ranges of error possibilities and analog a growing demand of new diagnostic approaches to handle also those highly complex systems as manufacturing systems which are following the Industry 4.0 vision. This is an ISA'95 compliant approach of a Big Data analytics methodology for analysis and observation in Industry 4.0 vision following manufacturing systems.

Keywords: Engineering · Big data · Cyber-physical systems · Context sensitivity · Diagnostics · Data mining · ISA-95

1 Introduction

By analyzing latest reports focusing on predictions of future data generation-consuming-traffic, a doubling of data growth every two years is foreseen [1]. This trend is also valid in the manufacturing domain. The Industry 4.0 vision [2] aims to establish an industrial infrastructure (the industrial internet of things (IIoT)) in which all things are able to exchange information over a network respecting the legacy ISA-95 compliant enterprise architecture. Given that the required connectivity and interoperability of the manufacturing things are guaranteed, the manufacturing environment appears as a very big data set that represents in a digital form the industrial system behind.

One future challenge is the exploitation growing complex data amounts in dynamic manufacturing environments. This work presents an approach to exploit the data through Big Data analysis for observation, analysis and diagnosis through identification, classification, filtering and analysis of data in Industry 4.0 following ISA'95 compliant manufacturing systems.

Published by Springer International Publishing Switzerland 2016. All Rights Reserved
L. Camarinha-Matos et al. (Eds.): DoCEIS 2016, IFIP AICT 470, pp. 116–123, 2016.
DOI: 10.1007/978-3-319-31165-4_12

At the same time this work presents an initial idea for a potential dissertation to reach the Ph.D. degree. Following described research question and hypothesis are base for this paper.

Research Question: Is an Industry 4.0 following Big Data observation, analysis and diagnosis approach based on classical Big Data analysis technologies useful to observe, analyze and diagnose Industry 4.0 vision following manufacturing systems as well as immigrated conventional manufacturing systems?

Hypothesis: Selected existing Big Data analysis technologies are adaptable and extendable to integrate Big Data observation, analysis and diagnosis functionalities into Industry 4.0 vision following manufacturing systems as well as in conventional manufacturing systems to observe, analyze and diagnose Industry 4.0 vision following manufacturing systems as well as immigrated conventional manufacturing systems.

The following chapters are structured as following: Sect. 2 describes relationship of this work to cyber-physical systems. Section 3 is a State of the Art (SotA) and describes related works which will be used as base for this approach. Section 4 describes the overall concept of the approach, Sect. 5 describes some application scenarios and Sect. 6 concludes the paper.

2 Relationship to Cyber-Physical Systems

The work describes an approach to implement ISA-95-compliant Big Data observation, analysis and diagnosis features in Industry 4.0 vision following manufacturing systems. The whole idea of Industry 4.0 is based on cyber-physical systems and the internet of things idea. This work will present how Big Data produced by cyber-physical systems in a service-based industrial internet of things are exploitable for observation, analysis and diagnosis use cases in the manufacturing domain.

3 State of the Art and Related Works

3.1 Analysis and Diagnosis in Manufacturing

Many analysis and diagnosis approaches are available in the manufacturing area. Behind classical manual diagnosis are often individual diagnosis solution used for manufacturing systems. Researcher are working on new approaches based on e.g. (automatic (predictive)) failure detection through fault tree analysis (FTA) [3] or self-learning approaches [4] handle also systems with a higher complexity (evolvable and emergent systems with changing physical and logical conditions and behaviors.

Self-learning approaches are observing systems to learn the behavior of a system and to detect based on knowledge base systems anomalies and failures. In deterministic finite automaton (DFA) [5] algorithms observing systems to build deterministic finite automats which represents a correct behavior of the system. In case that the system gets into a state which is not covered by the model (generated in the learning phase) will be made a defined reaction as e.g. sending an alarm or to start a troubleshooting routine [3]. Past results of this topic will influence the implementation of this approach. The relation is the idea of knowledge bases and the idea of building trees.

In the concept chapter will be shown that so called "Big Data profiles" are based on these ideas. The approach is able to identify anomalies which will be reviewed by humans and saved into a knowledge base so that time by time the knowledge will be increased. Big Data profiles describe among others a specific system state based on Big Data analyses but also possible progression which could end in this specified state. Fault tree analysis combined with the described deterministic finite automaton approach identifies errors very similar but without statistical Big Data analyses.

3.2 Data Mining and Big Data

Data mining is a broad interdisciplinary research area which covers the idea to extract for implicit, previously unknown, and potentially useful information from data [6]. Data Mining is the analysis step in the process for knowledge discovery. The process is mostly similar and can be divided into four steps [7]: (1) Focusing and selecting of the potential useful data, (2) pre-processing of the data (data cleaning and data completion), (3) transformation of data into a fitting format, (4) data mining analysis itself and (5) the evaluation/knowledge discovery of the data mining analysis results. For data mining there are a lot of algorithms and approaches available [8]. This approach aims to use latest outcomes of pattern recognition approaches as base for data mining.

One sector for Data Mining is Big Data mining. Big Data is one of the most promising technologies nowadays [9, 10]. The definition of Big Data is still in discussion and many suggestions were made. Gartner made a proposal in 2011 where was suggested to categorize Big Data through 3Vs (Volume of Data, Variety of Data and Velocity of Data) [11]. This definition is mostly accepted.

In the last years came up a range of technologies to deal with Big Data. One popular technology is the Googles "Map Reduce" Algorithm which is a programming model for parallel processing of big data sets [12]. This algorithm was implemented in the free Apache Hadoop framework which is an old solution but up to now the base/core element for many Big Data technologies. Hadoop provides basically the Hadoop Distributed File System (HDFS) and the Map Reduce Algorithm. Optionally it provides several extensions. This work will use past approaches of data mining and Big Data analysis and will extend, adapt and improve them to use these approaches in the manufacturing domain.

4   Overall Concept

This section describes a Big Data observation, analysis and diagnosis approach for Industry 4.0 vision following manufacturing systems. The approach aims to generate synergies between state-of-the-art technologies and approaches of diagnosis, pattern recognition, Big Data analysis and context extraction, to generate ISA-95 compliant functionalities for Big Data observation, analysis and diagnosis for Industry 4.0 vision following manufacturing systems.

Figure 1 shows the overall concept. The Approach is divided into a data storage part, a runtime part (where run-time modules will be deployed and executed) and an engineering part (where all engineering tools will be provided for Big Data analysis, observation and diagnosis). The approach provides functionalities for a Big Data analysis and diagnosis, and for continuous (automatic) Big Data observation. All parts and how these parts work together will be explained in the following.

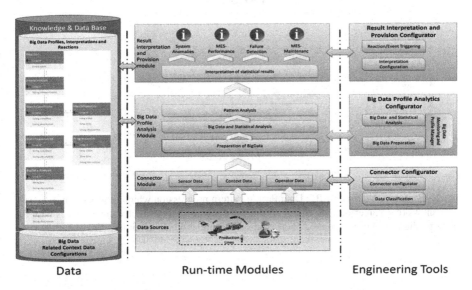

Fig. 1. Concept of a new Big Data analysis and observation approach for large-scale manufacturing systems.

4.1 Data Part

In the Data part will be stored all necessary data in a knowledge and data base. Saved will be data from data sources, related context information of data sources (location, which system, test mode y/n, etc.), Big Data Profiles for a continuous Big Data analysis and configurations (/settings) for engineering tools and run-time modules.

Input Data of data sources could be Big Data buckets (one-time/manual data downloads) for a manual analysis or could be streamed Big Data inputs (event-based of timeframe based for a continuous observation).

A pattern recognition defined in a Big Data profile can be interpreted in several ways which will be saved as interpretations in the knowledge and data base – and different interpretations are interesting for different Big Data analysis solution users (e.g. humans or systems which are connected to the Big Data analysis solution (or more precise: connected to the "Result interpretation and provision module")). Therefore "reactions" will be stored which are linked to interpretations. A reaction defines who will be how notified in case that a defined recognized pattern was found through a Big Data analysis (also called Big Data profile match).

Big Data profiles will be used to observe systems based on continuous Big Data analysis. Big Data Profiles are defining Big Data itself, statistical analyses and define conditions of a Big Data profile match to detect defined system states. Through saved pattern progressions (see Fig. 2) it is also possible to predict automatically future system behaviors and probable future Big Data profile matches.

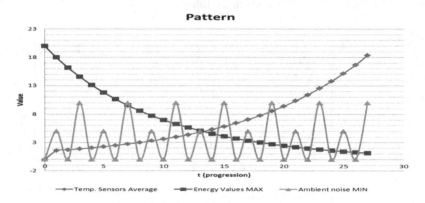

Fig. 2. Example Pattern of a Big Data Profile Pattern progression which describes an Anomaly – t27 shows (e.g.) an error state and t0–t26 shows the progression which can be used as base for future predictions. Further context data could specify under which conditions this pattern is valid or not.

4.2   Engineering Tools

This section describes the engineering tools for each module in the run-time part. The Engineering tools configure and use functionalities of run-time modules.

Connector Configurator - For the provision of data for a Big Data analysis will be used connector modules. Features/functionalities of a connector are (1.) the integration of conventional systems into a network of industry 4.0 vision following manufacturing systems (individual/application specific part to integrate proprietary interfaces) and (2.) the selection, filtering and preparation of usable system data for Big Data analysis, and (3.) the configuration of the communication with a Big Data Profile Analysis Module (data provision, event triggers, data streams, etc.).

Big Data Profile Analytics Configurator - The Big Data Profile Analytics Configurator is divided into (1.) a big data preparation, and (2.) a big data and statistical analysis configurator. These features will be used by the Big Data Monitoring and Profile Manager tool and for the Profile management. The big data preparation procedure will provide functionalities to manage input data from connectors. After selecting needed data follow the Big Data analysis. It will be provided a range of Big Data analysis algorithms which are depending on used technologies and the kind of data input (buckets or streams). The following statistical analysis step will provide functionalities to use several statistical methods (e.g. Gaussian distribution, runaways, averages, etc.) which are usable on Big Data analysis results. Provided will be also functionalities to check the progression of statistical results. These features are useful to identify e.g. causes of a progression and will be later used for automatic predictions.

Result Interpretation and Provision Configurator - The last engineering tool is the result interpretation and provision configurator. This tool provides functionalities to configure interpretation for Big Data Profiles and to configure interpretation related reactions. As already described could the match of a Big Data Profile have several interpretations. This tool will provide functionalities to manage, generate, edit or delete interpretations and makes it possible to connect them with profiles. In case of a Big Data profile match the interpretations will be triggered. A match of a profile could have e.g. following interpretations (simple examples): "power consumption to high", "performance is going down", "throughput is going down", "wear increases", etc. Different interpretations are interesting for different users: "the energy provider wants to know that the energy consumption goes up", "the MES system wants to know that the performance goes down", "the Factory manager wants to know that the throughput goes down", "the maintenance operator wants to know that the wear increases", etc. To notify users (other systems or user) this tool will provide functionalities to configure reactions.

4.3 Runtime Part

The modules in the runtime part of Fig. 1 (expect data sources) represent the engines of the Big Data analysis approach which will be executed for a (continuous) big data observation/analysis and diagnosis in a manufacturing environment. The modules of the runtime part will be described in the following.

Data Sources - Big Data solutions are using poly-structured input data (structured data, semi structured data as XML or HTML, and unstructured data as pictures or documents) as base for their Big Data analysis. Data Sources for Big Data solutions in the manufacturing area related to maintenance tasks are mainly data from systems settled in the levels 1–3 of the ISA-95. Those information are e.g. direct sensor data from the level 1; productions process monitoring data from (SCADA) systems of the level 2; or monitoring data of the whole production process as well as scheduling, quality management, maintenance and production tracking data from (MES) systems settled in the level 3. Other data are coming from context information sources to recognize also information about the context of a system as e.g. the location, the ambient temperature, etc. Context Data will be used to get additional information of big data sources in order to make Big Data Analyses sensitive and reactive to environmental happenings. Saved Big Data profiles will be the base for continuous Big Data analyses but not under all context conditions it is necessary to observe a system so that context related conditions will be used to check the validity of a big data profile.

Connector Module - The connector module establishes the communication between the Big Data Profile Analytics Module and data sources and is responsible for the data provision.

Big Data Profile Analytics Module - This module provides Big Data analyses features and executes procedures for an (automatic) continuous Big Data analysis. The run-time module is divided into (1.) a preparation of BigData part and (2.) into a Big

Data and Statistical Analysis part. The Big Data analysis part will execute the Big Data analysis based on Big Data analysis configurations which are defined in Big Data profiles. After the Big Data analysis will be sent the results to the Statistical Analysis step. In this step will be executed the statistical analysis based on defined Statistical Analysis configurations defined in Big Data profiles. The pattern analysis step will check if a saved Big Data profile matches with the current Big Data statistics analysis results and will trigger in that case the Result interpretation and provision module in case of a match.

Result interpretation and provision module - This module interprets configured Big Data Profile matches and informs users (systems, operators, etc.) which subscribed interpretation related events.

5   Application Scenarios

This work is very application driven and should bring Big Data observation, analysis and diagnosis into ISA'95 compliant Industry 4.0 vision following manufacturing systems. There are several possible application scenarios for such an approach in case that the concept will proof the hypothesis. It follows a short list of examples.

System Anomaly/Failure detection and prediction (detected and predicted system anomalies/failures.), *Process Optimization* (system parameter values which have a positive effect of the system can be identified), *Maintenance Plan Optimization and Individualization* (optimize and individualize maintenance plans based on Big Data analyses), *Energy Efficiency Optimization, Decision Support* (support strategic decisions, improvements) or *Security* (through system anomaly detection can be identified system manipulations through hacking activities).

Specially focused in this paper were the needs related to manufacturing systems but there are also several other potential use cases in other domains as supply chain management, product development and improvement or product individualization.

6   Conclusions

This paper presented an approach for implementing ISA 95-compliant Big Data observation, analysis and diagnosis in Industry 4.0 vision following manufacturing systems. During this project it is aimed to proof and validate that based on the hypothesis that such an approach is suitable for application in future Industry 4.0 vision following manufacturing systems. Results of this work will bring Big Data observation, analysis and diagnosis features into future smart factories and will represent on possibility to handle the growing data amounts and complexity in such systems. There are also visible critical points: Big Data analysis can recognize changes and identify context automatically but the interpretation is an application specific issue and will need mostly human support to define goals and to build a knowledge base. It is a future challenge to develop intelligent algorithms which are able make the interpretation of anomalies identified by Big Data analyses. In this approach the interpretation (the diagnosis) will be supported by human experts.

The validation and hypothesis proof of this approach will be made in a realistic flexible manufacturing system located in the University of Applied Sciences Emden/Leer. The system consists of conventional manufacturing components and will be transferred to an Industry 4.0 vision following service-based environment as base for a validation.

Acknowledgment. This work is partly supported by the ProSEco (Collaborative Environment for Eco-Design of Product-Services and Production Processes Integrating Highly Personalized Innovative Functions) project of European Union's 7th Framework Program, under the grant agreement no. NMP2-LA-2013-609143. This document does not represent the opinion of the European Community, and the European Community is not responsible for any use that might be made of its content.

References

1. Gantz, J., Reinsel, D.: The Digital Universe in 2020: Big Data, Bigger Digital Shadows, and Biggest Growth in the Far East Executive Summary: A Universe of Opportunities and Challenges. International Data Corporation (IDC) (2012)
2. Drath, R., Horch, A.: Industrie 4.0: hit or hype? [Industry Forum]. IEEE Ind. Electron. Mag. **8**, 56–58 (2014)
3. Maier, A.: Online passive learning of timed automata for cyber-physical production systems. In: 12th IEEE International Conference on Industrial Informatics (INDIN 2014), pp. 60–66 (2014)
4. Stokic, D., Scholze, S., Barata, J.: Self-learning embedded services for integration of complex, flexible production systems. In: IECON 2011-37th Annual Conference on IEEE Industrial Electronics Society, pp. 415–420 (2011)
5. Brookshear, J.G.: Theory of Computation: Formal Languages, Automata, and Complexity. Benjamin/Cummings Publishing Company, Redwood City (1989)
6. Leung, C.K.S., MacKinnon, R.K., Fan, J.: Reducing the search space for big data mining for interesting patterns from uncertain data. In: 2014 IEEE International Congress on Big Data (BigData Congress), pp. 315–322 (2014)
7. Ester, M., Sander, J.: Knowledge Discovery in Databases: Techniken und Anwendungen. Springer, Heidelberg (2013)
8. Grossman, R.L., Kamath, C., Kegelmeyer, P., Kumar, V., Namburu, R.: Data Mining for Scientific and Engineering Applications. Springer, New York (2013)
9. Manyika, J., Chui, M., Brown, B., Bughin, J., Dobbs, R., Roxburgh, C., et al.: Big data: the next frontier for innovation, competition, and productivity (2011)
10. Lohr, S.: The Age of Big Data. The New York Times, 11 February 2012
11. Beyer, M.: Gartner Says Solving 'Big Data' Challenge Involves More Than Just Managing Volumes of Data, 27 June 2011. https://www.gartner.com/newsroom/id/1731916
12. Dean, J., Ghemawat, S.: MapReduce: simplified data processing on large clusters. Commun. ACM **51**, 107–113 (2008)

Biomedical Applications

Development of Mixed Reality Systems to Support Therapies

Bruno Patrão[1(✉)], Paulo Menezes[1], and Paula Castilho[2]

[1] Faculty of Sciences and Technology, Institute of Systems and Robotics (ISR-UC),
University of Coimbra, Coimbra, Portugal
{bpatrao,paulo}@isr.uc.pt
[2] Faculty of Psychology and Education Sciences,
Cognitive and Behavioural Centre for Research and Intervention (CINEICC),
University of Coimbra, Coimbra, Portugal

Abstract. This work is focused on the development of Mixed Reality-based Systems suitable for their use in psychological therapies. Recently, on-going research about Immersive Virtual Environments has shown its applicability in psychological domains. Firstly, we intend to explore the bio signals data for emotion recognition. Secondly, we aim to develop an innovative system that allows the creation and manipulation of experiments oriented to psychological therapy. Finally, these experiments will be tested and evaluated in real scenarios by therapists with clinical patients. The results from this work will enable the assessment of the efficacy of these experiments and its improvement in order to apply in a generalized way to individuals with emotional, behavioural and psychological problems.

Keywords: Affective computing · Sensing technologies · Embodiment · Mixed reality · Computer graphics

1 Introduction

Virtual reality has been establishing itself as a powerful tool for the treatment of panic disorder and anxiety. Among these disorders, virtual reality has been used, for example, for the treatment of: fear of flying, driving, heights, public speaking, storms; claustrophobia; agoraphobia; arachnophobia; social phobia; panic disorder and post-traumatic stress due to traffic accidents.

Research carried out so far, where it uses virtual reality to exposure therapy, demonstrated the potential of this technology for the treatment of various types of phobias. The most common and effective treatment of certain phobias is through gradual exposure, i.e., the patient is exposed to conditions that stimulate gradually anxiety. Initially the anxiety increases although as the treatment progresses it tends to decreases. Another form of treatment is through imagery and recall of the dangerous situations and describing the experience to the therapist [1].

L. Camarinha-Matos et al. (Eds.): DoCEIS 2016, IFIP AICT 470, pp. 127–134, 2016.
DOI: 10.1007/978-3-319-31165-4_13

Blascovich et al. enumerated the methodological advantages of virtual-reality-based studies for social psychological research including increased safety and control, the treatment is more efficient and it becomes easier to schedule; it is more efficient and effective for therapists realize the concerns of patients, streamlining the diagnostic and treatment processes; unlimited number of repetitions of feared situations, and the therapist has better control in the environment in which the patient is exposed [2]. Virtual reality can be used to induce affect in the treatment of anxiety, particularly for exposure therapy. A meta-analysis showed effects comparable to clinical *in vivo* exposures [3].

For virtual reality to be effective in the treatment, it is necessary that the virtual environments are able to cause anxiety in patients, i.e. there is a need for patients to have a sense of being in fact experiencing a situation. In virtual reality, this is associated with the concepts of presence and immersion. The feeling of presence refers to the sense of involvement and excitement with the environment and the objects in it. When experiencing virtual environments, the sense of presence is a key concept, which is translated by the feeling of being present in another location or space [7]. Regarding the immersion, it is a psychological state that is characterized by a sense of involvement in the environment. This perception is created through images, sounds, and other *stimuli* that allow the interaction with an environment that provides a series of *stimuli* and experiences to the participant. The sense of presence is vital in any experience of this kind, as it will define if the participant will perceive or not the virtual world as real. One way to achieve the sense of presence is through multisensory stimulation. By increasing the number of senses stimulated in a virtual reality, it is possible to dramatically improve the feeling of presence of a user and consequently, his/her satisfaction and memories of the experiment.

1.1 Motivation

The general aim of this work is to develop a system capable of induce, identify and classify human emotions and physiological activation. We intend to explore two complementary parts: the induction of basic emotions related to threat cues (e.g. self-disgust, anxiety) which are consistently associated with psychopathology, and on the other side, the evaluation of physiological reaction, reflexive behaviours and identification of *stimuli* linked to emotion recognition. Thus, it will be possible to have a closed loop system capable of controlling the emotional state of the patient during a therapeutic process.

The first part will be based on the analysis of body motion (e.g. body language, eye tracking, facial expression) and measured bio-signals (e.g., heart rate, body temperature, skin conductance, respiratory rate). The second part will drive the development of a framework specially designed for the intended goals of therapeutic procedures, in which visual, auditory or haptic *stimuli* will be combined.

Finally, the developed solution will be tested and evaluated in real therapeutic scenarios, under the supervision of specialists and following the ethical and legal procedures. This work will enable the assessment of the efficacy of these technologies and how they can be improved and applied in a generalised way to individuals with psychological and emotional problems.

2   Cyber-Physical Systems

The recent availability of low cost immersive devices, the growing computing power of portable devices and cloud services create excellent opportunities for the application of virtual realty systems in therapeutic contexts, which is one of the areas that can have a real benefit with these systems.

The proposed tool intends to be used in therapeutic context but not only limited to it. Using the advantage of nowadays fast Internet connections it is possible to use this tool anywhere and the therapist can always remotely design or control the experience on the fly. Actually, there are two main advantages of using this kind of immersive technologies, first is that the therapist can manipulate all the environment and guide the patient through a specific situation without physically interfere, it can be applied to most of exposure based therapies. The second advantage is that the therapist can, in real time, observe the physiological responses of the patient and act in accordance to it. Furthermore, the therapist can have a virtual representation in the same virtual environment as the patient.

3   State of the Art

In a virtual reality environment, participants are exposed to digital contents representing real-world scenarios, people, objects, and events, which once combined with the full user's body tracking enables him/her immersion and natural interaction with artificial worlds. Virtual reality permits the participant to come in contact with immersion in scenarios and explore the environments through first-person perspectives as opposed to viewing the scene from stationary positions or third-person views, what generates different impacts on the user experience [4, 5]. Virtual reality has great potential as a method for the induction of affects and emotions; nevertheless this has been rarely used until now.

Rizzo et al. proposed virtual warfare scenarios to use therapeutically with soldiers with post-war trauma [6]. Also relevant, are studies conducted by Mel Slater et al., where they developed virtual environments for the study of neurological rehabilitation and in other psychological domains [8].

Other notable work is project EMMA (Engaging Media for Mental Health Applications) exploring the contribution of emotions have on "presence", which refers to the feeling of being immersed in a virtual environment. In this project a virtual urban park combined with multisensory *stimuli* (e.g., sounds, sights, lightning effects and other forms of affective *stimuli*) was used to induce changes such as anxiety or relaxation [9].

The success of these immersive systems depends on the "sense of presence" level that they can induce on a participant taking part in such experiences. The sense of presence refers to the sense of "being there", in the virtual environments created by the technology. Botvinick & Cohen in their work known as "rubber hand illusion" [10], found that a fake body part can be incorporated into human body representation through synchronous multi-sensory stimulation on the fake and corresponding real body part.

For instance, people experiencing virtual reality may have the illusion of being in a virtual place and, consequently, carry out actions as if the situation and events depicted were really happening. Accompanying these actions, we may observe physiological changes (e.g., heart rate, body temperature, skin conductance, and respiratory rate), reflexive behaviours (e.g., eye blinking or smiling at a virtual human character) and emotions arousal. As far as we know, few authors had combined virtual reality with physiological data.

4   Research Contributions and Innovation

The expected contributions of the current work are the analysis of body motion and bio signals data for emotion recognition and the development of an innovative framework that allows the creation and manipulation of experiments oriented to psychological therapy, to this end we have created different setups.

Figure 1 represents the diagram for the proposed system; the dashed boxes represent the on-going work. Furthermore, the therapist can manipulate some parameters of Stimuli System, such as, scene type and intensity of *stimuli* (input data) and receive the patient's biofeedback (output data).

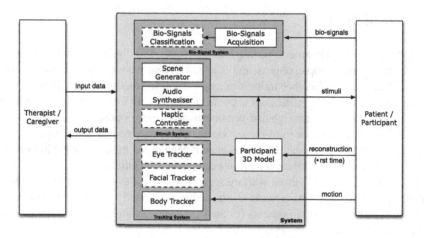

Fig. 1. Proposed system diagram.

The first prototype was designed to elicit fear and avoidance reactions. It consists in a virtual mirror room and, at a certain point, instructing the participant to touch the mirror triggers a set of grim visual and auditory events that culminate in the severance of the avatar's hand. To construct this prototype several steps had to be followed in order to produce the necessary components that upon integration result in the intended setup. These steps are described in the next subsections.

4.1   Model Acquisition System

First we start to reconstruct the user's 3D model so he/she can have a virtual representation (Fig. 2). The 3D reconstruction solves two problems by itself; in one hand is the user's height being used to place the point of view on the virtual world helping the spatial awareness, because environments are modulated in a 1:1 scale, in other hand, the user will be able to have a fastest acceptance of their virtual representation and consequently a better immersion feeling.

Fig. 2. User 3D reconstruction.

4.2 Bio-signals System

It is well known that physiological signals change in response to physical activities or emotional changes. There are morphological and physiological connections between these emotional states and autonomic nervous system, including its sympathetic and parasympathetic divisions. Sympathetic (SNA) and parasympathetic nervous activity (PNA) contributes to those responses, such as, the heart rate variability, skin conductance variability or body temperature. While the PNA controls the body's response in rest, SNA is responsible for the internal response in body's fight-or-flight, meaning the reaction that occurs in response to a perceived harmful event, attack, or threat to survival. However, it is very difficult to identify a direct correlation between physiological data and a particular emotion, but possible to detect the arousal of an emotion when presenting specific contextual *stimulus*, i.e., it is possible to detect fear when presenting a horror situation.

The following statements describe how we will use the physiological data in our system.

Electrocardiography. Essential tool to access the Heart Rate Variability (HRV) associated with changes in the Heart Rate (HR) strongly related to *stimuli* activation.

Electro-Dermal Activity. Measures electrical skin conductance varying with the wet level of sweat glands. These are controlled by the sympathetic system that affects EDA once it is active, especially when an individual is anxious or stressed.

Body Temperature. Varies depending on the place where it is measured. In our system the body temperature is measured upon the skin and upper torso.

Body Acceleration. When attached to a moving body, accelerometers are capable of accurately sensing it movements. We use it in several body parts (arms, torso, head) to detect psychomotor agitation.

Breathing Activity. The process involves the movement of the diaphragm, which is expressed in movements of expanding and compressing of the rib cage and abdominal area. This produces patterns that are used by the system to identify events.

Aiming to detect emotional arousal, the signals are acquired from a set of sensors, pre-processed to remove any noise and produce the features needed by the classification process.

4.3   Stimuli System

The majority of the development of immersive systems focuses on vision since it is the most dominant sense. Thus, the head mounted displays (HMD) are considered the core technologies for this end. By consequence the essential ingredient of Virtual Reality is a tracked HMD that lets user see new views of the visual world as he/she moves his head. Wearing an HMD, the user can look around and see the simulated virtual world just like in the real world.

Furthermore, objects producing sounds or the sound of objects interacting with other objects are part of our daily experience. Humans as many animals have the ability to detect the sound source location (i.e. where the sound is coming from). The synchronicity of the sounds and object-related events and coherency between object motion seen and the corresponding sound source displacement is something that we humans are very sensible to. In other words if synchronicity and coherency fail they are immediately detected, affecting the perception and immersion in the environment. For this end we are using the HMD Oculus Rift with audio phones combined with our graphical engine, OpenAIR.

4.4   Tracking System

With a level of importance similar to the sensory *stimuli* systems used, the tracking system is the cornerstone that transforms a simple wearable visualisation system into an immersive system, where the user can be active. An unobtrusive tracking mechanism is required to register any head and body motion and providing the data to the computer to make the required changes in viewpoint and position.

In order to have a full body tracker we are using a RGB-D sensor (Kinect) to collect all joints orientation and map them to the 3D model skeleton. This allows the user to interact with the virtual world and have the virtual representation mimicking, synchronously his/her movements, thus improving the immersion.

Additionally, there are reflexive behaviours when experiencing virtual environments very important to collect, such as, gaze orientation, eye blinking or smiling at a virtual character. In order to get this information, we are at the moment designing a prototype with cameras tracking the lower face and eye movements to be installed on the HMD covering that entire region.

5   Discussion of Results and Critical View

We started to develop a small virtual room with a mirror where the participant can freely move and look around. At the mirror, his/her reflection mimics all the movements creating a good way to adapt and feel immersed in the system. After a minute of adaptation the user is asked to touch his/her own reflection and when he/her does it a guillotine falls cutting the hand and left it bleeding. At this point different participant's reactions occur in their motor and physiological systems.

With this experiment in the virtual mirror prototype, our tests showed high peaks in Electro-dermal Activity (EDA) on most participants, confirming that physiological data has a direct relation with emotional responses and *stimuli* reactions (Fig. 3).

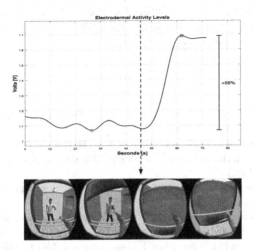

Fig. 3. Immersive Virtual Mirror Experiment (dashed line represents the moment when guillotine falls).

After these preliminary results, we decided to add more scenarios, and extend our data acquisition to include other bio-signals such as Heart Rate (HR), Respiratory Rate (RR), Body Temperature (BT), and Body Movements (BM). The addition of these bio-signals will enable a more complete analysis of the human physiological responses to

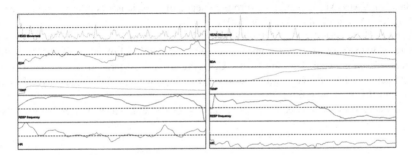

Fig. 4. Stressful environment: Haunted house (*left*). Relaxing environment: Zen garden (*right*).

emotional states or activation in an attempt to establish a correlation between them. Figure 4 shows these new acquired signals in two different situations, the left image corresponds to a set of stressful situations in a haunted house and the right one to a peaceful environment.

6 Conclusions

Being able to inhabit a virtual world through the same perspective as in real life unlocks the possibility for creating plausible enough experiences to be used as a tool in clinical therapy. Our tests have been confirming this claim by showing that different experiences can indeed elicit different physiological responses in the participant. These responses can be used by the therapist as an additional source of information about the evolution of the patient during the therapy sessions. This leads us to believe that our system can indeed help the aid of several psychopathologies.

References

1. Emmelkamp, P., Bouman, T., Scholing, A.: Anxiety Disorders: A Practitioner's Guide, 1st edn. Wiley, Manhattan (1992)
2. Blascovich, J., Loomis, J., Beall, A.C., Swinth, K.R., Hoyt, C.L., Bailenson, J.N.: Immersive virtual environment technology as a methodological tool for social psychology. Psychol. Inq. **13**(2), 103–124 (2002)
3. Powers, M.B., Emmelkamp, P.: Virtual reality exposure therapy for anxiety disorders: a meta-analysis. J. Anxiety Disord. **22**, 561–569 (2008)
4. Ochsner, K.N., Knierim, K., Ludlow, D.H., Hanelin, J., Ramachandran, T., Glover, G., Mackey, S.C.: Reflecting upon feelings: an fMRI study of neural systems supporting the attribution of emotion to self and other. J. Cogn. Neurosci. **16**(10), 1746–1772 (2004)
5. Ruby, P., Decety, J.: How would you feel versus how do you think she would feel? A neuroimaging study of perspective-taking with social emotions. J. Cogn. Neurosci. **16**(6), 988–999 (2004)
6. Rizzo, A., Pair, J., Graap, K., Manson, B., McNerney, P.J., Wiederhold, B., Wiederhold, M., Spira, J.: A virtual reality exposure therapy: application for Iraq war military personnel with post-traumatic stress disorder: from training to toy to treatment. In: Roy, M. (ed.) NATO Advanced Research Workshop on Novel Approaches to the Diagnosis and Treatment of Posttraumatic Stress Disorder, pp. 235–250. IOS Press, Washington, D.C. (2004)
7. Sanchez-Vives, M.V., Slater, M.: From presence to consciousness through virtual reality. Nat. Rev. Neurosci. **6**(4), 332–339 (2005)
8. Pan, X., Gillies, M., Barker, C., Clark, D.M., Slater, M.: Socially anxious and confident men interact with a forward virtual woman: an experimental study. PLoS ONE **7**(4), e32931 (2012)
9. Riva, G., Mantovani, F., Capideville, C.S., Preziosa, A., Morganti, F., et al.: Affective interactions using virtual reality: the link between presence and emotions. CyberPsychol. Behav. **10**(1), 45–56 (2007)
10. Botvinick, M., Cohen, J.: Rubber hands 'feel' touch that eyes see. Nature **391**, 756 (1998)

Brain-Computer Interfaces by Electrical Cortex Activity: Challenges in Creating a Cognitive System for Mobile Devices Using Steady-State Visually Evoked Potentials

Pedro Morais[1,2,3(✉)], Carla Quintão[2,3], and Pedro Vieira[2,3]

[1] Biomedical Engineering Doctoral Program, Universidade NOVA de Lisboa, Lisbon, Portugal
[2] Department of Physics, Faculty of Sciences and Technology,
Universidade NOVA de Lisboa, Lisbon, Portugal
{cmquintao,pmv}@fct.unl.pt
[3] Laboratory for Instrumentation, Biomedical Engineering and Radiation Physics,
Universidade NOVA de Lisboa, Lisbon, Portugal
jpe.morais@campus.fct.unl.pt

Abstract. The research field of *Brain-Computer Interfaces* (BCI) emerged in an attempt to enable communication between paralyzed patients and technology. Identifying an individual's mental state, through his brain's electric activity, a typical BCI system assigns to it a particular action in the computer. It is known that when the visual cortex is stimulated with a certain frequency, it shows activity with the same frequency. This *Steady-State Visually Evoked Potential* (SSVEP) activity can be used to achieve the aforementioned communication goal. In this work, we first analyze the spontaneous electrical activity of the brain, to distinguish two mental sates (concentration/meditation). Then, following an SSVEP type of approach, we divide the stimulating screen in four areas, each of which flickering at a distinct frequency. By observing the responding frequency from the occipital lobe of the subject, we can then estimate the 2 bit decision he made. We observe that such a setup is efficient for real time BCI, and can be easily integrated in mobile devices. Besides, the user is able to change voluntarily her/his decisions, interacting with the system in a natural manner.

Keywords: BCI · EEG · SSVEP · Mobile device · Tablet · Smartphone

1 Introduction

A biomedical cognitive control system must be able to interpret the electrical signals produced in our brain and distinguish different levels of activity. There have been various approaches taken in this area, such as event-related desynchronization and synchronization (ERD/ERS) [1], evoked potentials with latency of 300 ms (P300) [2, 3] and visual evoked potentials in stationary mode (SSVEP) [4, 5]. The ERD/ERS engaged in the study of alpha and beta waves, characterized by frequency between 8 Hz and 12 Hz and 12 Hz to 30 Hz, respectively, which can be observed, for example, during an imagination task motion. P300 in an evoked potential with a latency of about 300 ms, which appears

© IFIP International Federation for Information Processing 2016
Published by Springer International Publishing Switzerland 2016. All Rights Reserved
L. Camarinha-Matos et al. (Eds.): DoCEIS 2016, IFIP AICT 470, pp. 135–141, 2016.
DOI: 10.1007/978-3-319-31165-4_14

after a visual or auditory stimulus that requires attention and cause some surprise. SSVEP are elicited by retinal stimulation with a signal whose frequency can vary between 3.5 Hz and 75 Hz and consist of a continuous and periodic signal detected in the visual cortex with the same frequencies [6]. The publications in this area are many and varied, ranging from the presentation of tools to control different devices, to new technological solutions. Among the applications that resulted from the above described techniques, we highlight some technological advances such as: 1. Activation of a mobile robot using a BCI [7], wherein a robot is controlled through four imagined movement (foot, tongue, left arm and right arm) using ERD/ERS; 2. Application of dried EEG sensors to mobile BCIs [8], which are placed on the hair, and exhibit very similar results to traditional sensors that use saline solution or conductive gel for electrical contact; 3. Construction of a simple communication system using SSVEP based on BCI [9] where a user gives answers like "yes", "no", "good", "bad"; 4. Creation of an online BCI using static visual evoked potential [10] where through one EEG channel the user can write a word and perform a search on Google; 5. Characterization of stimuli based on P300 amplitude [11], wherein the study shows that there are several factors which influence the potential analyzed, for example, the effect of motivation as possible physiological influence on the amplitude P300 and, finally; 6. Development of a mobile phone based on BCI for communication on a day-to-day [12] through SSVEP stimulation.

In this context it is rather evident the importance of a transversal knowledge between neuroscience and computer science to reach new insights in BCI area, solving problems related to the acquisition, storage and retrieval of brain information, as well as creating new approaches to identify different actions on the same interface.

This work aims to answer the question: "How to develop a solution where an user can interact with mobile devices naturally changing voluntarily their mental task?". We propose a system consisting of two specific phases: First, the detection of an individual's state of concentration and second, the selection of the action to be taken through SSVEP after confirming the previous condition. To fulfill the first phase, the spontaneous elec-troencephalographic (EEG) signals were classified based on the traditional band frequency analysis (alpha band, between 8 to 13 Hz and beta band, between 13 and 30 Hz) [13, 14]. In the second phase, the SSVEP permit to distinguish between at least 4 different actions that the subject would like to performed.

2   Relationship to Cyber-Physical Systems

A cyber-physical system (CPS) comprises the junction of computing elements with nature physical processes. This approach provides the development of more specific applications such as process control, instrumentation, medical devices and smart structures. In the coming age of *internet of everything*, the development of this type of system contributes to a new era of products where everyone will be connected everywhere. To reach such solution is necessary a flexible architecture with new interfaces. The research for user-friendly interfaces is increasing and one of the aims will be to replace keyboard and mouse computers for more effective means of communication. The use of touch screens, commonly available in tablets and smartphones, is a clear example of this demand.

The project described in this paper presents a system that establishes the connection between machines and humans using brain-computer interfaces (BCI)/Electroencephalography (EEG). The EEG is used as a non-invasive electrophysiological monitoring approach to identifying behavior patterns of brain electrical activity, while BCIs promote a direct communication channel between the brain and an electronic device. Using brain electrical activity recorded at scalp level enables intelligent monitoring systems in real-time. The use of sensory channels becomes a new form of input in addition to providing information about the status and user intent. Using this type of information, systems can adapt dynamically contributing to the task that one wants to run. To develop new solutions in this field, it is essential to understand how the brain works and manages the information. Referring to EEG and other biological signals such as electrocardiography (ECG) and electromyography (EMG), opens up a new form of communication between humans and electronic devices. This approach presents several challenges such as the efficiency of embedded systems, the implementation of algorithms using brain electrical impulses and distribution architectures that add autonomy to the devices and increase the efficiency to the communication mechanisms. At the end, this solution must have a high degree of robustness, and should enable connection to the cloud, not confined to the local control devices.

3   Materials and Methods

To answer the challenges in creating this BCI solution it is used the EPOC equipment [15] to record the EEG signals. The system comprises 14 channels (AF3, F7, F3, FC5, T7, P7, O1, O2, P8, T8, FC6, F4, F8, AF4) and two reference electrodes located in P3 and P4 channels (Common Mode Sense active electrode and Driven Right Leg passive electrode - CMS/DRL). The device records the signals sequentially with a rate of 128 Hz, with a resolution of 14 bits per channel and have a frequency response between 0.16 and 43 Hz. In addition, it provides wireless data transmission using a frequency of 2.4 GHz, with a battery allowing for 8 h of continuous work. One of the greatest facilities of this device is the use of saline solution electrodes instead of the common conductive gel to establish the contact between the electrodes and the scalp. The acquisition control and the complete signal classification process were performed by two software platforms: *OpenVibe* and *EEGLAB*. The *OpenVibe* is an open source application, multi-platform, containing various pre-programmed modules for signal processing. In parallel, the *EEGlab*, a *Matlab* toolbox for EEG and event related potentials processing, is also used, being necessary to install the plug-in acquisition *BIOSIG* data for reading data in GDF format *(General Data Format)* [16].

3.1   Acquisition Protocol for Classification of the Mental State

The signals from the 14 channels are recorded while the subject is seated with a straight posture with the palms on the knees, avoiding any muscular movement. A complete test lasts a total of five minutes. The subject is asked to switch from one mental state (meditation) to the other (concentration), every 30 s, being notified with a beep. In meditation he should get away from any thought focusing his attention on the breath which should be long and deep. For the state of concentration he should countdown of 3 in 3 from 100, while he visualizes the results.

3.2 Signal Processing and Classification

A band pass analog filter from 0.16 Hz to 85 Hz, together with a 50 Hz notch filter, is applied to the signals. Since the meditation/concentration detection states is the main purpose of this phase of the project, we focalized our analysis on the prefrontal cortex electrodes (AF3, AF4), relating with attention activities [17] and in occipital electrodes (O1, O2), where the alpha rhythm, which measure the level of arousal of the subject, is particularly intense.

We used the *OpenVibe* application to implement the following steps: 1. The electrodes are selected; 2. A 4th order *Butterworth* digital band pass filter between the alpha band frequencies (8–12 Hz) is applied to the signals; 3. The signal is sub-divided into various time windows lasting 5 s with an overlap of 4.9 s; 4. The power spectral density is computed; 5. A moving average is applied.

The identification of meditation and concentration states was performed by real-time signal acquisition using the following criteria: if the alpha band spectrum in AF3 electrode is below a certain threshold (in our arbitrary units this threshold is 20) then the state will be classified as "concentration", else, the state will be classified as "meditation". If the alpha band spectrum exceeds the value 80 the corresponding EEG epoch is considered an artifact.

3.3 Choosing Actions Using SSVEP

In a second stage, after checking the "concentrate" state, it is necessary to identify an action by means of SSVEP. To fulfill this objective it is presented to the subject a screen divided into four areas, each one with an image formed by a black/white checkerboard that oscillates at a predefined frequency. The subject option is identified depending on the area/image where the subject focuses its attention.

The stimulation is made using the Psychophysics version 3, a *Matlab* toolbox, which allows imposing frequencies at 8.6 Hz, 10 Hz, 12 Hz and 15 Hz to the screen. The detection of the frequency emitted by the different screen areas is performed through the analysis of the power spectrum obtained from the electrodes O1 and O2, located in primary visual areas in the occipital lobe. This process allows the recognition of the action to be taken in a natural way.

In this stage the signals were filtered using a 4th order *Butterworth* digital band pass filter between 8–30 Hz, which comprises the alpha and beta bands. Once again, the analysis was performed on windows during 5 s with an overlap of 4.9 s. Finally, the spectrogram of those signals is generated and the results were compared with the frequencies of the visual stimuli in order to identify the action that should be taken.

4 Results

4.1 Meditation/Concentration

The states identification of meditation and concentration which corresponds to the 1st phase of the project was performed by real-time signal acquisition. One example is shown next, for electrodes AF3 and O1 [Fig. 1].

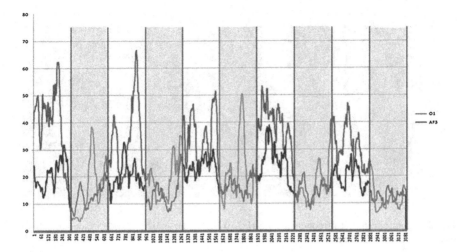

Fig. 1. Results of a 5 min essay in which the subject is switching from the meditation to concentration stage. Each state lasts 30 s starting with meditation. The graph represents the amplitude of the alpha waves in O1 and AF3 electrodes.

As expected, during the concentration periods the alpha activity decreases substantially to values that rarely exceed the value 20 (arbitrary units). It should be noted in this example that the two peaks seen in between 30.0–60.0 and 150.0–180.0 observed in O1 electrode, it is a state deconcentration unscheduled during the visual imagination of the subtraction results. Analyzing the values obtained, there is a clear distinction between the two states meditation/concentration, both on the front and occipital zone, although in the latter the difference is more pronounced.

These results demonstrate that it is possible to search and identify patterns of brain activity consistent with the classification of this information in real time.

4.2 SSVEP Results

After completing the 1st stage of the process, with the concentration state identification, it follows the 2nd phase results, corresponding to the choice of the action using SSVEP. The results obtained in the spectrogram show a clear identification of the 8.6 Hz, 10 Hz, 12 Hz and 15 Hz frequencies [Fig. 2] (the ones at which the four areas of the screen flick).

It was also realized that the checkerboard images in which the oscillating frequencies were applied, should be dispersed in the screen away from each other as much as possible. Besides, it is expected that the use of preset options will provide faster responses, compared to description of the desire. Taking an example of someone who wants to quench his thirst, using a grid with the image of a glass of water as one of the default options, will save significant time compared to writing this same intention.

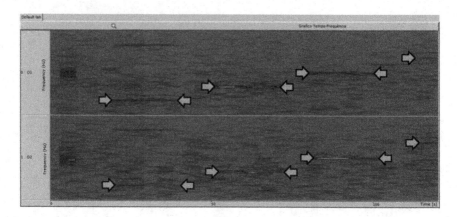

Fig. 2. Spectrogram with frequencies 8.6 Hz, 10 Hz, 12 Hz and 15 Hz identification using the electrodes O1 and O2 during the SSVEP stimulation.

5   Conclusions

The junction of the identification of a concentrated state with SSVEP will allow the implementation of a hybrid system with high accuracy. BCI approach allows the user to change voluntarily their mental task interacting with the system naturally. This work aims to contribute to the development of an autonomous system which allows monitoring in real-time mobile devices from the electrical brain activity. Associated with this application are the interfaces that will interact with the user. The flickering stimuli using checkerboard images can be replaced with icons that identify the user's intent with a clear differentiation of the actions to be taken. At the same time it should be evaluate the disruption that may be caused by peripheral vision of a concentrated individual, in order to remove this undesirable activity from the signal. We also believe that the use of higher frequencies will provoke a decreasing in visual fatigue caused by the oscillation of the image, making the system more comfortable. The presented approach should be tested in a statistical significant number of individuals to identify more precisely the degree of accuracy of the results.

References

1. Lopes da Silva, F.H., Pfurtscheller, G. (eds.): Event-Related Desynchronization. Handbook of Electroencephalogr and Clinical Neurophysiology, vol. 6, pp. 51–65. Elsevier, Amsterdam, Revised Edition (1999)
2. Fazel-Rezai, R., Abhari, K.: A region-based P300 speller for brain-computer interface. Can. J. Electr. Comput. Eng. **34**, 81–85 (2009)
3. Sellers, E., Arbel, Y., Donchin, E.: BCIs that uses P300 event-related potentials. In: Wolpaw, J., Wolpaw, E.W. (eds.) Brain-Computer Interfaces: Principles and Practice. Oxford University Press, Oxford (2012)

4. Allison, B., Faller, J., Neuper, C.H.: BCIs that use steady-state visual evoked potentials or slow cortical potentials. In: Wolpaw, J., Wolpaw, E.W. (eds.) Brain-Computer Interfaces: Principles and Practice. Oxford University Press, Oxford (2012)
5. Silberstein, R.B.: Steady state visually evoked potential (SSVEP) topography in a graded working memory task. Int. J. Psychophysiol. **42**, 219–232 (2001)
6. Pastor, M., Artieda, J., Arbizu, J., Valencia, M., Masdeu, J.: Human cerebral activation during steady-state visual-evoked responses. J. Neurosci. **23**(37), 621–627 (2003)
7. Barbosa, A.O.G., Achanccaray, D.R., Meggiolaro, M.A.: Activation of a mobile robot through a brain computer interface. In: 2010 IEEE International Conference on Robotics and Automation, pp. 4815–4821 (2010)
8. Chi, Y.M., Wang, Y.-T., Wang, Y., Maier, C., Jung, T.-P., Cauwenberghs, G.: Dry and noncontact EEG sensors for mobile brain-computer interfaces. IEEE Trans. Neural Syst. Rehabil. Eng.: Publ. IEEE Eng. Med. Biol. Soc. **20**, 228 (2011)
9. Sanchez, G., Diez, P.F., Avila, E., Leber, E.L.: Simple communication using a SSVEP-based BCI. J. Phys: Conf. Ser. **332**, 012017 (2011)
10. Liu, T., Goldberg, L., Gao, S., Hong, B.: An online brain-computer interface using non-flashing visual evoked potentials. J. Neural Eng. **7**, 036003 (2010). doi:10.1088/1741-2560/7/3/036003
11. Kleih, S.C., Nijboer, F., Halder, S., Kübler, A.: Motivation modulates the P300 amplitude during brain–computer interface use. Clin. Neurophysiol. **121**, 1023–1031 (2010)
12. Wang, Y.-T., Wang, Y., Jung, T.-P.: A cell-phone-based brain-computer interface for communication in daily life. J. Neural Eng. **8**, 025018 (2010). doi:10.1088/1741-2560/8/2/025018
13. Fisch, B., Spehlmann, R.: Fisch and Spehlmann's EEG Primer, 3rd edn. Elsevier, Amsterdam (1999)
14. Niedermeyer, E., da Silva, F.H.L.: Electroencephalography: Basic Principles, Clinical Applications, and Related Fields. Williams & Wilkins, Baltimore (1993)
15. Emotiv - EPOC neuroheadset (2012). http://emotiv.com/store/hardware/epocbci/epoc-neuroheadset/
16. Schlogl, A., Filz, O., Ramoser, H., Pfurtscheller, G.: GDF - a general dataformat for biosignals, Technical report (2004)
17. Rebollo, M.A., Montiel, S.: Atención y funciones ejecutivas. Revista de Neurologia, **42** (Supl 2), S3–7 (2006)

Automatic EOG and EMG Artifact Removal Method for Sleep Stage Classification

Ali Abdollahi Gharbali[1,2(✉)], José Manuel Fonseca[1,2], Shirin Najdi[1,2], and Tohid Yousefi Rezaii[3]

[1] Computational Intelligence Group of CTS/UNINOVA, 2829-516 Caparica, Portugal
[2] Faculdade de Ciências e Tecnologia, Universidade Nova de Lisboa, Campus da Caparica, Quinta da Torre, Monte de Caparica, 2829-516 Caparica, Portugal
{a.gharbali,s.najdi}@campus.fct.unl.pt,jmrf@fct.unl.pt
[3] Faculty of Electrical and Computer Engineering, University of Tabriz, Tabriz, Iran
yousefi@tabrizu.ac.ir

Abstract. In this paper, a new algorithm is proposed for artifact removing of sleep electroencephalogram (EEG) with application in sleep stage classification. Rather than other works which used artificial noise, in this study real EEG data contaminated with electro-oculogram (EOG) and electromyogram (EMG) are used for evaluating the proposed artifact removal algorithm's efficiency using classification accuracy. The artifact detection is performed by thresholding the EEG-EOG and EEG-EMG cross correlation coefficients. Then, the segments considered contaminated are denoised by normalized least-mean squares (NLMS) adaptive filtering technique. Using a single EEG channel, four sleep stages consisting of Awake, Stage1 + REM, Stage 2 and Slow Wave Stage (SWS) are classified. A wavelet packet (WP) based feature set together with artificial neural network (ANN) are deployed for sleep stage classification purpose. Simulation results show that artifact removed EEG allows a classification accuracy improvement of around 14 %.

Keywords: Sleep stage classification · Wavelet packet · Adaptive filtering · Artifact removing · Artifact detection

1 Introduction

Sleep is an essential part of a human's daily life that significantly affects his/her health, productivity, mental and moral states. Therefore, the diagnosis of sleep related disorders is of great importance in sleep research. Sleep scoring with the use of polysomnographic (PSG) recordings is one of the principal requirements of diagnosis procedure of sleep disorders. PSG recordings include electroencephalogram (EEG), electromyogram (EMG), electro-oculogram (EOG), electrocardiogram (ECG), snoring and other physiological signals to detect body movements. Currently sleep scoring is mainly done by a domain expert who visually analyses PSG data, diagnoses and prescribes treatment. However, this process is time consuming, tedious and highly subjective. Therefore,

© IFIP International Federation for Information Processing 2016
Published by Springer International Publishing Switzerland 2016. All Rights Reserved
L. Camarinha-Matos et al. (Eds.): DoCEIS 2016, IFIP AICT 470, pp. 142–150, 2016.
DOI: 10.1007/978-3-319-31165-4_15

several algorithms for automatic sleep scoring have been proposed to assist the expert and increase the reliability of the results [1].

EEG is widely adopted for the automatic detection of sleep stages and neuronal activity evaluation during sleep. However, EEG is usually contaminated with several artifacts such as power line noise, EMG, EOG and electrode movements. Removal or attenuation of the noise and unwanted signals is a prerequisite for most of the EEG signal processing applications. The presence of artifacts makes the EEG analysis difficult, since it may introduce spikes that can be confused with original EEG trend decreasing the reliability of the subsequent processing stages.

The basics for the artifact removing are diverse and are closely related to the specific application in which the algorithm is going to be used. A commonly used method for avoiding artifacts is rejecting the contaminated segments of the recorded EEG [2]. This method although simple, results in huge data loss. Instead, denoising the contaminated EEG segments would not only preserve the amount of data, but also would probably contribute to the increase of accuracy in the automatic sleep stage classification [3]. Following this idea, an extensive number of studies have tried to extract the clean EEG out of the contaminated recording in different research areas but still no optimal method is agreed upon [4].

Considering the fact that nowadays portable devices for patient monitoring and automatic sleep stage classification could be a helpful assistance for experts on the analysis of sleep signals, the main motivation for the current work is the lack of a systematic method for automatic artifact detection and cancellation which leads to an improvement in the automatic stage classification accuracy compared to the original acquired data. In view of this gap, the following research question emerges:

What would be a suitable methodology to detect, identify and remove the EOG and EMG artifacts from EEG in EEG-based sleep analysis, so that the overall artifact level is decreased and the automatic sleep stage classification accuracy is positively influenced?

In the rest of the paper we will try to answer this question. The paper is organized as follows: Sect. 2 explains how the proposed automatic artifact removal system can benefit cyber-physical systems. Section 3 gives a brief overview of how EEG artifact removal problem is treated in the literature. Section 4 provides detailed description of the proposed algorithm, and simulation results. Section 5 presents the results discussion and Sect. 6 finalizes the paper with the conclusions and future work directions.

2 Relationship to Cyber-Physical Systems

Cyber-physical systems (CPS) are an emerging technology that attracted a lot of attention in recent years. CPS can be regarded as the next generation computing systems that monitor, coordinate, control and integrate communication and computation abilities to interact with physical systems [5]. CPS are assumed to have a wide range of applications in several areas including defense, transportation, agriculture, energy, healthcare, etc.

In the healthcare area there are currently a limited number of CPS available (such as bedside monitors) and many of their components still work isolated. With the extensive use of CPS, all these components will be integrated to a networked closed loop system together with the human [6]. Wearable health monitoring systems, that allow the subject to continue normal life while his/her vital signs are recorded and analyzed, are a good example of such solutions. In most of the cases a whole night PSG signal is necessary for diagnosing a sleep related problem. This data is usually recorded in a hospital or sleep clinic while the change of sleeping environment and noise may distract the subject and induce significant alterations in the recorded data. Medical CPS like wearable sleep monitoring systems seem to be the perfect solution for recording whole night data while the subject is saved from any change in the sleeping environment. The medical data recorded this way is the main indicator of the subject's health status and its reliability is very important for further signal processing. Therefore, denoising of biological data is one of the most important requirements for improved medical CPS [7, 8]. This work is developed in such a context to support the relation between CPS and sleep monitoring.

3   State of the Art

EEG is mainly intended for recording cerebral activity, yet other extra electrical activities are also recorded. These extra activities are usually considered harmful artifacts that can be either physiological like EOG, EMG and ECG or extra physiological like power line interference. EOG measures are captured mainly by frontal electrodes, but they are strong enough to also affect other electrodes. About the EMG, the degree and type of contamination depends on the contracted muscle, the recording purpose and the environment [9].

Some of these artifacts are easily removed by a finite impulse response (FIR) or infinite impulse response (IIR) filter if their power spectrum doesn't have overlap with the EEG power spectrum. However, EMG and EOG, which have significant power spectrum overlap with EEG are not easy to remove requiring careful consideration [9].

The state of the art in EEG denoising is quite broad. The most classic methods used in the last years are regression (especially for ocular interferences), blind source separation or component base techniques [10] and Wiener and Bayes filtering methods [4]. Also adaptive filtering [4], wavelet denoising [11] and empirical mode decomposition (EMD) [12] are among the most widely used denoising techniques. A great range of studies pay particular attention to the improvement of existing methods or using more objective performance criteria [13].

In the context of sleep stage classification, Estrada et al. [14] proposed a denoising method based on applying a predefined threshold to the wavelet coefficients of noisy EEG with the objective of finding the best threshold defining rule and value. The noisy signal is constructed by adding white noise and 50 Hz power line sinusoidal noise. The coefficients that contribute to the noise components are zeroed out using a threshold discrimination filter. They use Mean Squared Error (MSE) and signal-to-noise ratio (SNR) as criteria for evaluating the performance of the proposed de-noising scheme.

4   Materials and Methods

Figure 1 shows an overview of the sleep stage classification with proposed sleep EEG artifact removal scheme.

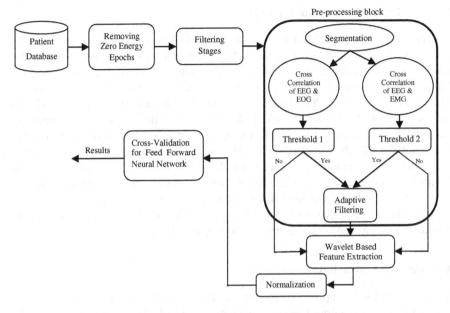

Fig. 1. Flowchart of the proposed algorithm.

4.1 Data

In this study, PSG records of 22 Caucasian males and females with the duration of nine hours each were used from the Sleep EDF Database [15]. Except for a slight difficulty in falling asleep, subjects were healthy and without any medication. All EEG (from Fpz-Cz and Pz-Oz electrode locations), EOG (horizontal) and submental chin EMG recordings were performed with a sampling rate of 100 Hz. All PSGs are divided into 30 s epochs and annotated by well-trained technicians according to the Rechtschaffen and Kales (R&K) manual. In this study, Pz-Oz EEG channel, horizontal EOG and submental chin EMG recordings of all the subjects are used.

4.2 Pre-processing

Visual inspection of the data reveals that occasionally there exist one or more consecutive epochs in which the energy of the signal is zero probably due to hardware failure. These epochs are removed in order to eliminate the confusion that they could introduce on the sleep stage classification algorithm.

Following the AASM manual [16] recommendations for the scoring of sleep, the EEG and EOG signals are filtered with a band-pass filter with lower cut-off of 0.3 Hz and higher cut-off of 35 Hz. In this work we use WP based decomposition and reconstruction as a filter for removing unwanted frequency band of physiological data [17].

4.3 Artifact Detection and Removal

Conventionally, it is assumed that the measured EEG is a linear combination of cerebral activity with one or more kind of artifacts. Therefore, in this paper, for detecting the EOG and EMG contamination, the filtered EEG, EOG and EMG recordings are divide into 1000-sample segments and then the cross correlation of each EEG segment is calculated with the corresponding EOG and EMG segment. If the absolute value of the EEG-EOG cross correlation coefficients or EEG-EMG cross correlation coefficients is more than threshold 1 or threshold 2 respectively, the corresponding segment will be fed to an artifact removal block which is based on NLMS adaptive filtering. Adaptive filtering [18] has been extensively used in EEG artifact removal algorithms. It uses a recorded reference of the artifact (in our case horizontal EOG and submental chin EMG) to adjust a vector of weights that models the contamination according to an optimization algorithm.

If the thresholding conditions for cross correlation coefficients are not satisfied, the relevant EEG segment will be copied to the output without any change.

4.4 Feature Extraction

In order to perform sleep stage classification, the output of the pre-processing block is used for feature extraction. There are five types of main brain waves that can be distinguished by their frequency range. These frequency bands are called Delta (0–3.99 Hz), Theta (4–7.99 Hz), Alpha (8–13 Hz) and Beta (> 13 Hz) [16]. Therefore, EEG is conventionally analyzed in the frequency domain. Moreover, EEG is a non-stationary signal and simultaneous time-frequency analysis can be quite useful. In this study, a WP tree with 7 decomposition levels and Daubechies order 2 (db2) mother wavelet is used for feature extraction. Different frequency bands of EEG including Delta, Theta, Alpha, spindle, Beta1 and Beta 2 are extracted according to the scheme proposed in [19]. The following statistical features are calculated for each epoch using the WP coefficients:

(1) Energy of the WP coefficients for each frequency band (F1–F6)
(2) Total Energy (F7)
(3) Mean of the absolute values of WP coefficients for all frequency bands (F8)
(4) Standard deviation of WP coefficients for all frequency bands (F9)
(5) Energy ratio of various frequency bands (F10 to F14)

F10 is the ratio between the energy in the Alpha band and the sum of the energy in the Delta and Theta bands. F11 is the ratio between the energy in the Delta band and the sum of the energy in the Alpha and Theta bands. F12 is the ratio between the energy in the Theta band and the sum of the energy in the Alpha and Delta bands. F13 is the ratio between the energy in the Alpha band and the energy in the Theta band and F14 is the ratio between the energy in the Delta band and the energy in the Theta band.

4.5 Normalization

Features are normalized to standardize their range. Since the range of values of raw EEG varies broadly, to avoid that features with larger numeric values dominate those with smaller numeric values affecting the accuracy of the classification technique, each feature (x_{ij}) is independently normalized by applying the following equation:

$$x'_{ij} = \frac{x_{ij} - \bar{x}_i}{\sigma_{x_i}} \tag{1}$$

where $\bar{x}_i$ and σ_{x_i} are the mean and the standard deviation of each independent feature vector x_i.

4.6 Classification

In this study ANN were used for the classification of sleep stages. The two-layer feed forward network consisting of 14 input neurons, 12 hidden neurons and 4 output neurons for discrimination between the four sleep stages *Wake, REM + S1, S2* and *SWS* is used. A sigmoid transfer function in the hidden layer and a linear transfer function in the output layer were selected. Levenberg-Marquardt training algorithm is chosen.

5   Experimental Results and Discussion

The performance of the proposed method was assessed using the six subjects selected from the dataset mentioned in Sect. 4.1. In our experiment for wavelet packet based filtering Daubechies order 20 (db20) was used as mother wavelet. In the artifact detection stage a threshold of 0.5 (*Threshold 1*) for EEG-EOG cross correlation coefficients and 0.25 (*Threshold 2*) for EEG-EMG cross correlation coefficients were selected. These thresholds are selected empirically considering highest classification accuracy. Three different result validation approaches including subjective and objective methods were applied.

The cross correlation coefficients for EEG-EOG and EEG-EMG which were detected by thresholding before and after applying the artifact removal algorithm are shown in

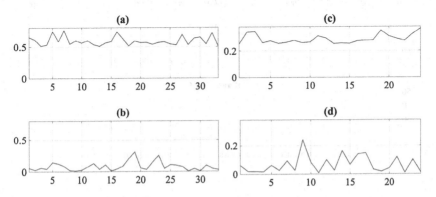

Fig. 2. Absolute Value of cross correlation coefficients, (a) EEG-EOG before artifact removal, (b) EEG-EOG after artifact removal, (c) EEG-EMG before artifact removal, (d) EEG-EMG after artifact removal algorithm.

Fig. 2. A significant reduction in the correlation coefficients is noticeable after artifact removal.

Figures 3 and 4 illustrate the cancellation of EOG and EMG artifacts from contaminated EEG segments. It can be seen that the artifacts can be correctly eliminated without distorting the original EEG.

After the completion of the artifact removal stage, the data is fed to feature extraction algorithm. For training the ANN, unlike the more conventional approaches in the literature, which import all the existing stages to the neural network, we used a quantity of training data to be selected out of each patient. This method is suitable for large databases helping on the reduction of the computational complexity of the classifier training stage.

Table 1. Results of the statistical analysis for comparison of each stage and overall accuracy.

	Wake (%)	REM + S1 (%)	S2 (%)	SWS (%)	Overall (%)
Raw	77.56	87.08	74.67	78.11	63.70
Filtered	79.44	78.75	83.26	90.74	70.60
Proposed method	87.08	87.25	87.38	90.93	77.80

In order to assess the effectiveness of our artifact removal algorithm, we studied the sleep stage classification accuracy for raw (after removing zero energy epochs), filtered and artifact removed data. Table 1 shows the results of statistical analysis for comparison of each stage and overall accuracy for all the above-mentioned data. The results are validated using repeated random sub-sampling method which is also known as Monte Carlo cross-validation technique. It is observed that there is an improvement in the performance of the classifier after filtering the data but the best performance is achieved by applying the proposed artifact removal algorithm.

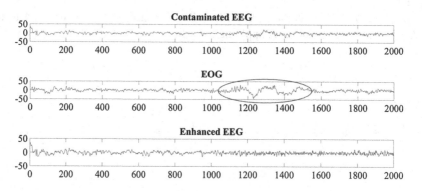

Fig. 3. EOG artifact cancelation from contaminated EEG.

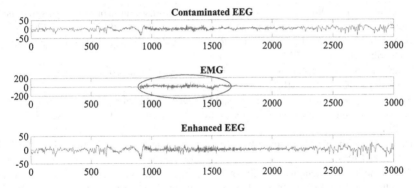

Fig. 4. EMG artifact cancelation from contaminated EEG.

6 Conclusions and Future Work

To the best of our knowledge there is a lack of EEG artifact removing studies in the sleep stage classification context that simultaneously removes the EOG and EMG artifact from EEG without rejecting epochs and also evaluates the performance of the classifier for de-noised data. This paper is a contribution in this regard. In conclusion our findings show that the proposed method for artifact cancelation is reliable for sleep stage classification giving a promising accuracy improvement.

Due to the database available at the moment for this research, EOG and EMG channels were utilized for cancellation of their contamination in EEG. The performance of proposed method can be improved in the future using other PSG channels like ECG.

Acknowledgment. This work was partially funded by FCT Strategic Program UID/EEA/00066/203 of UNINOVA, CTS.

References

1. Niedermeyer, E., da Silva, F.H.L.: Electroencephalography: Basic Principles, Clinical Applications, and Related Fields. Williams & Wilkins, Baltimore (1999)
2. Devuyst, S., Dutoit, T., Ravet, T., Stenuit, P., Kerkhofs, M., Stanus, E.: Automatic processing of EEG-EOG-EMG artifacts in sleep stage classification. In: IFMBE Proceedings, pp. 146–150 (2009)
3. Croft, R.J., Chandler, J.S., Barry, R.J., Cooper, N.R., Clarke, A.R.: EOG correction: a comparison of four methods. Psychophysiology **42**, 16–24 (2005)
4. Sweeney, K.T., Ward, T.E., McLoone, S.F.: Artifact removal in physiological signals–practices and possibilities. IEEE Trans. Inf Technol. Biomed. **16**, 488–500 (2012)
5. Wang, J., Abid, H., Lee, S., Shu, L., Xia, F.: A secured health care application architecture for cyber-physical systems. J. Control Eng. Appl. Inform. **13**(3), 101–108 (2011)
6. Philip, A., Broman, D., Lee, E.A., Torngren, M., Sunder, S.S.: Cyber-physical systems - a concept map. http://cyberphysicalsystems.org/

7. Milenković, A., Otto, C., Jovanov, E.: Wireless sensor networks for personal health monitoring: issues and an implementation. Comput. Commun. **29**, 2521–2533 (2006)
8. Haque, S.A., Aziz, S.M., Rahman, M.: Review of cyber-physical system in healthcare. Int. J. Distrib. Sens. Netw. **2014**, 1–20 (2014)
9. Urigüen, J.A., Garcia-Zapirain, B.: EEG artifact removal-state-of-the-art and guidelines. J. Neural Eng. **12**, 31001 (2015)
10. Romero, S., Mañanas, M.A., Barbanoj, M.J.: Ocular reduction in EEG signals based on adaptive filtering, regression and blind source separation. Ann. Biomed. Eng. **37**, 176–191 (2008)
11. Unser, M., Aldroubi, A.: A review of wavelets in biomedical applications. Proc. IEEE **84**, 626–638 (1996)
12. Safieddine, D., Kachenoura, A., Albera, L., Birot, G., Karfoul, A., Pasnicu, A., Biraben, A., Wendling, F., Senhadji, L., Merlet, I.: Removal of muscle artifact from EEG data: comparison between stochastic (ICA and CCA) and deterministic (EMD and wavelet-based) approaches. EURASIP J. Adv. Signal Process. **2012**, 127 (2012)
13. Fatourechi, M., Bashashati, A., Ward, R.K., Birch, G.E.: EMG and EOG artifacts in brain computer interface systems: a survey. Clin. Neurophysiol. **118**, 480–494 (2007)
14. Estrada, E., Nazeran, H., Sierra, G., Ebrahimi, F., Setarehdan, S.K.: Wavelet-based EEG denoising for automatic sleep stage classification. In: CONIELECOMP Proceedings of 2011 - 21st International Conference on Electronics, Communication and Computing, pp. 295–298 (2011)
15. The Sleep-EDF Database [Expanded]. http://www.physionet.org/physiobank/database/sleep-edfx/
16. The AASM Manual for the Scoring of Sleep and Associated Events - Rules, Terminology and Technical Specifications. http://www.aasmnet.org/scoringmanual/
17. Wiltschko, A.B., Gage, G.J., Berke, J.D.: Wavelet filtering before spike detection preserves waveform shape and enhances single-unit discrimination. J. Neurosci. Methods **173**, 34–40 (2008)
18. Haykin, S.S.: Adaptive Filter Theory. Prentice Hall, New Jersey (2002)
19. Ebrahimi, F., Mikaeili, M., Estrada, E., Nazeran, H.: Automatic sleep stage classification based on EEG signals by using neural networks and wavelet packet coefficients. In: Conference Proceedings of IEEE Engineering in Medicine and Biology Society, pp. 1151–1154 (2008)

Low Cost Inertial Measurement Unit for Motion Capture in Biomedical Applications

João Lourenço[1], Leonardo Martins[2]([✉]), Rui Almeida[1], Claudia Quaresma[3], and Pedro Vieira[3]

[1] Department of Physics, Faculdade de Ciências e Tecnologias, Universidade Nova de Lisboa, Quinta da Torre, 2829-516 Caparica, Portugal
jp.lourenco@campus.fct.unl.pt,rui.almeida@ngns-is.com
[2] UNINOVA, Institute for the Development of New Technologies, Quinta da Torre, 2829-516 Caparica, Portugal
l.martins@uninova.pt
[3] LIBPhys-UNL, Department of Physics, Faculdade de Ciências e Tecnologias, Universidade Nova de Lisboa, 2829-516 Monte da Caparica, Portugal
{q.claudia,pmv}@fct.unl.pt

Abstract. A low-cost inertial measurement unit has been developed for accurate motion capture, allowing real-time spatial position registration (linear and angular) of the user's whole-body. For this, we implemented a dedicated circuit for 9 degrees of freedom motion sensors, composed of an accelerometer, gyroscope and a magnetometer. We also applied signal processing and data fusion algorithms to prevent the inherent drift of the position signal. This drift is known to exist during the sensor integration process and the implemented algorithms showed promising results. This system is meant to be used in two specific biomedical applications. The first one is linked to the development of a low-cost system for gait analysis of the whole-body, which can be used in home-based rehabilitation systems. The second application is related to the real-time analysis of working postures and the identification of ergonomic risk factors for musculoskeletal disorders.

Keywords: Motion capture · Sensor fusion · Cyber-physical systems · Home rehabilitation · Work risk analysis

1 Introduction

Pervasive and ubiquitous environments have been spreading in our society, especially focusing on tracking user's location and activities by monitoring physiological signals, motion and orientation.

There are two main cyber-physical approaches to this field, using two different sensorial components, the first is based on computer vision devices for human motion capture, such as using single or a network of cameras, infrared cameras or other optical devices [1, 2], while the task of human movement analysis has also been influenced by

© IFIP International Federation for Information Processing 2016
Published by Springer International Publishing Switzerland 2016. All Rights Reserved
L. Camarinha-Matos et al. (Eds.): DoCEIS 2016, IFIP AICT 470, pp. 151–158, 2016.
DOI: 10.1007/978-3-319-31165-4_16

using inertial sensors, such as accelerometers and gyroscopes that are present in common smartphones or by using dedicated sensors connected to microcontrollers (which can also communicate to smartphones or other computation devices) [2, 3].

The remote analysis of the user's position and body posture can provide major benefits in the healthcare and manufacturing industries. Our studies will focus on the real-time analysis of two main problems. The first problem will analyse the position of each sensor to enable the gait analysis of the whole-body, which can in turn be used in low-cost and home-based rehabilitation systems. The second will use pattern recognition techniques to analyse static working postures, taking into account potential ergonomic risk factors of musculoskeletal disorders caused by poor work situations. From these open problems, there is a main research question that arises:

> *"How to design a low-cost system that is capable of accurate detection in real-time of the user's whole-body?"*

In this paper we propose a system that answers this question (a relation of this work to Cyber-Physical Systems is presented in Sect. 2). Section 3 gives a more detailed overview of the state of the art and the related literature to inertial measurement units and the associated applications of such systems. Section 4 provides a summarized description of our proposed physical system and the associated algorithms of signal processing, the 9 degrees of freedom data fusion and the implementation of a real-time data transmission protocol. The initial laboratory results are presented in Sect. 5, while Sect. 6 gives a summarized conclusion of our working progress and a preview of our future work plan.

2   Research Connection with Cyber-Physical Systems

Cyber-Physical Systems (CPS) have been emerging as the next computing revolution because they integrate the available communication and computational capacities to create new interactions between cyber and physical components [4]. These systems are merging the physical and virtual world and several applications have been identified in communication, transportation, infrastructure, energy, robotics, manufacturing and healthcare [4, 5].

Cyber-Physical frameworks have already been proposed for use in rehabilitation systems [6]. Such systems can be used for home-based physiotherapy services processes, such as the monitoring of users in senior care, patients with reduced mobility (for example with hip or knee replacement surgery), stroke and heart attack patients and other subjects that require occupational therapy to regain day to day skills.

CPS applications are already inside the healthcare system, which has been extensively reviewed [7], and require a special architecture that can handle the privacy of the data, that can also handle communication between the hospital/rehabilitation centre, the storage unit and the sensors. Finally they also need to handle the computational resources that arise from a feedback system that receives real-time data from various patients and a large network of sensors and requires a real-time classification or response back to the users [8].

In the manufacturing area, there have been several advances in hazard risk management and their approach to the preservation of worker's safety and health has been highly improved, by developing new work equipment features and definition of more secure tasks [9].

A big trend that has been emerging in this area is the definition of a sensing and smart factory, which uses information from a network of physical sensors and virtual databases to make real-time risk assessments [9], which has been extensively overviewed for the construction industry [10]. These sensors can also be used to monitor and track the assembly line, by using real-time information to guide the manual assembly process [11]. Right now, most studies in this area use visual based motion capture frameworks for to identify workers ergonomic risk factors [12], even adapting systems such as the Kinect^{TM} sensor for body kinematic measurement in the workplace [13]. Most of these studies lack the possibility of real-time management and will be hard to implement in a real setting, due to inherent problems in these systems: there is necessity of adaptation of the background (other objects can influence the field of view, the lightning can affect the measurements) and the user's to be completely inside the field of view of the cameras. To solve such a problem, some groups have been fusing data coming from physiological sensors (e.g. ECG, EEG, EMG) location sensors (e.g. GPS or Ultra-Wideband technology) and normal cameras to make a real-time monitor tasks in the workplace and analysis of ergonomic risks [14].

3   State of the Art

Motion capture sensorial systems are usually divided into two categories: optical sensors and non-optical sensors [2]. Optical systems use cameras to acquire motion information of the subject being studied. Due to their scalability and high sensitivity to the motion, they are regarded as the standard on several fields such as gait analysis, film production and video game animation. However, they have some disadvantages: they are usually very expensive, do not work on every lighting environment and require special software for different tasks. On the other hand, there are the non-optical systems, which are comprised of every other system e.g. magnetic arrays, microelectromechanical systems (MEMS)-based [2].

MEMS-based inertial motion capture systems use inertial sensors (e.g. gyroscopes and accelerometers) to acquire motion information, have been subject of great interest due to constant miniaturization of the sensors, ability to work on any environment and relative low cost when compared with other systems. They do, however, also have several drawbacks, namely the lack of information about the full 3D motion and the inherent noise in some sensors.

The MEMS-based sensors have been more recently implemented for real-time human motion tracking in various physical activities, such as boxing, golf swinging, and table tennis [15], but also have been able to actually identify some simple activities using fuzzy logic, such as walking, sitting or stair climbing [16]. These human activity measurements can be useful in the identification of bad postures in the workplace and the identification of ergonomic hazard risks in the workplace.

Both issues have been tackled simultaneously in two fronts, the first is using combinations of sensors to get the most information (e.g. combining 3D accelerometers with 3D gyroscopes to get information about the acceleration and angular velocity in 3 Dimensions). The second is by using Data Fusion algorithms to mitigate the effect of the measurement noise, for example using Kalman filtering in a limited frequency band [17] or other methodologies to filter the inherent noise over the whole frequency band of interest [18].

A final approach is by fusing both visual (both with and without using optical markers) and inertial sensors, in order to solve the intrinsic problems in each system [19]. Concerning the rehabilitation procedure of gait disorders, all the approaches, such as using just inertial sensors, or optical sensors or fusing both of them have been tested and studied [20–23].

4 Materials and Methods

In this section we provide a representation of the physical sensor architecture, with the implemented sensor elements and a description of the fusion and signal processing algorithms.

Our current prototype is using the LSM9DS1, a 9 degrees of freedom (gyroscope, accelerometer, magnetometer) digital sensor unit from STMicroelectronics, communicating the sensor output to a Raspberry Pi acting as signal processor, data logger and as gateway to a local network, which allows not only the sensor to be configured remotely but also the sensor output data to be accessed by any computer on the network using a software that allows the user to watch not only the data in text format, but also a 3D Model that mimics the sensor orientation in real-time. In Fig. 1 we show a simple scheme of the prototype and the experimental assembly of the prototype.

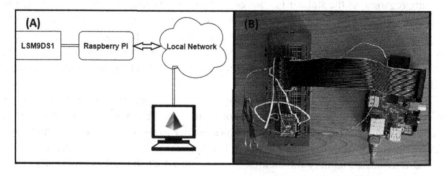

Fig. 1. (A) Prototype schematic (B) Experimental assembly of the prototype.

To address the noise issue described in Sect. 3, we implemented the Mahony's filter in Direction Cosine Matrix form in both Python and MATLAB following the algorithm in [24], a detailed discussion about the problem and filtering algorithms have been extensively discussed in [24, 25] respectively.

5   Initial Laboratory Results and Discussion

In Fig. 2 we present the results of the static test in which the sensor was idle for 15 min with a sampling rate of 119 Hz on both the gyroscope and accelerometer and 20 Hz on the magnetometer, it shows the difference between the uncorrected signal (A) and the corrected signal (B), the initial behaviour is the filter response. Alpha, beta and gamma correspond to the relative rotation (to the initial orientation) around the x,y,z axis respectively. We can notice that although there is an offset after the drift correction, the values tend to be stable after 15 min. This offset is a consequence of the filters signal response and can be easily removed after a few a few minutes of its stabilization. It is important to note, that this drift issue is still a problem, even in a recent study on reha-bilitation systems for gait disorders, which solved this problem by fusing the inertial sensors with information coming from an infrared camera [26].

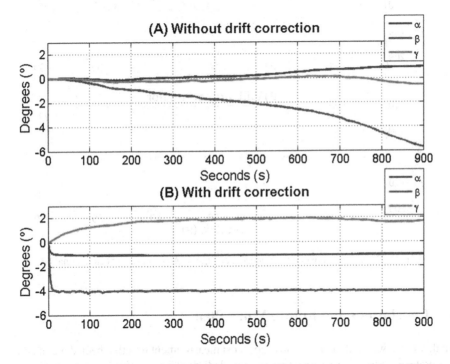

Fig. 2. Difference between the uncorrected signal (A) and the corrected signal (B)

For the next experiment, we tested the sensor for rapid and slow movements to check the filter response, using a goniometer to rotate the sensor around the z axis (gamma), while trying to maintain the other axis stabilized. The rotation sequence was the following: at 2.5 s, we moved from 0° (all the angles are relative to this starting point) to 24°, and then we moved to 165° with 2 additional steps of 5°. Afterwards we move to –100° (8 s), which followed to a fast rotation to 90° (11.5 s) then to –45°, then to 150°. At 15 s, we rotated back to 0°, then a fast rotation to 160°, with a slow rotation back to

90°, then again to 160°, with a gradual slow rotation to –25°. At 36 s, we rotated to 45°, back to –110°, then to 120° and a slow rotation to –125°, finalized by a rotation to 110°.

Figure 3 shows the respective test around the z axis (gamma) with a sampling rate of 60 Hz of both the gyroscope and accelerometer and 20 Hz on the magneto-meter, (A) represents the uncorrected signal and (B) the corrected one. Although the differences are barely noticeable in gamma plot, on the alpha and beta plots there is a significant effect of the drift, showing that we can also correct this problem when the sensor is moving.

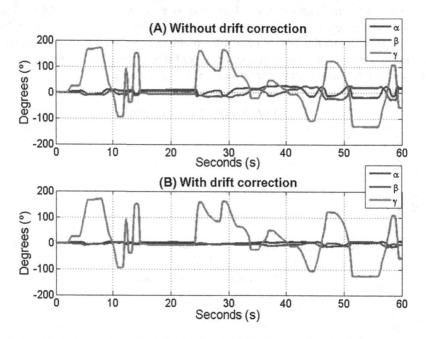

Fig. 3. Difference between the uncorrected signal (A) and the corrected signal (B)

6   Conclusions and Future Work Plan

In this work, we implemented a low-cost cost measurement unit that been developed for accurate motion capture, by implementing a dedicated circuit with 9 degrees of freedom sensors (accelerometer, gyroscope and magnetometer), connected to a Raspberry Pi, which transmits the data to a local network, allowing access by authorized computers. We implemented signal processing and data fusion algorithms to correct the inherent drift process that occurs even when the sensor is stopped, as shown in Fig. 2. As future work we can use an absolute position coordinate system (e.g. north east down system) instead of the existing coordinate system.

We are planning to use this system in two specific biomedical applications, as the first one is linked to the real-time analysis of working postures and the identification of ergonomic risk factors for musculoskeletal disorders. For this we are planning to start with a network of at least 3 sensors to test the analysis in the upper limbs. We will make a feedback system that uses the ISO 11226 recommendations for the holding time for upper arm elevation in a specific posture and the acceptable shoulder, forearm and hand positions in a static posture [27]. To analyse a dynamic working posture, we will use a similar framework to the project FAST ERGO_X [28], which is based on fuzzy rules. Instead of using camera sensors, we will be using inertial sensors, which will give the ability to make the analysis in real-time.

The second application is related to the development of a low-cost system for gait analysis of the whole-body, which can be used in home-based rehabilitation systems, we will use the same network of sensors for the upper limbs to validate them using the gold standard sensors used in rehabilitation centres.

Acknowledgments. Leonardo Martins is supported by a PhD Scholarship with the reference SFRH/BD/88987/2012 and is also partially funded by FCT Strategic Program UID/EEA/ 00066/203 of UNINOVA, CTS, funded by the Portuguese funding institution FCT - Fundação para a Ciência e a Tecnologia. We also acknowledge the technical support of the engineers of NGNS - Ingenious Solutions.

References

1. Moeslund, T.B., Hilton, A., Krüger, V.: A survey of advances in vision-based human motion capture and analysis. Comput. Vis. Image Underst. **104**, 90–126 (2006)
2. Corke, P., Lobo, J., Dias, J.: An introduction to inertial and visual sensing. Int. J. Rob. Res. **26**, 519–535 (2007)
3. Yang, C.-C., Hsu, Y.-L.: A review of accelerometry-based wearable motion detectors for physical activity monitoring. Sensors **10**, 7772–7788 (2010)
4. Geisberger, E., Cengarle, M.V., Keil, P., Niehaus, J., Thiel, C., Thönnißen-Fries, H.-J.: Cyber-Physical Systems - Driving force for innovation in mobility, health, energy and production (2011)
5. Khaitan, S.K., Member, S., Mccalley, J.D.: Design techniques and applications of cyberphysical systems: a survey. IEEE Syst. J. **9**, 350–365 (2015)
6. Ma, X., Tu, X., Huang, J., He, J.: A cyber-physical system based framework for motor. In: ACWR 2011 Proceedings of the 1st International Conference on Wireless Technologies for Humanitarian Relief, pp. 285–290 (2011)
7. Haque, S.A., Aziz, S.M., Rahman, M.: Review of cyber-physical system in healthcare. Int. J. Distrib. Sens. Netw. **2014**, 1–20 (2014)
8. Wang, J., Abid, H., Lee, S., Shu, L., Xia, F.: A secured health care application architecture for cyber-physical systems. Control Eng. Appl. Info. **13**, 101–108 (2011)
9. Lazaro, O., Moyano, A., Uriarte, M., Gonzalez, A., Meneu, T., Fernández-Llatas, C., Traver, V., Molina, B., Palau, C., Lopez, O., Sanchez, E., Ros, S., Moreno, A., Gonzalez, M., Antonio, J., Sepulcre, M., Gozalvez, J., Collantes, L., Prieto, G.: Integrated and personalised risk management in the sensing enterprise. In: Banaitiene, N. (ed.) Risk Management - Current Issues and Challenges, pp. 285–312. InTech, Rijeka (2012)
10. Pinto, A., Nunes, I.L., Ribeiro, R.A.: Occupational risk assessment in construction industry – Overview and reflection. Saf. Sci. **49**, 616–624 (2011)

11. Bader, S., Aehnelt, M.: Tracking assembly processes and providing assistance in smart factories. In: 6th International Conference on Agents and Artificial Intelligence, ICAART 2014, Proceedings, vol.1, ESEO, Angers, Loire Valley, France, 6–8 March 2014, pp. 161–168. SciTePress - Science and Technology Publications (2014)

12. Han, S., Lee, S.: A vision-based motion capture and recognition framework for behavior-based safety management. Autom. Constr. **35**, 131–141 (2013)

13. Dutta, T.: Evaluation of the Kinect™ sensor for 3-D kinematic measurement in the workplace. Appl. Ergon. **43**, 645–649 (2012)

14. Cheng, T., Migliaccio, G.C., Teizer, J., Gatti, U.C.: Data fusion of real-time location sensing and physiological status monitoring for ergonomics analysis of construction workers. J. Comput. Civ. Eng. **27**, 320–335 (2012)

15. Liu, H., Wei, X., Chai, J., Ha, I., Rhee, T.: Realtime human motion control with a small number of inertial sensors. Symp. Interact. **3D**(3), 133–140 (2011)

16. Liu, S., Chang, Y.: Using Accelerometers for Physical Actions. Telemed. e-Health **15**, 867–876 (2009)

17. Ligorio, G., Sabatini, A.: Extended Kalman filter-based methods for pose estimation using visual, inertial and magnetic sensors: comparative analysis and performance evaluation. Sensors **13**, 1919–1941 (2013)

18. Skaloud, J., Bruton, A.M., Schwarz, K.P.: Detection and filtering of short-term (1/f γ) noise in inertial sensors. J. Inst. Navig. **46**, 97–107 (1999)

19. Hol, J.D., Schön, T.B., Luinge, H., Slycke, P.J., Gustafsson, F.: Robust real-time tracking by fusing measurements from inertial and vision sensors. J. Real-Time Image Process. **2**, 149–160 (2007)

20. Ali, A., Sundaraj, K., Ahmad, B., Ahamed, N., Islam, A.: Gait disorder rehabilitation using vision and non-vision based sensors: a systematic review. Bosn. J. Basic Med. Sci. **12**, 193–202 (2012)

21. Zhou, H., Hu, H.: Human motion tracking for rehabilitation—a survey. Biomed. Signal Process. Control **3**, 1–18 (2008)

22. Zheng, H., Black, N.D., Harris, N.D.: Position-sensing technologies for movement analysis in stroke rehabilitation. Med. Biol. Eng. Comput. **43**, 413–420 (2005)

23. Tao, Y., Hu, H., Zhou, H.: Integration of vision and inertial sensors for home-based rehabilitation. IEEE Int. Conf. Robot. Autom. (2005)

24. Premerlani, W., Bizard, P.: Direction cosine matrix IMU: Theory (2009)

25. Madgwick, S.O.H., Harrison, A.J.L., Vaidyanathan, R.: Estimation of IMU and MARG orientation using a gradient descent algorithm. In: IEEE International Conference on Rehabilitation Robotics, pp. 179–185 (2011)

26. Ma, E., Popovic, M., Masani, K.: Wearable gait analysis using vision-aided inertial sensor fusion. In: Book of abstracts - IUPESM 2015 World Congress on Medical Physics and Biomedical Engineering, Toronto, 7–12 June 2015, p. 386 (2015)

27. International Organization for Standardization: ISO 11226 - Ergonomics—Evaluation of static working postures (2000)

28. Nunes, I.: FAST ERGO_X - a tool for ergonomic auditing and work-related musculoskeletal disorders prevention. Work. **34**, 133–148 (2009)

Intelligent Environments

Auto-Adaptive Interactive Systems for Active and Assisted Living Applications

João Quintas[1(✉)], Paulo Menezes[2], and Jorge Dias[2,3]

[1] Laboratory of Automatic and Systems, Instituto Pedro Nunes, Coimbra, Portugal
`jquintas@ipn.pt`
[2] Department of Electrical and Computer Engineering, University of Coimbra, Coimbra, Portugal
`{pm,jorge}@deec.uc.pt`
[3] Khalifa University of Science and Technology and Research, Abu Dhabi, UAE

Abstract. The objective of this work is of improving the efficacy, acceptance, adaptability and overall performance of Human-Machine Interaction (HMI) applications using a context-based approach. In HMI, we aim to define a general human model that may lead to principles and algorithms allowing more natural and effective interaction between humans and artificial agents. This is paramount for applications in the field of Active and Assisted Living (AAL). The challenge of user acceptance is of vital importance for future solutions, and still one of the major reasons for reluctance to adopt cyber-physical systems in this domain. Our hypothesis is that, we can overcome limitations of current interaction functionalities by integrating contextual information to improve algorithms accuracy when performing under very different conditions and to adapt interfaces and interaction patterns according user intentions and emotional states.

Keywords: Human-machine interaction · Context · Active and Assisted Living · Social agents · Adaptive systems

1 Introduction

There is an emerging trend focused in auto-adaptable and self-reconfiguring ambient intelligence systems in order to support smarter habitats. The associated technological challenges include active perception features, mobility in unstructured environments, understanding human actions, detect human behaviors and predict human intentions, access to large repositories of personal and social related data, adapt to changing context. In the case of social artificial agents, systems must incorporate features that allow an agent to be capable of delivering a sociable experience with the user. These can be classified as cyber-physical systems, as they are designed as a network of interacting elements with physical input and output instead of as standalone devices. These features are paramount for applications in the field of Active and Assisted Living (AAL). In AAL, the primary goal is to provide solutions that help people through ageing, by promoting active and healthy living. Part of being active and healthy includes socializing. In a vast number of cases, this activity is done in care centers or nursing homes.

© IFIP International Federation for Information Processing 2016
Published by Springer International Publishing Switzerland 2016. All Rights Reserved
L. Camarinha-Matos et al. (Eds.): DoCEIS 2016, IFIP AICT 470, pp. 161–168, 2016.
DOI: 10.1007/978-3-319-31165-4_17

In such scenarios, the demand for social stimulation as part of care service aggravates the need for scarce qualified human resources. Thus, technological solutions are seen as a benefice but the challenge of adoption remains associated to user acceptance. However, most of interactive approaches rely on explicit information (e.g. direct commands) to achieve expected behaviors from the system. How can we take into account implicit information to achieve human-like interaction skills in cyber-physical systems? Our hypothesis is that, we can overcome limitations of current interaction functionalities by integrating contextual information (i.e. implicit) to improve algorithms accuracy when performing under very different conditions, to select most adequate algorithms to provide a given functionality, and to adapt interfaces and interaction patterns according user intentions and emotional states. In Sect. 2 we present our contribution to the field of cyber-physical systems focusing in applications for the AAL domain. Section 3 describes a survey of relevant literature. In Sects. 4 and 5 we present our approach and initial results. Sections 6 and 7 conclude the paper summarizing the major findings and stating further work.

2 Contribution to Cyber-Physical Systems

Our contribution aims to develop automatically adaptive cyber-physical agents that will evolve according to the end-user needs and intentions. The overall goal is improving acceptance of interactive technologies, with special focus to robotics and cyber-physical systems, by users of assistive technologies, developed to address challenges from the Active and Assisted Living (AAL) domain.

From experience, it is common that the misunderstanding of functionalities and handling of new technology often leads to rejection or even to fear if the system is not behaving as expected. Thus, the challenge of user acceptance is of vital importance for future systems and still one of the major reasons for reluctance to deploy or introduce cyber-physical agents in AAL applications. The analysis of the results of relevant research projects allows us to identify common needs, which are repeatedly requested by end-users. In summary, users expect intuitive interaction with social agents. Therefore, Natural Interaction features (e.g. gestures, speech, etc.) and adaptation to the user's general profile, specific needs and intentions, were identified as high priority requirements. In the AAL domain, these features must address the needs of users with cognitive or physical skills degradation (e.g. elderly or impaired people) and aim to compensate these limitations, hence enriching user experience and accessibility.

Taking into account this list, we conclude that interaction aspects are a central point for end-users. Therefore, we can summarize the major features according Fig. 1.

To refine reasoning and to integrate context information about a person within his/her home environment in the broadest sense possible, namely the identification of activities of daily life (ADLs) in order to provide the correct services to attain end-user goals, models must be developed that are capable (1) of using a priori knowledge, either hardcoded or from experience, and (2) of evolving in time. However, there is not available yet a satisfactory implementation that is capable of reason and learn based in of context information in a distributed system, which involves sharing contextual information between different components of the system [1].

Fig. 1. Scientific and technological challenges related with HMI related features, as summarized from end-user needs

3 Related Works

The current state of the art in the domain of user interaction with cyber-physical agents is still mainly based on graphical user interfaces implemented on stationary (often wall-mounted) touch panels (or terminals), portable devices (like touch tablets or mobile phones) or by using TV-sets (often in combination with set-top-boxes) and their remote controls as front end for the elderly. The graphical user interfaces are often text or icon based. Speech input is becoming more and more stable but is still in the research and development phase since surrounding noise and sounds, speaker localization and optimum input with ambient microphones (by avoiding wearable microphones) are still a challenge and are addressed by several research projects (e.g. Companionable etc.). Recent approaches as well apply avatar technology to enhance (increase acceptance and entertainment value) and personalize user interaction [2]. Human-machine interaction (HMI) technologies and corresponding cognitive capabilities of agent systems have seen many developments in the last few decades – see [3] for an extensive survey. Solutions for multisensory active perception and attention allocation have greatly evolved [4], as has multimodal human emotion and dialogue analysis and human-like emotion and dialogue synthesis [5] and also human behavior analysis [6, 7]. In summary, we identified that existing cyber-physical systems that have been developed recently in order to provide assistance in domestic, professional and public environments are based on closed architectures, thus being operational only in specific contexts of usage, equipment and data. Furthermore, one drawback of existing implementations is commonly that perceptual models neglect the context and the spatial relation during the perception process. Our conclusions from the literature survey point us in the direction of proposing a human-machine interaction framework that considers integration as keystone to overcome existing challenges. This framework goes beyond the state-of-the-art approaches by taking into account contextual information in the interaction process.

4   Research Contribution and Innovation

Our main contribution aims to create a reference framework that targets improving the current state of the art of auto-adaptive interactive systems for Active Assisted Living applications. The innovative aspect of this framework is the use of contextual information to orchestrate the behaviors of the social agent (i.e. perception and actuation features involved during interaction). This paper extends the concepts and design presented from our previous work in [8].

4.1   Context-Based Human-Machine Interaction (CB-HMI) Framework

We consider the integration of contextual information a vital aspect in HMI approaches. In our approach we consider context recognition to be the process that leads to auto-adaption of the system. We call this process the *Contextual adaptation loop*. As depicted in Fig. 2, this is a periodic process that operates in the background of the system, while the user is interacting with it. It plays the role of detecting changes in the context and provides this information to the agent's main execution loop (e.g. perception-action).

The execution of this process in mainly related with the Interaction inputs, Environment analysis and Interaction control components depicted in Fig. 3, which invokes the adaption of the agent's interaction behavior (i.e. changes the orchestration of Interaction outputs components in Fig. 3). This process must take into consideration restrictions or rules that may apply for different situations. This information corresponds to *Context models* that are learnt and stored, in the Context recognition component, for future reference. The CB-HMI framework is designed to be modular, loosely coupled and to take advantage of context models to orchestrate different NI modules and user interfaces. The goal is to provide features allowing the agent to be able to adapt its behavior according to the user's expectations and intentions, as illustrated in Fig. 3.

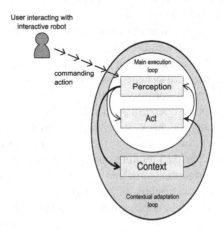

Fig. 2. Context verification process

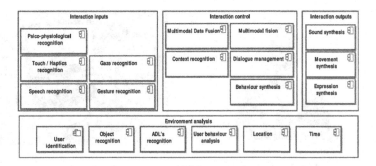

Fig. 3. Framework component diagram

The components are grouped into four main categories. *Interaction inputs* include the components related with user interaction inputs (i.e. Speech Recognition, Gaze recognition, Gesture recognition, Touch/Haptics recognition and Psyco-physiological recognition). These modules contribute to the overall system by perceiving the user's explicit state (i.e. intentions and emotions). *Environment analysis* is also related with perception mechanisms, but it contributes with implicit states. It refers mainly to what can be recognized and identified from the environment and from user habits profile (i.e. Object recognition, User identification, ADLs recognition, User behavior analysis, Time, Location). *Interaction control* components receive inputs from the Interaction inputs components and handle the data fusion and fission in order to update the agent's internal state and determine appropriate response (i.e. Multimodal data fusion, Multimodal data fusion, Context recognition, Behavior synthesis and Dialogue management). *Interaction outputs* are directly linked to the synthesis components generating the appropriate verbal and non-verbal behavior of the companion and communicate relevant information to the user via the graphical user interface (i.e. Expression synthesis, Movement synthesis) and vocal outputs (i.e. Sound synthesis).

5   Implementation and Results

To validate our approach we started the implementation of a cyber-physical social agent that took the form of a virtual assistant. Our objective at this stage was to create a working prototype in line with the state of the art to establish a baseline for future work. In this first version of the prototype we addressed the *Main execution loop* depicted in Fig. 2. The prototype included an initial implementation of some of the components listed in Fig. 3, namely: Speech recognition, Dialogue Management, Expression synthesis, Sound synthesis and Movement synthesis. The system operates on an all-in-one computer (Lenovo ThinkCentre Edge 93z All-in-One) and supports various interaction modalities. According to the preferences of the user and depending on the distance between the user and the hardware, interaction can be done either by speech (2–3 m) or via the Graphical User Interface (GUI) on the computer's touchscreen (arm length). The virtual assistant interface (Fig. 4a) closely simulates human conversational behavior through the use of synthesized voice and synchronized non-verbal behavior such as head

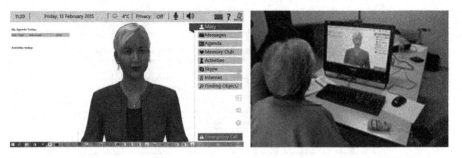

Fig. 4. Implemented prototype (a) and user testing (b)

nods, posture shifts, facial expressions and hand gestures. The interface is built on top of the Behavioral Markup Language (BML) to describe the physical realization of behaviors, such as speech and gesture and the synchronization constraints between these behaviors.

5.1 Experimental Validation

The current implementation of the system was validated in three different sites. Test trials were conducted in Switzerland, in the Netherlands and in Portugal. The criteria for selecting the participants stated that the target group should be 65+ years of age, living alone, and independent and or with light impairments related with ageing. In Table 1 we present detailed information regarding participant characteristics.

Table 1. Characteristics of trials participants

Topic		Switzerland	Netherlands	Portugal
	Participants	N = 7	N = 13	N = 12
Demographics	Average age	76,4	79,4	79,9
Technologic experience	Computers (1–5)	4	3,1	1
	Tablets (1–5)	2,6	3,8	1
Quality of life	Satisfaction daily life	87,5 %	100 %	NA
	Memory (1–10)	Qualitative	6,9	NA
Device used during trials	All-in-one	4	1 (used by N)	1 (used by N)
	Tablet	3	0	0

Evaluation Process: In all test sites, the common approach to introduce the system to the elderly was by gathering small groups of 2 to 3 and then the elderly interact with the system individually (Fig. 4b). These sessions worked as focus group. The Think aloud method and the System Usability Score (SUS) were used as tools to assess the overall impressions and usability of the system.

5.2   Results

System Usability Scores have been calculated based on the 10 questions of the validated questionnaire for SUS. In this tool, a score between 50-70 is positive, where above 68 is considered as 'above average'. Besides, a total SUS score below 70 is problematic since this value seems to be the threshold for a "good" usability. The overall results from the trials are depicted in Fig. 5.

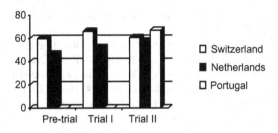

Fig. 5. SUS scores results

6   Discussion

In general all participants found the system very interesting and showed a lot of motivation in participation. The SUS scores, for the Swiss and Dutch users, were all above 68 and especially below 70 which means that the system is as "above average" and below the threshold for a "good usability". It seems to summarize what the participants reported and the results shown: the idea behind the system was very good but the system needs to be improved, to be more intuitive in its use. In Portugal, the overall of SUS scores were 67, which is still considered as positive but falls bellow "above average". This may be an indication that, for those users, the system requires a deeper adaptation. Evidence from the study revealed that the (cognitively impaired) older adults found it difficult to gain insights into the system's possibilities and cope with the limitations. Age related cognitive degradation, specifically age related memory changes and their effects on learning, make older adults feel mentally stressed when interacting with new technologies.

7   Conclusion and Further Work

The paper presented a framework that is being developed to improve the human-machine interaction process within cyber-physical systems. We identified HMI as a vital aspect in the adoption of Active and Assisted Living solutions. Moreover, we identified key HMI features that are commonly referred by end-users as a must, which are still imposing relevant scientific and technical challenges. To address these challenges, we implemented a virtual agent that aims to serve as artificial and interactive companion. The current version of the system was tested in three end-users organizations, in three different European countries. The results showed that the current version of the system can reach positive scores, but it is still considered to be in the borderline of usability. Future improvement will require further development of the proposed framework.

References

1. Quintas, J., Menezes, P., Dias, J.: Context-based perception and understanding of human intentions. In: 22nd IEEE International Symposium on Robot and Human Interactive Communication (2013)
2. Morandell, M.M., Hochgatterer, A., Wöckl, B., Dittenberger, S., Fagel, S.: Avatars@home: interFACEing the smart home for elderly people. In: Holzinger, A., Miesenberger, K. (eds.) USAB 2009. LNCS, vol. 5889, pp. 353–365. Springer, Heidelberg (2009)
3. Goodrich, M.A., Schultz, A.C.: Human-robot interaction: a survey. Found. Trends Hum.-Comput. Interact. **1**, 203–275 (2007)
4. Ferreira, J.F., Lobo, J., Bessiére, P., Castelo-Branco, M., Dias, J.: A Bayesian framework for active artificial perception. IEEE Trans. Syst. Man Cybern. Part B Cybern. **43**, 699–711 (2012)
5. Prado, J.A., Simplício, C., Lori, N.F., Dias, J.: Visuo-auditory multimodal emotional structure to improve human-robot-interaction. Int. J. Soc. Robot. **4**, 29–51 (2011)
6. Aliakbarpour, H., Khoshhal, K., Quintas, J., Mekhnacha, K., Ros, J., Andersson, M., Dias, J.: HMM-based abnormal behaviour detection using heterogeneous sensor network. In: 2nd Doctoral Conference on Computing, Electrical and Industrial Systems (2011)
7. Quintas, J., Almeida, L., Brito, M., Quintela, G., Menezes, P., Dias, J.: Context-based understanding of interaction intentions. In: 21st IEEE International Symposium on Robot and Human Interactive Communication (2012)
8. Menezes, P., Quintas, J., Dias, J.: The role of context information in human-robot interaction. In: 23rd IEEE International Symposium on Robot and Human Interactive Communication Workshop on Interactive Robots for Aging and/or Impaired People (2014)

Using Fuzzy Logic to Improve BLE Indoor Positioning System

Sérgio Onofre[1(✉)], Bernardo Caseiro[2], João Paulo Pimentão[2],
and Pedro Sousa[2]

[1] Holos SA, Caparica, Portugal
onofre@holos.pt
[2] Faculty of Sciences and Technology, NOVA University of Lisbon,
Monte de Caparica, 2829-516 Almada, Portugal
b.caseiro@campus.fct.unl.pt, {pim,pas}@fct.unl.pt
http://servrobot.holos.pt

Abstract. Accuracy and precision are key parameters in the definition of indoor positioning systems. We want to provide a mobile robot with the capacity to autonomously determining its location inside buildings, to allow it to autonomously navigate. The solution developed is based on spreading emitter beacons of Bluetooth Low Energy in the building and use location finding techniques to determine the robot's location. The main challenge is the capacity to obtain accurate readings of signal strength and the low repeatability of readings even under unchanged conditions. To improve the signal strength measurements it is necessary to deal with this imprecision. Our approach is based on the use of Fuzzy Logic to deal with the accuracy problem. Once better signal strength readings are achieved, using this method, approximate distances are calculated based on signal strength and the trilateration method is implemented to provide the location of the mobile robot.

Keywords: Fuzzy Logic · Indoor location techniques · Bluetooth low energy · Trilateration

1 Introduction

Indoor location is an emerging market that can be used for commercial purposes, public safety, and military. Commercially it can be applied to track children, people with special needs, help navigate blind people, locate equipment, mobile robots, etc. For military and public safety it will be useful to track prisoners, help policeman, soldiers and firemen navigate inside buildings [1].

Location and outdoors navigation was solved long ago with GPS. The problem occurs when you need accuracy or indoors location. GPS nowadays have an error of about 10 m, radio signals coming from distant satellites don't work well with obstructed paths and the high-frequency waves bounce around when they hit metal and walls. This technology is therefore not usable on indoor location issues [2].

Actually there are some companies searching and testing solutions using different technologies to find a solution for indoors location. In general, the most important components of the system are location sensor devices that produce metrics of the

© IFIP International Federation for Information Processing 2016
Published by Springer International Publishing Switzerland 2016. All Rights Reserved
L. Camarinha-Matos et al. (Eds.): DoCEIS 2016, IFIP AICT 470, pp. 169–177, 2016.
DOI: 10.1007/978-3-319-31165-4_18

relative position of a mobile terminal (MT) and a known reference point (RP). Most conventional positioning methods are Angle of arrival (AOA) to estimate direction; received signal strength indicator (RSSI), Time difference of arrival (TDOA) and Time of arrival (TOA) are used to estimate distance [1].

Our goal is to provide a mobile robot (ServRobot[1]) with a tool to know is location using Bluetooth Low Energy beacons signals.

2 Cyber Physical Systems

The cyber-physical systems (CPS) consider computational components and physical units to cooperate with humans using different technologies. CPS are intelligent, real-time, distributed, networked (wired/wireless) control systems and likely linked in a loop. They can be useful in several fields. CPS design consists of a network of cooperating elements with physical inputs and outputs. It connects computational and physical rudiments to obtain better effectiveness, functionality, security and flexibility [3].

This work is related to cyber-physical systems, by combining physical entities: Bluetooth Low Energy (BLE) beacons, with computational elements that through a BLE signal receptor connected to a computer, allows to process the input values and implement algorithms to build the desired system. The final goal is to locate an autonomous mobile robot in a wireless network in real time that could be adapted in the future to locate other objects like phones or even people. It consists on build a system in real-time with feedback loops where physical processes, in this case the received signal strength indicator from the BLE beacon affect computations and have influence on the robot next moves and upgrade others functionalities by knowing the robot location, it can improve the efficiency, safety, reliability and functionality of the mobile robot.

This project could help the research on efficient and functional indoor location systems contributing for the cyber physical systems development too.

3 State of the Art

3.1 Conventional Positioning Methods

In the next sections, several conventional positioning methods are presented.

RSSI – It is a process used to estimate the distance between the receiver and a beacon. It matches the probability distribution with the strength of the received signals, and estimates the distance using a statistical method. To start a sample of the signal strength is read in different locations, but the environment changes modify the path loss models that enables to repeat the same strength of the received signal. It is necessary an environment precise path loss model to use the location estimate system [4].

AOA – This method is used to measure the direction in a system and it can do it reading the direction of the received signal using two or more antennas on different positions [4].

TOA – It is a time-based positioning technique that measures the signal travel time between nodes [5].

TDOA – This method estimates the time difference of arrival between two signals moving between two reference nodes and the given node [5].

Table 1 compares the accuracy degradation of an indoor location system due to multipath fading and increase of the distance between devices for each positioning method.

Table 1. Comparison of position methods [5]

Method	Accuracy degradation	
	Multipath	Distance
TOA/TDOA	Slightly	None
AOA	Slightly	Slightly
RSSI	Severely	Slightly

3.2 Indoor Location Systems

There are different location determination systems and processes for indoor environments: ultrasound, infrared, electromagnetic, visual location, physical contact and radio frequency systems. We will focus on radio frequency systems.

Bluetooth based systems – Radio hardware (beacons) with Bluetooth low energy protocol. Beacons are spread over surface and the signals that they send can enable devices to determine their location. Some systems use beacons signal to incorporate on fingerprints database, others determine absolute or relative location by measuring distances, trilateration. Furniture and walls hinder signal propagation. Using three beacons is enough but a fourth one will decrease localization errors [6]. Indoor location is challenging because of the reflections and absorptions. Besides Bluetooth Low Energy there are Wi-Fi, RFID, ZigBee and Ultra-Wideband based systems. More details in [7–9]. Some developers combine multiple technologies like fusing data based on Wi-Fi and Bluetooth. Fusing helps to obtain more accurate results (Table 2).

Table 2. Radio frequency based systems [10]

	Frequency band	Power consumption	Cost	Accuracy (meters)
Wi-Fi 802.11	2.4 GHz	High	High	5 to 10
RFID	Multiple	Low	Very low	2 to 10
Bluetooth 4.0	2.4 GHz	Low	Low	1 to 2
ZigBee	Multiple	Low	Low	3 to 5
UltraWideBand	3.1 to 10.6 GHz	Low	Low	<2

3.3   Location Techniques in Indoor RF-Based Systems

There are two types of techniques to calculate the location using RF-based systems, geometric (traditional) and pattern recognition.

Geometric techniques – calculate positions based on distances between mobile transmitter-receiver and fixed receiver-transmitter. *Trilateration* – Provides an estimate of the device's location by using the received signal strength from different transmitters (non-linear). The distance estimation from the transmitters to the receiver is calculated using the relation between distance and signal strength:

$$Pr = Pt + 20 \log \left(\frac{\lambda}{4\pi}\right) + 10n \log \left(\frac{1}{d}\right) \tag{1}$$

Pt is the transmitters power (in dBm), Pr is the power at the receiver, λ is the wavelength, n is the path loss exponent (n = 2 is free space), d is the distance transmitter/receiver. The distance estimation generates a circle centered on each transmitter, the device will be located where circles coincide. The technique is called trilateration [11]. *Triangulation* - Uses geometric properties of triangles to estimate the target's location. More details in [11].

Pattern Recognition techniques – It consists in buildin a signal strength model. To estimate the mobile device's location is necessary to calculate the match with the signal strength model. It requires two phases: the offline phase, where a map is built after collecting the RF signals from defined location points. And the online phase, where an algorithm is used to find the match between the read signals and the previously collected signals to estimate the device location [5].

Traditional methods are K-Nearest Neighbors, Neural Networks and Probabilistic. More details in [12].

3.4   Fuzzy Logic

Fuzzy Logic is bets fit on under uncertainty, it deals automatically with tolerance for imprecise values. The variables are represented by membership functions, instead of the absolute true or false values. It reproduces the human mind capacity to employ approximate reasoning. In classic logic the decision is binary, true or false, whereas in Fuzzy Logic variables have a range between 0 and 1, of degree of membership towards a fuzzy set [13]. Imprecise inputs can be considered and used as "linguistic variables" denoting values such as big and small, low/medium/high are used in the model. Common membership functions used in Fuzzy sets are triangular, Gaussian, trapezoidal, etc., although triangular (see Fig. 1) is more widely used.

Figure 2 shows the block diagram of fuzzy logic mechanism. Fuzzification is a procedure that classifies numerical values into fuzzy sets, the rule base contains the IF-THEN rules that represent the linguistic variables. The evaluation relates the IF-THEN rules to the fuzzy sets to get a fuzzy output. The Fuzzification changes crisp data into linguistic values using linguistic variables, membership functions map every

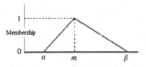

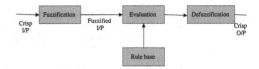

Fig. 1. Triangular Fuzzy membership function [13]

Fig. 2. Fuzzy Logic block diagram [13]

element of the input variables onto a membership rate from 0 to 1. The evaluation makes approximate reasoning [13].

The defuzzification procedure alters the fuzzy output back to the classical output with the control goal. There are three defuzzification techniques normally used, the most widely used is the Center of Gravity method (COG) is identical to the physical calculation of center of gravity in objects, the weighted average of the membership function of the center of the gravity of the area limited by the membership function curve is computed to be the crispest value of the fuzzy values. Besides COG there are Mean of Maximum and Height Methods [13].

4 Development

Our system development starts with the interaction between a BLE transmitter (beacon) and the BLE receptor (sniffer).

Bluetooth Low Energy Sniffer and Beacon
To receive the received signal strength indicator (RSSI) from the beacons a Bluetooth low energy sniffer (Adafruit bluefruit LE Sniffer nRF51822) is used. This device is a USB-to-BLE board that allows the computer to communicate with a BLE enabled device. The Gimbal U-Series 5 proximity beacon is a small, USB powered device that transmits a Bluetooth Low Energy signal that can be detected by other devices, in this case by the Bluetooth Low Energy sniffer. This signal enables the identification of the beacon by the MAC Address. The sniffer captures the RSSI value and with it it's possible to calculate the distance estimation between the devices.

Obtaining Distance from RSSI Values
The code used by the device Adafruit bluefruit LE Sniffer nRF51822 is written in python and allows to retrieve the RSSI and MAC address of nearby BLE beacons, this project uses that code as start point. To obtain the distance in meters from the received signal strength indicator (RSSI) it was necessary to measure a bunch of RSSI measurements at known distances, then we did a best fit curve to match the data points, this algorithm was developed by David Young in Java. In this work the code was adapted to python language [14].

$$distance = \begin{cases} \left(\frac{RSSI}{txPower}\right)^{10} for \ \frac{RSSI}{txPower} < 1 \\ 0.89976 \times \left(\frac{RSSI}{txPower}\right)^{7.7095} + 0.111 \ for \ \frac{RSSI}{txPower} \geq 1 \end{cases} \quad (2)$$

The variable txPower is the RSSI value at 1 meter of distance, it was calibrated and the obtained value was −87 db. The values 0.89976, 7.7095 and 0.111 are the three constants calculated when solving for a best fit curve to the measured data points. Each 0.95 s a distance reading is received, the main challenge is the capacity to obtain accurate readings of signal strength (and consequent approximate distance in meters) given the low repeatability of readings even under unchanged conditions. The direct signal strength measured by the Bluetooth signal receiver vary in each reading because of physical disturbing factors such as wall reflections and other devices transmitting the same frequency. To improve the measures, two normalizing techniques: "median and average filters" and "Fuzzy Logic" were implemented.

Median and Average Filters

After some tests comparing the real distance measured with a scale and the distance obtained at the program by the sniffer it was detected that the accuracy is bad and has low repeatability of readings. To try to improve the accuracy and the repeatability, a way to discard the absurd values on was developed. First a vector accumulates ten measurements and after that the values are sorted, the extremes are excluded and the four middle values are used to calculate the average. However, as the results were still far from good, since they were imprecise we decided to use another approach. Using Fuzzy Logic to deal with imprecision.

Fuzzy Logic Implementation

To deal with imprecise values we decided to use Fuzzy Logic. This could be a good way to improve the results. The tool used was an open source java library called jFuzzyLogic. This library imports a Fuzzy control language file with the input and output variables and the linguistic terms configured, then checks the rules to return an output variable based on inputs. We take the results of the median process and in the end it will return only one output variable with the final value. For each of the four inputs there are twelve linguistic terms possibilities to do the fuzzification ranging from "less 0.5 m" to "more 5 m". Then the system will check the rules block with 23 rules and makes the decision (Fuzzy inference). To finish the defuzzification is done and a real value is obtained as output. Considering ten direct distance measures read by the sniffer at 1 meter away from the beacon (real distance), the four middle values are selected as inputs for the FCL file and each of them is fuzzified. Then the system checks the rules block, makes the decision and does the deffuzification to obtain a real value as output. In our test the final distance output is 1.02 m, close to the real distance 1 m. The defuzzification technique, Center of Gravity method was used.

Distance Readings Results

Three different ways to obtain the distance between the BLE beacon and BLE sniffer were tested in an office room without objects between the devices. The first method uses direct distance readings each 0.95 s, the second method applying median and average filters after ten direct readings and third method implements a fuzzy logic inference system after ten direct readings. The BLE beacon was fixed on a power supply and the BLE sniffer connected to a computer was positioned at different distances measured by a scale at 0.5 m, 1 m, 1.5 m, 2 m, 2.5 m and 3 m. The test values were obtained ten seconds after fixating the BLE sniffer position to guarantee receiving a stabilized signal. The following graphics compare the measurements obtained by the three methods (Figs. 3 and 4 and Table 3):

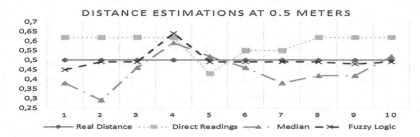

Fig. 3. Distance estimations at 0.5 m

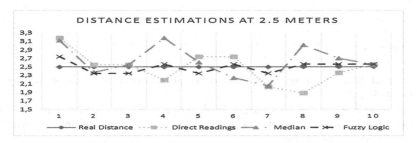

Fig. 4. Distance estimations at 2.5 m

Table 3. Average error for each method at different distances

	0.5 m	1 m	1.5 m	2 m	2.5 m	3 m
Direct readings	0.101	0.148	0.096	0.449	0.285	0.194
Median	0.082	0.235	0.252	0.300	0.306	0.243
Fuzzy Logic	0.028	0.115	0.136	0.050	0.118	0.193

5   Conclusions and Future Work

After doing these tests with the three methods some behaviors of the receiving signal were detected. The signal takes some time to stabilize after changing the distance between the beacon and the sniffer. It is necessary to wait some seconds until the values get closer to the real distance. Once the signal stabilizes the repeatability increases. In general as the distance between the beacon and the sniffer increases, the average error of distance estimations increases. That is an expected behavior because the disturbing factors like signal reflecting rise their negative influence as long as the direct signal wave power (the desired one) gets lower.

Comparing the direct readings average error and the estimations after applying median and average filter average error, the results of applying the filter are not as good as expected. That can be explained because even maintaining the sniffer at same fixed position when the distance estimation gets high or low it tends to still get higher or lower with time. The normalizing factor by applying median and average filter excludes the estimations that usually are far from real distance, but due to the trend to the estimation to get high or low with time, it doesn't improve the average error. In practice the average error obtained by direct readings method is similar to the average error applying median and average filter. Comparing the direct readings average error with the Fuzzy Logic inference system estimations average error, the implementation of Fuzzy Logic reduces the error between real distance and estimation, improving the accuracy and the repeatability. This normalizing method proved to be more satisfying that Median and Average method. It proved to be useful improving the estimation values and Fuzzy Logic inference system was adopted in this project. The next step will consist in implementing the trilateration algorithm after obtaining the estimated distance between three fixed BLE beacons in known positions at x, y coordinates and the BLE sniffer which is in an unknown position and can move.

References

1. Pahlavan, K., Li, X.: Indoor geolocation science and technology. IEEE Commun. Mag. **40**, 112–118 (2002)
2. New Indoor Navigation Technologies Work Where GPS Can't - IEEE Spectrum. URL consulted (2014). http://spectrum.ieee.org/telecom/wireless/new-indoor-navigation-techno logies-work-where-gps-cant
3. Reddy, Y.: Cyber-Physical Systems: Survey Cloud-based Cyber Physical Systems: Design Challenges and Security Needs (2015)
4. Park, C., Park, D., Park, J., Lee, Y., An, Y.: Localization algorithm design and implementation to utilization RSSI and AOA of ZigBee. In: 5th International Conference Future Information Technology, pp. 1–4 (2010)
5. Roxin, A., Gaber, J., Wack, M., Nait-Sidi-Moh, A.: Survey of wireless geolocation techniques. In: 2007 IEEE Globecom Work, pp. 1–9 (2007)
6. Akif, M., Mahallesi, E., Ankara, Y.: A bluetooth signal strength based indoor localization method
7. Kaemarungsi, K., Krishnamurthy, P.: Modeling of indoor positioning systems based on location fingerprinting (2004)
8. Liu, Y., Du, H., Xu, Y.: The research and design of the indoor location system based on RFID. 2011 Fourth International Symposium on Computational Intelligence and Design, pp. 87–90 (2011)
9. Fernández, S., Gualda, D., García, J. C., García, J. J., Ureña, J., Gutiérrez, R.: Indoor location system based on ZigBee devices and metric description graphs (2011)
10. Li, H.: Low-cost 3D bluetooth indoor positioning with least square. Wireless Pers. Commun. **78**, 1331–1344 (2014)

11. Liu, H., Member, S., Darabi, H., Banerjee, P., Liu, J.: Survey of wireless indoor positioning techniques and systems. IEEE Trans. Syst. Man Cybern. Part C Appl. Rev. **37**, 1067–1080 (2007)
12. Lin, T., Lin, P.: Performance comparison of indoor positioning techniques based on location fingerprinting in wireless networks, pp. 1–6 (2005)
13. Singh, A.K., Purohit, N., Varma, S.: Fuzzy logic based clustering in wireless sensor networks: a survey. Int. J. Electron. **100**, 121–141 (2012)
14. Young, D.: Understanding ibeacon distancing - Stack Overflow. URL consulted (2015). http://stackoverflow.com/questions/20416218/understanding-ibeacon-distancing

CMOS Indoor Light Energy Harvesting System for Wireless Sensing Applications: An Overview

Carlos Carvalho[1,2(✉)] and Nuno Paulino[2,3]

[1] Instituto Superior de Engenharia de Lisboa (ISEL – ADEETC),
Instituto Politécnico de Lisboa (IPL), Rua Conselheiro Emídio Navarro, nº1,
1949-007 Lisbon, Portugal
cfc@isel.ipl.pt

[2] UNINOVA/CTS, Instituto de Desenvolvimento de Novas Tecnologias,
Campus FCT/UNL, 2829-516 Caparica, Portugal
nunop@uninova.pt

[3] Departamento de Engenharia Electrotécnica, Faculdade de Ciências e Tecnologia,
Universidade Nova de Lisboa, Campus FCT/UNL, 2829-516 Caparica, Portugal

Abstract. This paper presents an overview of the PhD thesis "CMOS indoor light energy harvesting system for wireless sensing applications", whose main goal was designing a micro-power light energy harvesting system for indoor scenarios, addressing the challenges associated with this kind of environment. Light energy was taken in by using an amorphous silicon (a-Si) photovoltaic (PV) cell and conditioned using a switched-capacitor (SC) voltage converter, along with a maximum power point tracking (MPPT) capability. The MPPT method was the Fractional V_{OC}, put into practice by using an asynchronous state machine (ASM) which automatically establishes and controls the clock signals' frequency, thus controlling the switches of the voltage converter. To minimize the area of the SC section, MOSFET capacitors were used. A charge reusing scheme was proposed, so as to decrease the loss through the parasitic capacitance of their bottom plate. Laboratorial results, taken from a CMOS solid-state prototype, show that the proposed system can achieve better results than those in the present state of the art.

Keywords: Energy harvesting · CMOS integrated circuits · MPPT · PV cells · Power conditioning · Wireless sensor networks

1 Introduction

The capability of electronic systems to get energy out of the environment around them, to be self-powered, is a subject that has drawn increasing interest of the researchers for some time [1], regarding the context of wireless sensor networks (WSN) [2], and embedded systems [3]. This capability, known as energy harvesting, allows for electronic applications to operate without having to hardwire a connection to the power grid, and needing no regular replacement of batteries [1, 4, 5]. This is particularly important

© IFIP International Federation for Information Processing 2016
Published by Springer International Publishing Switzerland 2016. All Rights Reserved
L. Camarinha-Matos et al. (Eds.): DoCEIS 2016, IFIP AICT 470, pp. 178–194, 2016.
DOI: 10.1007/978-3-319-31165-4_19

for sensor networks deployed to inhospitable places, where an amount of energy, of any sort, obtainable from the environment, will suffice.

Besides a strong feature about ubiquity, energy harvesting is also interesting either ecologically and economically, by circumventing the use of batteries as the power of the main system. Firstly, the sensor node will not take part in the chain of chemical pollution caused by battery manufacturing or decommissioning. Secondly, by not using batteries, there is a decrease in costs, both in devices and replacement manpower.

In an indoor environment, it is preferable to have self-supplied sensors so as to avoid cord connections and the consequent costs with material, like cable duct. In addition, there is total flexibility to place the nodes where they are exactly desired.

Although energy can be taken in from light existing inside buildings, this specific source poses an upgraded challenge, since the level of indoor light energy is quite below the one obtainable in the outside. Moreover, the available indoor light can change significantly, as natural light from outside suffers an attenuation and is subjected to a mix with light coming from ceiling luminaries.

The current thesis presents an energy scavenging system fed by indoor light, whose function is to supply a sensor system, thus permitting a networking of the same kind as the one presented in [6]. In spite of the design being mainly to deal with typical light levels in indoors, the proposed system still proved to work using more intense levels, thus making it a highly flexible light energy harvesting system. Through minimizing size and cost, this system was fully integrated into a silicon die, except for the large value output storing capacitor and the scavenging PV cells.

To maximize the harvested energy coming from a PV cell, a DC-DC voltage converter is needed, such that it can follow the maximum power point (MPP) of the PV cells. Also aiming to decrease volume and cost, the proposed DC-DC voltage converter uses switched-capacitors (SC), putting aside inductors [7].

The system has to be as efficient as possible, because of the typically reduced light level existing indoors. Therefore, the MPPT algorithm was implemented by using analog circuitry so as to save power, discarding implementations that use a microcontroller, or similar. The MPPT method used is the Fractional Open Circuit Voltage (FV_{OC}) technique, implemented through the use of an ASM that automatically adapts to the environmental conditions, creating and adjusting the clock frequency of the signals that control the switches of the voltage converter circuit, making it to match its input impedance to the MPP of the PV module.

2 Contribution to Cyber-Physical Systems

The main purpose of the work contained in the research thesis under overview is concerned about building an energy harvesting system, or the energy harvesting section of a WSN node, such that the node can be supplied with light energy in an indoor environment. Such a node can be an embedded system that, more broadly, is a part of a cyber-physical system (CPS) with some sort of sensing utility.

For example, there are various areas and applications in which a CPS has proved to be effective. Among those areas, one can find health condition monitoring [8], forest surveillance and monitoring [9], environment and energy monitoring inside buildings [6], WSNs in automotive applications [10], structural health monitoring [11], or wireless networks for localizing and studying animals [12], just to mention a few.

In many contexts, the individual nodes, when deployed to a given scenario, are programmed with the required intelligence to establish a network with neighboring nodes, ultimately allowing for the establishment of a whole network among all the nodes, provided that the required layer levels of the protocol stack are implemented in the software. Although dealing with smaller and simpler computational structures than computers, the full set of requirements, typical for an operating system [13] and a computer network [14], can be implemented at a more basic level.

The system that has been designed and developed in the research thesis under overview is only focused about energy harvesting alone. Therefore, a natural improvement to the system that was implemented in this work is the addition of sensor and transceiver circuitry, so that the latter can give it full networking capability, allowing for communication links. Depending on the role that a node can be given inside its network, even bridging up between networks is also possible.

The aim of this thesis is not to develop a wireless node for a specific application, but to demonstrate that it is feasible to build the harvesting section, such that it can deal with the available light of indoors, proving that this concept can be put into practice. The proposed system can then be casted to operate with different kinds of sensors, such as: light, pressure, temperature, acceleration, etc. For example, the data to be transmitted could be the light intensity that the PV cell is experiencing. This information is embedded in the frequency of the phases controlling the switches in the DC-DC converter, i.e. the actual energy harvesting PV cell can be the sensor itself. Such a system could be used to monitor the amount of light inside an "intelligent house". By using a WSN based upon systems like these, the energy for illumination purposes could be more efficiently used, by lowering the costs both in money and in natural resources. The address of this kind of problem is mobilizing many researchers and [15] shows an example concerning this matter.

3   Literature Review

The literature review allowed obtaining a global overview about the main issues involved in the energy harvesting theme, namely regarding energy harvestable sources, PV cells, MPPT techniques, DC-DC voltage converters, and components able to store high amounts of energy.

3.1   Harvestable Sources of Energy

Around us, there are various energy sources that can be scavenged, so as to supply electronic circuits [4]. Energy can be obtained from several sources, like: light (natural [4] or artificial [16]), electromagnetic radiation [17], mechanical motion (vibrations, for example) [18], thermal gradients [19], etc. Each one of these sources has a reduced energy density, leading the electronic modules inside the node to work by making use of a highly reduced amount of energy and to have efficiencies as high as possible. Of all these, light has the highest density for low-power systems [4].

Mechanical Energy. Mechanical energy is obtained by wind, wave, vehicle motion, or any vibrations or movement. A significant aspect specific to scavenging this source of energy, is the need for rectifying circuits. There are various ways to convert mechanical motion into electricity. Converting movement to electricity is achieved by employing electromagnetics [4], piezoelectrics [18] or electrostatics [20]. Harvesters that exploit vibrations, have presently efficiencies that go from 25 % to 50 % [21].

Thermal Gradients. Harvesting thermal energy can be an option in situations where high temperature gradients exist. These gradients can be used as a source of heat, supplying a reasonable level of energy [1, 4, 19]. Thermoelectric generators (TEG) are silent and dependable, having no parts in motion. Currently, these transducers are able to convert about of 5 % to 6 % of the thermal source to electric energy. However, research has been undertaken, to mature new materials and modules in order to try to exceed the harvesting efficiency over 10 % [22].

Radio Frequency Electromagnetic Energy. In urban areas, radio frequency (RF) energy is everywhere, available around the clock, in day time and night time, being generated by sources such as mobile phone cell networks, Wi-Fi networks, radio and television broadcasting, and other sources of the same kind. Since the scavenged source is AC, rectifiers must be used in order to get stable DC supply [19, 23, 24]. Harvesters acting over the RF sources, presently have efficiencies reaching 50 % [21].

Human Generation. Using applications powered by human means is an idea that has interested researchers from some time [25]. Converting human motion into electricity is a subject of extensive interest [26]. In biomedical applications, implants are supplied by using the heat from the patient's metabolism, as the scavenged source [8].

Microbial Fuel Cells. Concepts behind the microbial fuel cell (MFC) come from the energy generated by the electrochemical reactions produced by bacteria when they are in activity [27]. A very hostile environment to humans, and where bacteria are plentiful, is wastewater. A WSN that does not require any human intervention after deployment is thus preferable. As such, monitoring and controlling wastewater treatment plants (WWTP) is addressed in [28], where a system is suggested. This system seeks for efficiency in the use of electric energy inside this kind of facilities.

Light Energy. Light is a visible electromagnetic wave. In indoors, the available light is much lower than outside, where the usual standard is AM1.5 ($1 \, kW/m^2$). The literature reports indoor irradiances that go from 1/10 of the maximum intensity of the Sun [29], to $0.833 \, W/m^2$ (100 lux, converted to W/m^2 using [30]) in a mildly illuminated room [31], or $10 \, W/m^2$, having the PV panel placed close to overhead lamps [16].

The main element for light energy scavenging is the PV cell. When illuminated, it behaves like the equivalent electric circuit in Fig. 1 (a), evidencing three main parameters, shown in Fig. 1 (b): maximum power point (MPP), open circuit voltage (V_{OC}) and short circuit current (I_{SC}). The model is established from the parameters of the corresponding circuit (I_1, R_P and R_S), but the model for the diode (D) is also important, given that it varies according to the amount of light.

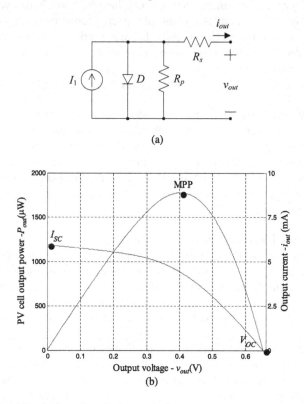

(a)

(b)

Fig. 1. PV cell: (a) Equivalent electric circuit; (b) Example of typical current and power curves.

PV technologies showing lower efficiencies and having lower manufacturing costs can be used, but resulting in a larger area for the PV cell, for the same magnitude of useful available power. The a-Si PV technology is an example of this, in which costs are lower, but a larger PV cell must be used to achieve the same output power. In [32], a cell of crystalline silicon (c-Si) is compared to another of a-Si. Here, when using various light sources to generate an equivalent amount of light, it was shown that in indoors, in opposition to c-Si PV cells, a-Si PV cells do not undergo a strong decrease in their output power, if the nature of light changes from sunlight to fluorescent lights. Thus, evidences indicate that a-Si PV cells are more suited for indoors.

3.2   PV Cell Technologies

PV cell technologies can be divided into three "generations". The first one uses c-Si structures, including the monocrystalline and polycrystalline ones. This technology has been evolving over time, improving both efficiency and capabilities. Although this is the first generation, it is not obsolete at all. The second one uses single junction components, while trying to optimize the use of the materials, aiming to maintain the efficiencies attained by the first generation. This generation encompasses, among other materials, a-Si. Lastly, the third generation uses multiple junctions and is showing good results and efficiently working cells at lower costs [33]. Organic cells and carbon nano tubes are examples of PV cells belonging to this generation [34].

3.3   DC-DC Voltage Converters

Energy harvested from environmental sources is not suited to directly power electronic circuits; it must be conveniently conditioned to be used. When working with a system based on photovoltaics, the converter circuit sets the output voltage of the cell, so as to attain optimal power operating conditions, using a MPPT algorithm.

Linear converters are not adequate because of their low efficiencies, thus switched converters must be used either based on inductors or on capacitors. Suitable inductors are not easily available in CMOS technology. As such, this system was based on SC.

SC DC-DC Voltage Converters. One can have a fully integrated SC voltage DC-DC converter, requiring no external components. In the converter, two types of capacitors exist: the output capacitor and the flying capacitors. According to the state of the switches, the flying capacitors connect between different nodes inside the circuit, and transfer charge. There are various SC voltage step-up topologies: the ladder [35], the Cockcroft–Walton [36], the Dickson charge pump [37], the Fibonacci [38], the Parallel-Series [39] and the Doubler [39] topologies. For the proposed system, the chosen topology was the Parallel-Series, aiming to double the input voltage. This is a very simple topology and that was the main reason for this choice.

3.4   Energy Storing Devices

The amount of energy that has been taken in, can be stored in a supercapacitor [40, 41] or in a rechargeable battery [7], allowing the node to be turned on during intervals when there is no energy available from outside. Each device requires special attention, because each of them involves very specific charging strategies [42].

Rechargeable Batteries. This device is a cell that stores energy, able to charge by inverting inside chemical reactions and is used when a large density of energy is needed. The lifetime of these devices is greatly affected by the number of charges and discharges, and trying to minimize the number of these cycles is an important issue. This topic relates to the quantity of time that such device can be operating, so that the accumulated charge can last greatly as possible. There are several types of batteries, such as the NiCd, NiMH, Li-ion and Li-polymer [43].

Supercapacitors. Supercapacitors have features making them different from regular capacitors. The electric model of a supercap isn't a simple high value capacitor, but an arrangement of branches having time constants of their own [44]. Supercapacitors can stand more charge and discharge cycles than batteries do. Appealing factors about supercapacitors are that they don't demand specific charging topologies, standing trickle charging and are inexpensive, when compared to batteries.

3.5 MPPT Methods

In order to maximize the energy harvested by the PV cells, a MPPT method should be used, maximizing the chance of obtaining as much energy as possible out of the PV cell which would be wasted, otherwise. This strategy enables augmenting the scavenged energy in about 65 % to 90 % [3]. Various methods can be found in [45], where two main types of tracking are identified: true MPPT and quasi-MPPT. The former are able to follow the real MPP of the cell, while the latter, also tracking the same point, are not concerned about the maximum itself. Being in the vicinity of it is enough. This leads to simpler implementations of the tracking circuits. Some examples of true MPPT methods are the "Hill Climbing" [7], the "Ripple correlation control (RCC)" [46], or the control based upon Neural Networks and Fuzzy Logic. As for the quasi-MPPT, one can find the "Fractional V_{OC}" [40], the "Fractional I_{SC}", the "DC link capacitor droop control", or the "One-cycle control (OCC) MPPT".

In the present work, the Fractional V_{OC} was chosen because of its simplicity and low energy budget, although involving the overhead of using pilot PV cells.

4   Proposed Energy Harvesting System

The proposed harvesting system presented in this research thesis is summarized in the block diagram depicted in Fig. 2.

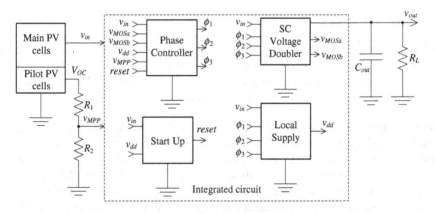

Fig. 2. Architecture of the proposed indoor light energy scavenging system.

The modules that make up the system, integrated into a 130 nm CMOS technology circuit, are: Phase Controller, SC Voltage Doubler, Start Up, and Local Supply.

4.1 SC Voltage Doubler

To reduce cost and volume, a SC voltage doubler was used. The circuit is a Parallel-Series SC step-up voltage converter (as by Sect. 3.3).

The concept of the step-up voltage converter circuit is depicted in Fig. 3. In addition, an equivalent switched parasitic capacitor (C_p), is included. This represents the load effect, created when the phase generator module is working, generating the phases ϕ_1 and ϕ_2. The PV cell has been linearized. So, the circuit shown in Fig. 1 (a) is now replaced by its Thévenin equivalent (v_S and R_S). C_{in} stands for the capacitance of the PV cell. The configurations that this circuit takes when in operation are shown next (Fig. 4):

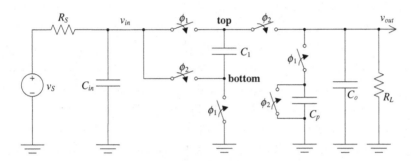

Fig. 3. Conceptual schematic of the Parallel-Series (doubler) voltage step-up converter.

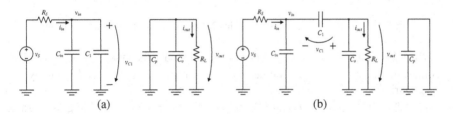

Fig. 4. SC step-up voltage converter: (a) Circuit during phase ϕ_1; (b) Circuit during phase ϕ_2.

In (1) and (2), the expressions of the steady state input and output voltages are presented, depending upon the elements that make up the circuit.

$$V_{IN} = \frac{T_{CLK}\left(4\left(C_o\,C_p + C_1\left(C_o + C_p\right)\right)R_L + \left(C_1 + 4\,C_o + C_p\right)T_{CLK}\right)v_S}{16\,C_1\,C_o\,C_p\,R_L\,R_S + 4\left(C_o\,C_p\,R_L + C_1\left(C_o + C_p\right)R_L + C_1\left(4\,C_o + C_p\right)R_S\right)T_{CLK} + \left(C_1 + 4\,C_o + C_p\right)T_{CLK}^2} \tag{1}$$

$$V_{OUT} = \frac{2\,C_1\left(4\,C_o\,R_L - T_{CLK}\right)T_{CLK}\,v_S}{16\,C_1\,C_o\,C_p\,R_L\,R_S + 4\left(C_o\,C_p\,R_L + C_1\left(C_o + C_p\right)R_L + C_1\left(4\,C_o + C_p\right)R_S\right)T_{CLK} + \left(C_1 + 4\,C_o + C_p\right)T_{CLK}^2} \tag{2}$$

Thus, V_{IN} and V_{OUT} are adjusted using the value of the clock frequency. Controlling this parameter, makes it possible to reach the MPP, occurring when the input impedance of the SC circuit is equal to the value of R_S, giving $V_{IN} = v_S/2$.

SC Voltage Doubler with Charge Reusing. To minimize the area needed by the switched capacitors, MOS transistor capacitors were used because they have the largest capacitance per unit of area, in the technology that was used. Unfortunately, this increases the value of the bottom plate parasitic capacitance [47], degrading efficiency. To overcome this problem and minimize the amount of charge lost through this capacitance, the approach that was adopted was to split the switched capacitor in two, and duplicating the SC circuit, as depicted in Fig. 5.

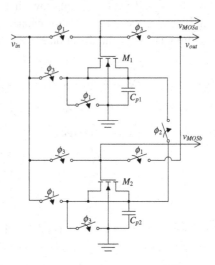

Fig. 5. SC step-up voltage doubler, using MOS capacitors with charge reusing.

When ϕ_1 is active, M_1 and C_{p2} connect in parallel with the input source (v_{in}) and M_2 is connected in series with this source, accomplishing an ideal voltage folding by two at v_{out}. During ϕ_1, C_{p1} is kept discharged, because both of its plates are connected to ground. When ϕ_3 is active, M_2 and C_{p1} swap role with M_1 and C_{p2}, leading to the same actions when ϕ_1 was active. Finally, when ϕ_2 is active, C_{p1} and C_{p2} are placed in parallel, achieving a charge redistribution between them. Since one of these two capacitances is charged to v_{in}, whilst the other is fully discharged, when they are tied together in parallel, the charge is uniformly divided between the two, and each capacitance gets half of the charge of the other capacitance that was firstly charged to v_{in}. The ideal voltage across this parallel arrangement is $v_{in}/2$. Thus, in ϕ_1 or ϕ_3, the parasitic capacitance that follows in the connection to v_{in}, will be then pre-charged to one half of its final voltage, thus requiring just half of the charge from the input voltage node. This procedure prevents wasting half of the input charge, which is inevitably committed to these capacitances, improving the efficiency of the converter.

4.2   Phase Controller

The Phase Controller creates the clock phase signals (ϕ_1, ϕ_2 and ϕ_3) required for the voltage doubler. The PV cell power changes according to light and temperature, so the phase controller generates these clock signals using the FV_{OC} MPPT method in order to automatically adjust the clock frequency to reach the MPP. This method uses the intrinsic characteristic of PV cells, in which there is a proportionality factor (k) between the open circuit voltage (V_{OC}) and the voltage at which the MPP occurs. Pilot PV cells, which are always in open circuit, or unloaded, are used to determine V_{OC}. The optimal voltage obtained from the pilot PV cell (v_{MPP}) is determined by computing the product of V_{OC} by k, using a voltage divider with resistors.

ASM Circuit. The clock signals are produced by an ASM, automatically and dynamically adjusting the frequency of the clock signals, thus obtaining the MPP condition out of the cell. The circuit that implements the MPPT algorithm, generates the clock phases, and establishes how to go from one state to the next, is depicted in Fig. 6.

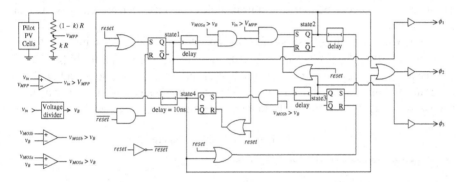

Fig. 6. Phase generator circuit using the Fractional V_{OC} MPPT.

The circuit has four states, which are determined by the output of four S-R latches. The states relate to the generation of the phase signals ϕ_1, ϕ_2, ϕ_3, and again ϕ_2, respectively. To go from a state to its successor, the *Set* signal of a given latch is turned on, switching its output from 0 to 1. In turn, this activates the *Reset* signal of the latch preceding it, causing its output to go from 1 to 0, effecting the state change.

4.3   Local Supply

The proposed system must create the power supply for the phase generator. The main output voltage cannot be used with this purpose, because at start-up, this voltage is 0 V and it will remain at about 0 V for a long time, until a significant charge has been delivered to the buffer capacitor. The answer is to have its own supply voltage, being not dependent from the output voltage. The Local Supply is a replica of the main SC module, with the capacitors and the switches scaled-down to a portion of the size of those used in the bigger SC circuit. The optimal ratio for both modules is 0.2, verified by a set of simulations. The output of this module is named v_{dd}.

4.4   Start-up

The start-up module guarantees that under very low ambient illumination, the system can start operating with success. Previously to the startup, the output of the Local Supply is 0 V. Next, the start-up module shunts the input voltage node to the output node of the Local Supply module. As soon as this voltage gets sufficiently high for the phase controller to start operating, the shunt is eliminated and the circuit begins its regular operation. This circuit also generates the *reset* voltage signal that will be used by the phase generator, in order to guarantee that it starts to operate in the correct state. More details about how this module works can be found in [48].

4.5   Manufactured Prototype

The total die area occupied by the various modules that have been presented is summarized in Table 1.

Using MiM capacitors, instead of MOS capacitors, would have an impact of about eight times [49], in the needed area, i.e. 2.03 mm^2, being a very significant issue.

The entire chip layout and a photograph of the prototype are depicted in Fig. 7. The integrated circuit was glued to a PCB and its pads were connected to the latter by using bond wires, as it is made evident from the photograph.

Table 1.   Partial and total layout areas.

Module	Area (mm^2)
Main MOS capacitances	0.1697
Local Supply	0.0839
Start-up	0.0189
Switches	0.0019
Fractional V_{OC} phase generator	0.0213
Total	0.2957
+ 5% (interconnections)	**0.31**

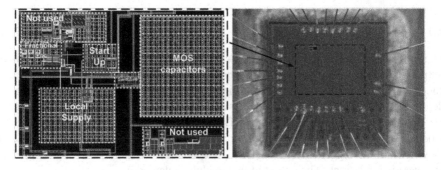

Fig. 7.   Layout of the integrated circuit and its photograph.

5 Experimental Results

The prototype of the proposed system was experimentally assessed to verify if it met the desired behavior and to assess the design parameters, by using the setup in Fig. 8.

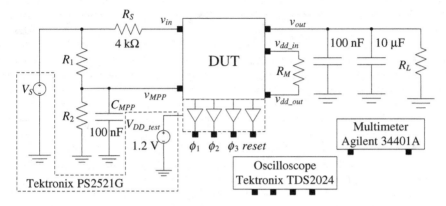

Fig. 8. Experimental test setup for the indoor light energy harvesting system.

It was confirmed that the whole proposed system could successfully start-up thanks to the start-up module. Subsequently to starting up, the system enters its steady state and it is feasible to evaluate the output and input voltages and currents, for different load values, allowing for the computation of the power efficiency of the DC-DC converter. The highest efficiency that could be measured for the prototype was 70.3 %.

The MPPT controller also showed the expected behavior, tracking the MPP and generating the required phase signals to double the input voltage, as seen in Fig. 9.

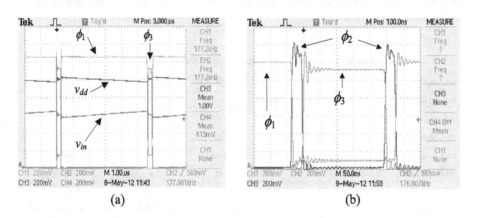

Fig. 9. (a) Behavior of the MPPT circuit; (b) Detail depicting the three phase signals.

By replacing the input voltage source with a-Si PV cells (main and pilot - 14 cm^2 and 2 cm^2, respectively), Fig. 10 (a) shows that this system is able to operate correctly with a Local Supply output of about 700 mV, just dissipating 1.43 μW, in a very reduced irradiance situation, which still permitted the system to correctly start-up (0.32 W/m²). In Fig. 10 (a), the spike that appears at the left of the image is the *reset* signal, generated by the Start Up circuit.

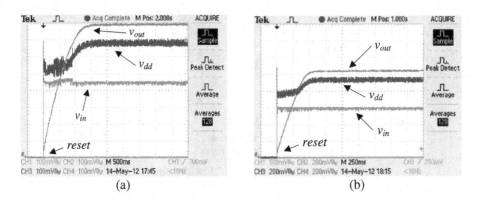

(a)   (b)

Fig. 10. Voltages amid start up, for irradiance of (a) 0.32 W/m² and (b) 4.97 W/m².

The phase generator module is even prepared to withstand lower irradiance levels, though the biggest part of the energy taken in is conveyed to power the phase generator. This means that almost no energy is delivered to the storage capacitor. The lowest irradiance that was registered, under which the system had started up and was operating correctly, was 0.18 W/m². The ASM circuit showed to operate correctly for a Local Supply output of only 453 mV, and dissipating 0.085 μW. This set of values were, to the best possible of the authors' knowledge and by the time the thesis was written, lower than those reported in the literature. The maximum efficiency of the converter is 70.3 % for an input power of 48 μW.

Table 2. Contrast with other state of the art publications.

Reference	[6]	[17]	[50]	[51]	This work
PV cell area (cm²)	42.5	N/A	16.5	N/A	14.0
Min. irradiance (W/m²)	2.50[a]	N/A	3.17	N/A	0.18
Min. input power (μW)	10625[b]	3300	180	5.0	8.7
Min. voltage supply (V)	2.500	5.000	2.800	1.000	0.453
Min. controller power (μW)	50	N/A	135	2.4	0.085
Efficiency (%)	75.0	40.0	91.8	87.0	70.3
Technology	Discrete	Discrete	Discrete	0.25 μm CMOS	0.13 μm CMOS
Storage device	Supercap	Supercap	Supercap	Battery	Supercap
Converter	Boost	Boost	Boost–Buck	Boost	Boost
Converter based device	COTS	Inductor	Inductor	Inductor	SC

[a]–Converting lux to W/m2 as performed in [30] and [50]
[b]–Computed by scaling (Min. irradiance) by (PV cell area)

6 Conclusions

The research thesis whose overview is presented in this paper showed the analysis, design and experimental assessment of an integrated CMOS energy harvesting system. This system uses a SC step-up voltage converter, developed to optimally work with a-Si PV cells (with an area of 14 cm^2), in indoor light situations.

The circuit was built in a 130 nm CMOS technology, occupying an area of 0.31 mm^2. The voltage converter uses MOSFET capacitors with a charge reusing strategy, so as to decrease the effect of the parasitic bottom plate capacitance loss. The phase controller module implements the MPPT Fractional V_{OC} algorithm, in order to extract, as much as possible, the power taken in by the PV cells. Laboratorial results showed that the proposed system could start-up from a 0 V condition, under an irradiance as low as 0.32 W/m^2. Afterwards, the system demands an irradiance of just 0.18 W/m^2 to stay operating.

A contrast between the proposed system and some state of the art papers is shown in Table 2, showing its merits over them, namely in terms of PV cell area, minimum irradiance, minimum voltage supply and minimum controller power, besides being based on SC and integrated using a 0.13 μm CMOS technology.

Acknowledgments. This work was supported by the Portuguese Foundation for Science and Technology (FCT/MCTES) (CTS multiannual funding) through the PIDDAC Program funds and by the grant SFRH/PROTEC/67683/2010, financially supported by the Instituto Politécnico de Lisboa. The authors wish to thank to Dr. Guilherme Lavareda, for his help on the subject of PV cells and for having produced the a-Si PV cells used to assess the proposed system.

References

1. Paradiso, J.A., Starner, T.: Energy scavenging for mobile and wireless electronics. IEEE Pervasive Comput. **4**(1), 18–27 (2005)
2. Kansal, A., Srivastava, M. B.: An environmental energy harvesting framework for sensor networks. In: Proceedings of the International Symposium on Low Power Electronics and Design (ISLPED 2003), pp. 481–486 (2003)
3. Raghunathan, V., Chou, P.H.: Design and power management of energy harvesting embedded systems. In: Proceedings of the International Symposium on Low Power Electronics and Design (ISLPED 2006), pp. 369–374 (2006)
4. Chalasani, S., Conrad, J.M.: A survey of energy harvesting sources for embedded systems. In: Proceedings of the IEEE Southeastcon 2008, pp. 442–447 (2008)
5. Jiang, X., Polastre, J., Culler, D.: Perpetual environmentally powered sensor networks. In: Fourth International Symposium on Information Processing in Sensor Networks (IPSN 2005), pp. 463–468 (2005)
6. Wang, W., Wang, N., Jafer, E., Hayes, M., O'Flynn, B., O'Mathuna, C.: Autonomous wireless sensor network based building energy and environment monitoring system design. In: Proceedings of the 2nd Conference on Environmental Science and Information Application Technology (ESIAT), pp. 367–372 (2010)

7. Shao, H., Tsui, C.-Y., Ki, W.-H.: The design of a micro power management system for applications using photovoltaic cells with the maximum output power control. IEEE Trans. Very Large Scale Integr. (VLSI) Syst. **17**(8), 1138–1142 (2009)

8. Hoang, D.C., Tan, Y.K., Chng, H.B., Panda, S.K.: Thermal energy harvesting from human warmth for wireless body area network in medical healthcare system. In: Proceedings of the International Conference on Power Electronics and Drive Systems (PEDS 2009), pp. 1277–1282 (2009)

9. Sun, Z.-J., Li, W.-B., Xiao, H.-F., Xu, L.: The research on solar power system of wireless sensor network node for forest monitoring. In: Proceedings of the International Conference on Web Information Systems and Mining (WISM 2010), vol. 2, pp. 395–398 (2010)

10. Thewes, M., Scholl, G., Li, X.: Wireless energy autonomous sensor networks for automobile safety systems. In: Proceedings of the 9th International Multi-conference on Systems, Signals and Devices (SSD 2012), pp. 1–5 (2012)

11. Park, G., Rosing, T., Todd, M., Farrar, C., Hodgkiss, W.: Energy harvesting for structural health monitoring sensor networks. J. Infrastruct. Syst. **14**(1), 64–79 (2008)

12. Gutiérrez, A., Dopico, N.I., Gonzalez, C., Zazo, S., Jimenez-Leube, J., Raos, I.: Cattle-powered node experience in a heterogeneous network for localization of herds. IEEE Trans. Ind. Electron. **60**(8), 3176–3184 (2013)

13. Farooq, M.O., Kunz, T.: Operating systems for wireless sensor networks: a survey. Sensors **11**(6), 5900–5930 (2011)

14. Cordeiro, C., Agrawal, D.: Ad Hoc and Sensor Networks, 2nd edn. World Scientific, Hackensack (2011)

15. Wang, W.S., O'Donnell, T., Ribetto, L., O'Flynn, B., Hayes, M., O'Mathuna, C.: Energy harvesting embedded wireless sensor system for building environment applications. In: Proceedings of the 1st International Conference on Wireless Communication, Vehicular Technology, Information Theory and Aerospace & Electronic Systems Technology (Wireless VITAE 2009), pp. 36–41 (2009)

16. Hande, A., Polk, T., Walker, W., Bhatia, D.: Indoor solar energy harvesting for sensor network router nodes. Microprocess. Microsyst. **31**(6), 420–432 (2007)

17. Dallago, E., Danioni, A., Marchesi, M., Nucita, V., Venchi, G.: A self-powered electronic interface for electromagnetic energy harvester. IEEE Trans. Power Electron. **26**(11), 3174–3182 (2011)

18. Kong, N., Ha, D.S.: Low-power design of a self-powered piezoelectric energy harvesting system with maximum power point tracking. IEEE Trans. Power Electron. **27**(5), 2298–2308 (2012)

19. Lhermet, H., Condemine, C., Plissonnier, M., Salot, R., Audebert, P., Rosset, M.: Efficient power management circuit: thermal energy harvesting to above-IC microbattery energy storage. IEEE J. Solid-State Circ. **43**(1), 246–255 (2008)

20. Torres, E.O., Rincón-Mora, G.A.: Electrostatic energy-harvesting and battery-charging CMOS system prototype. IEEE Trans. Circ. Syst. I: Regul. Pap. **56**(9), 1938–1948 (2009)

21. Kumar, S.S., Kashwan, K.R.: Research study of energy harvesting in wireless sensor networks. Int. J. Renew. Energy Res. (IJRER) **3**(3), 745–753 (2013)

22. Lu, X., Yang, S.-H.: Thermal energy harvesting for WSNs. In: Proceedings of the IEEE International Conference on Systems Man and Cybernetics (SMC 2010), pp. 3045–3052 (2010)

23. Colomer, J., Miribel-Catala, P., Saiz-Vela, A., Samitier, J.: A multi-harvested self-powered system in a low-voltage low-power technology. IEEE Trans. Ind. Electron. **58**(9), 4250–4263 (2010)

24. Fernandes, J.R., Martins, M., Piedade, M.: An energy harvesting circuit for self-powered sensors. In: Proceedings of the 17th International Conference on Mixed Design of Integrated Circuits and Systems (MIXDES), pp. 205–208 (2010)
25. Starner, T.: Human-powered wearable computing. IBM Syst. J. **35**(3–4), 618–629 (1996)
26. Zeng, P., Chen, H., Yang, Z., Khaligh, A.: Unconventional wearable energy harvesting from human horizontal foot motion. In: Proceedings of the 26th Annual IEEE Applied Power Electronics Conference and Exposition (APEC 2011), pp. 258–264 (2011)
27. Meehan, A., Gao, H., Lewandowski, Z.: Energy harvesting with microbial fuel cell and power management system. IEEE Trans. Power Electron. **26**(1), 176–181 (2011)
28. Chen, Y., Twigg, C.M., Sadik, O.A., Tong, S.: A self-powered adaptive wireless sensor network for wastewater treatment plants. In: IEEE International Conference on Pervasive Computing and Communications Workshops (PERCOM Workshops 2011), pp. 356–359 (2011)
29. Rabaey, J., Burghardt, F., Steingart, D., Seeman, M., Wright, P.: Energy harvesting - a systems perspective. In: Proceedings of IEEE International Electron Devices Meeting (IEDM 2007), pp. 363–366 (2007)
30. Randall, J.F., Jacot, J.: Is AM1.5 applicable in practice? Modelling eight photovoltaic materials with respect to light intensity and two spectra. Renew. Energy **28**(12), 1851–1864 (2003)
31. Weddel, A.S., Merret, G.V., Al-Hashimi, B.M.: Photovoltaic sample-and-hold circuit enabling MPPT indoors for low-power systems. IEEE Trans. Circ. Syst. I: Regul. Pap. **59**(6), 1196–1204 (2012)
32. Wang, W.S., O'Donnell, T., Wang, N., Hayes, M., O'Flynn, B., O'Mathuna, C.: Design considerations of sub-mW indoor light energy harvesting for wireless sensor systems. ACM J. Emerg. Technol. Comput. Syst. (JETC) **6**(2), 7–9 (2010). Article 6
33. Chen, C.J.: Physics of Solar Energy. Wiley, Hoboken (2011)
34. El Chaar, L., Lamont, L.A., El Zein, N.: Review of photovoltaic technologies. Renew. Sustain. Energy Rev. **15**(5), 2165–2175 (2011)
35. Lin, P.M., Chua, L.O.: Topological generation and analysis of voltage multiplier circuits. IEEE Trans. Circ. Syst. **34**(2), 517–530 (1977)
36. Cockcroft, J.D., Walton, E.T.S.: Experiments with high velocity positive ions. (I) further developments in the method of obtaining high velocity positive ions. Proc. R. Soc. Lond. Ser. A (Containing Papers of a Mathematical and Physical Character) **136**(830), 619–630 (1932)
37. Dickson, J.: On-chip high-voltage generation in NMOS integrated circuits using an improved voltage multiplier technique. IEEE J. Solid-State Circ. **11**(6), 374–378 (1976)
38. Cabrini, A., Gobbi, L., Torelli, G.: Voltage gain analysis of integrated fibonacci-like charge pumps for low power applications. IEEE Trans. Circ. Syst. II: Express Briefs **54**(11), 929–933 (2007)
39. Starzyk, J., Jan, Y.-W., Qiu, F.: A DC-DC charge pump design based on voltage doublers. IEEE Trans. Circ. Syst. I: Fundam. Theor. Appl. **48**(3), 350–359 (2001)
40. Brunelli, D., Moser, C., Thiele, L., Benini, L.: Design of a solar-harvesting circuit for batteryless embedded systems. IEEE Trans. Circ. Syst. I: Regul. Pap. **56**(11), 2519–2528 (2009)
41. Simjee, F., Chou, P.H.: Everlast: long-life, Super-capacitor-operated Wireless Sensor Node. In: Proceedings of the 2006 International Symposium on Low Power Electronics and Design (ISLPED 2006), pp. 197–202 (2006)
42. Jeong, J., Jiang, X., Culler, D.: Design and analysis of micro-solar power systems for Wireless Sensor Networks. In: Proceedings of the 5th International Conference on Networked Sensing Systems (INSS 2008), pp. 181–188 (2008)

43. Sudevalayam, S., Kulkarni, P.: Energy harvesting sensor nodes: survey and implications. IEEE Commun. Surv. Tutorials **13**(3), 443–461 (2011)
44. Zubieta, L., Bonert, R.: Characterization of double-layer capacitors for power electronics applications. IEEE Trans. Ind. Appl. **36**(1), 199–205 (2000)
45. Esram, T., Chapman, P.L.: Comparison of photovoltaic array maximum power point tracking techniques. IEEE Trans. Energy Convers. **22**(2), 439–449 (2007)
46. Esram, T., Kimball, J.W., Krein, P.T., Chapman, P.L., Midya, P.: Dynamic maximum power point tracking of photovoltaic arrays using ripple correlation control. IEEE Trans. Power Electron. **21**(5), 1282–1291 (2006)
47. Seeman, M.D., Sanders, S.R.: Analysis and optimization of switched-capacitor DC–DC converters. IEEE Trans. Power Electron. **23**(2), 841–851 (2008)
48. Carvalho, C., Paulino, N.: Start-up circuit for low-power indoor light energy harvesting applications. Electron. Lett. **49**(10), 669–671 (2013)
49. Carvalho, C., Paulino, N.: A MOSFET only, step-up DC-DC micro power converter, for solar energy harvesting applications. In: Proceedings of the 17th International Conference on Mixed Design of Integrated Circuits and Systems (MIXDES), pp. 499–504 (2010)
50. Tan, Y.K., Panda, S.K.: Energy harvesting from hybrid indoor ambient light and thermal energy sources for enhanced performance of wireless sensor nodes. IEEE Trans. Ind. Electron. **58**(9), 4424–4435 (2011)
51. Qiu, Y., van Liempd, C., op het Veld, B., Blanken, P.G., Van Hoof, C.: 5 μW-to-10 mW input power range inductive boost converter for indoor photovoltaic energy harvesting with integrated maximum power point tracking algorithm. In: IEEE International Solid-State Circuits Conference Digest of Technical Papers (ISSCC), pp. 118–120 (2011)

Control and Fault Tolerance

Initial Study on Fault Tolerant Control with Actuator Failure Detection for a Multi Motor Electric Vehicle

Bruno dos Santos[(✉)] and Rui Esteves Araújo

INESC TEC, Faculty of Engineering,
University of Porto, 4200-465 Porto, Portugal
{bruno.laranjo,raraujo}@fe.up.pt

Abstract. This study presents a scheme to detect and isolate faults in over-actuated electric vehicles. Although this research work is still emerging, it already provides a view of the main challenges on the problem and discusses some possible approaches that can be useful to overcome the key difficulties. This paper intends to present a fault detection algorithm based on Unknown Input Observer (UIO). The residuals are built through the difference of signals between the measured outputs and the output estimations from the observer. The main idea is to detect fault in the electric motors and steering wheel actuator. The algorithm is presented and tested with some fault scenarios using the co-simulation tool between Simulink/MATLAB and the high-fidelity model from Carsim software.

Keywords: Fault tolerant control · Fault detection and isolation · Unknown input observer · Electric vehicles

1 Introduction

The road map on road transport published by ERTRAC, EPOSS and SMARTGRIDS in 2010 had the expectation that the adaptation of conventional vehicles to electric power train was the first milestone [1]. At the present stage, most compact electric vehicle is based on today's drive train platforms with one central electric motor. However the potential of novel power drive concepts is not completely achieved with these first generation vehicles. Multiple electric motors mounted close to the wheels or even in the wheel opens the development of new compact vehicle drive train architectures.

System redundancy could be explored in order to have high safety and robustness requirements. The system redundancy in multi-motor electric vehicles should be exploited. If an actuator fails, it is necessary to redistribute the control effort among the healthy actuators such that stability is retained.

Some accidents are prevented by using mechanism that detect and inform the driver about a fault. Between July 2005 and December 2007 6.8% of all accidents are something related with some vehicle malfunction [2]. In the next section will be focused on the formulation of the Research Questions that need to be answer with this PhD work.

© IFIP International Federation for Information Processing 2016
Published by Springer International Publishing Switzerland 2016. All Rights Reserved
L. Camarinha-Matos et al. (Eds.): DoCEIS 2016, IFIP AICT 470, pp. 197–205, 2016.
DOI: 10.1007/978-3-319-31165-4_20

1.1 Research Question

Research Question 1: How can we improve the architecture of the fault tolerant control with Fault Diagnosis for four-wheel independently driven electric vehicle?
The key idea to be explored is to replace the standard control allocation by the concept of "smart control allocation" in order to improve the overall structure. The main objective of the smart control allocation is to ensure that the system adapts in the best way when an actuator fails.

A general overview of control architecture is shown in the Fig. 1. The reconfigurable allocator uses information provided by the fault diagnosis to accommodate faults by performing an appropriate reconfiguration. The fault diagnosis system uses the signals from the Hardware and IMU (Inertial Measurement Unit) sensors with Unknown Input Observer (UIO) algorithm to detect faults. The fault diagnosis system output will be a set of fault flags. This information will be interpreted as the total fault indicator vector, which will be an input for the reconfigurable allocator. The relationships between actuators states, fault type and remedial actions will be defined in the fault code table in order to specify the constraint bounds of the control allocation algorithm.

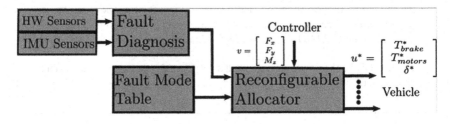

Fig. 1.   Proposed FTC structure in the PhD.

Research Question 2: How should we exploit the fault tolerant detection subsystem to implement an active Fault Tolerant Control (FTC) for multi-motor electric vehicles?
If we started by investigating how to use of the information provided by the Fault Detection (FD) scheme, we could help to improve the active FTC performance. One of the biggest challenges of this thesis is the FD capability to send useful information to the controller. To be useful, the information sent to the controller has to be accurate and needs to be on time. Although some losses in accuracy can be compensated by using robust control techniques, also the delay issues will be crucial. The response time or detection test time needs to be the slowest possible in order to maximize the fault reaction time.

2   Contribution to Cyber-Physical Systems (CPS)

From the automotive electronic point of view, the modern vehicles contain a large number of CPS's such as Anti-Lock Brake Systems, Traction Control, Electronic Stability Program, etc. The same happens for the electric vehicles. The tight interaction between all the computational devices like the battery management system, electric drives of motors and the physical components brings many challenges in this field. For safety and efficiency reasons, these spatially distributed components require a sophisticated monitoring and control. For this purpose, a paradigm change for Electrical and Electronic (E/E) architectures becomes necessary to increase the flexibility between elements [3].

Functional safety is taking more importance in electric vehicles development, due the increased complexity and the wide employment of electrical and electronic systems. In contrast with the past, it will be more important for automotive industry to put in place a normative process to assure the reliability of the safety-related systems, as opposed to regulate themselves with different rules [4].

In order to ensure functional safety the automotive functional safety standard was launched (ISO 26262) [5] to solve the problem of components faults in electrical and electronic system in vehicles. This standard is an adaptation of the IEC 61508 and overcome all the lifecycle of electrical/electronic components. Some metrics introduced by the norm can be useful to quantify the fault tolerant control performance.

3 State of the Art

The literature presents numerous FDI applications for aeronautical and aero-space systems, chemical process, nuclear plants, power system, automotive and electronic systems, see for example [6] or [7]. Also a FDI for the ROboMObil, a concept car, was designed and tested in real operating conditions on DLR (Deutschen Zentrums für Luft- und Raumfahrt) institute. They used the well-known parity equation to detect faults in sensors and actuators [8]. Also, the work started by Rongrong Wang in his PhD thesis at Ohio State University [9] demonstrates the significance of having a FDI scheme to the same type of vehicles.

Only few papers about fault tolerant control of over actuated electric vehicles consider the fault diagnosis in their studies, others assuming that the faults are perfectly identified [10]. However, this is not a realistic hypothesis since fault diagnosis is a very challenging topic, especially for non-linear systems.

In this paper is presented an FDI system based on a Linear Parameter Varying (LPV) UIO scheme. The idea of this approach is to represent the nonlinearities inherent in the model by a LPV approach. Then, the way to use the proposed LPV model is to develop a bank of UIOs that generates a set of residuals which will be sensitive to a pre-established set of faults while insensitive to the others which are considered as disturbances.

4 Vehicle Model

The vehicle modeling is composed by several physical quantities represented in a three coordinate system. The dynamics behavior is based on Euler and Newton laws of motion. The system axis (x, y) is fixed in the vehicle Center of Gravity (CoG), and the movement is based on CoG's velocities and forces:

$$\dot{v}_x = v_y r + \frac{F_x}{m}, \quad \dot{v}_y = -v_x r + \frac{F_y}{m}, \quad \dot{r} = \frac{M_z}{I_z} \quad (1)$$

Where v_x is the longitudinal speed, v_y is the lateral speed and r is the yaw rate. Through a simplified Euler-Newton equation system it is possible to represent the vehicle behavior as a function of the longitudinal force F_x the lateral force F_y and the yaw moment M_z applied in the CoG. These virtual forces are the forces that act to move the vehicle. The virtual inputs are related with the tire forces by:

$$v = \begin{bmatrix} F_x & F_y & M_z \end{bmatrix}, v = BF_{xy} \quad (2)$$

$$v = \begin{bmatrix} 1 & 0 & 1 & 0 & 1 & 0 & 1 & 0 \\ 0 & 1 & 0 & 1 & 0 & 1 & 0 & 1 \\ -\frac{l_s}{2} & l_f & \frac{l_s}{2} & l_f & -\frac{l_s}{2} & -l_r & \frac{l_s}{2} & -l_r \end{bmatrix} F_{xy}$$

$$F_{xy} = \begin{bmatrix} F_{xFL} & F_{yFL} & F_{xFR} & F_{yFR} & F_{xRL} & F_{yRL} & F_{xRR} & F_{yRR} \end{bmatrix}$$

$$\begin{bmatrix} F_{xi} & F_{yi} \end{bmatrix} = \begin{bmatrix} \cos(\delta_i) & -\sin(\delta_i) \\ \sin(\delta_i) & \cos(\delta_i) \end{bmatrix} \begin{bmatrix} F_{ci} \\ F_{li} \end{bmatrix}, i \in \{FL, FR, RL, RR\}$$

Where F_{ci} and F_{li} are the centripetal and longitudinal tire forces applied to the ground. In order to define the tire forces, the Pacjeka semi empiric model can be employed [11]. In this section we assume that the forces are in the linear part of the tire function and also that the side slip angle for each tire is small and the difference between the torque input and the longitudinal force is zero ($F_{li} - T_i r_i \approx 0$). At this stage of the work we assume that the torque signal sent to the in-wheel actuator is a combination of the motor and brake system signals. So the tire force components are:

$$F_{ci_{front}} = N_i \cdot D_i \left(\delta_i - \frac{v_y - l_f r}{v_x} \right), F_{ci_{rear}} = N_i \cdot D_i (\frac{v_y - l_r r}{v_x}) \quad F_{li} = \frac{T_i}{R} \quad (3)$$

where T_i is the input torque to each in-wheel motor, δ_i is the steering angle in each wheel ($\delta_i = 0$ for the rear wheels and for the front wheels the steering angle is equal to the command input). Also N_i is the normal force applied in the wheel; D_i is the cornering stiffness; l_f and l_r are the distance from the CoG to the front and rear axis.

5   Fault Detection System Through LPV UIO

In order to detect actuator faults, the command inputs from the driver are sent to the force observer. The virtual force observer is obtained applying the Eqs. (1, 2, 3). This estimator determine the value of each virtual forces that are delivered to the vehicle. As a result, the state reconstruction is carried out by a bank of UIOs using the virtual force provided by the force observer with different outputs. With the purpose to achieve isolation of faults, the inputs (individual virtual forces) need to be divided into groups and be selected the output related with the fault in consideration.

5.1   LPV UIO

In this work, it is considered that only one actuator fault is detected by each UIO. More details about the UIO observer can be found in [12]. For a LPV system like:

$$\begin{cases} \dot{x}(t) = A(\rho)x(t) + B(\rho)u(t) + D(\rho)d(t) + B_s(\rho)f_s(t) \\ \qquad\qquad\qquad y(t) = Cx(t) \end{cases} \tag{4}$$

where $x(t) \in \mathbb{R}^n$; $u(t) \in \mathbb{R}^m$; $d(t) \in \mathbb{R}^d$; $y(t) \in \mathbb{R}^p$; and $f_s(t) \in \mathbb{R}^1$; are the state vector, input vector, perturbation vector, output vector and fault vector, respectively. The distribution matrix for the fault is represented by the corresponding column number (s^{th}) of B (B_s). The parameter $\rho(t) = [\rho_1, ..., \rho_r] \in \mathbb{D}$ is a time varying function. The requirements to LPV system are: all the matrices are affine in ρ; the parameter ρ is accessible; as well as the ρ and $\dot{\rho}$ are bounded. We are going to first transform the system into is polytopic form:

$$A(\rho) = \sum_{i=1}^{v} h_i(\rho(t))A_i, \text{ with: } h_i(\rho) > 0 \text{ and } \sum_{i=1}^{r} h_i(\rho(t)) = 1 \tag{5}$$

where v is the number of vertices limits of ρ. The new system is described by:

$$\begin{cases} \dot{x}(t) = \sum_{i=1}^{v} h_i(\rho)(A_i x + B_i u + D_i d + B_{si} f_s) \\ \qquad\qquad\qquad y(t) = Cx \end{cases} \tag{6}$$

This structure allows the usage of linear UIO for the transformed system:

$$\begin{cases} \dot{z} = \sum_{i=1}^{v} h_i(\rho)(F_i z + G_i v + K_i y) \\ \qquad\qquad \hat{x} = z - Hy \end{cases} \tag{7}$$

The goal is that estimation error ($e(t)=x(t)-\hat{x}(t)$) converges to zero under any initial conditions and any input. Replacing the estimate error expression from Eq (7), it follows that $e=Tx - z$, where $T=I-H\,C$ (note that the index i is omitted). Solving this equation for the time derivative of the error gives:

$$\begin{aligned} \dot{e} = Fe + (TB - G)u + (TD)d \\ + (TB_s)f_s + (TA - KC + FHC - F)x \end{aligned} \tag{8}$$

In order to design the UIO, is necessary to meet the conditions:

$(C1)G = TB(cancel\,u(t));$ $\qquad\qquad$ $(C2)TD = 0(cancel\,d(t))$

$(C3)F = TA - LC(F\,is\,''Hurwitz'');$ $\qquad\quad$ $(C4)K = L - FH$

$(C5)rank(TB_S) = rank(B_s)$ $\qquad\qquad\qquad$ $(C6)T = I - HC$

Thus, if all conditions are respected the UIO error dynamics becomes $\dot{e} = Fe + TB_s f_s$ and the residual correspond to $r = Ce(t)$.

For the disturbance decoupling, first it is necessary to check if $rank(CD) = rank(D)$ to guaranteed a solution for the system $HCD = D$ (from the condition (C2) and (C6)). One generalized solution for H can be expressed by [13]:

$$H = D(CD)^+ + H_0(I - (CD)(CD)^+) \tag{9}$$

where (X^+) is the generalized inverse matrix given by $(X)^+ = ((X)^T(X)^{-1})(X)^T$, since X is full rank. H_0 is arbitrary matrix of appropriate dimensions and give some freedom on the design. The eigenvalues of F can be arbitrary located by choosing a suitable matrix L, only if the pair (TA, C) is observable. For LPV systems the condition $(rank(CD_i) = rank(D_i))$ for the linear decoupling is not sufficient, it is necessary that $(rank(C[D_1, ...,D_v]) = rank[D_1,...,D_v])$ in order of to obtain $(HCD_\rho = D_\rho)$. Also the polytopic stability of (L_i) is subject to the global stability of $F = TA_\rho - L_\rho C$ where (X_ρ) are all matrices from the "LPV space".

UIO Design. The system in Eq. (1) can be written into a LPV for the (x,y) dynamics. The LPV system has following parameters:

$$x = [v_x, v_y]^T; \quad \rho = r; \quad A_\rho = \begin{bmatrix} 0 & \rho \\ -\rho & 0 \end{bmatrix}; \quad B_\rho = \frac{1}{m}I_2; \quad C = I_2; \quad u = [F_x, F_y]^T.$$

The residual in yaw rate dynamics is the difference between the models in a non-fault scenario with the read values. The UIO algorithm design is presented in Table 1.

Table 1. UIO Algorithm design for planar motion;

UIO	Faults in F_x	Faults in F_y
1 - Define Fault Distribution	$B_{fx} = \frac{1}{m}[1\ 0]^T$	$B_{fy} = \frac{1}{m}[0\ 1]^T$
2 - Perturbation assumption	$f_x \gg d_x$; perturbation matrix $D_x = [0\ 1]^T$	$f_y \gg d_y$; perturbation matrix $D_y = [1\ 0]^T$
3 - Polytopic Transformation $(\rho = h_1 \cdot r + h_2 \cdot (-r))$	$h_1 = \frac{\rho+r}{2r}$; $h_2 = \frac{-\rho+r}{2r}$	
4 - $rank([D_1,\ D_2]) = rank(C[D_1,\ D_2])$		
5 - The design matrix H_{0i} is chosen in order to guaranties $rank(TB_f) = rank(B_f)$	$H_{01} = H_{02} = \begin{bmatrix} 0 & 0 \\ 0 & 1 \end{bmatrix}$;	$H_{01} = H_{02} = \begin{bmatrix} 1 & 0 \\ 0 & 0 \end{bmatrix}$;
6 - $T = I - HC$; $G = TB$;	$T = \begin{bmatrix} 1 & 0 \\ 0 & 0 \end{bmatrix}$	$T = \begin{bmatrix} 0 & 0 \\ 0 & 1 \end{bmatrix}$
7 - Check if $rank(TB_f) = rank(B_f)$ and if $(TA,\ C)$ is observable		
8 - $F = TA - LC$; $K = L + FH$	$L = \begin{bmatrix} l_1 & 0 \\ 0 & l_2 \end{bmatrix}$	$L = \begin{bmatrix} l_1 & 0 \\ 0 & l_2 \end{bmatrix}$

5.2   Simulations

The proposed LPV-UIO scheme has been simulated in different case scenarios. The nonlinear vehicle model was simulated through the Carsim software in co-simulation with Simulink/MATLAB where the proposed observer was implemented. The vehicle parameters that have been used to perform simulations are on the Table 2.

Table 2.   Vehicle parameter used in the simulations

Parameters	Symbol	Value	Parameters	Symbol	Value
Vehicle total mass	m	1100 kg	Wheel radius	R	0.3 m
Tire cornering stiffness	D	0.2179 rad^{-1}	Wheel moment of inertia	I_w	1.1 kgm^2
Distance from CoG to front axle	l_f	1.2 m	Distance from CoG to rear axle	l_r	1.3 m
Distance from CoG to front ground	h	0.37 m	Track width	l_s	1.5 m
Moment of inertia about vertical axis	I_z	996 kgm^2			

In order to demonstrate the capacity to detect faults in actuators it is defined two fault scenarios: The first, a J-turn maneuver with steer fault in constant acceleration with faults in two of the four motors. In this simulations only abrupt fault are considered (step). The fault amplitude is for the steering test a part of the steering input. For the wheel test it is always considered the total failure of the motor in-wheel unit. The inputs of torque to wheel and steering angle reference from the driver are the system input.In the simulations, when a fault occurs, the same fault has to be signalized by an alteration in the residual for each component. The residual amplitude has to be greater than a certain threshold in order to signalize that there is a fault.

Steer Fault. The vehicle starts with an initial longitudinal speed of *20 m/s* with no steer input. At the time *t=1 s* the steering angle is set in *0.05 rad*. This steering is kept constant for the entire test. In this simulation, several faults input have been tested. From the Fig. 2 it is possible to see that the vehicle global trajectory is affected by the steering fault. At the moment when steering input is set to *0.05 rad*, the residuals have a slightly change that are neglected by the observer. When at *t=5 s* fault is injected into the system, all residuals reflects the actuator faults. Obviously the steering fault impact has more influence in the residual provided by the second UIO and from M_z. The steering component has more impact in these residuals.

In-wheel Motor Faults. In the second test, the vehicle starts with equal initial speed of the first test v_x=20 m/s. In this case the steering remains δ =0° rad for the entire test. At t=5 s in-wheel motor faults combinations are injected in the system. The residuals for this test are shown in the Fig. 3. The first point to notice is that the faults in wheels are signalized with low amplitude signals. It is because the faults have low contribution to the trajectory. Also, it is necessary to point out that the observer gain is very low, approximately 10^{-3}. However the trajectory deviation is not large. Also, in the residual

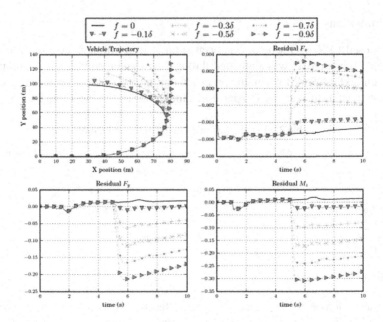

Fig. 2. Results for the steer fault test.

in F_x the value in a non-fault case is not zero due to the longitudinal friction forces that were not modeled in Eq. (1).

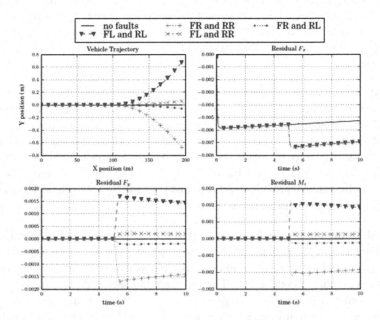

Fig. 3. Results for the in-wheel motor fault test.

6   Conclusions and Future Work

This work considers the design of UIO for use in fault detection for an over-actuated electric vehicle. Although research work is in its first steps, it already provides an overview on the problem while discussing some possible approaches and proposes a method to overcome the key difficulties. Simulation results show a good performance in order to detect several types of faults. Future work, currently under way, includes find a way that does not use the information about lateral velocity as well as implement a validation of the proposed scheme on the prototype vehicle.

Acknowledgments. This work is partially funded by National Funds through FCT - Science and Technology Foundation, through the scholarships SFRH/BD/90418/2012.

References

1. Ertrac, EPoSS, SmartGrids: European Roadmap Electrification of Road Transport (2010)
2. NHTSA: National Motor Vehicle Crash Causation Survey Report to Congress (2008)
3. Sankavaram, C., Kodali, A., Pattipati, K.R.: An integrated health management process for automotive cyber-physical systems. In: International Conference on Computing and Network Communications{ICNC} 2013, San Diego, CA, USA, January 28–31, 2013, pp. 82–86 (2013)
4. Boules, N.: Reinventing the Automobile: The Cyber-Physical Challenge (2008)
5. ISO: ISO/DIS 26262 - Road vehicles - Functional safety. International Organization for Standardization/Technical Committee 22 (ISO/TC 22), Geneva, Switzerland (2011)
6. Isermann, R.: Model-based fault-detection and diagnosis – status and applications. Annu. Rev. Control **29**, 71–85 (2005)
7. Zhang, Y., Jiang, J.: Bibliographical review on reconfigurable fault-tolerant control systems. Annu. Rev. Control **32**, 229–252 (2008)
8. Ho, L.M., Ossmann, D.: Fault detection and isolation of vehicle dynamics sensors and actuators for an overactuated X-by-Wire Vehicle, pp. 6560–6566 (2014)
9. Wang, R.: Fault-tolerant control and fault-diagnosis design for over-actuated systems with applications to electric ground vehicles (2013)
10. Lopes, A., Araujo, R.E.: Fault-tolerant control based on sliding mode for overactuated electric vehicles (2014)
11. Pacejka, H.: Tire and Vehicle Dynamics. Elsevier Science, Woburn (2012)
12. Ichalal, D., Mammar, S.: On unknown input observers for LPV systems. IEEE Trans. Ind. Electron. **62**, 5870–5880 (2015)
13. Darouach, M., Zasadzinski, M., Xu, S.J.: Full-order observers for linear systems with unknown inputs. IEEE Trans. Automat. Contr. **39**, 606–609 (1994)

Fault Analysis of Three-Level NPC Inverters in Synchronous Reluctance Motor Drives

Diogo M.B. Matos, Jorge O. Estima[✉], and Antonio J. Marques Cardoso

CISE – Electromechatronic Systems Research Centre, University of Beira Interior,
Covilhã, Portugal
jestima@ieee.org

Abstract. A performance analysis of a synchronous reluctance motor (SynRM) drive, operating under different fault conditions, with a three-level NPC inverter, controlled by a seven-segment Space Vector Modulation (SVM) technique, is presented in this paper. Considering the voltage source inverter, open-circuit faults of different types are introduced and their effects are studied regarding the SynRM and the inverter performance evaluation. The healthy and faulty operating conditions comparison will take into account the evaluation of some variables, such as the motor power factor, electromagnetic torque, efficiency, total waveform distortion values, currents RMS values, and total waveform oscillation values, obtained from simulation results.

Keywords: Open-circuit faults · Synchronous reluctance motor · Inverter failures · Three-level NPC inverter failures

1 Introduction

During the last years, the interest in the use of synchronous reluctance motors (SynRMs) has increased quite a lot due to their high efficiency, simple structure and rugged characteristics. Moreover, these motors are capable to operate in high-speed applications and in high-temperature environments, enhancing their potential for high-performance variable speed drives [1–4].

Likewise, there has been a growing use of multilevel inverters (MI) such as the neutral point clamped (NPC) converters, since comparing with the traditional two-level converters, they provide higher output voltage waveform quality, lower harmonic content, lower *dv/dt* transients and lower switching losses. In addition, due to the availability of more voltage levels, switching-states or voltage-levels redundancy, and space vectors redundancy, actually, these converters are used in many critical applications in which it is necessary to maximize the reliability of the system through fault-tolerant capability [5–7]. Note also that, traditionally, the use of MI has been associated to high- and medium-voltage applications, whereas, nowadays, it is known that its use is also viable for low-power and low-voltage applications [8, 9]. Despite all that, it is worth mentioning that the more complex structure associated to the MIs may result also in additional problems which do not occur with traditional converters. The MIs comprise

© IFIP International Federation for Information Processing 2016
Published by Springer International Publishing Switzerland 2016. All Rights Reserved
L. Camarinha-Matos et al. (Eds.): DoCEIS 2016, IFIP AICT 470, pp. 206–216, 2016.
DOI: 10.1007/978-3-319-31165-4_21

a greater number of components that result in higher costs. At the same time, the increase of the semiconductors may result in a reduction of the reliability of the system, since a failure in a semiconductor device may cause severe damages on the drive system. In general, power device failures can be classified as open-circuit faults and short-circuit faults. Typically, short-circuit faults are more destructive, requiring special care. Open-circuit faults do not necessarily cause the system shutdown and can remain undetected for an extended period of time. This may lead to secondary faults in the converter or in the remaining drive components, resulting in the total system shutdown and high repairing costs. Moreover, in the particular case of the NPC inverter, issues related to the unbalanced voltage of the dc-link capacitors must be also addressed. Depending on the operating conditions, different voltages values at the capacitors' terminals may occur, which cause undesirable distortion at the inverter output [5, 7].

Several studies have been presented regarding to the most diverse aspects related to the NPC inverters. On one hand, issues related to the capacitors voltage unbalance in some operating conditions has been addressed in [5, 7, 10–12], assuming nowadays as a solved problem. On the other hand, subjects such as control techniques and fault-tolerant solutions have also been addressed, observing significant contributions in these fields in the last years [13–16]. Regarding the fault-tolerant capability, this topic is particularly interesting for NPC inverters, since it is considered that NPC converters present limited fault tolerance.

Although many studies have been published on these fields, few studies exist with respect to what happens after the occurrence of a failure. Besides, usually the analyses are carry out mainly for PMSM drives and induction motor drives operating under faulty conditions, being provided much less attention to the SynRM performance under the same conditions. Hence, some of these issues are discussed in this paper when semiconductor open-circuit faults occur. These faults are introduced in a three-level NPC converter, controlled by a SVM technique, allowing the control of the machine speed. The SynRM performance assessment will take into account the evaluation of some variables, such as the motor power factor, electromagnetic torque, efficiency, total waveform distortion (TWD) values, currents RMS values, and total waveform oscillation values (TWO), obtained from simulation results.

2 Relationship to Cyber-Physical Systems

Cyber-Physical Systems (CPS) is a new concept that has been developed in recent years as a result of the integration of computer systems, networking and physical processes. It is expected that the advances in the CPS will enable capability, adaptability, scalability, resiliency, safety, security, and usability, which will exceed widely the embedded systems of today.

By definition, CPS are related to the integration of computation, networking and the control of physical processes, where physical processes affect computations and vice versa. In this context, electric drives as the one presented in this work, can be easily integrated in a CPS. These applications are strictly related to the control of physical processes, which is a fundamental part of a CPS. Moreover, nowadays, CPS based on electric drives are widely used in industry. Some examples can be pointed out, such as the case of electric power plants and complex transformation industries, where a great number of electric drives are connected to a monitoring network, providing constant data exchange to a main computational platform, allowing for precise control and monitoring of the entire production/transformation processes.

3 SynRM Model

The SynRM model dq equivalent circuits, including iron losses and saturation in the synchronous reference frame, are presented in Fig. 1. An extra resistor R_c connected in parallel with the magnetizing branch in both d and q axes is used to take into account the iron losses effect.

The effect of magnetic saturation was also considered in the SynRM model by modeling the d and q axes inductances as dependent on the supply current. It is worth noting that the saturation effect in the d axis will be distinct from the one in the q axis, since the magnetic paths around the rotor present different reluctances. For this reason, dq axes saturation behavior will be different with the current variation.

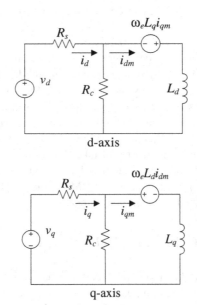

d-axis

q-axis

Fig. 1. SynRM dq equivalent circuits in the synchronous reference frame.

Taking this into account, the SynRM voltage equations in the synchronous reference frame are expressed by:

$$v_d = R_s i_d - \omega_e L_q i_{qm} + L_d \frac{di_{dm}}{dt} \tag{1}$$

$$v_q = R_s i_q - \omega_e L_d i_{dm} + L_q \frac{di_{qm}}{dt} \tag{2}$$

where i_d and i_q are the dq axes supply currents, v_d and v_q are the dq axes supply voltage components, i_{dm} and i_{qm} are the dq axes magnetizing currents, L_d and L_q are the dq axes inductance components (which depend on the i_{dm} and i_{qm} currents), R_c and R_s are the iron losses resistance and the stator resistance, and finaly, ω_e represents the electrical frequency. The electromagnetic torque expression and the mechanical moving equation are given by:

$$T_e = \frac{3}{2}p \left(L_d - L_q\right) i_{dm} i_{qm} \tag{3}$$

$$T_e = J \frac{d\omega_m}{dt} + B_m \omega_m + T_L + T_k. \tag{4}$$

where T_e is the developed electromagnetic torque, p the number of pole pairs, T_L is the load torque, J represents the system moment of inertia, B_m is the viscous friction coefficient, ω_m is the mechanical speed of the motor and T_k is the constant friction coefficient.

4 Control of Three-Level NPC Inverter

The power circuit of the three-level NPC inverter is shown in the Fig. 2. Each inverter leg is composed of four IGBT switches (S_{x1} to S_{x4}, where $x = a,b,c$) with antiparallel diodes, and two clamping diodes (D_{x1} and D_{x2}), which are connected between the pairs of upper and lower IGBT switches (S_{x1}/S_{x2} and S_{x3}/S_{x4}) and the middle point 0. The middle point is available since on the DC side of the inverter, the DC bus capacitor is split into two. Note, however, that the correct neutral point potential at the middle point O is only achieved if the capacitor voltages are balanced.

The operating states of the switches in each NPC inverter leg can be characterized by the three switching states shown in Table 1. The switching state P indicates that the upper two switches of the leg x are turned on and the inverter terminal voltage V_{x0} is $+V_{dc}/2$. In contrast, the switching state N means that the two lower semiconductor devices are turned on, leading to a V_{x0} value equal to $-V_{dc}/2$. In turn, the switching state O signifies that the two inner switches are turned on, and therefore, V_{x0} is connected to the middle point through the clamping diodes. In an ideal operating condition this means that V_{x0} is equal to zero.

Regarding the control, only a good sequence of switching states enables a suitable operation of the drive. This implies an appropriate selection of the reference vector $\vec{V}_{ref}$, in order to be generated just the required voltage for the operation of the SynRM. The generation of the reference vector is, however, associated to the voltage vectors that can be generated by the inverter on a sampling period Ts.

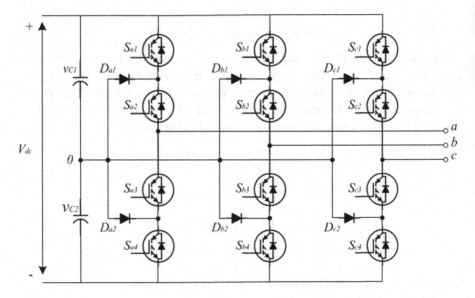

Fig. 2. Circuit diagram of three-level neutral-point clamped inverter.

Table 1. Definition of the NPC converter switching states.

Switching State	Device Switching Status				Inverter Terminal Voltage V_{x0}
	S_{x1}	S_{x2}	S_{x3}	S_{x4}	
P	On	On	Off	Off	$+V_{dc}/2$
O	Off	On	On	Off	0
N	Off	Off	On	On	$-V_{dc}/2$

Considering the switching states P, O, and N for each leg, the inverter has $3^3 = 27$ different switching state combinations that result in the voltage vectors distribution throughout a space vector diagram, as shown in Fig. 2. Some of the switching states are redundant, since they generate the same voltage vector.

The SVM technique will modulate the reference vector using the three nearest vectors, based on the "volt-second balancing" principle [17]. In order to accomplish this, the modulator needs to know the position of the reference voltage vector on the space vector diagram to identify the closest vectors.

In the present work, a seven-segment SVM scheme was used. An overview of the applied SVM control is shown in Fig. 3. As it can be seen, first, it is identified the region, sector and the modulation index related to the reference vector needed for the SynRM operation. After that, it is calculated the dwell times for the application of vectors considering the modulation period T_s. Lastly, it is used a lookup table to generate the switching states based on the calculated above (Fig. 4).

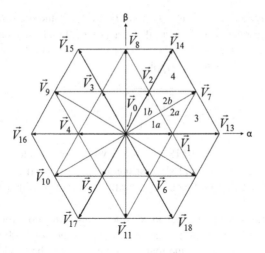

Fig. 3. Schematic diagram of the implemented SVM technique.

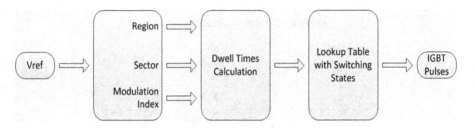

Fig. 4. Space vector diagram of the NPC inverter.

5   Simulation Results

The MATLAB/Simulink programming language was used for the modeling and simulation of the SynRM variable speed drive. The machine mathematical model was connected to a three-level NPC power converter controlled by a SVM rotor field oriented control algorithm.

The SynRM performance was assessed by considering three distinct operating modes: normal operation conditions, an inverter single semiconductor open-circuit failure in IGBT S_{a1} and in IGBT S_{a2}.

The SynRM phase currents evaluation was performed by calculating of the RMS and the total waveform distortion (TWD) values. The TWD coefficient expresses the SynRM currents' harmonic content and it is given by:

$$TWD = \frac{\sqrt{I_{RMS}^2 - I_1^2}}{I_1} \times 100\%$$ (5)

where I_{RMS} and I_l represent the current's total RMS value and the fundamental component RMS value, respectively. In turn, the analysis of the electromagnetic torque was performed by obtaining the Total Waveform Oscillation (TWO) factor, which is directly related to the amount of the torque ripple:

$$TWO = \frac{\sqrt{T_{e_{RMS}}^2 - T_{e_{DC}}^2}}{\left|T_{e_{DC}}\right|} \times 100\,\% \tag{6}$$

where T_{eRMS} and T_{eDC} are the RMS and mean values of the electromagnetic torque, respectively. Other computational simulations were performed in order to obtain the motor power factor (PF) and efficiency (η) for the different considered operating conditions.

The evaluation and comparison of all considered operating conditions was performed for the same values of speed and torque. A reference speed of 1200 rpm was assumed together with a constant load torque that was equivalent to the SynRM rated value. In Table 2 are presented the SynRM parameters used in this study.

Table 2. SynRM dataplate parameters.

Parameters	Values
Power	2.2 kW
Voltage	400 V
Current	5.69 A
Speed	1500 rpm
N° of pole pairs	2
Moment of inertia	0.0018 kg.m^2

5.1 Normal Operating Conditions

The SynRM phase a current and electromagnetic torque spectrograms and their corresponding time-domain waveforms under normal operating conditions are shown in Fig. 5 (a) and (b).

As it can be seen, the SynRM phase currents are virtually sinusoidal, presenting low noise and a well-defined fundamental component. As a result, a TWD value equal to 1.59 % together with a RMS value of 5.69 A were obtained for each motor phase.

Regarding the SynRM electromagnetic torque, it can be observed that it presents only low amplitude high-frequency noise and that it is basically constant, resulting in a low TWO value equal to 1.97 %.

As far as the motor efficiency and power factor is concerned, the values of 89.67 % and 0.684, were, respectively, obtained.

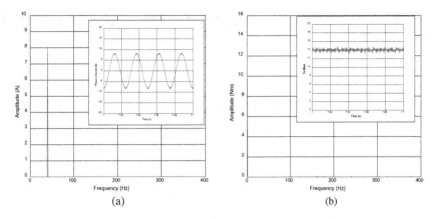

Fig. 5. Spectrogram and time-domain waveform of the SynRM: (a) phase a current; (b) electromagnetic torque (under normal operating conditions).

5.2 Single Semiconductor Open-Circuit Failure in IGBT S_{a1}/IGBT S_{a2}

With the aim to introduce semiconductor single open-circuit faults, the corresponding gate signals for the IGBT were set to zero. It is worth noting that in both cases the antiparallel diode remains connected. Furthermore, several simulations were performed and it was concluded that for these specific operating conditions, the failures in IGBT S_{a1} or IGBT S_{a2} result in very similar effects. Thus, the results will be presented for only one of these situations (fault in IGBT S_{a1}). The failures also lead to the discharge of the lower capacitor and, conversely, a charge of the upper capacitor to a value approximately twice of the initial one, which results in the loss of the three voltage levels and, at the same time, an overvoltage of healthy semiconductors.

Under these faulty conditions, the NPC power converter becomes unbalanced, resulting in different current waveforms. Figure 6 (a), (b), and (c) show these waveforms together with the corresponding spectrogram. Analyzing the data, it can be easily observed that the machine phase currents present a very distorted waveform. Consequently, the TWD values increase considerably, as shown in Table 3. Moreover the SynRM losses are negatively affected due to the existence of a large DC component and a high second order harmonic.

Table 3. RMS and TWD values for an inverter failure in IGBT S_{a1}.

Parameters	Phase a	Phase b	Phase c
RMS (A)	9.04	9.04	6.13
TWD (%)	100.73	70.75	109.02

By analyzing the RMS values, it can be also concluded that the NPC converter is operating under unbalanced conditions. Since the power switch IGBT S_{a1} is opened, and considering the system steady-state operation after the failure, the phase *a* current cannot assume positive values.

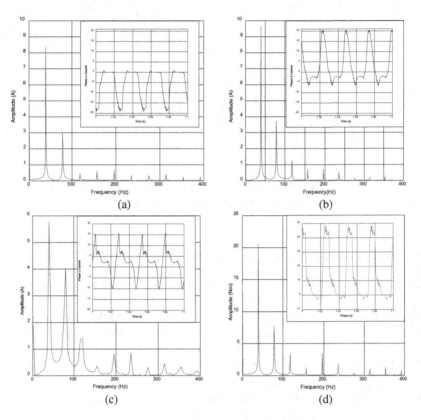

Fig. 6. Spectrogram and time-domain waveform of the SynRM: (a) phase *a* current; (b) phase *b* current; (c) phase *c* current; (d) electromagnetic torque (for an open-circuit fault in IGBT T1).

The SynRM electromagnetic torque waveform together with its corresponding spectral analysis is depicted in Fig. 6(d). These results demonstrate that the electromagnetic torque presents a strong oscillation/ripple. Contrary to the results obtained for the healthy case, the electromagnetic torque is not smooth anymore, being clearly visible high frequency harmonic components multiple of the fundamental frequency, resulting in a TWO value of 130.32 %. This fact proves that under these operating conditions, the SynRM shaft suffers strong mechanical efforts which may endanger the drive operation.

Regarding the machine efficiency and power factor, the values of 81.77 % and 0.436, were, respectively, obtained for this specific operating mode. Due to the increase of RMS values, the copper losses are also higher, resulting in a large SynRM input power and lower efficiency. On the other hand, under these circumstances the machine apparent power is also increased, resulting in a low power factor value.

6   Experimental Implementation Plan

Currently, further work is being carried out in order to build the setup for the experimental validation. A 2.2 kW 1500 rpm SynRM with similar characteristics to the one used in the simulation results will be used. A Semikron three-level NPN converter will be acquired, which already includes the power switch modules and their corresponding drive circuits and protections. The drive control algorithm will be implemented into a dSPACE DS1103 system. An incremental encoder with a resolution of 2048 pulses per turn will be used to measure the motor shaft position. The dSPACE platform provides a dedicated software that allows to build a real-time user interface that can be used to introduce converter faults by disabling the semiconductors gate pulses.

7   Conclusions

This paper has presented a performance evaluation of a SynRM variable speed drive fed by a three-level NPC converter under distinct faulty conditions. The drive healthy operation was assessed as well as the case where power converter open-circuit failures in IGBT Sa1 and IGBT Sa2 were considered. The SynRM performance assessment was based on the evaluation of some variables like the motor power factor, electromagnetic torque, efficiency, TWD, currents RMS values, and TWO values.

It has been concluded that when the three-level NPC converter presents a single semiconductor open-circuit failure, the SynRM supply currents will be negatively affected due to their large harmonic distortion. As a consequence, there is a noticeable increase of their TWD values, being observed a strong electromagnetic torque pulsating waveform. The motor efficiency and power factor are also reduced due to the increase of the RMS values of the two remaining healthy phases.

As future work, the analysis of the distinct patterns and behavior of the different variables under these abnormal operating conditions will enable the development of appropriate fault diagnostic techniques. Finally, after identifying the damaged devices, fault-tolerant remedial strategies can be then implemented in order to improve the drive's performance.

Acknowledgment.   The authors acknowledge the financial support from the Portuguese Foundation for Science and Technology (FCT) under grant no. SFRH/BD/102345/2014 and grant no. SFRH/BPD/87135/2012.

References

1. Lipo, T.A.: Synchronous reluctance machines – a viable alternative for AC drives? Electr. Mach. Power Syst. **19**, 659–671 (1991)
2. Yahia, K., Matos, D.M.B., Estima, J.O., Cardoso, A.J.M.: Modeling synchronous reluctance motors including saturation, iron losses and mechanical losses. In: Proceedings of the 22nd International Symposium on Power Electronics, Electrical Drives, Automation and Motion, Ischia, Italy, pp. 595–600 (2014)

3. Matos, D.M.B., Estima, J.O., Yahia, K., Cardoso, A.J.M.: Modeling and implementation of MTPA control strategy for SynRM variable speed drives. In: International Review of Electrical Engineering, vol.9, no. 6 (2014)
4. Matos, D.M.B. Estima, J.O., Cardoso, A.J.M.: Performance evaluation of synchronous reluctance motor drives under inverter fault conditions. In: 10th IEEE International Symposium on Diagnostics for Electric Machines, Power Electronics and Drives. Guarda (2015)
5. Franquelo, L.G., Rodríguez, J., León, J.I., Kouro, S., Portillo, R., Prats, M.M.: The age of multilevel converters arrives. IEEE Industr. Electron. Mag. **2**(2), 28–39 (2008)
6. Rodriguez, J., Bernet, S., Steimer, P.K., Lizama, I.E.: A survey on neutral-point-clamped inverters. IEEE Trans. Industr. Electron. **57**(7), 2219–2230 (2010)
7. Abu-Rub, H., Holtz, J., Rodriguez, J., Baoming, G.: Medium voltage multilevel converters – State of the art, challenges and requirements in industrial applications. IEEE Trans. Industr. Electron. **57**(8), 2581–2596 (2010)
8. Teichmann, R., Malinowski, M., Bernet, S.: Evaluation of three-level rectifiers for low-voltage utility applications. IEEE Trans. Industr. Electron. **52**(2), 471–481 (2005)
9. Teichmann, R., Bernet, S.: A comparison of three-level converters versus two-level converters for low-voltage drives, traction, and utility applications. IEEE Trans. Industr. Appl. **42**(3), 855–865 (2005)
10. Lin, L., Zou, Y., Wang, Z., Jin, H.: Modeling and control of neutral point voltage balancing problem in three-level NPC PWM inverters. In: Proceedings of 36th IEEE PESC, Recife, Brazil, pp. 861–866 (2005)
11. Holtz, J., Oikonomou, N.: Neutral point potential balancing algorithm at low modulation index for three-level inverter medium voltage drives. IEEE Trans. Industr. Appl. **43**(3), 761–768 (2007)
12. Park, J.-J., Kim, T.-J., Hyun, D.-S.: Study of neutral point potential variation for three-level NPC inverter under fault condition. In: Proceedings of the 34th IEEE IECON, pp. 983–988 (2008)
13. Ceballos, S., Pou, J., Robles, E., Gabiola, I., Zaragoza, J., Villate, J.L., Boroyevich, D.: Three-level converter topologies with switch breakdown fault-tolerance capability. IEEE Trans. Industr. Electron. **55**(3), 982–995 (2008)
14. Li, S., Xu, L.: Strategies of fault tolerant operation for three-level PWM inverters. IEEE Trans. Industr. Electron. **21**(4), 933–940 (2006)
15. Li, J., Huang, A.Q., Liang, Z., Bhattacharya, S.: Analysis and design of active NPC (ANPC) inverters for fault tolerant operation of high-power electrical drives. IEEE Trans. Industr. Electron. **27**, 519–533 (2012)
16. Ceballos, S., Pou, J., Robles, E., Zaragoza, J., Martin, J.L.: Performance evaluation of fault-tolerant neutral-point-clamped converters. IEEE Trans. Industr. Electron. **57**(8), 2709–2718 (2010)
17. Wu, B.: High-Power Converters and AC Drives. IEEE Press/Wiley Interscience, Hoboken (2006)

Analysis of Lift Control System Strategies Under Uneven Flow of Passengers

Kyaw Kyaw Lin[1], Sergey Lupin[1], and Yuriy Vagapov[2(✉)]

[1] National Research University of Electronic Technology, Zelenograd, Moscow 124498, Russia
u002020@edu.miet.ru, lupin@miee.ru
[2] Glyndwr University, Plas Coch, Mold Road, Wrexham, LL11 2AW, UK
y.vagapov@glyndwr.ac.uk

Abstract. Modern embedded microcontrollers used in lift automation can provide the system operation under both static and dynamic strategies. The control system operating under dynamic strategy can change the lift system controls under different passenger behaviours, varying to improve the efficiency of service. However, it is extremely challenging to determine the system conditions leading to the decision to switch the control strategies. The present paper investigates the influence of the passengers' flow parameters on lift control systems strategies using a hybrid mathematical model combining agent-based and discrete event methods. It has been shown that the operation of the lift control systems, in the context of decision leading conditions, can be effectively assessed through a model analysis. The proposed model has been simulated using AnyLogic systems to estimate the impact of uneven flow of passengers on performance of a ten floor lift system. The results of simulations have been used to determine the optimum intervals between the calls and lift arrival time which lead to strategies that minimise passenger's waiting time.

Keywords: Lift control · Dynamic control · AnyLogic · Lift modelling

1 Introduction

Due to rapid development of microelectronics, modern engineering control systems including lift control utilise specialised microcontrollers [1, 2] or universal microprocessors [3] replacing analogue or relay control circuits dominated in the past few decades. Implementation of digitally based controllers significantly expanded computational resource of control systems and range of their control algorithms while reducing cost of system design.

Reengineering of lift control systems is often based on a simple reproduction of old control circuit algorithm using a digital control. This approach does not use advanced features and functions available in the digital platform. Alternatively, intensive employment of computational power of modern microcontrollers can make control more complex and "intelligent". However, the intelligent control requires more information about the controlled object in order to bring new features into control system. Therefore, the systems should be modernised by installation of additional sensors [4].

© IFIP International Federation for Information Processing 2016
Published by Springer International Publishing Switzerland 2016. All Rights Reserved
L. Camarinha-Matos et al. (Eds.): DoCEIS 2016, IFIP AICT 470, pp. 217–225, 2016.
DOI: 10.1007/978-3-319-31165-4_22

2   Lift as Cyber-Physical System

Kim and Kumar [5] defined cyber-physical system (CPS) as an "engineered system in which computing, communication, and control technologies are tightly integrated." This definition covers a wide range of engineering systems including lifts.

In lift systems, electro-mechanical elements represent the physical component of CPS definition whereas lift control is related to the cyber aspect of CPS. Modern lift control comprises a variety of computational technologies providing effective and efficient operation of electro-mechanical elements. Passengers are able to interact to a lift control system using simple communication (buttons) and sensors. The service provided by a lift is actually physical vertical motion of passengers in up and down directions. Therefore the research in the area of lift systems including analysis, modelling and optimisation of lift control algorithms is directly linked to CPS [6].

3   Lift System

A conventional lift control provided operation of the system under static algorithm. It means that the control strategy is stable or unchanged during the system operation. However, modern digital controllers used to automate lift system can provide the operation under both static and dynamic control strategies. The lift control system operating under dynamic strategy can change the system behaviour if the passengers flow rate is vary [7]. In order to provide a dynamic strategy it is extremely challenging to determine the system condition leading to the decision to switch the control algorithm.

If lift is considered as a queuing system, the time of passenger service T_{serv} can be represented as:

$$T_{serv} = T_{dis} + T_{wait} + T_{mov} \tag{1}$$

where T_{dis} is the time of passenger dislocation from flat to lift's hall, T_{wait} is the passenger waiting time of lift arrival, and T_{mov} is the time of passenger carrying to a destination floor. Therefore, in terms of lift service efficiency, the primary task of lift system is to minimise T_{serv}.

Architectural design of lift systems also aims to reduce T_{serv}. It has been observed that the optimal lift allocation in a building affects on the service time reduction. Goetschalckx and Irohara [8] present an algorithm to choose the optimal floor topology based on minimising the vertical and horizontal motions (T_{dis}) of passengers. Matsuzaki et al. [9] suggest the genetic algorithm for optimisation of the quantity and capacity of lifts using the same approach. Markos and Dentsoras [10] discuss an approach to determine the location of passenger lifts in a commercial building.

Many lift manufactories offer various effective solutions aimed to decrease the time T_{mov}. Company Otis [11], one of the world leaders in the lift industry, proposes models that can operate at a speed faster than 2.54 m/s. However, even so high speed will not significantly decrease the time T_{serv} without effective management strategies ensuring the reaction of control system on the passengers' queries flow rate [12].

4   Lift Control Strategies

To reduce the number of parameters affecting the efficiency of control the system analysis employed single strategies. The list of available single lift control strategies is given below:

- Simple management (SM);
- Unilateral collective control with one button (up) (U1U);
- Unilateral collective control with one button (down) (U1D);
- Unilateral collective control with two buttons (up) (U2U);
- Unilateral collective control with two buttons (down) (U2D);
- Bilateral collective control with one button (B1);
- Bilateral collective control with two buttons (B2).

The analysis is focused on lift control strategies operating under uneven flow of passengers. Most of people are often following a scheduled behaviour which generates variable passenger flow rate in a residential building. If the daylight distribution of passengers flow is known or determined using various sensors then that can be used to change the control strategies in order to increase the efficiency of lift control system.

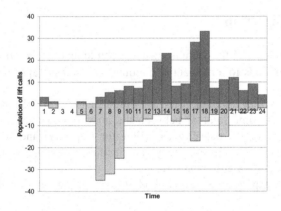

Fig. 1. Population of lift calls versus time in a residential building for a day (24 h). Blue chart is the calls to move the lift cabin up; green chart is the calls to move the lift cabin down (Colour figure online).

Figure 1 shows a typical example of number of calls versus time in a residential building. The positive values are corresponding to "up" calls while the negative values are corresponding to "down" calls. It can be seen that during the morning peak time the majority of calls are "down" calls whereas the majorly of "up" calls belong to evening time [13].

5   Lift Modelling

AnyLogic system [14] was used for modelling and simulation of the lift system. The suggested model combines agent-based and discrete event methods and has been created

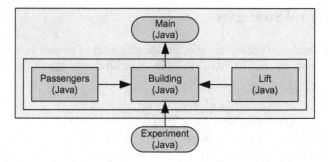

Fig. 2. Structure of the lift system model in AnyLogic having five classes: Passengers, Building, Lift, Experiment and Main.

using an object-oriented programming technology. The model contains five classes: *Building*, *Passengers*, *Lift*, *Experiment* and *Main* as shown in Fig. 2. Agent-based behavior of the model is simulated using state diagrams (statecharts). Statechars of the classes describe performance of agents in accordance of the control algorithms [15].

6   Model Simulation

The model implements a lift system in a ten-floor building ($n_{floor} = 10$). The flow rate of passengers corresponds to a residential building – passengers appear on the floors at equal probability.

Time parameters in simulation are measured in sec. Variable parameters in the experiment are: the time interval between the arrivals of passengers *IBP* (Interval Between Passengers), lift capacity *EC* (Elevator Capacity) and lift efficiency *EE* (Elevator Efficiency). The lift efficiency is defined as an average value of occupancy rate of the lift cabin during moving time.

The following expression is used to find the average waiting time during the simulations:

$$T_{wait} + T_{mov} = \frac{1}{N} \sum_{i=1}^{N} \left(t_{out}^i - t_{inp}^i \right) \qquad (2)$$

where t_{inp} is arrival time of the passenger in the call floor; t_{out} is the time when the passenger leaves the lift on the destination floor, and N is a quantity of passengers.

In the analysed examples the lift stop time (T_{stop}) was equal to 5 s and the time of moving between floors (T_{floor}) equals 3 s.

Let's estimate average value of T_{wait} for a low rate down stream of passengers under U1D control strategy. We consider that after service of current passenger, lift will stay on the first floor.

$$T_{wait}^{down} = \frac{T_{floor} \left(n_{floor} - 1 \right)}{2} + T_{stop} = \frac{3 \left(10 - 1 \right)}{2} + 5 = 18.5 \text{ s} \qquad (3)$$

The waiting time T_{wait} can be adjusted using knowledge of passengers flow peaks where control algorithm has to be changed according a new behaviour of lift system.

In the case of downstream scenario having morning peak the waiting time T_{wait} can be decreased due to the following control strategy: after servicing of a querying passenger, lift does not stay on the first floor; it will move in a position in the middle of building – new park zone. For ten floors building it corresponds to sixth floor ($n_{floor} = 5$ for the equation below). In this case the waiting time equals:

$$T_{wait}^{down} = \frac{T_{floor}\left(n_{floor} - 1\right)}{2} + T_{stop} = \frac{3\,(5 - 1)}{2} + 5 = 11\,\text{s} \tag{4}$$

This control strategy brings the benefits of 7.5 s comparing to (3). Let's call such control strategy as "down modification".

Similar estimation of T_{wait} can be used for a low rate upstream of passengers and U1D control strategy. After service of a passenger, lift will stay on the destination floor:

$$T_{wait}^{up} = \frac{T_{floor}\left(n_{floor} - 1\right)}{2} + T_{stop} = \frac{3\,(10 - 1)}{2} + 5 = 18.5\,\text{s} \tag{5}$$

In the case of upstream scenario having evening peak of passenger flow the waiting tame T_{wait} can be decreased using the following control strategy: after the servicing of a querying passenger, lift does not stay at the destination floor; it returns to the first (ground) floor. In this case the waiting time equals:

$$T_{wait}^{up} = 0 \tag{6}$$

It brings the benefits of 18.5 s comparing to (5). This control strategy is called "up modification".

Let's estimate T_{wait} for low rate mixed stream of passengers with equal probability of up and down directions queries.

$$T_{wait} = (T_{wait}^{up} + T_{wait}^{down})/2 \tag{7}$$

For "down modification":

$$T_{wait} = (18.5 + 11)/2 = 14.75\,\text{s} \tag{8}$$

For "up modification":

$$T_{wait} = (0 + 18.5)/2 = 9.25\,\text{s} \tag{9}$$

The value of T_{wait} for not modified U1D strategy under up and down streams of passengers is exactly the same:

$$T_{wait}^{down} = T_{wait}^{up} = 18.5\,\text{s} \tag{10}$$

In previous work [16] authors compared the efficiency of various strategies while this work aims to assess the impact of accuracy in determination of the peak of passenger flow. Figure 3 presents six versions of U1D lift control strategy (B–G) widely used in civil buildings. Each version includes up and down modifications acting in accordance to the shown schedule. The version C provides operation of up and down modifications within proposed upper and down peaks of passengers flow. It means that the lift control system has determined the peaks of uneven flow of passengers correctly. The version B simply switches the strategies every 12 h according to the passenger's stream, while the versions D–G provide the same operation as the version C with 30 min time shift. The versions D–G simulate low accuracy determination of the flow peaks switching the modified control strategies earlier or later.

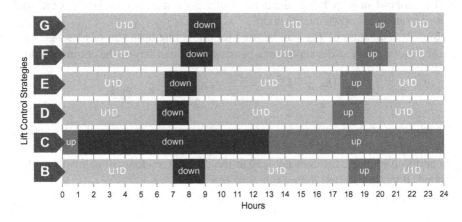

Fig. 3. The schedule of modified lift control strategies. U1D is the unilateral collective control with one button (down) lift control strategy, up is "up modification" of U1D control strategy, down is "down modification" of U1D control strategy.

In order to conduct the simulation IBP was varying to change the passenger flow rate. The parameters chosen for simulation are: IBPpeak = 30 s, IBPnormal = 210 s; down peak has been predicted from 7 to 9 h, upper peak – from 18 to 20 h.

Table 1 presents results of simulation of control strategies U1D (not modified) and versions B and C in terms of lift cabin capacity (column EC). It can be seen that for all experiments versions B and C show better waiting time Twait comparing to U1D, while version B has better lift efficiency (EE) compared to C.

Table 2 shows the results of simulation of all analysed strategies. Since the strategies C-D simulate accuracy of peak flow determination the results demonstrate the influence of accuracy of peak detection on the lift efficiency and the waiting time. It can be seen that the mistakes in peak time determination bring stronger influence on the time waiting but not on the elevator efficiency.

Table 1. Results of simulation of control strategies U1D and versions B and C.

EC	U1D		B		C	
	T_{wait}	EE	T_{wait}	EE	T_{wait}	EE
$IBP_{peak} = 30$ s, $IBP_{normal} = 210$ s						
2	13.4 s	0.30	11.4 s	0.28	10.3 s	0.24
3	13.1 s	0.20	11.2 s	0.18	10.3 s	0.16
4	12.9 s	0.15	11.4 s	0.14	10.7 s	0.12
$IBP_{peak} = 30$ s, $IBP_{normal} = 150$ s						
2	12.9 s	0.3	10.9 s	0.28	10.1 s	0.24
3	13 s	0.20	11.7 s	0.19	10.7 s	0.16
4	12.7 s	0.15	11.8 s	0.14	10.7 s	0.12
$IBP_{peak} = 30$ s, $IBP_{normal} = 90$ s						
2	11.8 s	0.30	10.5 s	0.28	9.8 s	0.24
3	12.1 s	0.20	11 s	0.19	9.3 s	0.16
4	12.1 s	0.15	11.5 s	0.14	9.2 s	0.12
$IBP_{peak} = 40$ s, $IBP_{normal} = 160$ s						
2	12.2 s	0.28	8.1 s	0.27	7.7 s	0.22
3	11.7 s	0.19	9 s	0.18	7.9 s	0.15
4	12.1 s	0.14	8.9 s	0.13	7.9 s	0.11
$IBP_{peak} = 50$ s, $IBP_{normal} = 150$ s						
2	10.8 s	0.28	7.9 s	0.26	7.7	0.22
3	11.1 s	0.19	7.7 s	0.17	7.7	0.15
4	11.1 s	0.14	7.8 s	0.13	7.3	0.11

Table 2. Results of simulation of control strategies B - G. ($IBP_{pick} = 30$ s, $IBP_{normal} = 210$ s)

EC	B		D		E		F		G	
	T_{wait}	EE	T_{wait}	EE	T_{wait}	EE	T_{wait}	EE	T_{wait}	EE
2	11.6 s	0.28	12.5 s	0.29	12.5 s	0.29	12.4 s	0.28	13.2 s	0.29
3	11.3 s	0.18	12.8 s	0.19	12.1 s	0.19	12.3 s	0.19	12.7 s	0.19
4	11.3 s	0.14	12.4 s	0.14	11.8 s	0.14	12.7 s	0.14	13.3 s	0.15

7 Conclusion

It has been shown that the agent-based modelling in AnyLogic system can be effectively used for assessment of the impact of changes in control strategies on the lift operation. The modelling and simulation can provide the solution of the time interval between the calls led to switching of the control strategies in order to minimise passenger's waiting time. The variable passenger flow rate obtained from the simulation can be used as an input parameter for analysis of various lift control strategies.

References

1. Yang, X., Zhu, Q., Xu, H.: Design and practice of an elevator control system based on PLC. In: IEEE Workshop on Power Electronics and Intelligent Transportation System, pp. 94–99. IEEE Press, New York (2008)
2. Bernard, P., Deaconu, I.-D., Popescu, S.V., Ghita, C., Deaconu, A.-S., Chirila, A.-I.: PLC controlled elevator drive system. In: 9th International Symposium on Advanced Topics in Electrical Engineering, pp. 166–169. IEEE Press, New York (2015)
3. Munoz, D.M., Llanos, C.H., Ayala-Rincon, M., van Els, R.H.: FPGA implementation of dispatching algorithms for local control of elevator systems. In: IEEE International Symposium on Industrial Electronics, pp. 1997–2002. IEEE Press, New York (2008)
4. Kwon, O., Lee, E., Bahn, H.: Sensor-aware elevator scheduling for smart building environments. Build. Environ. **72**, 332–342 (2014)
5. Kim, K.-D., Kumar, P.R.: Cyber-physical systems: a perspective at the centennial. Proc. IEEE **100**, 1278–1308 (2012)
6. Klein, R., Xie, J., Usov, A.: Complex events and actions to control cyber-physical systems. In: 5th ACM International Conference on Distributed Event-based System, pp. 29–38. ASM, New York (2011)
7. Irmak, E., Colak, I., Kaplan, O., Kose, A.: Development of a real time monitoring and control system for plc based elevator. In: 14th European Conference on Power Electronics and Applications, pp. 1–8. IEEE Press, New York (2011)
8. Goetschalckx, M., Irohara, T.: Formulations and optimal solution algorithms for the multi-floor layout problem with elevators. In: IIE Annual Conference and Expo: Industrial Engineering's Critical Role in a Flat World, pp. 1446–1452. Institute of Industrial Engineers, Norcross (2007)
9. Matsuzaki, K., Irohara, T., Yoshimoto, K.: Heuristic algorithm to solve the multi-floor layout problem with the consideration of elevator utilization. Comput. Ind. Eng. **36**(2), 487–502 (1999)
10. Markos, P., Dentsoras, A.: Floor circulation index and optimal positioning of elevator hoistways. In: Setchi, R., Jordanov, I., Howlett, R.J., Jain, L.C. (eds.) KES 2010, Part II. LNCS, vol. 6277, pp. 331–340. Springer, Heidelberg (2010)
11. Otis. http://www.otisworldwide.com
12. Ding, B., Zhang, Y.-M., Peng, X.-Y., Li, Q.-C., Tang, H.-T.: A hybrid approach for the analysis and prediction of elevator passenger flow in an office building. Autom. in Constr. **35**, 69–78 (2013)
13. Fernandez, J.R., Cortes, P.: A survey of elevator group control systems for vertical transportation: a look at recent literature. IEEE Control Syst. **35**(4), 38–55 (2015)
14. AnyLogic. http://www.xjtek.com

15. Huang, Y.-S., Chen, J.-R., Lee, S.-S., Weng, Y.-S.: Design of elevator control systems using statecharts. In: 10th IEEE International Conference on Networking, Sensing and Control, pp. 322–327. IEEE Press, New York (2013)
16. Lupin, S., Lin, K.K., Davydova, A., Vagapov, Y.: The impact of sensors' implementation on lift control system. In: IEEE East-West Design and Test Symposium EWDTS 2014, pp. 217–219. IEEE Press, New York (2014)

Scalar Variable Speed Motor Control for Traction Systems with Torque and Field Orientation Filter

Paulo Mendonça[1] and Duarte M. Sousa[1,2(✉)]

[1] DEEC, Instituto Superior Técnico, Technical University of Lisbon, Lisbon, Portugal
{paulo.mendonca,duarte.sousa}@tecnico.ulisboa.pt
[2] INESC-ID, Lisbon, Portugal

Abstract. Scalar traction control systems can be used in trains, trams and electrical vehicles equipped with induction motors. To increase the versatility and efficiency of such systems compared to conventional solutions, they must enable dynamic links to different types of power grids, i.e., be supplied by multi-frequency and multi-voltage power sources. Behind these solutions, there are control systems based on vectorial controllers. In order to set systems with the above features, in this work it is intended to build a scalar traction control system, in which the speed is controlled via a scalar controller. The system controls the motor speed and also the position of the rotor and stator fields. Furthermore, it allows to clearly setting the operation conditions, i.e., avoiding situations that would change from braking to traction and vice-versa due to operational and/or functional disturbances. To achieve the required speed reference follow up, the proposed solution includes a torque and field orientation filter.

Keywords: Variable speed · Motor control · Electric traction

1 Introduction

Nowadays are commercialized several types of electrical vehicles, both road and rail vehicles, but in the near future it is expected that even more electrical vehicles will be manufactured. Besides the electrical vehicles, there are several industrial systems and solutions based on different types of electrical motors. The main purpose of this is to build a command variable speed system, for a rail traction system based on an induction motor, in which the speed is controlled instead of the torque and the controller is scalar and not vectorial. In addition, the proposed system allows avoiding undesirable situations, as for instance, changing from braking to traction and vice-versa due to operational and/or functional disturbances.

To build this system, which is a part of a larger traction system that is being developed in a PhD thesis, it was used as core systems a three arm inverter, a triphasic induction motor and a constant DC voltage source [1]. The inverter uses IGBT semiconductors, and they are controlled via a PWM SVM controller, which can be disenabled by the main control system. This drive system must be adaptable to the different power grids used in the railway sector, i.e., be designed considering the requirements of multi-frequency and multi-voltage power supplies. Technically, should also constitute a step

L. Camarinha-Matos et al. (Eds.): DoCEIS 2016, IFIP AICT 470, pp. 226–234, 2016.
DOI: 10.1007/978-3-319-31165-4_23

forward of other requirements and features, such as energy efficiency, energy recovery, service interruptions, interoperability of transport networks, disturbance and failures in power systems, for instance. The solution to synthesize be an energy efficient and versatile solution in the face of conventional solutions, namely dynamically adapt to different feeding systems (multi-voltage and multi-frequency). This Ph.D. thesis has the objective of building energy efficient and versatile multi voltage traction system for a rail vehicle when compared to conventional solutions.

The main control system controls the voltage input of the induction motor. The direct voltage is kept always at zero, and the quadrature voltage module is regulated according to the speed, via a hysteretic controller and a proportional controller. The control angle is calculated and set according to the desired speed.

In many speed control systems, it is usual that the fluxes of the stator and rotor are, sometimes, in undesirable positions [2–6]. In these cases, the motor breaks when it is submitted to a traction effort. In this paper, this situation is analyzed and presented a solution, by predicting the position of the two vector fields and not allowing undesirable flux positions. This way the fluxes are always in the correct position and this flux filter is combined with a torque limitation filter. With this type of solution it is not only guaranteed a better controller reference follow up, but also, energy savings.

For testing and validating the proposed solution, a model of the system was simulated [2]. The main results illustrating the operation principles and the added value of the proposed solution are presented and discussed in this paper.

2   Contribution to Cyber-Physical Systems

Over the past decades, the operation of the railways have been changed in response to technological, social and economic challenges. As push factors of these changes, it is important to highlight the increasing number of rail operators sharing the same infrastructure, the advent of the smart grids and the adoption of cyber solutions. In this context, apart from the interoperability, security and safety of electric traction systems, remote control and cyber monitoring of railway operation are topics that have to be investigated and developed. To obtain efficient solutions that address these challenges have also to be made on valid cybernetic solutions.

It should be also referred that the operation of trams and of light rail vehicles can be monitored remotely in real time. These systems should also be able to optimize its operation, which can be set according to traffic or weather conditions, for instance. In addition, these systems have potential to incorporate new features, as for instance, acting as a backup energy storage system. Under this context, in this Ph.D. work is described an approach that intends to contribute to control the speed of these type of electric vehicles. The proposed approach is integrated in a more complex solution that allows moving up the interaction of urban rail systems with smart electrical grids [1].

Considering the operating principle of the proposed solution, the main contribution of this work to "Technological Innovation for Cyber-Physical Systems" consist on the development of a scalar variable speed motor control for traction systems that will allow implementing a full and remote automated driving.

3   System Description

3.1   Power System

The design used for the main power electronics system was a three arm inverted feed by a 600 V DC power source, which corresponds to the feeding voltage used in the Lisbon tramways. The inverter uses six IGBT semiconductors with parallel diodes [7, 8].

The inverter feeds a 100 horsepower induction motor, which is a typical traction motor power in light railway.

The mathematical model used is the same as the one implemented in the simulation model. The induction motor electrical model is simulated based on the following equations:

$$
\begin{aligned}
V_{qs} &= R_s i_{qs} + d\varphi_{qs}/dt + \omega\varphi_{ds} \\
V_{ds} &= R_s i_{qs} + d\varphi_{ds}/dt - \omega\varphi_{qs} \\
V'_{qr} &= R'_r i'_{qr} + d\varphi'_{qr}/dt + \left(\omega - \omega_r\right)\varphi'_{ds} \\
V'_{dr} &= R'_r i'_{dr} + d\varphi'_{dr}/dt - \left(\omega - \omega_r\right)\varphi'_{qs}
\end{aligned}
\tag{1}
$$

Equation system (1), describes the voltage model in dq0 coordinates, where ω is the reference frame angular velocity and ω_r is the electrical angular velocity.

Equation system (2) is used to describe the torque behavior.

$$
T_e = 1.5p\left(\varphi_{ds} i_{qs} - \varphi_{qs} i_{ds}\right)
\tag{2}
$$

Equation system (3) is used to describe the flux model.

$$
\begin{aligned}
\varphi_{qs} &= L_s i_{qs} + L_m i'_{qr} \\
\varphi_{ds} &= L_s i_{ds} + L_m i'_{dr} \\
\varphi'_{qr} &= L'_r i'_{qr} + L_m i_{qs} \\
\varphi'_{dr} &= L'_r i'_{dr} + L_m i_{ds}
\end{aligned}
\tag{3}
$$

Equation system (4) describes the inductance relationship.

$$
\begin{aligned}
L_s &= L_{ls} + L_m \\
L'_r &= L'_{lr} + L_m
\end{aligned}
\tag{4}
$$

This model will give the dq fluxes components for the field orientation filter.

3.2   Main Controller

The main controller sets the direct voltage to zero and controls the quadrature voltage according to the desirable speed. The angle is calculated according to the desired speed, according to Eq. (5). The voltage provided by the controller is converted into gate pulses trough a space vector controller.

Together with the main controller, the system has also two filters that are described in the following subsection. These two systems are responsible for limiting the torque, not allowing torques greater than 1000 Nm in the motor, and fluxes in undesirable positions.

To control the motor speed, it is assumed that the motor speed is measured by a sensor, which output is used as feedback in the control chain (Fig. 1). The real speed is compared with the reference signal and multiplied by a gain. This result is limited between 500 V and −500 V, which are the desirable limits to the voltage applied to the motor. In practice, this is a proportional controller that controls the quadrature voltage module. The direct voltage is kept always at zero.

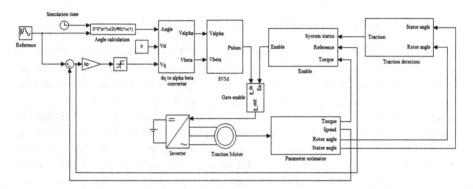

Fig. 1. Model of the proposed system

The voltage angle θ is calculated according to Eq. (5).

$$\theta = P_P \left(2\pi \frac{t}{60} \right) N_{ref} \tag{5}$$

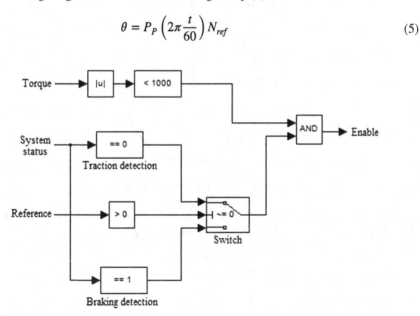

Fig. 2. Model of the "*Gate Enable*" block

In Eq. (5), P_P is the number of pole pairs in the machine, which is two in this case, t is the time and N_{ref} is the reference speed.

The block "*dq to alfa beta converter*", converts the voltage in dq0 coordinates to $\alpha\beta$ coordinates. The block "*SVM*" converts the $\alpha\beta$ voltage components to PWM signals using a space vector modulator controller. So that, a control of the machine supply voltage is performed. The PWM signals command the six IGBT gates via a "*Gate Enable*" block, shown in Fig. 2, which is responsible for enabling and disenabling the command signals. This block is commanded by the "Enable" block, so when the enable signal is 0, all gates commands are put to 0 also.

3.3   Torque and Field Orientation Filter

This filter prevents the torque to be greater than 1000 Nm. Every time that the torque exceeds this value the gate signals are disenabled, which means in practice that all signals are set to zero. With this filter it is implemented a supervision of the torque avoiding to reach high torque values and thus preventing big current and torque surges.

The positions of the fluxes, the torque and the speed are calculated within the "*parameter estimator*" block, according to Eqs. (1), (2), (3) and (4). It is important to refer that the accuracy of this model and of the prediction implemented depends on the knowledge of the motor parameters, as for instance, the real inductance and inductor resistance. The angle values are inputs of the "*Traction Detection*" block that predicts if the motor is braking or in traction. This filter contributes for a best energy usage and better reference follow up. Figure 3 shows the flux position and its effect.

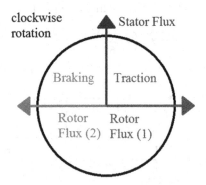

Fig. 3.   Fluxes positioning and effect

Inside the "*traction detection*" block is the equation set (6) and (7). If the statement (6) OR (7) = TRUE is valid, than the system is braking.

$$\begin{cases} \varphi_r + \pi \geq \varphi_s \\ \varphi_s \geq \varphi_r \\ \varphi_r \geq 0 \\ \varphi_r \leq \pi \end{cases} \tag{6}$$

$$\begin{cases} \varphi_s \geq \varphi_r \\ \varphi_r \leq 0 \\ \varphi_r \geq \pi \end{cases} \tag{7}$$

After the selection had been made, than the enable block decides to enable or disenable the gate command block according to Fig. 4.

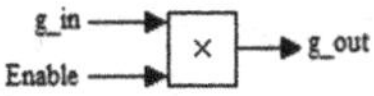

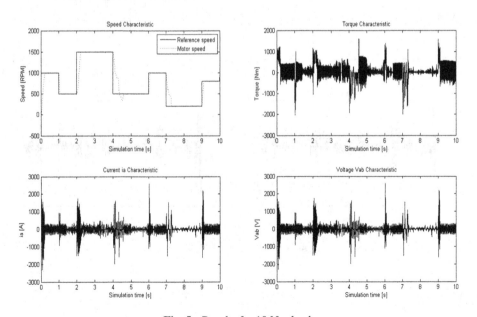

Fig. 4. Model of the *"Enable"* block

4 Results

In order to validate the proposed solution the system was tested by simulation with variable speed steps and loads, i.e., submitting the motor to different torques. It was tested the influence of the flux and torque filter, as well. The filter is responsible for not allowing unwanted flux positions and torques higher than 1000 Nm.

In the Fig. 5 it can be observer a simulated result using a 10 Nm load, Fig. 6 has the same simulation using a 100 Nm load, Fig. 7 has a 300 Nm load and simulation Fig. 8 has a 100 Nm load and has no filter applied.

Fig. 5. Results for 10 Nm load.

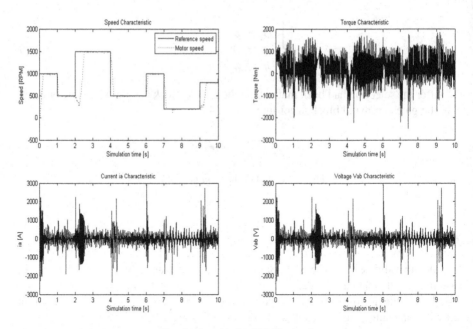

Fig. 6. Results for 100 Nm load.

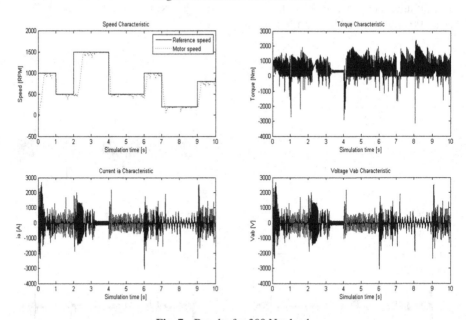

Fig. 7. Results for 300 Nm load.

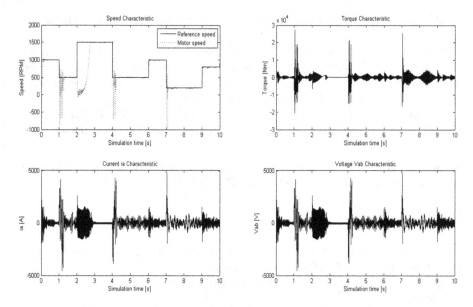

Fig. 8. Results for 100 Nm load, without using the filters described in subchapter 3.3.

In Figs. 5, 6, 7 and 8, it can be observed a good system follow up of the reference speed and without any static error. At higher speeds it can be observed some ripple and transients during some speed transitions. This can be due to simulation errors or a controller unable of guarantee every time smooth transitions. As it can be observed, at higher speeds the ripple decreases due to the machine characteristics, as expected, since the better performance of the system will be achieved in this speed range. All of this ripple and transients will be reduced inside the rail vehicle car, because the transmission has a 1:6 reduction factor, in most light rail vehicles, and mechanic filters like silent blocks and other shock absorbing systems.

5   Conclusions and Further Work

In this work it was made a speed controller for rail traction, using a scalar controller. With the proposed system, it is achieved an effective control of the speed and, at the same time, the control system rejects states where the stator and rotor fluxes are in a traction position when the vehicle is in braking and states where the fluxes are in braking position when the vehicle is in traction. In addiction the system prevents high levels of torque. To implement this option, filters are used constituting an important feature of this system, giving that the filters are compatible with a scalar control system and allow to achieve the required performance. This system is an alternative to the field orientation systems available and widely used.

For testing the results consistency, laboratorial test will be performed for the proposed system. It is expected to confirm the transient states and the load ripple observed in the simulations, since in a real rail vehicle there are mechanic filters like silent blocks and other shock absorbing systems that contribute to overcome these technical issues.

Considering the results obtained and the characteristics of the proposed system, being the speed control robust, in the future it is expected to implement a full automated driving, although it can be conventionally driven. It has the advantage of being energy efficient when it blocks high torques and undesirable flux positioning.

As further work, an interesting development would be to build a similar controller using the same concept, but instead of controlling the speed, controlling the torque.

Acknowledgements. This work was supported by national funds through FCT – Fundação para a Ciência e a Tecnologia, under project UID/CEC/50021/2013.

References

1. Mendonça, P., Sousa, D.M., Silva, J.F., Pinto, S.: An approach to recover braking energy of a tram. In: IEEE PEDG2014, Galway-Ireland (2014)
2. Kenny, B.H., Lorenz, R.D.: Stator- and rotor-flux-based deadbeat direct torque control of induction machines. IEEE Trans. Ind. Appl. **39**(4), 1093–1101 (2003)
3. Marques, G.D., Sousa, D.M.: Sensorless direct slip position estimator of a DFIM based on the air gap pq vector—sensitivity study. IEEE Trans. Industr. Electron. **60**(6), 2442–2450 (2013)
4. Marques, G.D., Sousa, D.M., Iacchetti, M.F.: An open-loop sensorless slip position estimator of a DFIM based on air-gap active power calculations—sensitivity study. IEEE Trans. Energy Convers. **28**(4), 959–968 (2013)
5. Iacchetti, M.F., Marques, G.D., Perini, R., Sousa, D.M.: Stator inductance self-tuning in an air-gap-power-vector-based observer for the sensorless control of doubly fed induction machines. IEEE Trans. Industr. Electron. **61**(1), 139–148 (2014)
6. Marques, G.D., Sousa, D.M., Iacchetti, M.F.: Air-gap power-based sensorless control in a DFIG connected to a DC link. IEEE Trans. Energy Convers. **30**(1), 367–375 (2015)
7. Sun, J.: Dynamics and Control of Switched Electronic Systems. Advanced Perspectives for Modeling Simulation and Control of Power Converters. Springer, London (2012). Chap. 2, Vasca, F., Lannelli, L. (eds.). ISBN 978-1-4471-2884-7
8. José Fernando, S.: Electrónica Industrial, Fundação Calouste Gulbenkian, Instituto Superior Técnico (1998)

Optimization and Decision Support

Variation-Aware Optimisation for Reconfigurable Cyber-Physical Systems

Rui Policarpo Duarte[1(✉)] and Christos-Savvas Bouganis[2]

[1] Universidade Autónoma de Lisboa, 1050-293 Lisbon, Portugal
rpduarte@ual.pt
[2] Department of Electrical and Electronic Engineering,
Imperial College London, London, SW7 2AZ, UK
ccb98@imperial.ac.uk

Abstract. Cyber-Physical Systems are present in many industries such as aerospace, automotive, health-care and transportation, and over time they have become critical and require high levels of resiliency and fault tolerance. Often they are implemented on reconfigurable logic due to IP design reutilisation, high-performance, and low-cost. Nevertheless, the continuous technology shrinking and the increasing demand for systems that operate under different power profiles with high-performance has led to implementations operating below the maximum performance offered by a particular technology. Design tools are conservative in the estimation of the maximum performance that can be achieved by a design when placed on a device, accounting for any variability in the fabrication process of the device. This work takes a new view on the performance improvement of circuit designs by pushing them into the error prone regime, as defined by the synthesis tools, and by investigating methodologies that reduce the impact of timing errors at the output of the system. In this work two novel error reduction techniques are proposed to address this problem. One is based on reduced-precision redundancy and the other on an error optimisation framework that uses information from a prior characterisation of the device. Both of these methods allow to achieve graceful degradation in performance whilst variation increases.

Keywords: FPGA · Error minimisation · Over-clocking · Reduced-precision redundancy · Bayesian optimisation

1 Introduction

The constant fabrication process scaling has led to devices operating faster, consuming less power, but with increased variability in their fabrication. Hence, transistors are not created equally on the same device. This negatively impacts the performance of the device. Besides variations introduced by the physical constraints, transistors are also affected by other parameters such as voltage and temperature. Therefore, the newly fabricated devices are more susceptible to such variations [1] as the technology is further reduced. Modern devices are limited by their worst performing transistor for a given family of devices.

© IFIP International Federation for Information Processing 2016
Published by Springer International Publishing Switzerland 2016. All Rights Reserved
L. Camarinha-Matos et al. (Eds.): DoCEIS 2016, IFIP AICT 470, pp. 237–252, 2016.
DOI: 10.1007/978-3-319-31165-4_24

To guarantee error-free operation of the implemented designs, once synthesised, the synthesis tools are prudent in determining the maximum clock frequency of digital circuits for a set of devices to operate on an error-free regime. As such, there is a significant gap between the maximum clock frequency reported by the models used in synthesis tools, and the actual maximum clock frequency that the actual device can operate, where the circuit will be placed on.

This work investigates mechanisms to increase the throughput of arithmetic units on Field-Programmable Gate Arrays (FPGAs) under Process-Voltage-Temperature (PVT) variation without changing the algorithm being implemented, while investigating the tradeoff in throughput, circuit area and timing errors.

Many times Cyber-Physical Systems (CPSs) are implemented on FPGAs because of the advantages they offer as high-performance, low-power, and highly specialised embedded blocks. Moreover, FPGAs are well positioned to tackle the aforementioned research problems because of their reconfigurability capabilities which is essential for the characterisation process that no other competitive technologies offer.

Examples of candidate applications for such CPSs that demand real-time performance, under different operating conditions (i.e. high-performance and low-power) are: Synthetic Aperture Radar (SAR) [2] for real-time, or high-performance; and medical capsule robots [3], and Electroencephalogram (EEG) [4] for low-power.

2 Contribution to Cyber-Physical Systems

CPSs can be seen as the combination of control theory and computer engineering, leading to control and computing co-design. They are usually comprised of sensors, actuators, and (feedback) controllers. The complexity of modern CPSs requires the usage of computational platforms to process large volumes of data. Nevertheless, these systems are discrete and suffer from delays (constant and variable) and also in variations in the system's response, which can make them unstable. Moreover, data dependent control systems execute different branches of software, inducing different responses by the system. To minimise the influence of software, the most sensitive parts of these systems are implemented in hardware, usually in computational data-paths for increased performance and reduced delays. Hence, real-time CPSs use dedicated hardware (co-)processors in reconfigurable logic which ensure high-performance, constant throughput and flexibility to instantiate customised sub-systems. Nevertheless, as any other silicon device, reconfigurable logic is sensitive to degradation, and variation of its operating conditions. Consequentially, the work here presented makes important contributions to the implementation of more resilient CPSs on reconfigurable devices, by promoting their graceful degradation.

In a nutshell, this research contributes to the advancement of CPSs by investigating alternatives to reduce, or mitigate, timing errors in applications that tolerate some errors in its calculations, and proposes methods to close an existing gap in this research area. Additionally, a possible by-product of this research is the promotion of security in circuit designs by turning the replication of the CPS unfeasible (i.e. Physical Unclonable Functions).

3 Background and Related Work

3.1 Sources of Variation

The performance of transistors is affected by variation in their sizes, supply voltages, temperature, cross-talk and jitter [7]. Contributions [5, 6] present how the performance of devices from the same family degrades with the aging of the device.

3.2 Variation-Aware Methods for Throughput Increase in FPGAs

Several methods have been proposed to minimise, or recover from, timing errors caused by the aforesaid sources of variation. They operate on different levels of the design, namely placement and routing, Register-Transfer Level (RTL) and algorithm level.

Variation-aware placement and routing [8] makes use of a model created from a characterisation of the fabric where the design is going to be deployed. The main benefit of this method is to instruct the placement and routing tool [9] to assign the critical-paths to the fastest elements on the device, thus reducing the critical-path delay, increasing the designs' overall clock frequency. The drawback is the necessity to characterise the design before deploying the design, which is currently only available on FPGAs [10].

In order to increase the clock frequency of a datapath, on the RTL level, typically the Digital Signal Processing (DSP) designer either reduces its word length or introduces additional pipeline levels. Word length optimisation while offering reductions in area and delay, it also reduces the precision of the results produced. Extensive work has been published on the aforementioned topics, offering techniques to implement and optimise DSP designs [11–13].

3.3 Error Recovery Methods

Since the early days of computing, engineers have been concerned with faults and errors from different natures [14]. Most of the mechanisms and methodologies proposed to mitigate them rely on extra circuitry and processing time. Usually a compromise is achieved in terms of the minimum requirements by the application to work, amount of resources and the time to produce results. Recently [15] has proposed a method to adapt the circuit's voltage according to the level of errors, thus trading off power for accuracy. However, some circuits don't admit errors of any kind in their computations. Hence, circuits that require a deterministic output in their calculations have to rely on other methodologies to recover from errors [16], such as Razor [17]. Other alternatives for fault tolerance techniques, their benefits and their limitations have been presented in [18]. Reduced-Precision Redundancy RPR was originally presented in [19] as a mechanism to contain errors in designs under voltage over-scaling, for low-power, based on the assumption that DSP design can tolerate some errors in their calculations, trading off precision for power [20, 21].

It relies on a smaller implementation of the original system computed in parallel using truncated operands. It compares both outputs and verifies if the magnitude of the difference is below a user defined limit. The method selects the original output if this result is below the specified threshold (T), and the approximated result otherwise. Reduced-Precision Redundancy (RPR) has been compared with other error recovering schemes, such as Triple-Modular Redundancy (TMR) [22], in terms of the tradeoff between errors and resources.

3.4 Resource Optimisation Through Bayesian Inference

A method to optimise linear projection implementations through inference was first introduced in [23] and later extended in [24, 25]. Here, the problem is to discover a projection matrix to compute the best approximation for the original data, minimising resources and reconstruction Mean-Square Error (MSE) of the projected data in the original space. One of the improvements, compared to other works is the avoidance of exhaustive search for solutions.

Being the factors F from a linear projection of data X and the basis matrix Λ, searching for a possible solution for Λ and F is an ill-conditioned problem, for which solutions from heuristic methods are suboptimal. The framework [25] uses a Bayesian conception of the factor analysis model rather than the Karhunen-Loeve Transformation (KLT) algorithm to find the elements of the Λ matrix that minimise the area cost, rather than a constant area cost consideration in the KLT algorithm. The framework receives the problem data, the area models and its parameters and iteratively computes the basis for all projection vectors. As a result it returns the basis matrix and the assignment for the logic elements.

3.5 Summary

Here are some of the most important concepts present in the design of DSP designs for FPGAs focused on performance optimisation, and also methods to mitigate timing errors, or to achieve graceful degradation.

Table 1 summarises the different techniques that can be applied to minimise, or minimise, timing errors. For each technique it shows how it actuates and its limitations.

Table 1. Summary of the existing techniques for acceleration of computations, in order to mitigate timing errors, and their limitations.

Technique	Effect	Limitations
Deep pipelining	Break the critical-path by inserting registers	Unsuitable for some streaming algorithms, or algorithms using recursion
Word length optimisation	Reduce the critical path by processing less bits from truncated operands	Penalty in the quality of results
Razor [17]	Check if the output matches the shadow register	Temporal redundancy is unsuitable for streaming applications
Reduced- Precision Redundancy [19]	Check if the error is within a threshold	Requires extra latency. Unsuitable for algorithms using recursion

4 Reduced Precision Redundancy Framework

RPR as originally proposed in [19] replaces the result produced by the arithmetic unit with an approximate result, which is computed in parallel, in order to control the propagation of errors. This tool operates similarly as existing RPR implementations as it wraps the original combinatorial unit with a redundant circuitry. Yet, it separates itself from other architectures as it has characteristics that are distinct for high-throughput and applications intolerant to latency, and depends on a novel RPR architecture with zero latency cost.

This tool reuses the concept of substituting a set of Most Significant Bytes (MSBs) from the original arithmetic unit with an approximate result, in case of error, nevertheless it introduces new mechanisms to: (a) identify timing errors and (b) generate approximations. Typically, RPR architectures use some of the MSBs of arithmetic operators, as usually they are the critical paths.

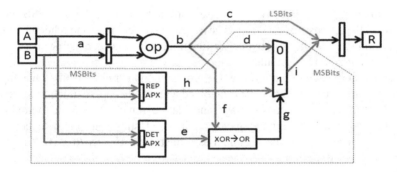

Fig. 1. The new RPR architecture on a generic operator.

4.1 Architecture

The two main novelties in this architecture are: (a) the use ROMs to have the MSBs of the approximations, rather than the simplified operator; and (b) a bit-wise testing (XOR → OR) as a substitute of a subtraction before the comparison with a user-defined limit to identify errors in computations.

Figure 1 presents the new architecture in a combinatorial unit op. Its inputs are A, B; and R is its output. Label a represents the inputs and b the original operator's output. The other labels refer to the paths added by the RPR. They are explained in more detail in [26].

The use of ROMs and bit-wise output testing (XOR → OR) minimises the delay between the input output ports of the new RPR unit. As long as the difference between the delay of the approximation's output and the original result inflates, it favours the increasing of the clock frequency.

As a results of potential realizations with the new RPR scheme, a taxonomy is presented to support the identification of key elements' word-lengths, such as: original unit, error-detection and error-correction ROMs. The structure is given by: iWL and oWL as the input and output word-lengths, and Ori, Det and Rep depict the Original, Detection and Replacement correspondingly: Ori iWL : Ori oWL / Det iWL : Det oWL / Rep iWL : Rep oWL.

Moreover, to discriminate the many RPR architectures, the succeeding prefix is included to the previous taxonomy, which is followed hereafter:

- LUT-SUB - existing RPR but without registers; approximate result is computed with a truncated arithmetic unit implemented with Look-Up Tables (LUTs)/Logic Elements (LEs); errors are identified by the absolute difference between the output of the original unit and the approximation;
- ROM-XOR - proposed RPR; detection approximation is obtained from a ROM, and errors are detected via a bitwise testing of the MSBs with the original unit.

4.2 Approximation Functions

The approximation functions aim to precisely produce the original unit's MSBs in order to decrease the approximation errors. The novelty when compared to other variations is the opportunity to use any approximation function, rather than depending on a truncated arithmetic unit computed simultaneously.

The new architecture employs two approximations simultaneously. One approximation in error detection (*DET APX*) combined with another approximation for replacement (*REP APX*) of erroneous results. Since data is removed from an approximation computed via truncated input arguments, a bit-wise testing between the expected and approximation results will be identified as a mismatch. Hence, *REP APX* is used to correct those MSBs.

4.3 ROM-XOR RPR Arithmetic Operators

The tool here presented supports any arithmetic unit. The details about how they are supported by this tool can be found in [26]. The tool is indifferent to synthesis of the operator and can be adapted to any arithmetic operator of any word-length.

Adder. In this operator the linear approximation function examined was: $apx(a,b,k) = a + b + k$. Here a and b are the truncated input operands and k is an offset to balance the truncation of the input operands. The new RPR tool searches for values of k which minimise the objective function regarding the predicted and the approximation MSBs.

Multiplier. The RPR multiplier is similar to the RPR adder, being the arithmetic unit the only change. In this case, a new approximation function is investigated to attenuate the objective function. The approximation function is given by: $apx(a,b,m,l) = (a + m) * (b + m) + l$. Here a and b are the truncated operands, and m and l are the approximation function coefficients.

4.4   Performance Evaluation

To assess this work, it was compared against standard arithmetic units, with and without RPR. It employs the default implementation for the synthesis of the combinatorial operators.

To obtain precise results, the device was kept at constant temperature of 20 °C, by using a Peltier element on top of the FPGA device and a 1, 2 V core voltage from an independent power source. All tests were conducted on a EP3C16F484C6 Cyclone III FPGA from Altera [27], on a DE0 board from Terasic [28].

Multiplier. Three 8-bit multipliers were used in the performance comparison: no RPR, LUT-SUB RPR and ROM-XOR RPR. Figure 2 presents the variance and mean of the difference between the expected value and the value obtained from the board for all multipliers. The top clock frequency of the ROM-XOR-RPR scheme is near the clock frequency of the multiplier without redundancy, and it surpasses any other multiplier being extremely over-clocked, e.g. 300 MHz. Going beyond 340 MHz the ROM-XOR-RPR multiplier has similar variance values as the other multipliers. Nevertheless, its mean error is still near zero.

Linear Projection Designs. Karhunen-Loeve transformation (KLT) [29], or linear projection, is normally utilised to compress data. In this assessment, a 8:16/5:3/5:3 ROM-XOR RPR, a LUTSUB RPR with an approximation obtained from the 5 MSBs and a threshold equivalent to the 2 MSBs, is compared against the implementation of the linear projection without any redundancy.

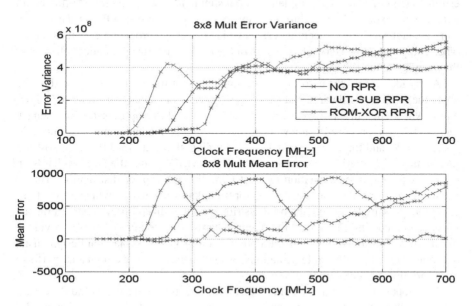

Fig. 2. Error variance (top) and mean error (bottom) of an 8-bit unsigned LUT-based multipliers for different clock frequencies.

The circuit to compute the linear projection corresponds to a fused Multiply-Accumulate (MAC) for each projected dimension. In the implementation all inputs are encoded with 9-bit sign-magnitude. The output of the multiplier is 16 bits unsigned. The word-length of the output increases with the number of accumulation stages (log_2).

Figure 3 presents the results for the 3 implementations at 270 MHz. The top row corresponds to the anticipated error-free result. The rows below relate to the results for: NO RPR, LUT-SUB RPR and ROM-XOR RPR. For all images the PSNR is computed from the projected data reconstruction, on the FPGA, into the original space in software. It's clear that the ROM-XOR RPR provides linear projections circuits with less errors and produce the smallest reconstruction Peak Signal-to-Noise Ratio (PSNR) for all images.

The results demonstrate that a small impact in the top clock frequency in the error-free regime is paid back when in the error prone regime.

5   Optimisation of Linear Projection Designs for Over-Clocking

In the circuit to implement the Linear Projection, or KLT, design, the data path holds the most critical paths. The main focus of this work is on over-clocking multiplier circuits, as they're the components with the largest delay in the data path of the design. In the KLT algorithm, the calculation of the projection matrix Λ and its hardware mapping onto FPGAs are often considered as two independent steps in the design process. However, considerable area savings can be achieved by coupling these two steps as shown in [23, 25]. The Bayesian formulation presented considers the subspace estimation and the hardware implementation simultaneously, allowing the framework to efficiently explore the possibilities of custom design offered by FPGAs. This framework generates Linear Projection designs which minimise errors and circuit resources, when compared to the standard approach of the KLT transform application followed by the mapping to the FPGA.

A key idea from [25] is to inject information about the hardware (i.e. in this case about the required hardware resources of a Constant Coefficient Multiplier (CCM)) as a prior knowledge in the Bayesian formulation of the above optimisation problem. In more detail, the proposed framework in [25] estimates the basis matrix Λ, the noise covariance Ψ, and the factors using Gibbs sampling algorithm [30] from the posterior distribution of the variables, having injecting knowledge about the required hardware recourses for the implementation of the CCMs through a prior distribution. Thus, a probability density function is generated for the unknown Λ matrix, which is used to for generation of samples, where the prior distribution tunes this posterior distribution, and thus accommodating the impact the required hardware resources. [31–34] provide an extension of the above work for the optimisation of Linear Projection designs using different arithmetic units implementations, to combat the effects of circuit area, performance variation and error minimisation.

The proposed work aims to support other arithmetic unit architectures and PVT variation in the characterisation, error modelling, and generation of designs to implement. The framework selects the multipliers used for the implementation of each dot product in along with the coefficients of the Λ matrix that define the lower dimension space.

Variation-Aware Optimisation for Reconfigurable CPS 245

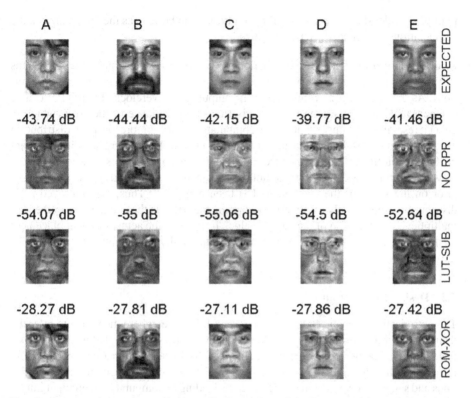

Fig. 3. Images of reconstructed faces (A–E) in the original space without timing errors (EXPECTED), obtained from different multiplier implementations (NO RPR, LUT-SUB RPR, ROM-XOR RPR) at 270 MHz. On top of each face there's the corresponding reconstruction error.

5.1 Defining the Objective Function

In this work, the objective function is based on the Mean Square Error (MSE) of the reconstructed data when are projected back to the original space, as well as on arithmetic errors at the output of the embedded multipliers that are generated when the design is over-clocked, due to PVT variation. In order to simplify the optimisation process and avoiding the formation of a multi-objective function, in this work it is assumed that the errors at the output of the embedded multipliers are uncorrelated for consecutive inputs. As such, the objective function is formed only by the errors due to dimensionality reduction and the variation of the error at the output of the multiplier when it is stimulated by a random input.

Error Models. The proposed framework builds a database of the errors that can be observed at the output of the multipliers when one of the multiplicands is fixed, modelling the constant coefficient of the Λ matrix. The process is repeated for a set of frequencies, as well as for a set of multiplicands resembling in this way the possible values of the coefficients in the the Λ matrix. The observed error variance at the output of the multiplier under the different operating conditions can be seen as the uncertainty in the computations which needs to be minimized.

Prior Distribution. The prior distribution $p(\cdot)$ in the framework aims to favour designs that are known (due to the previous error profiling) to perform poor under certain conditions. As the expectation of the error at the output of an overclocked multiplier can be compensated, the focus is on the minimisation of the variance of the error, which in effect it resembles the uncertainty in the computation. Furthermore, the prior distribution can also favour certain coefficients in terms of how well the corresponding constant coefficient multiplier "fits" in the FPGA device in terms of resources, power, and so on. However, in this work, the above has not been taken into account and thus an informative prior on the value of the coefficients has been employed. Thus, the employed prior distribution captures only information regarding the errors at the output of the embedded multipliers, as a function of the targeted clock frequency, the actual physical placement on the FPGA device, the utilised core voltage, and finally the operating temperature of the device.

5.2 Design Exploration

The optimisation problem falls under Bayesian Inference, where the aim is to infer the distribution of the parameters (i.e. the coefficients of the Λ matrix) that would approximate (in MSE manner) the targeted data. In this work we selected to use Gibbs sampling [30], a sampling methodology that breaks the joint distribution to conditional distributions and samples the parameters one at a time, leading to computational efficient implementations. The sampled design points (effectively the coefficients of the Λ matrix) converge to designs that minimise the objective function U. To further improve the computational complexity, the proposed framework samples each dimension (i.e. column) of the Λ matrix sequentially. The designer provides the dimensionality of the low-dimensional space K, as well as the targeted operating conditions of the device, which define the prior distribution $p(\cdot)$ and effectively guide the framework to select coefficients that perform well under the targeted operating conditions.

5.3 Performance Evaluation

The proposed framework is evaluated for a set of Linear Projection problems when various operating conditions such as process variation, voltage, and temperature are targeted. The performance of the resulting design from the proposed framework was compared to standard design (i.e. baseline) that follows the utilises the KLT algorithm for the computation of the coefficients for the Λ matrix, thus it is unaware of the actual operating conditions and the variance of the error that is expected at the output of the multipliers. The DEO board from Terasic was used for the experiments, which hosts a Cyclone III EP3C16 FPGA. The control of the core voltage that was supplied to the FPGA was done through the PL303QMD-P [35] power supply from TTI. To fine control the temperature of the device, a thermoelectric cooler was placed on top of the device and it was calibrated using a digital thermometer from Lascar Electronics [36] with a

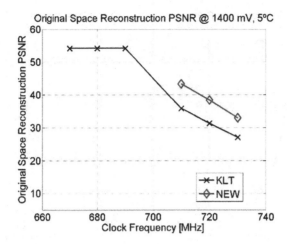

Fig. 4. Performance comparison between the design generated by the proposed framework (NEW) and the reference design (KLT) for a number of operating frequencies when the supply voltage is 1400 mV and the device temperature is kept at 5 °C.

deviation below 1 °C. It should be noted that all the reported results have been collected by running the actual system.

As a case study the problem of generating a Linear Projection design that projects data from Z^6 to Z^3 is utilised. The characterisation of the embedded multipliers under various conditions as well as the estimation of the projection matrix was performed utilising a different set of data to the ones that were used of the computation of the reconstruction error of the system (i.e. evaluation of the design). The utilised metric for the performance of a design is the PSNR of the reconstructed data in the original space.

Optimisation for Maximum Performance. This scenario captures the case where maximum performance is required from the system, which requires an increased FPGA core voltage and the application of an active cooling method for the device. In our test, the device was kept at 5 °C and supplied with 1400 mV, instead of the 1200 mV specified as the maximum supply voltage by the manufacturer. Further than that, the device was clocked with a clock frequency that was double the the maximum frequency specified by the synthesis tool for the normal working conditions. Figure 4 shows the obtained results for a number of frequencies. The results demonstrate that the proposed methodology can generate designs that result in a gain of 10 dB in PSRN of the reconstruction compared to the baseline design utilising the KLT algorithm. Moreover, fixing the targeted PSNR, the designs generated by the proposed framework can be clocked 20 MHz higher than the designed based on KLT.

Optimisation for Low Voltage. This case investigates the gains providing by the proposed framework when a low-power system is targeted, by utilising a low core voltage and without any active cooling component. Figure 5 (left) demonstrates the performance achieved by the reference design that utilises the KLT algorithm when the device operates at 35 °C under a set of FPGA core voltages (0.9 V to 1.2 V) and a number

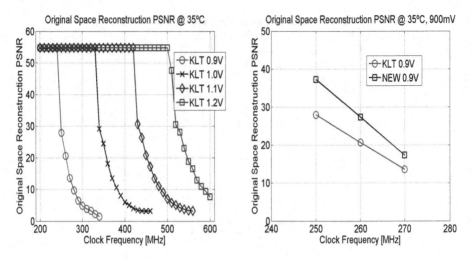

Fig. 5. Performance of the KLT Linear Projection application under different core voltages (left), and a comparison between the two methods at 900 mV (right).

of clock frequencies. The figure shows that as the core voltage drops, the maximum frequency where the design operates without computational errors decreases. Figure 5 (right) focuses on a subset of these results, and in particular the design point with core voltage of 900 mV. The figure depicts also the performance achieved by the design generated by the proposed framework. The results show that an improved PSNR (around 10 dB) is achieved by the design generated by the proposed framework, compared to the baseline design for the same clock frequency. Moreover, for a similar PSNR, a higher clock frequency of up to 10 MHz is achieved by the new design compared to the reference design.

Optimisation for Device Temperature Tolerance. This test scenario investigates the case where the generated design operates under various temperature conditions. A common design methodology to address the problem is to design for the worst case condition. The proposed framework has been extended in order to address the following case in order to generate designs that would achieve a good performance under various temperatures.

In this work, in order to capture the performance of the multiplier under a set of temperatures, a weighted average of the characterisation errors was utilised. In this test case, the design is assumed to operate in temperatures: 20, 35 and 50 °C with a proportional time spend in each temperature captured by the following weights: $\alpha_{20} = 0.3$, $\alpha_{35} = 0.5$ and $\alpha_{50} = 0.2$. Please note that these weights are also used in the weighted average of the characterisation errors. In practice, the proposed framework generates circuit designs per clock frequency, covering all the temperatures within the expected range. They are identified with **NEW WAVG** in the results.

Figure 6 (top-left) depicts the performance of the reference Linear Projection circuit (KLT) as a function of the utilised frequency for the targeted temperatures of the device, when the core voltage is 1200 mV. The rest of Fig. 6 focuses on the comparison between the reference (KLT design), the design produced by the proposed framework using the average weight approach (NEW WAVG), and the design generated by the proposed framework assuming that highest operating temperature (i.e. worst case scenario), for a set of frequencies and temperatures.

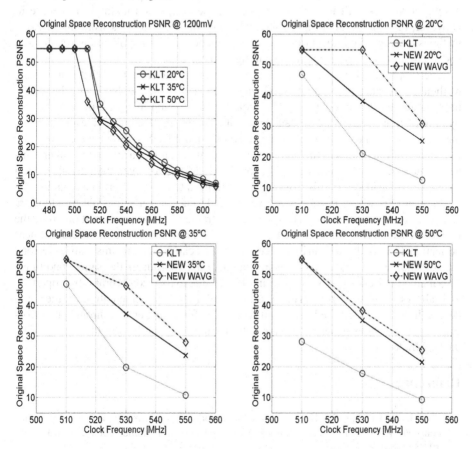

Fig. 6. Performance of the Linear Projection application for a set of device temperatures (top-left), and a comparison between the three methods at 20 °C (top-right), 35 °C (bottom-left) and 50 °C (bottom-right).

The figure shows that the designs generated by the proposed framework outperform the designs based on KLT across all operating temperatures and frequencies, providing at the same time significant improvements on the reconstruction of the data. Moreover, the NEW WAVG designs perform significantly better than the NEW ones for the two out of the tree operating temperatures, as they incorporate information about the performance of the device at these temperatures, performing slightly worst than NEW at 50 °C, as NEW has specifically optimised for this temperature.

6 Conclusions

The constant scaling in the fabrication process has led to devices exhibiting an increase in their process variation. Hence, when the maximum throughput offered by traditional design techniques isn't enough, over-clocking the design is a method to increase it. However, this makes the design susceptible to produce errors. This work investigated methods to assess the impact of errors on arithmetic units and applications, on devices under variation, as well as methods to mitigate those errors.

The proposed RPR scheme fulfils the need for a generic method that can provide resilience to a data path without introducing extra latency, neither having to change the implementation of the algorithm. The (non-trivial) solution imagined proved to work by accelerating the units in the data path while controlling the errors. Despite the fact that the novel architecture requires twice the LEs and 2 ROMs, tests have showed that the quality of the results at the output of the RPR unit can't be achieved by other mitigation methods for the same operating conditions. Moreover, even though only timing errors were considered in this work, it can be utilised to mitigate permanent faults.

In scenarios where resources are scarce and hence it's not feasible to add extra resources to mitigate errors, the optimisation framework uses information from the previous characterisation to create an error model. From that model it generates linear projection designs, through an inference method that can produce results with less errors, when compared to traditional implementations operating under the same conditions. This method is suitable to be adopted in FPGAs, due to its reconfigurability properties, as it allows to have a prior characterisation and a later implementation on the same device. It was also identified that when accounting for timing errors, throughput, errors and area, the designs generated by the optimisation framework were the ones offering the best trade-off.

References

1. Sedcole, P., Cheung, P.Y.K.: Parametric yield modeling and simulations of FPGA circuits considering within-die delay variations. ACM Trans. Reconfigurable Technol. Syst. **1**, 10:1–10:28 (2008)
2. Nascimento, J., Bioucas Dias, J.: Vertex component analysis: a fast algorithm to unmix hyperspectral data. IEEE Trans. Geosci. Remote Sens. **43**, 898–910 (2005)
3. Beccani, M., Tunc, H., Taddese, A., Susilo, E., Volgyesi, P., Ledeczi, A., Valdastri, P.: Systematic design of medical capsule robots. IEEE Design Test **32**, 98–108 (2015)
4. Ke, L., Li, R.: Classification of EEG signals by multi-scale filtering and PCA. In: IEEE International Conference on Intelligent Computing and Intelligent Systems, ICIS 2009, vol. 1, pp. 362–366, November 2009
5. Stott, E.A., Wong, J.S., Sedcole, N.P., Cheung, P.Y.K.: Degradation in FPGAs: measurement and modelling. In: FPGA, pp. 229–238 (2010)
6. Wong, J.S.J., Sedcole, P., Cheung, P.Y.K.: Self-measurement of combinatorial circuit delays in FPGAs. ACM Trans. Reconfigurable Technol. Syst. **2**, 10:1–10:22 (2009)
7. Sedcole, P., Wong, J.S., Cheung, P.Y.K.: Characterisation of FPGA clock variability. In: Proceedings of the IEEE Computer Society Annual Symposium VLSI, ISVLSI 2008, pp. 322–328 (2008)

8. Guan, Z., Wong, J.S., Chaudhuri, S., Constantinides, G., Cheung, P.Y.: A two-stage variation-aware placement method for FPGAs exploiting variation maps classification. In: 2012 22nd International Conference on Field Programmable Logic and Applications (FPL), pp. 519–522, August 2012

9. Betz, V., Rose, J.: VPR: a new packing, placement and routing tool for FPGA research. In: Field-Programmable Logic and Applications, pp. 213–222 (1997)

10. Wong, J.S.J., Cheung, P.Y.K.: Timing measurement platform for arbitrary black-box circuits based on transition probability. IEEE Trans. Very Large Scale Integr. (VLSI) Syst. **21**, 2307–2320 (2013)

11. Constantinides, G.A., Cheung, P.Y.K., Luk, W.: Synthesis and Optimization of DSP Algorithms. Kluwer Academic Publishers, Norwell (2004)

12. Deschamps, J., Bioul, G., Sutter, G.: Synthesis of Arithmetic Circuits: FPGA, ASIC, and Embedded Systems. Wiley, New York (2006)

13. Moore, R.E.: Automatic error analysis in digital computation. Technical report, Space Div. Report LMSD84821, Lockheed Missiles and Space Co., Sunnyvale, CA, USA (1959)

14. Neumann, J.: Probabilistic logics and the synthesis of reliable organisms from unreliable components. Automata Stud. **34**, 43–98 (1956)

15. Krause, P.K., Polian, I.: Adaptive voltage over-scaling for resilient applications. In: Proceedings of the Design, Automation & Test in Europe Conference & Exhibition (DATE), pp. 1–6 (2011)

16. Roberts, D., Austin, T., Blauww, D., Mudge, T., Flautner, K.: Error analysis for the support of robust voltage scaling. In: Proceedings of the Sixth International Symposium Quality of Electronic Design, ISQED 2005, pp. 65–70 (2005)

17. Ziesler, C., Blaauw, D., Austin, T., Flautner, K., Mudge, T.: Razor: a low-power pipeline based on circuit-level timing speculation (2003)

18. Sharma, U.: Fault tolerant techniques for reconfigurable platforms. In: Proceedings of the 1st Amrita ACM-W Celebration on Women in Computing in India, A2CWiC 2010, pp. 60:1–60:4. ACM, New York (2010)

19. Shim, B., Shanbhag, N.: Reduced precision redundancy for low-power digital filtering. In: 2001 Conference Record of the Thirty-Fifth Asilomar Conference on Signals, Systems and Computers, vol. 1, pp. 148–152 (2001)

20. Hegde, R., Shanbhag, N.R.: Energy-efficient signal processing via algorithmic noise-tolerance. In: Proceedings of the 1999 International Symposium on Low Power Electronics and Design, ISLPED 1999, pp. 30–35. ACM, New York (1999)

21. Huang, J., Lach, J., Robins, G.: A methodology for energy-quality tradeoff using imprecise hardware. In: DAC (2012)

22. Hentschke, R., Marques, F., Lima, F., Carro, L., Susin, A., Reis, R.: Analyzing area and performance penalty of protecting different digital modules with Hamming code and triple modular redundancy. In Proceedings of the 15th Symposium on Integrated Circuits and Systems Design, pp. 95–100, IEEE Computer Society, Washington, DC (2002)

23. Bouganis, C.-S., Pournara, I., Cheung, P.Y.K.: Efficient mapping of dimensionality reduction designs onto heterogeneous FPGAs. In: Proceedings of the 15th Annual IEEE Symposium Field-Programmable Custom Computing Machines, FCCM 2007, pp. 141–150 (2007)

24. Bouganis, C.-S., Park, S.-B., Constantinides, G.A., Cheung, P.Y.K.: Synthesis and optimization of 2D filter designs for heterogeneous FPGAs. ACM Trans. Reconfigurable Technol. Syst. **1**, 24:1–24:28 (2009)

25. Bouganis, C.-S., Pournara, I., Cheung, P.: Exploration of heterogeneous FPGAs for mapping linear projection designs. IEEE Trans. Very Large Scale Integr. (VLSI) Syst. **18**(3), 436–449 (2010)

26. Duarte, R., Bouganis, C.-S: Zero-latency datapath error correction framework for over-clocking DSP applications on FPGAs. In: 2014 International Conference on ReConFigurable Computing and FPGAs (ReConFig), pp. 1–7, December 2014
27. Altera: Cyclone III device handbook. http://www.altera.co.uk/
28. Terasic Technologies: Terasic DE0 board user manual v. 1.3 (2009)
29. Hotelling, H.: Analysis of a complex of statistical variables into principal components. J. Educ. Psychol. **24**, 417–441 (1933)
30. Geman, S., Geman, D.: Stochastic relaxation, Gibbs distributions, and the Bayesian restoration of images (PAMI). IEEE Trans. Pattern Anal. Mach. Intell. **6**, 721–741 (1984)
31. Duarte, R., Bouganis, C.: High-level linear projection circuit design optimization framework for FPGAs under over-clocking. In: 2012 22nd International Conference on Field Programmable Logic and Applications (FPL), pp. 723–726, August 2012
32. Duarte, R.P., Bouganis, C.-S.: A unified framework for over-clocking linear projections on FPGAs under PVT variation. In: Goehringer, D., Santambrogio, M.D., Cardoso, J.M., Bertels, K. (eds.) ARC 2014. LNCS, vol. 8405, pp. 49–60. Springer, Heidelberg (2014)
33. Duarte, R.P., Bouganis, C.-S.: Over-clocking of linear projection designs through device specific optimisations. In: 21st Reconfigurable Architectures Workshop (RAW 2014), pp. 9–60 (2014)
34. Duarte, R.P., Bouganis, C.-S.: Pushing the performance boundary of linear projection designs through device specific optimisations (abstract only). In: Proceedings of the 2014 ACM/SIGDA International Symposium on Field-programmable Gate Arrays, FPGA 2014, p. 245. ACM, New York (2014)

Virtual Reference Feedback Tuning of MIMO Data-Driven Model-Free Adaptive Control Algorithms

Raul-Cristian Roman[1(✉)], Mircea-Bogdan Radac[1], Radu-Emil Precup[1], and Emil M. Petriu[2]

[1] Politehnica University of Timisoara, Timisoara, Romania
raul-cristian.roman@student.upt.ro,
{mircea.radac,radu.precup}@upt.ro
[2] University of Ottawa, Ottawa, Canada
petriu@uottawa.ca

Abstract. This paper proposes a new tuning approach by which all Model-Free Adaptive Control (MFAC) algorithm parameters are computed using a nonlinear Virtual Reference Feedback Tuning (VRFT) algorithm. This new mixed data-driven control approach, which results in a mixed data-driven tuning algorithm, is advantageous as it offers a systematic way to tune the parameters of MFAC algorithms by VRFT using only the input/output data of the process. The proposed approach is validated by a set of MIMO experiments conducted on a nonlinear twin rotor aerodynamic system laboratory of equipment position control system. The mixed VRFT-MFAC algorithm is compared with a classical MFAC algorithm whose initial parameter values are optimally tuned.

Keywords: Model-Free Adaptive Control · Optimization · Twin Rotor Aerodynamic System · Virtual Reference Feedback Tuning

1 Introduction

The Virtual Reference Feedback Tuning (VRFT) technique for data-driven controllers has been proposed and applied in [1] to Single Input-Single Output (SISO) systems, in [2] to Multi Input-Multi Output (MIMO) systems, and next extended in [3, 4] to nonlinear systems. The main feature of VRFT, as that of all data-driven controller tuning techniques, consists in collecting the input/output (I/O) data of an unknown open-loop process to compute the controller parameters. A shortcoming of this technique is that it does not guarantee the control system (CS) stability.

The main features of Model-Free Adaptive Control (MFAC) algorithms are [5, 6]: only the online I/O data of the process is used, and the CS stability is guaranteed through reset conditions imposed to the Pseudo-Partial-Derivatives (PPD) matrix. As pointed out in [7], the MFAC algorithm applied to TRAS acts as a pure integrator when the PPD matrix is almost constant during adaptation revealing an insufficient parameterization of the controller.

© IFIP International Federation for Information Processing 2016
Published by Springer International Publishing Switzerland 2016. All Rights Reserved
L. Camarinha-Matos et al. (Eds.): DoCEIS 2016, IFIP AICT 470, pp. 253–260, 2016.
DOI: 10.1007/978-3-319-31165-4_25

Given that the MFAC and VRFT have complementary features, this paper proposes a new data-driven control approach that combines these data-driven techniques resulting in new mixed data-driven algorithms. The mixed VRFT-MFAC algorithm has been applied to a class of nonlinear SISO systems in [7]. The mixed approach aims to control the azimuth and pitch motions of a Cyber-Physical System (CPS) represented by Twin Rotor Aerodynamic System (TRAS) system. Our mixed data-driven algorithm is efficient versus the current specialized literature [1–6] in finding the optimal parameters of MFAC algorithms as it has only eight parameters in the MIMO setting.

The paper is organized as follows: the relation to CPSs is submitted in the next section, the mixed MFAC-VFRT approach and algorithm are presented in Sect. 3, the 4th section offers the experimental validation on the TRAS laboratory equipment, and the conclusions are drawn in Sect. 5.

2   Relation to Cyber-Physical Systems

The nonlinear state-space model that describes the MIMO TRAS process is specified as follows in [8], and the variables are defined in [9]:

$$\dot{\Omega}_h = [l_t F_h(\omega_t)\cos\alpha_v + \Omega_h f_h + u_2 k_{vh}]/J_h,$$
$$\dot{\Omega}_v = \{l_m F_v(\omega_m) + \Omega_v f_v + g[(A - B)\cos\alpha_v - C\sin\alpha_v]$$
$$- (\Omega_h^2/2)(A + B + C)\sin 2\alpha_v\}/J_v, \tag{1}$$
$$\dot{\alpha}_h = \Omega_h, \ \dot{\alpha}_v = \Omega_v, \ \dot{\omega}_h = (u_1 - \omega_h/k_{Hh})/I_h, \ \dot{\omega}_v = (u_2 - \omega_v/k_{Hv})/I_v,$$
$$y_1 = \alpha_h, y_2 = \alpha_v.$$

The relation to CPSs is threefold from both the process and the controller points of view. First, Eq. (1) shows that TRAS is a nonlinear MIMO mechatronics process, its complexity associated with the inclusion of communication, electronics, hardware and software makes it a representative example of CPS. Second, the digital control algorithms are offered for this process, the controllers are implemented on a PC which communicates with the process by means of an adequate interface. Third, the mix of two data-driven approaches (namely, MFAC and VRFT) is advantageous as the process model is not needed in the controller tuning. This is especially important in case of CPSs as our approach outperforms the traditional model-based approaches that are strongly dependent on the process model, and accurate models are extremely difficult to obtain for CPSs.

3   Mixed MFAC-VRFT Control Approach

The control law and estimation mechanism specific to MFAC algorithms are expressed in the state-space form

$$\mathbf{u}(k) = \mathbf{u}(k-1) + \frac{\rho \hat{\boldsymbol{\Phi}}^T(k)[\mathbf{y}^*(k+1) - \mathbf{y}(k)]}{\lambda + ||\hat{\boldsymbol{\Phi}}(k)||^2},$$

$$\hat{\boldsymbol{\Phi}}(k) = \hat{\boldsymbol{\Phi}}(k-1) + \frac{\eta[\Delta \mathbf{y}(k) - \boldsymbol{\Phi}(k-1)(\mathbf{u}(k-1) - \mathbf{u}(k-2))](\mathbf{u}^T(k-1) - \mathbf{u}^T(k-2))}{\mu + ||\mathbf{u}(k-1) - \mathbf{u}(k-2)||^2}, \qquad (2)$$

where $\mathbf{y}^*(k+1) = [\,y_1^*(k+1) \quad y_2^*(k+1)\,]^T$ is the tracking reference input vector, $\mathbf{y}(k) = [\,y_1(k) \quad y_2(k)\,]^T \in \mathbf{R}^{2\times1}$ is the controlled output vector, $\mathbf{u}(k) = [\,u_1(k) \quad u_2(k)\,]^T \in \mathbf{R}^{2\times1}$ is the control signal vector, T indicates matrix transposition, $\hat{\boldsymbol{\Phi}}(k)$ is the estimated PPD matrix, $0 < \eta < 1$ is the first step size constant, $\mu > 0$ is a weighting factor parameter, $\rho > 0$ is the second step size constant. $\lambda \geq 0$ in (2) is a weighting parameter that appears in the following optimization problem whose solving is the objective of MFAC:

$$\mathbf{u}^*(k) = \arg \min_{\mathbf{u}(k)} J_{MFAC}(\mathbf{u}(k)),$$

$$J_{MFAC}(\mathbf{u}(k)) = ||\mathbf{y}^*(k+1) - \mathbf{y}(k+1)||^2 + \lambda ||\Delta \mathbf{u}(k)||^2. \qquad (3)$$

An equivalent model to (2) is

$$\mathbf{u}(k) = \mathbf{g}(\hat{\boldsymbol{\Phi}}(k), \mathbf{u}(k-1), \mathbf{y}^*(k+1), \mathbf{y}(k), \boldsymbol{\theta}),$$

$$\hat{\boldsymbol{\Phi}}(k) = \mathbf{h}(\hat{\boldsymbol{\Phi}}(k-1), \mathbf{u}(k-1), \mathbf{u}(k-2), \mathbf{y}(k), \mathbf{y}(k-1), \boldsymbol{\theta}), \qquad (4)$$

where $\mathbf{g}, \mathbf{h} \in \mathbf{R}^{2\times1}$ are nonlinear functions of their arguments, $\boldsymbol{\theta} = [\rho \; \eta \; \lambda \; \mu]^T$ is the parameter vector. Introducing the additional state vector $\mathbf{z}(k) = \mathbf{u}(k-1)$ and replacing $\hat{\boldsymbol{\Phi}}(k)$ from the second equation in (4) in the first equation in (4), the state-space model (4) is transformed into

$$\boldsymbol{\chi}(k) = \mathbf{F}(\boldsymbol{\chi}(k-1), \mathbf{U}(k), \boldsymbol{\theta}), \qquad (5)$$

with the extended state vector $\boldsymbol{\chi}(k) = [\mathbf{u}(k) \; \mathbf{z}(k) \; \hat{\boldsymbol{\Phi}}(k)]^T$, the extended control signal vector $\mathbf{U}(k) = [\mathbf{y}^*(k+1) \; \mathbf{y}(k) \; \mathbf{y}(k-1)]^T$, and

$$\mathbf{F}(\boldsymbol{\chi}(k-1), \mathbf{U}(k), \boldsymbol{\theta}) = \begin{bmatrix} \mathbf{g}(\hat{\boldsymbol{\Phi}}(k-1), \mathbf{u}(k-1), \mathbf{z}(k-1), \mathbf{y}(k), \mathbf{y}(k-1), \mathbf{y}^*(k+1), \boldsymbol{\theta}) \\ \mathbf{u}(k-1) \\ \mathbf{h}(\hat{\boldsymbol{\Phi}}(k-1), \mathbf{z}(k), \mathbf{z}(k-1), \mathbf{y}(k), \mathbf{y}(k-1), \boldsymbol{\theta}) \end{bmatrix}. \qquad (6)$$

Starting with the initial conditions $\hat{\boldsymbol{\Phi}}(1), \mathbf{u}(1), \mathbf{z}(1) = \mathbf{u}(0)$, $\mathbf{u}(k)$ is expressed recurrently as

$$\hat{\Phi}(2) = \mathbf{h}(\hat{\Phi}(1), \mathbf{u}(1), \mathbf{u}(0), \mathbf{y}(2), \mathbf{y}(1), \boldsymbol{\theta}),$$

$$\mathbf{u}(2) = \mathbf{g}(\hat{\Phi}(1), \mathbf{u}(1), \mathbf{u}(0), \mathbf{y}(2), \mathbf{y}(1), \mathbf{y}^*(3), \boldsymbol{\theta}),$$

$$\mathbf{u}(3) = \mathbf{g}(\hat{\Phi}(2), \mathbf{u}(2), \mathbf{u}(1), \mathbf{y}(3), \mathbf{y}(2), \mathbf{y}^*(4), \boldsymbol{\theta})$$

$$= \mathbf{g}(\mathbf{h}(\hat{\Phi}(1), \mathbf{u}(1), \mathbf{u}(0), \mathbf{y}(2), \mathbf{y}(1), \boldsymbol{\theta}), \mathbf{g}(\hat{\Phi}(1), \mathbf{u}(1), \mathbf{u}(0), \mathbf{y}(2), \mathbf{y}(1), \mathbf{y}^*(3), \boldsymbol{\theta}),$$

$$\mathbf{u}(1), \mathbf{y}(3), \mathbf{y}(2), \mathbf{y}^*(4), \boldsymbol{\theta})$$

$$= \mathbf{g}(\hat{\Phi}(1), \mathbf{u}(1), \mathbf{u}(0), \mathbf{y}(3), \mathbf{y}(2), \mathbf{y}(1), \mathbf{y}^*(3), \mathbf{y}^*(4), \boldsymbol{\theta}), \tag{7}$$

$$\dots$$

$$\mathbf{u}(k) = \mathbf{g}(\hat{\Phi}(1), \mathbf{u}(1), \mathbf{u}(0), \mathbf{y}(k), \mathbf{y}(k-1), \dots, \mathbf{y}(2), \mathbf{y}(1), \mathbf{y}^*(k+1), \mathbf{y}^*(k), \dots, \mathbf{y}^*(4),$$

$$\mathbf{y}^*(3), \boldsymbol{\theta})$$

$$= \mathbf{g}(\hat{\Phi}(1), \mathbf{u}(1), \mathbf{u}(0), \mathbf{y}^*(k+1) - \mathbf{y}(k), \mathbf{y}^*(k) - \mathbf{y}(k-1), \dots, \mathbf{y}^*(3) - \mathbf{y}(2), \mathbf{y}(1), \boldsymbol{\theta}).$$

Using the notation $\mathbf{e}(k) = \mathbf{y}^*(k+1) - \mathbf{y}(k)$ for the tracking error, $\mathbf{u}(k)$ in (7) can be considered to emerge from the I/O nonlinear recurrent controller

$$\mathbf{u}_{\boldsymbol{\theta}_e}(k) = C_{\boldsymbol{\theta}_e}(\boldsymbol{\theta}_e, \mathbf{u}(k-1), \dots, \mathbf{u}(k-n_{uc}), \mathbf{e}(k), \dots, \mathbf{e}(k-n_{ec})), \tag{8}$$

with $\boldsymbol{\theta}_e = \{\hat{\Phi}(1), \boldsymbol{\theta}^T\}$. Using the controller model (8), the model reference objective function (o.f.) specific to nonlinear VRFT is [10].

$$J_{MR}(\boldsymbol{\theta}) = \sum_{k=1}^{N} \left\| \mathbf{y}_{\boldsymbol{\theta}}(k) - \mathbf{y}^d(k) \right\|^2, \tag{9}$$

where $\mathbf{y}_{\boldsymbol{\theta}}(k+1) = \mathbf{f}(\mathbf{y}(k), \dots, \mathbf{y}(k-n_y), \mathbf{u}_{\boldsymbol{\theta}}(k), \dots, \mathbf{u}_{\boldsymbol{\theta}}(k-n_u))$ is the nonlinear process model (with $\mathbf{y}_{\boldsymbol{\theta}}(k+1)$ – the process output), uc and ec are the known orders of the fixed structure controller parameterized by the vector $\boldsymbol{\theta}$, $\mathbf{r}(k)$ is the reference input to the closed-loop CS, $\mathbf{y}^d(k) = \mathbf{m}(\mathbf{y}^d(k-1), \dots, \mathbf{y}^d(k-n_{ym}), \mathbf{r}(k-1), \dots, \mathbf{r}(k-n_{rm}))$ is the output of the user-selected nonlinear reference model $\mathbf{m}$ of orders ym and rm if the input is set as $\mathbf{r}(k)$. Herein, $\mathbf{m}$ is assumed to be invertible.

Assuming that an I/O data pair $\{\mathbf{u}(k), \mathbf{y}(k)\}$, $k = 0 \dots N$, is available form an open loop experiment on the stable process. Then a virtual reference input $\bar{\mathbf{r}}(k)$ is calculated as $\bar{\mathbf{r}}(k) = \mathbf{m}^{-1}(\mathbf{y}(k))$ such that the reference model output and the closed-loop CS output have similar trajectories. By abuse of notation, $\mathbf{m}^{-1}(\mathbf{y}(k))$ results in $\bar{\mathbf{r}}(k)$, which, set as input to $\mathbf{m}$, gives $\mathbf{y}(k)$. The virtual reference tracking error is calculated as $\bar{\mathbf{e}}(k) = \bar{\mathbf{r}}(k) - \mathbf{y}(k)$. The controller that achieves $\mathbf{u}(k)$ caused by $\bar{\mathbf{e}}(k)$ is the one that carries out reference model tracking, and the parameters of this controller are calculated by minimizing the o.f. [10]

$$J_{VRFT}(\boldsymbol{\theta}) = \frac{1}{N} \sum_{t=1}^{N} \left\| C_{\boldsymbol{\theta}}(\boldsymbol{\theta}, \bar{\mathbf{e}}(k)) - \mathbf{u}(k) \right\|^2. \tag{10}$$

As shown in [10], MIMO VRFT does not require any time-varying filter to make $J_{MR}(\theta)$ and $J_{VRFT}(\theta)$ approximately equal, as is usually the case in classical VRFT. The two o.f.s in (9) and (10) can be made approximately equal for a rich parameterization of the controller which can be, for example, a neural network [10, 11].

Our mixed data-driven control approach is based on considering that $r(k)$ specific to VRFT equals $y^*(k + 1)$ specific to MFAC:

$$r(k) = y^*(k + 1). \tag{11}$$

Therefore, the MFAC algorithm structure is included in a closed-loop CS. Figure 1 shows the feedback control structure with the MFAC algorithm tuned by VRFT.

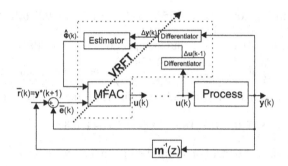

Fig. 1. Structure of mixed MFAC-VRFT CS

4 Experimental Results

Two tuning approaches are proposed to highlight the performance of the mixed VRFT-MFAC control approach, an indirect one in which the VRFT framework is used and the o.f. in (10) is minimized, and a direct one in which a process model is used and a meta-heuristic Gravitational Search Algorithm (GSA) optimizer [12–15] minimizes the following o.f. in the MIMO setting:

$$\tau^* = \arg \min_\tau J(\upsilon), \quad J_\varepsilon(\tau) = \frac{1}{N} \sum_{k=1}^{N} ((y_1^*(k, \tau) - y_1(k, \tau))^2 + (y_2^*(k, \tau) - y_2(k, \tau))^2), \tag{12}$$

where $\tau = [\hat{\Phi}_{11}(1)\ \hat{\Phi}_{12}(1)\ \hat{\Phi}_{21}(1)\ \hat{\Phi}_{22}(1)\ \rho\ \eta\ \lambda\ \mu]^T$ is the parameter vector of VRFT-MFAC and MFAC algorithms and τ^* is the optimal value of τ.

The MFAC objective [5] is to solve the optimization problem in (3).

Use is made in MFAC of the PPD matrix $\Phi(k) = [\phi_{ij}(k)]_{i,j \in \{1,2\}}, \|\Phi(k)\| \le b$, which appears in the compact form dynamic linearization (CFDL) MFAC process model [5]

$$\Delta y(k + 1) = \Phi(k)\Delta u(k), \tag{13}$$

this matrix should be diagonally dominant and fulfill

$$|\phi_{ij}(k)| \le b_1,\ b_2 \le |\phi_{ii}(k)| \le a\,b_2,\ i,j \in \{1,2\},\ i \ne j,\ a \ge 1,\ b_2 > b_1(2a+1), \qquad (14)$$

The digital reference model is

$$\mathbf{m}(z) = \begin{pmatrix} \frac{0.00115z^{-1}+0.00115z^{-2}}{1-1.997z^{-1}+0.999z^{-2}} & 0 \\ 0 & \frac{0.00092z^{-1}+0.00091z^{-2}}{1-1.947z^{-1}+0.949z^{-2}} \end{pmatrix}, \qquad (15)$$

which is the discretized version of a diagonal transfer matrix of second-order normalized t.f.s.

The values of the initial parameters of the mixed VRFT-MFAC algorithm obtained by a GSA that minimizes the o.f. given in (10) are

$$\boldsymbol{\tau} = [746.99\ \ 0.064\ \ 0.048\ \ 1757\ \ 46.42\,0.101\ \ 70.46\ \ 907.6]^T, \qquad (16)$$

where, according to (14), the lower and the upper bounds of the PPD matrix were set as $\begin{pmatrix} 373.49 & 0 \\ 0 & 878.5 \end{pmatrix}$ and $\begin{pmatrix} 1120.49 & 0.129 \\ 0.097 & 2635.51 \end{pmatrix}$. The initial parameters of the MFAC algorithm were also obtained by a GSA that minimizes the o.f. given in (12):

$$\boldsymbol{\tau} = [835\ \ 0.1\ \ 0.1\ \ 1838\ \ 49.34\,0.7\,6.544\ \ 0.874]^T, \qquad (17)$$

where the lower and the upper bounds of the PPD matrix were set as $\begin{pmatrix} 417.5 & 0 \\ 0 & 919 \end{pmatrix}$ and $\begin{pmatrix} 1252.5 & 0.2 \\ 0.2 & 2757 \end{pmatrix}$. Different bounds were imposed to the PPD matrices specific to the two algorithms such the matrix elements can vary according to [5, 6] and the stability is fulfilled.

One of the aims of this paper is to analyze if the performance indices of the CS with mixed VRFT-MFAC algorithm is similar to that of the with MFAC algorithm. The CS performance is assessed through ten experimental trials concerning J_ε. The averages and variances of these o.f.s are next computed in order to highlight more accurately the algorithms and CS performance.

The values of J_ε obtained by the application of both data-driven algorithms are presented in Table 1. A sample of experimental results is exemplified in Fig. 2.

Table 1. The values of the o.f. in the MIMO experimental scenario

	VRFT-MFAC	MFAC
Average of J_ε	0.0143	0.0138
Variance of J_ε	$3.6600 \cdot 10^{-6}$	$5.7020 \cdot 10^{-6}$

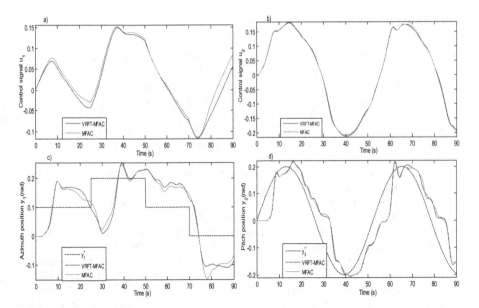

Fig. 2. Experimental results expressed as CS responses in the MIMO scenario: control signals u_1 and u_2, process outputs y_1 and y_2 and references trajectories y_1^* and y_2^*

5 Conclusions

This paper has given a new MIMO model-free data-driven controller tuning approach, referred as mixed VRFT-MFAC approach, which optimally tunes all parameters of MFAC algorithms by VRFT. The mixed VRFT-MFAC algorithm that results from this approach has been implemented on the nonlinear TRAS laboratory equipment.

The comparison with the classical MFAC algorithm (Table 1 and Fig. 2) shows, in correlation with [16, 17], the small differences between what VRFT-MFAC and MFAC exhibit. Therefore, the mixed VRFT-MFAC approach is a time-saving controller tuning solution with similar performance to MFAC.

The proposed approach is useful when controlling processes whose identification is difficult or impossible, and such processes are specific to CPSs. Further research will be focused on the constraints concerning the reference model choice. The constraints related to the optimization problems come from the special features of CPS processes.

Acknowledgements. This work was supported by grants of the Romanian National Authority for Scientific Research, CNCS – UEFISCDI, project numbers PN-II-RU-TE-2014-4-0207 and PN-II-ID-PCE-2011-3-0109, and by a grant from the NSERC of Canada.

References

1. Campi, M.C., Lecchini, A., Savaresi, S.M.: Virtual reference feedback tuning: a direct method for the design of feedback controllers. Automatica **38**(8), 1337–1346 (2002)
2. Formentin, S., Savaresi, S.M., Del Re, L.: Non-iterative direct data-driven tuning for multivariable systems: theory and application. IET Control Theor. Appl. **6**(9), 1250–1257 (2012)
3. Campi, M.C., Savaresi, S.M.: Virtual reference feedback tuning for non-linear systems. In: IEEE Conference on Decision and Control and European Control Conference, pp. 6608–6613. Seville, Spain (2005)
4. Campi, M.C., Savaresi, S.M.: Direct nonlinear control design: the virtual reference feedback tuning (VRFT) approach. IEEE Trans. Autom. Control **51**(1), 14–27 (2006)
5. Hou, Z., Jin, S.: Data-driven model-free adaptive control for a class of MIMO nonlinear discrete-time systems. IEEE Trans. Neural Netw. **22**(12), 2173–2188 (2011)
6. Hou, Z., Jin, S.: A novel data-driven control approach for a class of discrete-time nonlinear systems. IEEE Trans. Control Syst. Technol. **19**(6), 1549–1558 (2011)
7. Roman, R.-C., Radac, M.-B., Precup, R.-E., Petriu, E.M.: Data-driven model-free adaptive control tuned by virtual reference feedback tuning. Acta Polyt. Hung. **13**(1), 83–96 (2016)
8. Two Rotor Aerodynamical System, User's Manual. Inteco Ltd., Krakow, Poland (2007)
9. Roman, R.-C., Radac, M.-B., Precup, R.-E., Petriu, E.M.: Data-driven optimal model-free control of twin rotor aerodynamic systems. In: IEEE International Conference on Industrial Technology, pp. 161–166. Seville, Spain (2015)
10. Yan, P., Liu, D., Wang, D., Ma, H.: Data-driven controller design for general MIMO nonlinear systems via virtual reference feedback tuning and neural networks. Neurocomputing **171**, 815–825 (2016)
11. Esparza, A., Sala, A., Albertos, P.: Neural networks in virtual reference tuning. Eng. Appl. Artif. Intell. **24**(6), 983–995 (2011)
12. Precup, R.-E., David, R.-C., Petriu, E.M., Preitl, S., Radac, M.-B.: Gravitational search algorithms in fuzzy control systems tuning. In: 18th IFAC World Congress, pp. 13624–13629. Milano, Italy (2011)
13. David, R.-C., Precup, R.-E., Petriu, E.M., Radac, M.-B., Preitl, S.: Gravitational search algorithm-based design of fuzzy control systems with a reduced parametric sensitivity. Inf. Sci. **247**, 154–173 (2013)
14. Precup, R.-E., David, R.-C., Petriu, E.M., Preitl, S., Radac, M.-B.: Fuzzy logic-based adaptive gravitational search algorithm for optimal tuning of fuzzy controlled servo systems. IET Control Theor. Appl. **7**(1), 99–107 (2013)
15. Precup, R.-E., David, R.-C., Petriu, E.M., Preitl, S., Radac, M.-B.: Novel adaptive charged system search algorithm for optimal tuning of fuzzy controllers. Expert Syst. Appl. **41**(4), 1168–1175 (2014)
16. Radac, M.-B., Roman, R.-C., Precup, R.-E., Petriu, E.M.: Data-driven model-free control of twin rotor aerodynamic systems: algorithms and experiments. In: IEEE Multi-Conference on Systems and Control, pp. 1889–1894. Antibes, France (2014)
17. Roman, R.-C., Radac, M.-B., Precup, R.-E.: Data-driven model-free adaptive control of twin rotor aerodynamic systems. In: IEEE 9th International Symposium on Applied Computational Intelligence and Informatics, pp. 25–30. Timisoara, Romania (2014)

Normalization Techniques for Multi-Criteria Decision Making: Analytical Hierarchy Process Case Study

Nazanin Vafaei[✉], Rita A. Ribeiro, and Luis M. Camarinha-Matos

CTS/UNINOVA and Faculty of Sciences and Technology,
NOVA University of Lisbon, 2829-516 Caparica, Portugal
nv@ca3-uninova.org, {rar, cam}@uninova.pt

Abstract. Multi-Criteria Decision Making (MCDM) methods use normalization techniques to allow aggregation of criteria with numerical and comparable data. With the advent of Cyber Physical Systems, where big data is collected from heterogeneous sensors and other data sources, finding a suitable normalization technique is also a challenge to enable data fusion (integration). Therefore, data fusion and aggregation of criteria are similar processes of combining values either from criteria or from sensors to obtain a common score. In this study, our aim is to discuss metrics for assessing which are the most appropriate normalization techniques in decision problems, specifically for the Analytical Hierarchy Process (AHP) multi-criteria method. AHP uses a pairwise approach to evaluate the alternatives regarding a set of criteria and then fuses (aggregation) the evaluations to determine the final ratings (scores).

Keywords: Normalization · AHP · MCDM · Rank reversal · Cyber-physical systems (CPSs)

1 Introduction

Everybody makes decisions in their daily lives, as for example: "Should I take an umbrella today"? "Where should I go for lunch"? To make decisions we need access to information (or data) and to reach a decision we need to combine the data to obtain a final score for each candidate decision alternative (e.g. combining food prices and service of restaurants to recommend). The aim of Multi-Criteria Decision Making (MCDM) methods is to rate and prioritize a set of alternatives that best satisfy a given set of criteria [1]. Criteria are a set of requirements or independent attributes that have to be satisfied by several alternatives. Each criterion may be measured in different units, for example, degrees, kilograms or meters; but they all have to be normalized to obtain dimensionless classifications, i.e. a common numeric range/scale, to allow aggregation into a final score. Hence, data normalization is an essential part of any decision making process because it transforms the input data into numerical and comparable data, allowing using MCDM methods to rate and rank alternatives [2, 3].

In this work, the main research question that we address is: *Which normalization technique is more suitable for usage with the AHP method?*

© IFIP International Federation for Information Processing 2016
Published by Springer International Publishing Switzerland 2016. All Rights Reserved
L. Camarinha-Matos et al. (Eds.): DoCEIS 2016, IFIP AICT 470, pp. 261–269, 2016.
DOI: 10.1007/978-3-319-31165-4_26

The motivation for carrying out this work includes four interconnected issues: (a) the importance of data normalization for decision problems where we need to fuse or aggregate data to obtain a final score per alternative; (b) the reduced number of research studies available in this topic; (c) continuation of previous work on suitability of normalization techniques for well-known MCDM methods (e.g. TOPSIS) [4]; (d) contributing to advances in Cyber Physical Systems [5] research, where huge amounts of available data from heterogeneous sensors need to be fused (aggregated) to determine a combined view. Specifically, in this study we focus on the well-known AHP method because it is a well-known and widely used MCDM method [6–13] but we plan to perform the same study for data fusion problems as well as other MCDM methods in the future, to determine which technique is more suitable for any decision problem that requires combining (fusing) data.

The Analytic Hierarchy Process (AHP) was introduced by Saaty [6, 7] to solve unstructured problems in economics, social sciences, and management [8]. AHP has been used in a vast range of problems from simple ones (e.g. selecting a school) to harder ones (e.g. in allocating budgets and energy domains) [8]. When applying the AHP method, the decision maker is able to structure the decision problem and break it down into a hierarchical top-down process. Then, he/she performs a pairwise matrix comparison of criteria using a [1–9] scale (corresponding to semantic interpretations such has "A is much more important than B" regarding a criterion). After normalization, the priorities are determined using either Eigen vectors or a simplified version with weighted sum (SAW) [9, 10].

AHP involves five main steps [13]: Step 1: Decompose the problem into a hierarchical structure; Step 2: Employ pairwise comparisons. A pairwise comparison is the process of comparing the relative importance, preference, or likelihood of two elements (objectives) with respect to another element (the goal). Pairwise comparisons are carried out to establish priorities. Decision elements at each hierarchy level are compared pairwisely and then the reciprocal matrix is completed; Step 3: Determine the logical consistency and if >10 % revise the pairwise classifications until the consistency index is below 10 %. In the implementation of AHP, we may face with inconsistent judgment of input data that it may cause some bad effects on decision process. For example, A1 may be preferred to A2 and A2 to A3, but A3 may be preferred to A1. So, Saaty [7] defined a measure of deviation from consistency that is called a consistency index, as: $C.I. = (\lambda max - N)/(N - 1)$, where N is the dimension of the matrix and λ is the largest eigenvalue of the matrix A. Then, Saaty calculated a consistency ratio (C.R.) as the ratio of the C.I. to a random index (R.I.) which is the average C.I. of sets of judgments (from a 1 to 9 scale) for randomly generated reciprocal matrices [7]. Step 4: Estimate the relative weights by combining the individual subjective judgments. We can use the eigenvalue method to estimate the relative weights of the decision elements. In order to estimate the relative weight of the decision elements in a matrix, we can use $A.W = \lambda max.W$ where W is the weight of criterion [13]. Step 5: Determine the priority of alternatives by doing aggregation on relative weights which is obtained by combining the criterion priorities and priorities of each decision alternatives relative to each criterion. Since in our work we discuss the suitability of normalization techniques for the AHP method, we focus on step 4 and 5.

In this work we propose an assessment approach for evaluating five common normalization techniques (see Table 1), using an illustrative example solved with AHP method [1, 2]. We choose AHP because it is a well-known and widely used MCDM method [6–13] but we plan to perform the same study for other MCDM methods in the future. Our novel assessment approach calculating Pearson correlation for global weight of alternatives and Spearman correlation for rank of alternatives which are borrowed from [14] to determine mean values in order to ensure a more robust evaluation and selection of the best normalization technique in AHP. The novelty of this study is making adaptation between assessment process and AHP in order to find best normalization technique for AHP method. The next section presents the experimental study performed.

Table 1. Normalization techniques.

Normalization technique	Condition of use	Formula
Linear: Max (N1) [14]	Benefit criteria	$n_{ij} = \frac{r_{ij}}{r_{max}}$
	Cost criteria	$n_{ij} = 1 - \frac{r_{ij}}{r_{max}}$
Linear: Max-Min (N2) [14]	Benefit criteria	$n_{ij} = \frac{r_{ij} - r_{min}}{r_{max} - r_{min}}$
	Cost criteria	$n_{ij} = \frac{r_{max} - r_{ij}}{r_{max} - r_{min}}$
Linear: sum (N3) [14]	Benefit criteria	$n_{ij} = \frac{r_{ij}}{\sum_{i=1}^{m} r_{ij}}$
	Cost criteria	$n_{ij} = \frac{1/r_{ij}}{\sum_{i=1}^{m} 1/r_{ij}}$
Vector normalization (N4) [2]	Benefit criteria	$n_{ij} = \frac{r_{ij}}{\sqrt{\sum_{i=1}^{m} r_{ij}^2}}$
	Cost criteria	$n_{ij} = 1 - \frac{r_{ij}}{\sqrt{\sum_{i=1}^{m} r_{ij}^2}}$
Logarithmic normalization (N5) [2]	Benefit criteria	$n_{ij} = \frac{\ln(r_{ij})}{\ln(\prod_{i=1}^{m} r_{ij})}$
	Cost criteria	$n_{ij} = \frac{1 - \frac{\ln(r_{ij})}{\ln(\prod_{i=1}^{m} r_{ij})}}{m - 1}$

2 Relationship to Cyber-Physical Systems

Cyber-physical systems (CPS) involve merging computation and physical processes, often denoted as embedded systems [15]. In most CPS, the physical inputs and outputs are typically designed as a network of interacting elements. This conceptual model is tied to the notion of robotics and sensor networks and their usage has been increasing day by day [16]. But CPS also inherits ideas from the areas of embedded and real-time systems. CPS have a broad scope of potential application in areas such as reactive interventions (e.g., collision avoidance); precision operations (e.g., robotic surgery and nano-level manufacturing); operation in dangerous or inaccessible environments (e.g., search and rescue, firefighting, and deep-sea exploration); complex systems coordination (e.g., air traffic control, war fighting); efficiency (e.g., zero-net energy buildings);

and augmentation of human capabilities (e.g., healthcare monitoring and service delivery) [16], to name a few. There are some discussions on the relationship between Cyber-Physical Systems and Internet of Things [17–19]. Camarinha and Afsarmanesh [5] mention that "there is a growing convergence between the two areas since CPSs are becoming more Internet-based". For example, in smart car parking, data from the parking space is transferred to the car drivers with the help of CPS and IoT technologies. Data is collected from sensors, which are installed in the parking lot, and transferred to the data center to be processed with MCDM methods, to determine the ranking of alternatives (best parking spaces). The best parking spaces are provided to the car drivers to support them making more informed decisions. In the illustrative example section, we will compare several normalization techniques for usage with the AHP method to rank alternatives and support car drivers. The smart car parking example shows a robust relationship between cyber physical system (CPS), Internet of Thing (IoT) and multi-criteria decision making (MCDM) concepts.

3 Normalization

There are several definitions for data normalization, depending on the study domain. For example, in Databases, data normalization is viewed as a process where data attributes, within a data model, are organized in tables to increase the cohesion and efficiency of managing data. In statistics and its applications, the most common definition is the process of adjusting values measured on different scales to a common scale, often prior to aggregating or averaging them [19]. Many other definitions exist, depending on the context or study domain (see for example [20]). Here we focus on normalization techniques for MCDM. In general, normalization in MCDM is a transformation process to obtain numerical and comparable input data by using a common scale [4]. After collecting input data, we must do some pre-processing to ensure comparability of criteria, thus making it useful for decision modeling. Furthermore, in MCDM, normalization techniques usually map attributes (criteria) with different measurement units to a common scale in the interval [0-1] [21, 22]. Several studies on the effects of normalization techniques on the ranking of alternatives in MCDM problems have shown that certain techniques are more suitable for specific decision methods than others [14, 23–28].

Chakraborty and Yeh [23] analyzed four normalization techniques (vector, linear max-min, linear max and linear sum) in the MCDM simple additive weight (SAW) method. They used a ranking consistency index (RCI) and calculated the average deviation for each normalization technique and concluded that the best normalization technique for SAW is the vector normalization. Further, the same authors analyzed the effects of those normalizations for order preference by similarity to ideal solution method (TOPSIS) by calculating ranking consistency and weight sensitivity of each normalization and proved that vector normalization technique is the best for implementing in TOPSIS method [24]. The authors [24] defined weight sensitivity as a method to analyze sensitivity level of different normalization procedures under different problem settings. They assumed same weights for attributes and then they increased their weights to find the sensitivity of the alternatives (normalization techniques) [24].

Also, the result was further validated by Vafaei et al. [4], who used Pearson and Spearman correlation coefficients to also conclude that the best normalization technique for TOPSIS method is the vector normalization.

In this work, we selected five (shown in Table 1) of the most promising normalization techniques [2, 14] and analyzed their effect on the AHP method. In Table 1, each normalization method is divided in two formulas, one for benefit and another for cost criteria, to ensure that the final decision objective (rating) is logically correct, i.e. when it is a benefit criterion for high values it will correspond to high normalized values (maximization - benefit) and when it is a cost criterion high values will correspond to low normalized values (minimization - cost).

Summarizing, the aim of this study is to identify which normalization technique is best suited for the AHP method.

4 Experimental Study with a Smart Car Parking Example

Here, we discuss the suitability of five normalization techniques for AHP with an illustrative example for smart car parking. This illustrative case consists of 3 criteria (C1, C2, C3), which correspond to time to park, distance, and size of the parking space, and 7 alternatives (A1, A2, ..., A7), which correspond to candidate location sites for parking. Finding the best place for parking the car is the goal; C1 and C2 are cost criteria, where low values are better, and C3 is a benefit criterion, where high values are desirable. Following the AHP method we defined three pairwise comparison matrices for each criterion (example in Table 2) and then one pairwise comparison matrix between criteria. To these four matrices we applied the five normalization techniques, separately, to determine the ranking of alternatives and compare results. The pairwise decision matrix for criteria "time to park", after steps 1, 2, 3 of AHP, is shown in Table 2.

Table 2. Pairwise comparison matrix with respect to the time.

	A1	A2	A3	A4	A5	A6	A7
A1	1	1/3	1/2	3	1/3	2	1
A2	3	1	1	4	1	3	1
A3	2	1	1	2	1/2	3	2
A4	1/3	1/4	1/2	1	1/4	1	1/3
A5	3	1	2	4	1	3	1
A6	1/2	1/3	1/3	1	1/3	1	3
A7	1	1	1/2	3	1	1/3	1

We started by testing the sum-based normalization (N3 in Table 1), the usual normalization technique for AHP [7], because it ensures column sum per alternative is equal to one that is defined by Saaty [7]. The other normalization techniques do not include this characteristic and the sum of the normalized values can be bigger than 1; hence, for comparison purposes we opted for re-normalizing the other four using N3. For illustrating the alternatives rating procedure we show the calculation for vector normalization of alternative A1 and the final results for all alternatives are shown in the Tables 3 and 4:

$$P_{11} = \frac{x_{11}}{\sqrt{\sum_{j=1}^{7} x_{1j}}} = \frac{1}{\sqrt{(1^2) + (3^2) + (2^2) + \left(\frac{1^2}{3}\right) + (3^2) + \left(\frac{1^2}{2}\right) + (1)^2}} = 0.7974$$

$$AverageP1 = \frac{0.7974 + 0.8390 + 0.8091 + 0.5991 + 0.8227 + 0.6524 + 0.7583}{7} = 0.7540$$

$$A_{11} = \frac{AverageP1}{Sum} = \frac{0.7974}{4.8050} = 0.1659$$

$$AverageA1 = \frac{0.1659 + 0.1814 + 0.1659 + 0.1304 + 0.1769 + 0.1393 + 0.1598}{7} = 0.1605$$

Table 3. Normalization results for vector normalization technique for cost criteria.

	P1	P2	P3	P4	P5	P6	P7	Average
P1	0.7974	0.8390	0.8091	0.5991	0.8227	0.6524	0.7583	0.7540
P2	0.3922	0.5169	0.6182	0.4655	0.4681	0.4786	0.7583	0.5283
P3	0.5948	0.5169	0.6182	0.7327	0.7341	0.4786	0.5165	0.5988
P4	0.9325	0.8792	0.8091	0.8664	0.8670	0.8262	0.9194	0.8714
P5	0.3922	0.5169	0.2365	0.4655	0.4681	0.4786	0.7583	0.4737
P6	0.8987	0.8390	0.8727	0.8664	0.8227	0.8262	0.2748	0.7715
P7	0.7974	0.5169	0.8091	0.5991	0.4681	0.9421	0.7583	0.6987
Sum	4.8051	4.6247	4.7730	4.5946	4.6508	4.6829	4.7437	4.6964

Table 4. Re-normalization results for vector normalization technique for cost criteria.

	A1	A2	A3	A4	A5	A6	A7	Average
A1	0.1659	0.1814	0.1695	0.1304	0.1769	0.1393	0.1598	0.1605
A2	0.0816	0.1118	0.1295	0.1013	0.1007	0.1022	0.1598	0.1124
A3	0.1238	0.1118	0.1295	0.1595	0.1578	0.1022	0.1089	0.1276
A4	0.1941	0.1901	0.1695	0.1886	0.1864	0.1764	0.1938	0.1856
A5	0.0816	0.1118	0.0495	0.1013	0.1007	0.1022	0.1598	0.1010
A6	0.1870	0.1814	0.1828	0.1886	0.1769	0.1764	0.0579	0.1644
A7	0.1659	0.1118	0.1695	0.1304	0.1007	0.2012	0.1598	0.1485
Sum	1	1	1	1	1	1	1	1

The global weights of alternatives and ranking results for the four tested normalization techniques are shown in Table 5. We discarded the logarithmic normalization technique from our results because we obtained negative and infinite data (due to the characteristics of pairwise matrices), hence it is not usable (appropriate) for the AHP method. As it can be seen in Table 5, there is consensus on which normalization techniques is better for alternatives A2, A3, A4 and A5 (i.e. they all have the same ranking), but for the other alternatives there was no consensus. Since, it is not possible to distinguish which is the best normalization technique just by looking at the results, we used the evaluation approach proposed in [4] to make the assessment. Hence, we calculated Pearson correlation and mean r_s values [4] with the global weights of alternatives and Spearman correlation with the ranks of alternatives to assess the suitability of the four tested normalization techniques for the AHP method. Table 6 displays that there exists complete consensus between Pearson and Spearman correlation's results and it is clear that the best normalization technique is N1 (linear: max) because it has the highest mean r_s value (P = 0.9606 & S = 0.9524) and the worst one is N3 (linear sum) with the lowest mean r_s value (P = 0.9029 & S = 0.8413).

Table 5. Global weight (G) and Ranking (R) of alternatives for the smart parking example.

	N1		N2		N3		N4	
	G	R	G	R	G	R	G	R
A1	0.1972	2	0.1925	2	0.1505	4	0.1693	2
A2	0.0681	6	0.0634	6	0.0762	6	0.1165	6
A3	0.1143	5	0.1161	5	0.0993	5	0.1297	5
A4	0.2469	1	0.2658	1	0.2876	1	0.1755	1
A5	0.0460	7	0.0291	7	0.0749	7	0.1101	7
A6	0.1765	3	0.1869	3	0.1598	2	0.1450	4
A7	0.1509	4	0.1462	4	0.1517	3	0.1538	3

Table 6. Pearson correlation between global weights and Spearman correlation between ranks of alternatives for each normalization technique.

	N1		N2		N3		N4		Mean r_s		Rank	
	P	S	P	S	P	S	P	S	P	S	P	S
N1			0.9961	1	0.9171	0.8571	0.9687	1	0.9606	0.9524	1	1
N2	0.9961	1			0.9273	0.8571	0.9458	0.9524	0.9564	0.9365	2	2
N3	0.9171	0.8571	0.9273	0.8571			0.8643	0.8095	0.9029	0.8413	4	4
N4	0.9687	1	0.9458	0.9524	0.8643	0.8095			0.9263	0.9206	3	3

* P = Pearson
**S = Spearman

From the example we can conclude that linear max (N1) is the best normalization technique for the AHP method and linear sum (N3) is the worst one. It is interesting to note that the single normalization used in AHP (linear sum- N3) is the worst one from this comparison study. Although N1 is elected as the most suitable normalization technique it required a re-normalization with N3 because the sum of the normalized values has to be 1. Therefore, we may conclude that a combination of max-normalization (N1) with linear-sum (N3) seems the most appropriate for AHP.

5 Conclusion

Normalization is the first step of any decision making process to transform data in different units into a common scale and comparable units. In this study we tested five common normalization techniques to assess their suitability for the AHP MCDM method. The tests showed that the logarithmic normalization technique (N5) is not usable in the AHP method because it can result in zero or infinite values in the normalized data, which is not acceptable to use in the method. Further, since AHP requires the columns of the pairwise matrices to sum up 1, the techniques: linear max, linear max-min and vector normalization techniques had to be re-normalized with linear sum (N3) before being compared. To assess the suitability of the normalization techniques for AHP we used Pearson and Spearman correlation and mean r_s values; the results showed that the best normalization technique is N1 (linear: max) combined with N3 (linear-sum) to ensure the sum is 1, while the worst one is N3 alone.

In a previous work we did the same assessment study for TOPSIS and in the future we plan to extend it to other well-known MCDM methods, with the aim to support decision makers by recommending the most suitable normalization techniques for usage with each MCDM method.

Acknowledgements. This work was partially funded by FCT Strategic Program UID/EEA/00066/203 of Computational Intelligence Group of CTS/UNINOVA.

References

1. Triantaphyllou, E.: Multi-criteria decision making methods. In: Multi-criteria Decision Making Methods: A Comparative Study, vol. 44, pp. 5–21. Springer, New York (2000)
2. Jahan, A., Edwards, K.L.: A state-of-the-art survey on the influence of normalization techniques in ranking: improving the materials selection process in engineering design. Mater. Des. **65**(2015), 335–342 (2014)
3. Nayak, S.C., Misra, B.B., Behera, H.S.: Impact of data normalization on stock index forecasting. Int. J. Comput. Inf. Syst. Ind. Manage. Appl. **6**(2014), 257–269 (2014)
4. Vafaei, N., Ribeiro, R.A., Camarinha-Matos, L.M.: Data normalization techniques in decision making: case study with TOPSIS method. Int. J. Inf. Decis. Sci. (2016, in press)
5. Camarinha-Matos, L.M., Afsarmanesh, H.: Collaborative systems for smart environments: trends and challenges. Collab. Syst. Smart Networked Environ. **434**(2014), 3–14 (2014)
6. Saaty, T.L.: A scaling method for priorities in hierarchical structures. J. Math. Psychol. **15** (3), 234–281 (1977)
7. Saaty, T.L.: The Analytic Hierarchy Process. McGraw-Hill, New York (1980)
8. Cheng, C.-H., Yang, K.-L., Hwang, C.-L.: Evaluating attack helicopters by AHP based on linguistic variable weight. Eur. J. Oper. Res. **116**(1999), 423–435 (1999)

9. Gaudenzi, B., Borghesi, A.: Managing risks in the supply chain using the AHP method. Int. J. Logist. Manage. **17**(1), 114–136 (2006)
10. Zahedi, F.: The analytic hierarchy process – a survey of the method and its applications. Interfaces (Providence) **16**(4), 96–108 (1986)
11. Tuzkaya, G., Onut, S., Tuzkaya, U.R., Gulsun, B.: An analytic network process approach for locating undesirable facilities: an example from Istanbul, Turkey. J. Environ. Manage. **88** (2008), 970–983 (2007)
12. Saaty, T.L., Vargas, L.G.: Models, Methods, Concepts & Applications of the Analytic Hierarchy Process. Institute for Operations Research and the Management Sciences, Maryland (2006)
13. Tzeng, G.-H., Huang, J.-J.: Multiple Attribute Decision Making: Methods and Applications. Taylor & Francis Group, Boca Raton (2011)
14. Celen, A.: Comparative analysis of normalization procedures in TOPSIS method: with an application to Turkish deposit banking market. INFORMATICA **25**(2), 185–208 (2014)
15. Shi, J., Wan, J., Yan, H., Suo, H.: A survey of cyber-physical systems. In: 2011 International Conference on Wireless Communications and Signal Processing (WCSP), pp. 1–6 (2011)
16. Wiki1: Cyber-physical system. https://en.wikipedia.org/wiki/Cyber-physical_system. Accessed 02 November 2015
17. Camarinha-Matos, L.M., Tomic, S., Graça, P. (eds.) Technological Innovation for the Internet of Things, vol. 394. Springer, Heidelberg (2013)
18. Camarinha-Matos, L.M., Tomic, S., Graça, P. (eds.): DoCEIS 2013. IFIP AICT, vol. 394. Springer, Heidelberg (2013)
19. Jeschke, S.: Everything 4.0? In: Drivers and Challenges of Cyber Physical Systems (2013)
20. Wiki2: Normalization (statistics). https://en.wikipedia.org/wiki/Normalization_%28statistics %29. Accessed 15 October 2015
21. Wiki3: Normalization. https://en.wikipedia.org/wiki/Normalization. Accessed 15 October 2015
22. Pavlicic, D.M.: Normalization affects the results of MADM methods. Yugosl. J. Oper. Res. **11**(2011), 251–265 (2011)
23. Etzkorn, B.: Data normalization and standardization. http://www.benetzkorn.com/2011/11/data-normalization-and-standardization/. Accessed 28 April 2015
24. Chakraborty, S., Yeh, C.-H.: A simulation based comparative study of normalization procedures in multiattribute decision making. In: International Conference on Artificial Intelligence, Knowledge Engineering and Data Bases, pp. 102–109 (2007)
25. Chakraborty, S., Yeh, C.-H.: A simulation comparison of normalization procedures for TOPSIS. In: Computers Industrial Engineering, pp. 1815–1820 (2009)
26. Chakraborty, S., Yeh, C.-H.: Rank similarity based MADM method selection. In: International Conference on Statistics in Science, Business and Engineering (ICSSBE 2012) (2012)
27. Milani, A.S., Shanian, A., Madoliat, R., Nemes, J.A.: The effect of normalization norms in multiple attribute decision making models: a case study in gear material selection. Struct. Multidiscip. Optim. **29**(4), 312–318 (2004)
28. Wang, Y.-M., Luo, Y.: Integration of correlations with standard deviations for determining attribute weights in multiple attribute decision making. Math. Comput. Model. **51**(2010), 1–12 (2010)

Wireless Technologies

A WLS Estimator for Target Localization in a Cooperative Wireless Sensor Network

Slavisa Tomic[4($\boxtimes$)], Marko Beko[1,3], Rui Dinis[2,5], and Milan Tuba[6]

[1] Universidade Lusófona de Humanidades e Tecnologias, Lisbon, Portugal
mbeko@uninova.pt
[2] Instituto de Telecomunicações, Lisbon, Portugal
rdinis@fct.unl.pt
[3] CTS, UNINOVA – Campus FCT/UNL, Caparica, Portugal
[4] ISR/IST, LARSyS, Lisbon, Portugal
s.tomic@campus.fc.unl.pt
[5] DEE/FCT/UNL, Caparica, Portugal
[6] Faculty of Computer Science, Megatrend University, Belgrade, Serbia
tuba@ieee.org

Abstract. This paper addresses target localization problem in a cooperative 3-D wireless sensor network (WSN). We employ non-traditional methodology which merges distance and angle measurements, respectively withdrawn from the received signal strength (RSS) and angle-of-arrival (AoA) information. Based on RSS measurement model and effortless geometry, a novel non-convex estimator according to the weighted least squares (WLS) criterion is obtained, which closely approximates the maximum likelihood (ML) estimator for small noise. It is shown that the devised estimator is appropriate for distributed implementation. Following the squared range (SR) approach, we propose a suboptimal SR-WLS estimator according to the generalized trust region sub-problem (GTRS) framework, to estimate the locations of all targets in the WSN. According to our simulations, the new estimator has excellent performance in a great variety of considered settings, in which the effectiveness of fusing two radio measurements is confirmed.

Keywords: Angle-of-arrival (AoA) · Generalized trust region sub-problem (GTRS) · Received signal strength (RSS) · Wireless sensor network (WSN)

1 Introduction

Accurate information about sensor's location is a key component in many practical applications. Wireless localization algorithms usually rely upon range measurements extracted from angle-of-arrival (AoA), received signal strength (RSS), time-of-arrival (ToA) information, or a combination of them [1].

In [2], a hybrid methodology which fuses distance and angle measures has been considered. Both estimators proposed in [2] deal with the non-cooperative target localization problem in a 3-D space: linear least squares (LS) and optimization based. The former one is a fairly simple estimator, and the later one uses Davidson-Fletcher-Powell algorithm [3]. The authors in [4] derived an LS and a maximum likelihood (ML) estimator

© IFIP International Federation for Information Processing 2016
Published by Springer International Publishing Switzerland 2016. All Rights Reserved
L. Camarinha-Matos et al. (Eds.): DoCEIS 2016, IFIP AICT 470, pp. 273–283, 2016.
DOI: 10.1007/978-3-319-31165-4_27

for a hybrid scheme that fuses RSS difference (RSSD) and AoA measurements. The authors in [4] employed non-linear constrained optimization to estimate the unknown location from multiple RSS and AoA measurements. In [5], a selective weighted LS (WLS) estimator for mixed RSS/AoA localization problematic was presented. The target position was established by employing loaded distances from the two *closest* anchor measures, together with the serving base station AoA measurement. Another WLS estimator for a non-cooperative localization problem for the case where the transmitted power is not known was proposed in [6]. Nonetheless, similar as the method proposed in [5], the WLS method has been derived without requiring information about the statistical properties of RSS and AoA measurements, which might lead to significant performance degradation in practice. Also, the authors in [6] investigated a small-scale wireless sensor network (WSN) only, with extremely low noise power.

All of the above approaches examine non-collaborative localization problem only, where the location of a single target, which communicates with anchors exclusively, is established at a time. Contrarily to the above mentioned approaches, here, the target localization problem in an extensive WSN is considered, where the number of anchors is insufficient and the communication range of all sensors is limited (*e.g.*, to prolong sensor's battery life). Hence, only some targets can directly communicate with anchors and sensor cooperation is required in order to acquire adequate amount of information to carry out the localization. Through employing the RSS spreading model and straightforward geometry, we first develop a new local non-convex estimator based on the WLS criterion that closely approximates the local ML one in the case where noise is low. Next, by following the squared range (SR) approach, we propose a suboptimal SR-WLS estimator based on the generalized trust region sub-problem (GTRS) framework, which is possible to solve *exactly* with a bisection method [7]. To the best of the authors' knowledge, distributed localization algorithms for hybrid RSS/AoA systems in cooperative WSNs are yet to be published.

2   Relationship to Cyber-Physical Systems

Cyber-physical systems (CPSs) have recently attracted much attention from both the educational and engineering public. These schemes are displaying an enormous potential in interacting with the physical world and creating its management more efficient. Integrating the computation and communication competencies into the modules of the physical surrounding [8] enables this. CPSs represent nowadays the new peer group of networks and embedded structures.

The placement of CPS raises numerous difficulties, where the most significant topic that has activated massive quantity of study is localization [8]. Basically, localization goals at estimating the location of CPS modules. In CPSs where data are closely related with the surroundings and the location at which they are produced, localization is a fundamental task.

Collaboration of a large number of scattered sensors in a WSN can be seen as a CPS. In such systems, localization is of a crucial importance, since a system may be configured to react locally to the variations within sensor records; hence, accurate determination of the location where the deviations arise is the key. Furthermore, services based on location-awareness represent a key component of countless wireless structures nowadays. By exploring the synergies between computational and physical components we can form smart environments which offer improved safety and efficiency in everyday life, *e.g.* smart parking, assistance for elderly or people with disabilities, monitoring of storage conditions and goods, *etc.*

3 Problem Statement

We examine a large-scale WSN with N anchors and M targets, that may also be viewed as a connected graph, $\mathcal{G}(\mathcal{V}, \varepsilon)$, with $|\mathcal{V}| = M + N$ vertices and $|\varepsilon|$ edges (connections), where $|\cdot|$ represents the cardinality of a set. The set of targets and the set of anchors are respectively denoted as $\mathcal{T}(|\mathcal{T}| = M)$ and $\mathcal{A}(|\mathcal{A}| = N)$, and their locations are denoted by $x_1, x_2, \cdots, x_M$ and $a_1, a_2, \cdots, a_N$, $(x_i, a_j \in \mathbb{R}^3, \forall i \in \mathcal{T}$ and $\forall j \in \mathcal{A})$, respectively. Due to limited communication range, R, two sensors, i and j, can interchange data if and only if they are inside the communication range of each other. The sets of all target/anchor and target/target edges are defined as $\varepsilon_A = \{(i,j): \ \|x_i - a_j\| \le R, \forall i \in \mathcal{T}, \ \forall j \in \mathcal{A}\}$ and $\varepsilon_T = \{(i,k): \ \|x_i - x_k\| \le R, \forall i, k \in \mathcal{T}, \ i \ne k\}$, respectively.

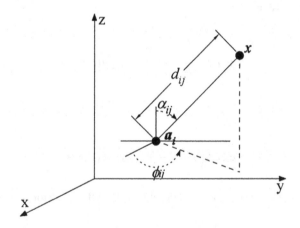

Fig. 1. Representation of a target and anchor locations within 3-D.

In favor of ease of expression, we describe a matrix $X = [x_1, \cdots, x_M] (X \in \mathbb{R}^{3 \times M})$ as the matrix of all unknown target locations. These unknown locations are estimated with the means of a hybrid methodology which merges distance and angle measures.

Throughout this work, we imply that the distance measurements are extracted from the RSS measurements, because ranging based on RSS does not oblige additional hardware [1]. The path loss between two sensors is given by:

$$L_{ij} = L_0 + 10\gamma \log_{10} \frac{\|x_i - \widehat{a}_j\|}{d_0} + n_{ij}, \forall (i,j) \in \varepsilon_A \cup \varepsilon_T \tag{1}$$

(see [9], [10]) where L_0 represents the path loss value at a reference distance d_0 $((\|x_i - \widehat{a}_j\| \geq d_0))$, γ is the path loss exponent (PLE), $\widehat{a}_j$ is the j-th neighbor of the i-th target $(a_j$ if $j \in A$, x_j if $j \in T)$, and n_{ij} is the log-normal shadowing considered as a Gaussian random variable, $i.e.$, $n_{ij} \sim \mathcal{N}(0, \sigma_{n_{ij}}^2)$. The target/target path loss measures are assumed to be symmetric[1].

Figure 1 gives a representation of a target and anchor locations within 3-D. Looking at Fig. 1, $x_i = [x_{i1}, x_{i2}, x_{i3}]^T$ and $a_j = [a_{j1}, a_{j2}, a_{j3}]^T$ represent the coordinates of the i-th target (not known) and the coordinates of the j-th anchor (known) respectively, whereas d_{ij}, ϕ_{ij} and α_{ij} respectively stand for the range, azimuth and elevation angle amongst the i-th target and the j-th anchor. According to the measurement model (1) the ML distance between two sensors is obtained as follows [1]:

$$\widehat{d_{ij}} = d_0 10^{\frac{L_{ij} - L_0}{10\gamma}}, \forall (i,j) \in \varepsilon_A \cup \varepsilon_T. \tag{2}$$

By applying simple geometry, azimuth and elevation angle measurements can be modeled respectively as [2]:

$$\phi_{ij} = \tan^{-1}\left(\frac{x_{i2} - a_{j2}}{x_{i1} - a_{j1}}\right) + m_{ij}, \forall (i,j) \in \varepsilon_A, \tag{3}$$

and

$$\alpha_{ij} = \cos^{-1}\left(\frac{x_{i3} - a_{j3}}{\|x_i - a_j\|}\right) + v_{ij}, \forall (i,j) \in \varepsilon_A, \tag{4}$$

According to (1), (3) and (4), we can obtain the ML estimator as [11]:

$$\widehat{X} = \arg\min_{X} \sum_{i=1}^{3|\varepsilon_A| + |\varepsilon_T|} \frac{1}{\sigma_i^2}[\theta_i - f_i(X)]^2, \tag{5}$$

where $\theta = [L^T, \phi^T, \alpha^T]^T$ $(\theta \in \mathbb{R}^{3|\varepsilon_A| + |\varepsilon_T|})$, with $L = [L_{ij}]^T, \forall (i,j) \in \varepsilon_A \cup \varepsilon_T, \phi = [\phi_{ij}]^T, \forall (i,j) \in \varepsilon_A, \alpha = [\alpha_{ij}]^T, \forall (i,j) \in \varepsilon_A$, and

---

[1] This assumption is made without loss of generality; it is readily seen that, if $L_{ij}^T \neq L_{ji}^T, \forall (i,j) \in \varepsilon_T$, then it is enough to replace $L_{ij}^T \leftarrow (L_{ij}^T + L_{ji}^T)/2$ and $L_{ji}^T \leftarrow (L_{ij}^T + L_{ji}^T)/2$ when solving the localization problem.

$$f(X) = \begin{bmatrix} \vdots \\ L_0 + 10\gamma \log_{10} \frac{\|x_i - \widehat{a_j}\|}{d_0} \\ \vdots \\ \tan^{-1}\left(\frac{x_{i2} - a_{j2}}{x_{i1} - a_{j1}}\right) \\ \vdots \\ \cos^{-1}\left(\frac{x_{i3} - a_{j3}}{\|x_i - a_j\|}\right) \\ \vdots \end{bmatrix}, \sigma = \begin{bmatrix} \vdots \\ \sigma_{n_{ij}} \\ \vdots \\ \sigma_{m_{ij}} \\ \vdots \\ \sigma_{v_{ij}} \\ \vdots \end{bmatrix}.$$

The non-convex LS estimator in (5) does not have a closed-form solution. Nonetheless, we will show that it can be transformed into GTRS framework and solved efficiently.

4 The Proposed SR-WLS Estimator

Assuming that the initial location estimations of the targets, $\widehat{X}^{(0)}$, are given, the ML in (5) can be solved locally per target, resorting merely to the data collected from its neighbors and employing an iterative methodology. Consequently, target i updates its location estimation in each iteration, t, by solving the below local ML estimator:

$$\widehat{x}_i^{(t+1)} = \arg\min_{x_i} \sum_{j=1}^{3|\varepsilon_{A_i}| + |\varepsilon_{T_i}|} \frac{1}{\sigma_j^2} [\theta_j - f_j(x_i)]^2, \forall i \in \mathcal{T} \ominus, \quad (6)$$

where $\varepsilon_{A_i} = \{j : (i,j) \in \varepsilon_A\}$ and $\varepsilon_{T_i} = \{k : (i,k) \in \varepsilon_T\}$ represent the set of all anchor and all target neighbors of the target i respectively.

Given $\widehat{X}^{(0)}$, for the case where noise power is low, out of (1) we can write:

$$\lambda_{ij}\|x_i - \widehat{a_j}\|^2 \approx d_0^2, \forall i \in \mathcal{T}, \ \forall j \in \varepsilon_{A_i} \cup \varepsilon_{T_i}, \quad (7)$$

where $\lambda_{ij} = 10^{\frac{L_0 - L_{ij}}{5\gamma}}$. Similarly, from (3) and (4) we respectively get:

$$c_{ij}^T(x_i - a_j) \approx 0, \forall i \in \mathcal{T}, \ \forall j \in \varepsilon_{A_i}, \quad (8)$$

and

$$k^T(x_i - a_j) \approx \|x_i - a_j\| \cos(\alpha_{ij}), \forall i \in \mathcal{T}, \ \forall j \in \varepsilon_{A_i}. \quad (9)$$

where $c_{ij} = \left[-\sin(\phi_{ij}), \cos(\phi_{ij}), 0\right]^T$ and $k = [0, 0, 1]^T$. To grant more significance to *nearby* connections, bring into play weights, $w = \left[\sqrt{w_{ij}}\right]$, where $w_{ij} = 1 - \widehat{d}_{ij} \Big/ \sum_{\forall i \in \mathcal{T}, \forall j \in \varepsilon_{A_i} \cup \varepsilon_{T_i}} \widehat{d}_{ij}$, and substitute $\|x_i - \widehat{a}_j\|$ in (9) with $\widehat{d}_{ij}$ described in (2). According to the WLS criterion and (7), (8) and (9) every target updates its location through minimization of the below estimator:

$$\widehat{x}_i^{(t+1)} = \arg \min_{x_i} \sum_{j \in \varepsilon_{A_i} \cup \varepsilon_{T_i}} w_{ij} \left(\lambda_{ij}\|x_i - \widehat{a}_j\|^2 - d_0^2\right)^2 + \sum_{j \in \varepsilon_{A_i}} w_{ij} \left(c_{ij}^T(x_i - a_j)\right)^2$$
$$+ \sum_{j \in \varepsilon_{A_i}} w_{ij} \left(k^T(x_i - a_j) - \widehat{d}_{ij}\cos\left(\alpha_{ij}^A\right)\right)^2$$

(10)

The above WLS problem is not convex and does not have a closed-form solution. Still, (10) may be expressed as a quadratic programming problem of which the *global* solution is found effortlessly [7]. Applying the replacement $y_i = \left[x_i^T, \|x_i\|^2\right]^T, \forall i \in \mathcal{T}$, Eq. (10) is modified as:

$$y_i^{(t+1)} = \arg \min_{y_i} \|W(Ay_i - b)\|^2$$

subject to

$$y_i^T \mathbf{D} y_i + 2l^T y_i = 0,$$

(11)

where $W = I_3 \otimes \text{diag}(w)$, and $\otimes$ denotes the Kronecker product, I_q represents the q dimensional identity matrix, and

$$A = \begin{bmatrix} \vdots & \vdots \\ -2\lambda_{ij}\widehat{a}_j^T & \lambda_{ij} \\ \vdots & \vdots \\ c_{ij}^T & 0 \\ \vdots & \vdots \\ k^T & 0 \\ \vdots & \vdots \end{bmatrix}, b = \begin{bmatrix} \vdots \\ d_0^2 - \lambda_{ij}\|\widehat{a}_j\|^2 \\ \vdots \\ c_{ij}^T a_j \\ \vdots \\ k^T a_j + \widehat{d}_{ij}\cos(\alpha_{ij}) \\ \vdots \end{bmatrix}, D = \begin{bmatrix} I_3 & 0_{3\times 1} \\ 0_{1\times 3} & 0 \end{bmatrix}, l = \begin{bmatrix} 0_{3\times 1} \\ \frac{1}{2} \end{bmatrix},$$

i.e., $A \in \mathbb{R}^{3|\varepsilon_{A_i}| + |\varepsilon_{T_i}| \times 4}$, $b \in \mathbb{R}^{3|\varepsilon_{A_i}| + |\varepsilon_{T_i}| \times 1}$, $W \in \mathbb{R}^{3|\varepsilon_{A_i}| + |\varepsilon_{T_i}| \times 3|\varepsilon_{A_i}| + |\varepsilon_{T_i}|}$, and $0_{p\times q}$ is the $p \times q$-dimensional matrix of all zeros.

Both the objective and the restraint functions in (11) are quadratic. This type of problem is identified as GTRS [7]; it may be solved *exactly* by a bisection method [7].

Assuming that $\mathcal{C}$ is the set of colors of the nodes (in order to coordinate the network [12]), Algorithm 1 encapsulates the suggested distributed SR-WLS algorithm. Lines 5–7 are carried out concurrently by all targets $i \in \mathcal{C}_c$, which might reduce the realization time of the approach. Information interchange occurs exclusively at Line 7, when targets send their location updates $\hat{x}_i^{(t+1)}$ to their neighbors. Because $\hat{x}_i^{(t+1)} \in \mathbb{R}^3$, one concludes that the new approach obliges a transmission of a maximum $3 \times T_{\max} \times M$ real values. The worst case computational complexity of the proposed method is linear, *i.e.*, $T_{\max} \times M \times \mathcal{O}\left(N_{\max} \times \max_i\{3|\varepsilon_{\mathcal{A}_i}| + |\varepsilon_{\mathcal{T}_i}|\} \right)$, where $N_{\max}$ denotes the maximum allowed number of iterations of the bisection method. In the further text, we will refer to Algorithm 1 as "SR-WLS".

Algorithm 1. The proposed distributed SR-WLS algorithm

Require: $\hat{X}^{(0)}, T_{\max}, \mathcal{C}, a_j, \forall j \in \mathcal{A}$
1. **Set:** $t \leftarrow 0$
2. **replicate**
3.   **for** $c = 1, \ldots, \mathcal{C}$ **do**
4.   **for all** $i \in \mathcal{C}_c$ (simultaneously) **do**
5.   Gather $\hat{a}_j, \forall j \in \mathcal{E}_{\mathcal{A}_i} \cup \mathcal{E}_{\mathcal{T}_i}$
6. $\hat{x}_i^{(t+1)} \leftarrow$ solve (11)
7.   Broadcast $\hat{x}_i^{(t+1)}$ to $\hat{a}_j, \forall j \in \mathcal{E}_{\mathcal{A}_i} \cup \mathcal{E}_{\mathcal{T}_i}$
8.   **end for**
9.   **end for**
10. $t \leftarrow t + 1$
11. **until** $t < T_{\max}$

---

5   Performance Evaluation

A set of simulation results is presented here with the intention of assessment of the performance of the new approach in the view of the estimation precision and convergence. All of the considered approaches were solved via MATLAB. In order to demonstrate the benefit of fusing two radio measurements versus traditional localization systems, we include also the performance results of the proposed method when only RSS measurements are employed, called here "SR-WLS$_{RSS}$".

In order to produce the radio measures, models (1), (3) and (4) are employed. A random arrangement of sensors within a box with the edge span $B = 20$ m in each Monte Carlo (M_c) trial was considered. If not declared otherwise, the PLE was set to $\gamma = 3$, the reference distance to $d_0 = 1$ m, the reference path loss to $L_0 = 40$ dB, and the maximum number of steps in the bisection procedure was $N_{max} = 30$. In practice however, it is almost impossible to perfectly estimate the value of the PLE. Thus, in order to account for a realistic measurement model mismatch and test the robustness of the considered approaches to flawed knowledge of the PLE, the true PLE for each connection was chosen from a uniform distribution on an interval $\gamma_{ij} \in [2.7, 3.3], \forall (i,j) \in \varepsilon_A \cup \varepsilon_T$. Finally, the zero estimate of the targets' locations, $\widehat{X}^{(0)}$, was assumed to be at the intersection of the big diagonals of the cube area. The main performance criterion used is the normalized root mean square error (NRMSE), defined as

$$\text{NRMSE} = \sqrt{\sum_{i=1}^{M_c} \sum_{j=1}^{M} \frac{||x_{ij} - \widehat{x}_{ij}||}{MM_c}},$$

where $\widehat{x}_{ij}$ represents the estimation of the real location of the j-th target, x_{ij}, in the i-th M_c trial.

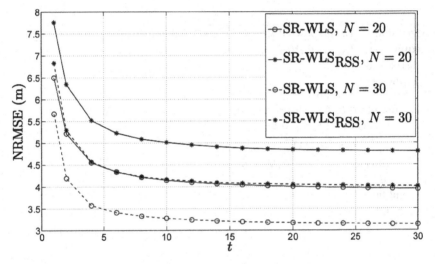

Fig. 2. NRMSE set against t assessment, when $M = 50, R = 6.5\,m, \sigma_{n_{ij}} = 3\,dB, \sigma_{m_{ij}} = 6°, \sigma_{v_{ij}} = 6°, \gamma_{ij} \in [2.7, 3.3], \gamma = 3, \mathbf{B} = 20\,m, L_0 = 40\,dB, d_0 = 1\,m, M_c = 500$.

Figures 2 and 3 illustrate the NRMSE versus t performance when $R = 6.5\,\text{m}, \sigma_{n_{ij}} = 3\,\text{dB}, \sigma_{m_{ij}} = 6°$, and $\sigma_{v_{ij}} = 6°$, for $M = 50, N = 20$ and $N = 30$, and $N = 20, M = 50$ and $M = 60$, respectively. From these figures, we can notice that the output from considered approaches improves as t grows, as anticipated. Also, it can be seen from Fig. 2 that the performance of all algorithms improves considerably as more anchors are included into the network. This behavior is anticipated, because when N is increased more reliable information is introduced in the network. Furthermore, Fig. 3 reveals that both methods need a somewhat elevated number of repetitions to converge for augmented M. Nonetheless, the estimation precision of the considered algorithms improves when more targets are added in the network. Moreover, from Figs. 2 and 3 one can perceive that all major changes in the performance for the considered algorithms occur in the first few iterations ($t \leq 15$), and that the performance gain is insignificant after-wards. This result is very important because it shows that our approach necessitates a low number of signal broadcast, which might augment the exploitation productivity of the spectrum, a valuable resource for wireless communications. It also shows that our

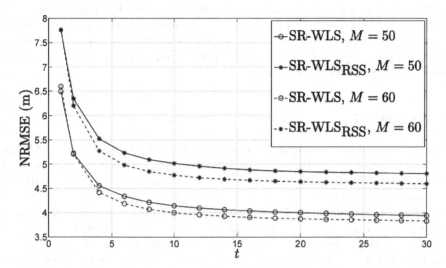

Fig. 3. NRMSE versus t comparison, when $N = 20, R = 6.5\,\text{m}, \sigma_{n_{ij}} = 3\,\text{dB}, \sigma_{m_{ij}} = 6°, \sigma_{v_{ij}} = 6°, \gamma_{ij} \in [2.7, 3.3], \gamma = 3, \text{B} = 20\text{m}, L_0 = 40\,\text{dB}, d_0 = 1\,\text{m}, M_c = 500.$

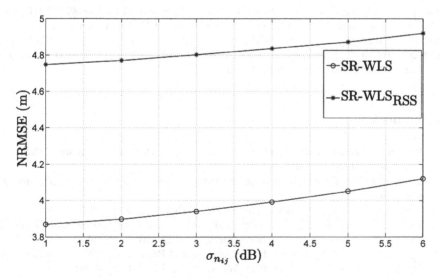

Fig. 4. NRMSE versus $\sigma_{n_{ij}}$ (dB) assessment, for $N = 20, M = 50, R = 6.5\,\text{m}, \sigma_{m_{ij}} = 1°, \sigma_{v_{ij}} = 1°, \gamma_{ij} \in [2.7, 3.3], \gamma = 3, T_{\max} = 30, B = 20\,\text{m}, L_0 = 40\,\text{dB}, d_0 = 1\text{m}, M_c = 500.$

282 S. Tomic et al.

algorithm is energy efficient; the communication stage is considerably more demanding (in terms of energy) than the data processing one [1, 12].

Figure 4 illustrates the NRMSE versus $\sigma_{n_{ij}}$ (dB) assessment, when $N = 20$, $M = 50, R = 6.5$ m, and $T_{max} = 30$. In this figure, one can perceive the performance degradation of the proposed algorithm as the quality of the RSS measurement reduces, as projected. Nonetheless, we can see from the figure that the deterioration in the performance is lower than 15% for the proposed algorithm, which is relatively low for the considered noise range. Finally, Fig. 4 confirms the effectiveness of measurement fusion, showing a gain of almost 1 m when the hybrid system is employed in comparison with the traditional RSS system.

6 Conclusion

Here, the hybrid RSS/AoA target localization problem in a collaborative 3-D WSN was addressed. By fusing information from RSS and AoA, we devised a new problem formulation based on a nonconvex WLS. We showed that the derived novel estimator is suitable for distributed implementation, and we presented a novel non-centralized procedure founded on the GTRS framework. This new algorithm is light in terms of computational complexity, and it can provide accurate localization in a variety of scenarios, having exhibited excellent performance both in view of the estimation precision and convergence. Moreover, the simulation results confirmed the effectiveness of combining RSS and AoA radio measurements in comparison with conventional RSS localization system, displaying a notable enhancement of the estimation precision.

Acknowledgments. This work was partially supported by Fundação para a Ciência e a Tecnologia under Projects PEst-OE/EEI/UI0066/2014, and PEst-OE/EEI/LA0008/2013 (IT pluriannual founding and HETNET), PEst-OE/EEI/UI0066/2011 (UNINOVA pluriannual founding), ADIN PTDC/EEI-TEL/2990/2012, COPWIN PTDC/EEI-TEL/1417/2012 and PTDC/EEITEL/6308/2014-HAMLeT, as well as the grants SFRH/BPD/108232/2015, SFRH/BD/91126/2012 and Ciência 2008 Post-Doctoral Research grant.

References

1. Patwari, N.: Location Estimation in Sensor Networks. Ph.D. thesis, University of Michigan, Ann Arbor, MI, USA (2005)
2. Yu, K.: 3-D localization error analysis in wireless networks. IEEE Trans. Wirel. Commun. 6(10), 3473–3481 (2007)
3. Fletcher, R.: Practical Methods of Optimization. Wiley, Chichester (1987)
4. Wang, S., Jackson, B.R., Inkol, R.: Hybrid RSS/AOA emitter location estimation based on least squares and maximum likelihood criteria. In: IEEE QBSC, pp. 24–29, June 2012
5. Gazzah, L., Najjar, L., Besbes, H.: Selective hybrid RSS/AOA weighting algorithm for NLOS intra cell localization. In: IEEE WCNC, pp. 2546–2551, April 2014
6. Chan, Y.T., Chan, F., Read, W., Jackson, B.R., Lee, B.H.: Hybrid localization of an emitter by combining angle-of-arrival and received signal strength measurements. In: IEEE CCECE, pp. 1–5, May 2014

7. Beck, A., Stoica, P., Li, J.: Exact and approximate solutions of source localization problems. IEEE Trans. Sig. Process. **56**(5), 1770–1778 (2008)
8. Koubaa, A., Jamaa, M.B.: Taxonomy of Fundamental Concepts of Localization in Cyber-Physical and Sensor Networks. Technical report, CISTER, Research Center in Real-Time and Embedded Computing Systems, February 2013
9. Rappaport, T.S.: Wireless Communications: Principles and Practice. Prentice-Hall, Upper Saddle River (1996)
10. Sichitiu, M.L., Ramadurai, V.: Localization of wireless sensor networks with a mobile beacon. In: IEEE MASS, pp. 174–183, October 2004
11. Kay, S.M.: Fundamentals of Statistical Signal Processing: Estimation Theory. Prentice-Hall, Upper Saddle River (1993)
12. Tomic, S., Beko, M., Dinis, R.: Distributed RSS-based localization in wireless sensor networks based on second-order cone programming. Sensors **14**(10), 18410–18432 (2014)

Effective Over-the-Air Reprogramming for Low-Power Devices in Cyber-Physical Systems

Ondrej Kachman[✉] and Marcel Balaz

Institute of Informatics, Slovak Academy of Sciences,
Dubravska cesta 9, 84507 Bratislava, Slovakia
{ondrej.kachman,marcel.balaz}@savba.sk

Abstract. Cyber-physical systems often include sensor devices in their structure. These devices may require firmware updates once deployed and these updates must be energy efficient for battery powered, physically inaccessible sensors. The problem of energy saving reprogramming can be split to four tasks– making old and new firmware versions more similar, generating small delta files using differencing algorithms, propagating delta files and applying updates at the end devices. This paper describes existing approaches dealing with this problem, analyzes their power consumption and introduces new optimizations for differencing algorithms. A new approach is presented that requires no external flash memory, device reboot or complex update agent at the sensor device.

Keywords: Reprogramming · Low-power devices · Over-the-air programming · Firmware similarity · Differencing algorithm · Update agent · Power consumption

1 Introduction

As the self-programming flash memories and wireless communication entered the world of embedded systems, over-the-air reprogramming became possible. It was especially examined for wireless sensor networks (WSNs), where hundreds of small sensor devices may be deployed in a large area. Firmware preloaded on devices in WSN is usually developed under test conditions and may not work properly once the devices are deployed. Over-the-air reprogramming made it easier to update firmware on the devices, but transferring the whole firmware image to many devices is energy inefficient as sending 1 bit wirelessly can consume the same energy as 1000 instructions [1]. This is critical for wireless low-power devices powered by batteries that require firmware update.

The problem of effective over-the-air reprogramming motivates the research question of this paper. How do over-the-air updates influence energy consumption of the updated devices and what can be done to improve the energy efficiency of these updates. The task is also to identify partial problems in this area and their solutions. We present our contribution to differencing algorithms and show how it can help to perform firmware updates effectively.

© IFIP International Federation for Information Processing 2016
Published by Springer International Publishing Switzerland 2016. All Rights Reserved
L. Camarinha-Matos et al. (Eds.): DoCEIS 2016, IFIP AICT 470, pp. 284–292, 2016.
DOI: 10.1007/978-3-319-31165-4_28

2 Relationship to Cyber-Physical Systems

Cyber-physical systems (CPS) are a promising direction in automated interaction between the physical and the virtual world. Based on information collected from physical world, CPS can properly respond with programmed actions. Interaction between the two worlds is provided by intelligence of CPS, which may be created by 3D displays, sensors, actuators, cameras and other devices. This links CPS with previously mentioned WSNs and low-power devices. Sensors and actuators often create an important part of CPS and face the same challenges [2], for example limited battery life, network traffic processing with limited computing capabilities or the adaptation to environment. Over-the-air reprogramming for small devices within CPS should be effective in terms of energy consumption, speed and security.

Energy can be wasted by excessive amount of update data disseminated through the network to end devices. Many protocols have been developed for WSNs [3–5]. However, WSNs usually consist of the same or similar devices and serve to collect data. CPS may consist of many devices with different resources and tasks that may be networked by different technologies, for example WLAN, Bluetooth or GSM [6]. When embedding over-the-air programming capabilities into the low-power device's firmware, developers must consider the right networking technology and protocols in order to reduce the amount of energy consumed by the firmware updates. Another way of energy wasting can be caused by the firmware code. Modern microcontrollers usually have firmware code stored in a program flash memory. Firmware updates are applied from delta files that encode the differences between older and newer versions. Some approaches proposed alteration to compilers [7, 8] and linkers [9] in order to make the versions more similar. Compiler alterations led to more instructions, thus worse execution times and some linker alterations resulted in fragmented program memory. Calls to functions in different parts of flash memory can lead to worse energy consumption as activation of different flash regions requires additional energy [10]. When considering many different devices and platforms in CPS, compiler and linker alterations would be required for every different platform. Developers can instead work with the product of compilers– object files, which usually have the standard executable and linkable (ELF) format, and work with sections and relocation entries [11].

Update speed is important for CPS too. Ineffective update dissemination can slow down network traffic and long downtime of important devices may cause CPS to malfunction. The problem of network traffic can be addressed by effective protocols, Quality of Service mechanisms (QoS) and reduced amount of update data using delta files generated by the differencing algorithms that compare the old and new firmware versions. The delta files are later processed by the end devices and their decoding followed by application of the update may greatly influence the update speed, so their structure and encoding must be taken into consideration by the firmware developers.

The security problem of over-the-air updates concerns CPS as well as any other systems of collaborating computational devices. A device with rogue firmware could infect the network and compromise the system. Network protocols disseminating the update data must implement some security mechanisms and the delta files should include an integrity check to prevent them from tampering.

3   State of the Art and Related Work

The area of effective over-the-air reprogramming can be split into different problems. Based on our analysis of the existing research in this area, we split it to four partial problems: (1) Firmware version similarity, (2) Delta generating differencing algorithms, (3) Update dissemination, and (4) Update application.

Overview of over-the-air update procedures is shown in Fig. 1. The following subsections analyze state of the art for each of the listed problems. One more subsection on flash memory energy consumption modeling is included as a base for our energy analysis of over-the-air updates.

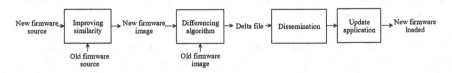

Fig. 1. Over-the-air updates overview

3.1   Firmware Version Similarity

Researchers agree, that firmware updates are more energy effective when data that must be sent through the network are minimal, so the devices save energy on the network communication. First step towards the smaller update data is through improvement in the similarity of two consecutive firmware versions. Considering multiplatform CPS, ELF files are the best way to improve firmware similarity. Authors of [12] identified 4 ELF file properties that caused larger delta file generation – branches, global variables, indirect addressing and relative jumps. Preserving addresses allocated for functions, variables and constants between the firmware versions results in a smaller difference between the linked files. Some of these problems can be addressed by changes to object file relocation entries [11, 13]. Tools altering the ELF files to improve the similarity of firmware images can help the differencing algorithms to perform significantly better.

3.2   Delta Generating Differencing Algorithms

The differencing algorithms that generate delta files (also called deltas or patches) for low-power devices must generate delta as small as possible and encode it in such a way, that the end device can decode it and apply the update easily. Basic approaches used block level comparison [14], which was fast but resulted in unnecessarily large deltas. These deltas transferred whole blocks of fixed size that have changed. Soon, byte level differencing algorithms and delta files were introduced [15–17]. These algorithms detected common sequences in the firmware images and new data. The difference information were encoded into delta files using usually two basic operations – COPY and ADD, each with different attributes. Basic format encoded into file is following:

- <COPY > <old address > <new address > <number of bytes>
- <ADD > <new address > <number of bytes > <first byte > … < last byte>

Every attribute may cost different amount of bytes. Cost for every attribute of an operation adds up to operation's total cost within a delta file, usually in bytes or words.

3.3 Update Dissemination

The delta dissemination stage greatly depends on the network protocol. One of the most notable protocols for WSNs updates is Deluge [4], included in TinyOS, running on many wireless sensor nodes. As mentioned in Sect. 2, CPS consist of different platforms and the network topology may be dynamic. Standard protocols can be used on the higher layers of the CPS. Supported by the QoS mechanisms, update data should reach end devices quickly. If these low power devices cannot support standard the communication protocols (like the most common TCP/IP protocol) due to their computational restraints, they should be provided with border routers that would communicate with them using some lightweight protocol [2].

3.4 Update Application

Updates are applied at the end devices using update agents [18]. The update agents decode the received delta file and execute decoded operations. Depending on implementation, the new firmware image can be reconstructed in an external memory and loaded into the program memory after reboot [11], or it can be applied on the fly, when the update agent edits the program memory directly and runs new firmware right after all the operations from the delta file are executed [12]. This saves the requirement of reboot and firmware loading, but may result in device malfunction if the update is not applied properly or interrupted. The Rollback option from external memory should be therefore present on these devices if possible.

3.5 Flash Memories Energy Consumption Modeling

Creating an accurate model for energy consumption of flash memories is not an easy task. There are many parameters that must be taken into account. The energy model proposed and validated in [10] is focused on embedded flash memories. The flash memory controller activates different regions of flash. These regions are created by 2^k bytes and activation of each region consumes E_k energy. Formal representation:

$$i \rightarrow j = \sum_{k=0}^{N(i,j)} E_k \qquad (1)$$

i and j represent memory address. Term $N(i,j)$ represents the largest changed region – $2^{N(i,j)}$ bytes, $N(i,j) = \log_2(i \oplus j)$. The authors of [19] created and validated an instruction-level energy model for embedded systems. They define a memory access energy consumption of a process as E_{memory}:

$$E_{memory} = E_{cntr} + E_{Flash} + E_{SRAM} \qquad (2)$$

E_{cntr} is energy consumed by the memory controller. E_{flash} is energy consumed by the flash memory write, read and erase operations. E_{SRAM} is energy consumed by write and read operations in RAM memory. This basic model serves us for theoretical evaluation of our differencing algorithm optimizations.

4   Research Contribution and Innovation

In our research of update energy consumption, apart from delta file size, we also take into account encoding of a delta file and update application. We propose optimizations of the differencing algorithms and delta file encoding for updates without external flash. We also create an energy consumption model that can evaluate our solution.

4.1   Differencing Algorithms for Updates Without Use of an External Memory

Low-power devices can be updated without use of an external memory. Self-programmability allows the device to rewrite its program memory on the fly. Some low-power devices used in CPS for control may not have an external memory, so they have to be updated without it if needed. On the fly update does not require copying of identical firmware parts like reconstruction in the external memory. Update application in device's program memory therefore requires different delta file encoding.

Delta file encoding. Updates executed on the fly require different delta file encoding. The traditional way that the reconstructs firmware mixes ADD and COPY operations.

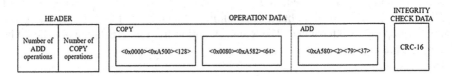

Fig. 2. Delta file format for updates without external memory usage

Now, COPY operations must be carried out first before new data are added. We propose the delta file to be split into three parts – header, operation data and integrity check data (Fig. 2). The header consists of two numbers, the number of COPY and the number of ADD operations. The operation data encode addresses and bytes for each operation. Integrity check data encode CRC-16 code that is checked prior to the firmware update in order to prevent corrupted delta files from updating firmware.

Optimizing differencing algorithms. All analyzed differencing algorithms work with full firmware images. For two images of length n and m, the space complexity in the worse cases is $O(n \times m)$ [15], in the better cases $O(n + m)$ [13, 17]. If we apply xor operation on these images and only compare non-matching sections of a new image to an old image, it is possible to improve algorithm's space complexity, which also results in better execution time. Let k be the number of non-matching bytes between n and m bytes of compared images. Space complexity is reduced to $O(n + k)$ as k is less than m.

Non-matching segments are built using COPY and ADD operations. The generated operations may be sometimes so small that they can be merged into preceding or following operation. For ADD operation with cost x bytes (without data bytes) and COPY operation with cost y bytes, we propose following optimizations:

- If COPY copies at most y bytes, it can be merged into preceding ADD
- If two consecutive COPY operations copy less than $2 \times$ bytes, they can be merged into ADD operation
- If result of COPY operations merging are two consecutive ADD operations, they can be merged into one ADD operation

4.2 Modeling Energy Consumption of the Update

We define energy consumed by flash write operation as $\lambda_{flash(write)}$. This is energy consumed by the erase and write operations for one page. Let s be the flash page size, $addr$ the address we start writing data to and n the number of bytes we write. NAND memories can only be written one page at time, so the number P of pages we must write in order to store n bytes is:

$$P = \lfloor (addr + n) \div s \rfloor - \lfloor addr \div s \rfloor + 1 \tag{3}$$

Using this formula, we can also determine the number of pages we read from, but flash memory can read bytes directly. If energy needed to perform each read operation is $\lambda_{flash(read)}$, then reading n bytes and writing them during COPY and ADD operations will result in flash memory energy consumption:

$$E_{flash} = n \cdot \lambda_{flash(read)} + P \cdot \lambda_{flash(write)} \tag{4}$$

E_{cntr} can be calculated based on Eq. (1). We omit small region activations, for example reading a file byte by byte, and take into account only jumps from location we read to page we write and back ($2P$ jumps total). This usually causes activation of more regions. If jump between two addresses i and j requires activation of R regions, jumping between these addresses during writing of P pages will consume energy:

$$E_{cntr} = 2P \sum_{k=0}^{R} E_k \tag{5}$$

4.3 Experimental Results and Discussion

We implemented the differencing algorithm called Delta Generator (DG) with proposed optimizations that generates delta files in the proposed format. The evaluation was done for 7 firmware change cases on ATmega32U4 microcontroller with 32 KB of self-programming flash memory. The page size of this memory is 128 bytes. We compare our solution to R3diff differencing algorithm [13], one of the best differencing algorithms that generate deltas for firmware updates in an external memory. For this purpose we created tool that calculated the amount of required read and write operations encoded in deltas based on Eq. (3). We also calculated the approximate number of region activations (Eq. (5)) that update process required. Results of our analysis are shown in Table 1.

Table 1. Comparison of R3diff [13] differencing algorithm to proposed Delta Generator (DG)

Change case	Changed bytes	Delta size (bytes)		Read bytes		Written pages		Regions activated	
		R3diff	DG	R3diff	DG	R3diff	DG	R3diff	DG
1	2	15	12	6	4	1	1	28	28
2	2668	966	954	1370	1361	45	37	922	772
3	1222	100	106	628	623	18	12	348	236
4	3054	1522	1448	1621	1674	138	121	2990	2636
5	3150	2051	1986	1675	1722	140	120	3036	2618
6	648	1193	970	563	408	120	85	2936	2198
7	3136	2057	1990	1672	1709	138	122	3026	2724

The experimental results show upgrade over existing solution – R3diff. The amount of read operations is roughly the same, but the proposed optimizations reduced the amount of page writes. The calculation of activated flash regions is not precise, byte by byte readings were excluded from the calculations. We were not yet able to determine exact power consumption of read and write operations for the flash memory used. The exact numbers could better show how not only size, but also encoding influences the energy efficiency of over-the-air updates. We also have not yet evaluated similarity improving techniques. We provided basic model that is a base for future improvements.

5   Conclusion and Further Work

We identified partial problems in the area of over-the air updates. In addition to this, we created the basic energy consumption model of reprogrammed memories and investigated the effect of delta file encoding on update energy performance. The current model is basic. It should be upgraded to a more detailed model and experimentally evaluated with real data. The executed experiments have shown good results. The vision is to create a precise model that takes into account similarity improvements, differencing algorithms, energy consumption of wireless interfaces, energy consumption of update agents and updated firmware. These are the main problems that influence effectivity of over-the-air updates. Perfection of solutions to these problems in combination with good energy consumption model can lead to fast, secure and energy effective updates for low-power devices in cyber-physical systems.

Acknowledgment. This work has been supported by Slovak national project VEGA 2/0192/15.

References

1. Levis, P., Culler, D.: Maté: a tiny virtual machine for sensor networks. In: Proceedings of the 10th International Conference on Architectural Support for Programming Languages and Operating Systems (ASPLOS-X), pp. 85–95. ACM Press, New York (2002)
2. Wu, F.-J., Kao, Y.-F., Tseng, Y.-C.: From wireless sensor networks towards cyber-physical systems. Pervasive Mob. Comput. **7**, 397–413 (2011). Elsevier B.V., Philadelphia
3. Stathopoulos, T., Heidemann, J., Estrin, D.: A Remote Code Update Mechanism for Wireless Sensor Networks. Technical report, Center for Embedded Networked Sensing (CENS). University of California (2003)
4. Hui, J. W., Culler, D.: The dynamic behavior of a data dissemination protocol for network programming at scale. In: Proceedings of the 2nd International Conference on Information Processing in Sensor Networks (IPSN 2008), pp. 81–94. ACM Press, New York (2004)
5. Aschenbruck, N., Bauer, J., Bieling, J., Bothe, A., Schwamborn, M.: Selective and secure over-the-air programming for wireless sensor networks. In: 21st International Conference on Computer Communications and Networks (ICCCN), pp. 1–6. Munich (2012)
6. Shi, J., Wan, J., Yan, H., Suo, H.: A Survey of cyber-physical systems. In: International Conference on Wireless Communications and Signal Processing (WCSP). Nanjing (2011)
7. Zhang, Y., Yang, J., Li, W.: Towards energy-efficient code dissemination in wireless sensor networks. In: International Symposium on Parallel and Distributed Processing (IPDPS), pp. 1–5. Miami (2008)
8. Huang, Y., Zhao, M., Xue, C. J.: WUCC: Joint WCET and update conscious compilation for cyber-physical systems. In: 18th Asia and South Pacific Design Automation Conference (ASP-DAC), pp. 65–70. Yokohama (2013)
9. Koshy, J., Pandey, R.: Remote incremental linking for energy-efficient roprogramming for sensor networks. In: Proceedings of the Second European Workshop on Wireless Sensor Networks, pp. 354–365. Istanbul (2005)
10. Pallister, J., Eder, K., Hollis, S. J., Bennet, J.: A high-level model of embedded flash energy consumption. In: International Conference on Compilers, Architecture and Synthesis for Embedded Systems (CASES), ACM Press, New York (2014)
11. Dong, W., Liu, Y., Chen, C., Bu, J., Huang, C., Zhao, Z.: R2: incremental reprogramming using relocatable code in networked embedded systems. IEEE Trans. Comput. **62**, 1837–1847 (2013)
12. Shafi, N. B., Ali, K., Hassanein, S.: No-reboot and zero-flash over-the-air programming for wireless sensor networks. In: 9th Annual IEEE Communication Society Conference on Sensor, Mesh and Ad Hoc Communications and Networks (SECON), pp. 371–379. Seoul (2012)
13. Dong, W., Mo, B., Huang, C., Liu, Y., Chen, C.: R3: optimizing relocatable code for efficient reprogramming in networked embedded systems. In: IEEE INFOCOM Proceedings, pp. 315–319. Turin (2013)
14. Jeong, J., Culler, D.: Incremental network programming for wireless sensors. In: 1st Annual IEEE Communications Society Conference on Sensor and Ad Hoc Communications and Networks (SECON), pp. 25–33. Santa Clara, California (2004)
15. Hu, J., Xue, C. J., He, Y., Sha, E. H.-M.: Reprogramming with minimal transferred data on wireless sensor network. In: 6th International Conference on Mobile Adhoc and Sensor Systems (MASS 2009), pp. 160–167. Macau (2009)
16. Panta, R.K., Bagchi, S., Midkiff, P.: Efficient incremental code update for sensor networks. ACM Trans. Sens. Netw. (TOSN) **7**, 30 (2011). ACM Press, New York

17. Mo, B., Dong, W., Chen, C., Bu, J., Wang, Q.: An efficient differencing algorithm based on suffix array for reprogramming wireless sensor networks. In: International Conference on Communications (ICC), pp. 773–777. Ottawa (2012)
18. Jurković. G., Sruk, V.: Remote firmware update for constrained embedded systems. In: 37th International Convention on Information and Communication Technology, Electronics and Microelectronics (MIPRO), pp. 1019–1023. Opatija (2014)
19. Bazzaz, M., Salehi, M., Ejlali, A.: An accurate instruction-level energy estimation model and tool for embedded systems. IEEE Trans. Instrum. Meas. 62, 1927–1934 (2013)

Electromagnetic Interference Impact of Wireless Power Transfer System on Data Wireless Channel

Elena N. Baikova[1], Stanimir S. Valtchev[1], Rui Melício[2,3(✉)], and Vítor M. Pires[4]

[1] EST FCT, Universidade Nova, Monte da Caparica, Lisbon, Portugal
[2] IDMEC/LAETA, Instituto Superior Técnico, Universidade de Lisboa, Lisboa, Portugal
ruimelicio@gmail.com
[3] Departamento de Física, Escola de Ciências e Tecnologia,
Universidade de Évora, Évora, Portugal
[4] EST Setúbal, Instituto Politécnico de Setúbal, Setúbal, Portugal

Abstract. This paper focuses on measurement and analysis of the electromagnetic fields generated by wireless power transfer system and their possible interaction on data transmission channel. To measure the levels of electromagnetic fields and spectrum near the wireless power transfer equipment the measurement system in the frequency range 100 kHz to 3 GHz was used. Due to the advances in technology it becomes feasible to apply the wireless power transfer in the electric vehicles charging. Currently, in the Faculty of Science and Technology of the University Nova high power wireless power transfer systems are in development. Those systems need to be controlled by several microcontrollers in order to optimize the energy transmission. Their mutual communication is of extreme importance especially when high intensity fields will generate highly undesired influence. The controllers are supposed to communicate with each other through radio frequency data channels. The wireless power transfer system with the electromagnetic interference may influence or completely disrupt the communication which will be a severe problem.

Keywords: Wireless power transfer · Wireless data transmission · Electromagnetic field · Electromagnetic interference · Experimental results

1   Introduction

Wireless power transfer (WPT) is a promising technology which attracts attention of researchers and manufacturers. In the area of low power the applications are already wide spread: laptops, mobile phones, PDA, wireless headphones, implants or razors. In the same time the high power equipment is also eager to get rid of the wires too, e.g. the industry of intelligent machining systems, robots, the forklift trucks, and electric/hybrid cars. Recently the great deal of attention has been focused on the wireless charging systems for electric vehicles (EV) [1, 2].

Among the existent WPT technologies one of the best results was obtained by Kurs et al. using magnetically coupled resonators [3]. The magnetic resonance technology is

L. Camarinha-Matos et al. (Eds.): DoCEIS 2016, IFIP AICT 470, pp. 293–301, 2016.
DOI: 10.1007/978-3-319-31165-4_29

proven to be the most suitable to achieve efficient energy transfer for the EV wireless charging [3–5]. The magnetic coupling system is more advantageous since it doesn't need an accurate parking position of the vehicle as in the case of inductive coupling.

Communications play a key role in wireless charging systems. The data transmission during wireless power transfer allows a transmitter unit to detect and identify receivers and optimize the energy transfer process by increasing or decreasing the transferred power. The simultaneous power and data exchange by proper choice of the modulation strategy are proposed in the inductive powering systems [6, 7] and in the resonant systems [8, 9].

In this paper, an application of wireless communications technologies in high power WPT systems is discussed focusing on the electromagnetic compatibility problems in simultaneous wireless energy and data transfer system. To the best of authors' knowledge, the wide overview on the electromagnetic compatibility (EMC) problems of wireless power transfer simultaneously with the wireless data transmission has not yet been reported [10], so this paper is the contribution in the field of EMC issues on wireless power and data transmission.

2   Technological Innovation for Cyber-Physical Systems

Nowadays there is a global integration increase of renewable energy systems in the existing electric grid, i.e., power system. Renewable technologies are dependent on weather conditions and are demanding regarding integration into the electric grid. This issue that needs to be fully solved involves the development of smart grid systems [11].

Cyber-physical systems (CPS) can be described as smart systems that includes software and hardware, namely components for sensing, monitoring, gathering, actuating, computing, communication and controlling physical infrastructures, completely integrated and directly interacting to sense the alterations in the state of the surrounding environment [12, 13]. In CPS computing and physical systems interact tightly in real-time. Smart electric grid, smart buildings, healthcare, wireless charging systems for electric vehicles are examples of emerging CPS.

The WPT system for the electric vehicles charging proposed in this paper is controlled by two microcontrollers in order to optimize the energy transmission. The mutual communication between the physical sub-systems, i.e. microcontrollers with sensing components, which allows to control and management of EV charging process, is the example of CPS as an integration of computation/communication with physical processes. In this case the bi-directional data transmission through a Radio Frequency (RF) channel in the WPT system make possible to directly interact with events in the physical world, i.e. to monitor and collect data from physical processes of charging EV battery.

3   WPT Simultaneously with the Wireless Data Transmission

The wireless power and data transmission system with resonant coils tuned at one single working frequency is analyzed in [6–8]. The transmitted working frequency is modulated at the same time by the data transmission. In order not to be influenced by the energy transfer the data transmission can be treated by different technical methods, e.g. amplitude shift keying (ASK), frequency shift keying (FSK) and phase shift keying (PSK) modulation.

The feasibility of the power and data receiver implementation applying amplitude modulation with a single antenna is proposed in [6]. The system of wireless power and data transmission, as described in [8], operates at a frequency of 13.56 MHz and allows simultaneous transmission of energy and data at speeds up to 1 Mb/s. The principle of operation is the modulation of the carrier frequency by the information signal.

However, the efficiency of energy transfer in this method is low because the transferred average power is reduced by the modulation signal. The modulation signal cannot be too low since must be protected from the energy transferred, i.e., noise.

The contactless power and data transmission system that uses two pairs of resonant coils is shown in [14, 15]. The adopted solution is based not only on separated coils for power and information transmission, but also on specially shaped multiple frames. The special geometry of data coils reduces the mutual coupling between the power and data coils. This way the voltage induced by electromagnetic field of WPT system can be reduced.

In the systems generally described, the power and data transfer are transmitted through the near electromagnetic field which operates across relatively short distances, i.e., up to a few meters. Unfortunately, the near field operation limits the communication speed and energy transfer efficiency.

To overcome these limitations a more efficient power and data transfer system is proposed to implement. In this system a bidirectional communication between the transmitter and receiver is created. For the coordination and optimization of the transmitter and receiver of the WPT system, a set of microcontrollers is in development. It is assumed that this system will transmit not only high density energy but also high-density (high speed) data through RF data channels.

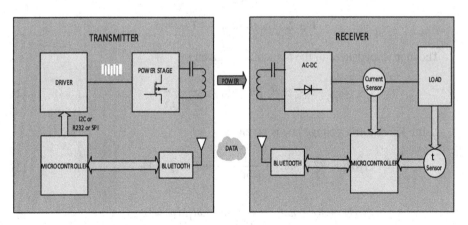

Fig. 1. Wireless Power Transfer System: Transmitter and Receiver Block Diagrams.

Currently, in the Faculty of Science and Technology (FCT), University Nova wireless power and data transmission system based on magnetic resonant coupling are being constructed. The simultaneous wireless powering and data communication transmitter and receiver block diagram is shown in Fig. 1.

The system shown in Fig. 1 is based on magnetic resonant coupling. The transmitter consists of a driver, a power stage, a transmitting coil, and a microcontroller. The receiver side includes a receiving coil, AC-DC rectification, current and temperature sensing and another microcontroller. The load is the EV battery.

The circuit representation of the two-coil WPT system is shown in Fig. 2. The schematic is composed of two resonant circuits corresponding to the two coils. These coils are connected together via a magnetic field, characterized by coupling coefficient k_{12}. The coupling coefficient k_{12} is given by:

$$k_{12} = \sqrt{\frac{M_{12}}{L_1 L_2}} \tag{1}$$

where M_{12} is the mutual inductance between the coils with inductances L_1 and L_2.

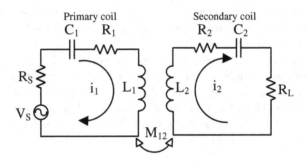

Fig. 2. The two-coil WPT system.

The loop impedances of the two coils are given by:

$$Z_{1,2} = R_{S,L} + R_{1,2} + j\omega L_{1,2} + \frac{1}{j\omega C_{1,2}}. \tag{2}$$

Using Kirchhoff's voltage law is given by:

$$I_1\left(R_s + R_1 + j\omega L_1 + \frac{1}{j\omega C_1}\right) - j\omega I_2 M_{12} = V_S \tag{3}$$

$$I_2\left(R_L + R_2 + j\omega L_2 + \frac{1}{j\omega C_2}\right) - j\omega I_1 M_{12} = 0. \tag{4}$$

4 Wireless Data Technology for WPT System

A data transmission system with higher frequencies, i.e., on the order of GHz allows increasing the data transmission rate. In the case of exchange of information on advanced wireless transmission systems, several communication technologies could be adopted, such as Bluetooth, Wi-Fi, ZigBee [1, 9, 16].

The development and research using well known and well supported technology allows obtaining more reliable solutions. So one of important factors for the wireless data technology will be the high level of standardization and interoperability between devices from different manufacturers. According to [17], in general, there are five potential wireless networks that are being considered for intra- and inter-vehicles applications, namely Bluetooth, UWB, RFID, ZigBee and Wi-Fi. However, only the Bluetooth was implemented in several vehicles in recent years and is now a widely used technology in many vehicles [16, 18].

ZigBee, UWB and Wi-Fi have not yet been implemented in all vehicles, and investigations in this field are still developing [17]. The ZigBee fills a gap provided by other technologies, including wireless sensor interconnection for the control. It is expected that the ZigBee can be used in monitoring and control applications, related to temperature and moisture measurement, as well as heating, air conditioning and lighting control [19].

In order to generalize the communications in WPT systems, it is of any interest to use a well-known, developed and supported technology. In this way one of the important factors for choosing the wireless data transmission technology is the high level of standardization and interoperability between devices from different manufacturers [19].

Given these considerations and the fact that Bluetooth technology is widely used in the automotive industry [17, 19] it's considered a good compromise in terms of the data rate/efficiency/cost. According to these considerations, this technology was adopted since it is quite suitable for data transmission in WPT system.

5 Experimental Results

The analysis was concentrated on the EM radiation caused by converter operating at 20 kHz switching frequency, at which the studied WPT system operates. In principle, the electromagnetic interferences from a relatively low frequency, as expected in case, probably will be not so strong. The Bluetooth technology operates at the frequency of 2.4 GHz which is enough higher than the first harmonic of the power transfer frequency (20 kHz). In order to verify this hypothesis some measurements and analysis were done.

The test equipment used to perform measurements was composed by an experimental setup of WPT system and the measurement system. The WPT system consists of a high frequency power source, transmitting resonant coil, receiving resonant coil and electronic programmable load. The measurement equipment consist of a basic unit, measuring instrument Narda SRM-3000 Selective Radiation Meter, 3-axis E-field antennas in 100 kHz to 3 GHz frequency domain and portable field strength meter, PMM 8053A.

The first series of measurements was taken in the FCT Laboratory. Some disturbance was observed at the frequency 1.9 MHz which may be caused by the heavy consumption of nearby located smartphone. Because of this, in order to prevent external electromagnetic interference, the second series of measurements was made inside a Faraday cage. The experimental setup of WPT system and the measurement equipment in the Faraday cage are shown in Fig. 3.

The first measurement of second data series was executed when the WPT system was non-energized. The electric field in the absence of interference of the WPT system is shown in Fig. 4. As it may be observed, there are interferences at frequencies between 100 kHz and 200 kHz. This can be explained by the nearby fluorescent lamp, whose electronic ballast was working. The electronic ballast operates in the range of frequencies from 25 kHz to 133 kHz.

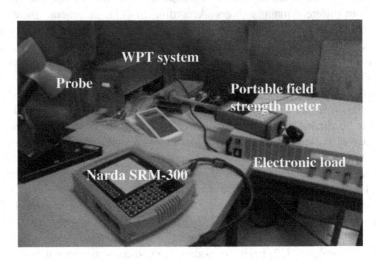

Fig. 3. WPT system - Experimental setup.

The harmonic generation from WPT system operating at 20 kHz frequency is shown in Fig. 5. The used measurement equipment was capable to analyze frequencies starting from 100 kHz, so the amplitudes of the harmonics were measured starting from the 5th harmonic, corresponding to 100 kHz.

To obtain the 1st harmonic and the 3rd harmonic amplitudes, the minimum squares method was used. Using the method of minimum squares the experimental data were adjusted with a coefficient of determination $R^2 = 0.9685$. The high value obtained for R^2 value indicates that the trend line quite precisely fits the data. The amplitude of the harmonics including the 1st harmonic and the 3rd one generated by the WPT system is shown in Fig. 6.

In order to estimate the electromagnetic interference impact on data wireless channel, it is necessary to evaluate the bit error rate (BER), the packet error rate (PER), or throughput of the communication system.

The most usual method to evaluate receiving quality of data wireless channel, including the effect of interference and disturbance in industrial wireless communication, is to check the PER. The PER is the percentage of the number of packets that failed to be received correctly to the number of whole packets transmitted. The PER is also one of the factors determining system throughput and latency [20].

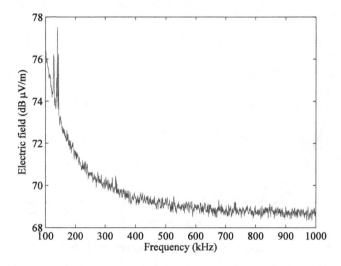

Fig. 4. Electric field in the absence of interference of the WPT system.

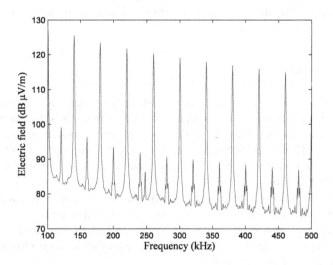

Fig. 5. Harmonic generation from the WPT system operating at 20 kHz frequency.

The Bluetooth communication performance was evaluated using the free software Iperf3 [21] and Wireshark [22]. Initially Bluetooth communication between two laptops placed in the operating area of WPT system was established. One of the laptops was used as a server and another one as a client. The server created data streams and sent them to the client using the software Irerf3.

To capture frames and determine the PER a network protocol analyzer, Wireshark, was used. It can capture and analyze packets to determine those ones that were lost by transmission errors.

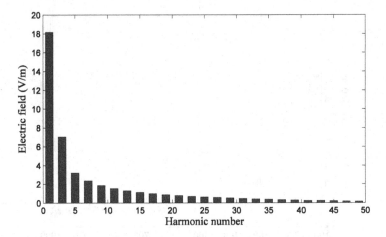

Fig. 6. Amplitude of the harmonics generated by the WPT system.

The results confirmed that there were no packets with transmission errors, so it can be considered that the studied WPT system doesn't influence the data wireless channel operating at 2.4 GHz frequency.

6 Conclusions

This paper presents a study related to the electromagnetic interference impact of Wireless Power Transfer system on data wireless channel. The wireless data transmission is important to improve the efficiency of the wireless power transfer. The consideration about choosing the wireless data technology was also presented. From the described technologies, the Bluetooth can be considered one of the solutions for the data transmission in WPT systems.

To prevent external electromagnetic interference the measurements of the electromagnetic fields generated by WPT system were taken in the Faraday cage. The levels of electromagnetic fields due to the WPT system are present. In order to estimate the possible impact of the electromagnetic interference on data wireless channel the evaluation of packet error rate was made.

The results of experiments confirmed that there is no significant impact from Wireless Power Transfer system operating at 20 kHz frequency on data transfer channel operating at 2.4 GHz frequency.

References

1. Li, S., Mi, C.: Wireless power transfer for electric vehicle applications. IEEE J. Emerg. Sel. Top. Power Electron. **3**(1), 4–17 (2015)
2. Valtchev, S.S., Baikova, E.N., Jorge, L.R.: Electromagnetic field as the wireless transporter of energy. Electron. Energ. Facta Universitatis, Nis, Serbia **25**(3), 171–181 (2012)

3. Kurs, A., Karalis, A., Moffatt, R., Joannopoulos, J.D., Fisher, P., Soljačić, M.: Wireless power transfer via strongly coupled magnetic resonances. Science **317**, 83–86 (2007)
4. Imura, T., Okabe, H., Hori, Y.: Basic experimental study on helical antennas of wireless power transfer for electric vehicles by using magnetic resonant couplings. In: IEEE Vehicle Power and Propulsion Conference, pp. 9936–9404. Dearborn, USA (2009)
5. Chunbo, Z., Kai, L., Chunlai, Y., Ma, R., Hexiao, C.: Simulation and experimental analysis on wireless energy transfer based on magnetic resonances. In: IEEE Vehicle Power and Propulsion Conference, pp. 1–4. Harbin, China (2008)
6. Wu, C.M., Sun, J.S., Itoh, T.: A simple self-powered AM-demodulator for wireless/data transmission. In: 42th European Microwave Conference, pp. 325–328. Amsterdam, Holland (2012)
7. Tibajia, G.V., Talampas, M.C.R.: Development and evaluation of simultaneous wireless transmission of power and data for oceanographic devices. In: IEEE Sensors, pp. 254–257. Limerick, Ireland (2011)
8. Hmida, G.B., Ghairani, H., Samet, M.: Design of a wireless power and data transmission circuits for implantable biomicrosystem. Biotechnology **6**(2), 153–164 (2007)
9. Yokoi, Y., Taniya, A., Horiuchi, M.,Kobayashi, S.: Development of kW class wireless power transmission system for EV using magnetic resonant method. In: 1st International Electric Vehicle Technology Conference, pp. 1–6. Yokohama, Japan (2011)
10. Obayashi, S., Tsukahara, H.: EMC Issues on Wireless Power Transfer. In: International Symposium on Electromagnetic Compatibility, pp. 601–604. Tokyo, Japan (2014)
11. Blaabjerg, F., Ionel, D.M.: Energy devices and systems – state-of-the art technology, research and development, challenges and future trends. Electr. Power Compon. Syst. **43**(12), 1319–1328 (2015)
12. Foundations for innovation in cyber-physical systems - workshop summary report. National Institute of Standards and Technology (2013)
13. Batista, N.C., Melicio, R., Mendes, V.M.F.: Layered smart grid architecture approach and field tests by ZigBee technology. Energy Convers. Manage. **88**, 49–59 (2014)
14. Bieler, T., Perrottet, M., Nguyen, V., Perriard, Y.: Contactless power and information transmission. IEEE Trans. Ind. Appl. **38**(5), 1266–1272 (2002)
15. Rathge, C., Kuschner, D.: High efficient inductive energy and data transmission system with special coil geometry. In: 13th European Conference on Power Electronics and Applications. Barcelona, Spain (2009)
16. Wireless connectivity guide. Texas Instruments. www.ti.com
17. Green, R.J., Rihawi, Z., Mutalip, Z.A., Leeson, M.S.: Networks in automotive systems: the potential for optical wireless integration. In: 14th International Conference on Transparent Optical Networks, pp. 1–4. Coventry, UK (2012)
18. Woodings, R.W., Cypress, M.G.: Avoiding interference in the 2.4-GHz ISM Band. http://www.eetimes.com/document.asp?doc_id=1273359
19. Ramteke, A., Gurmule, A., Sonkusare, K.: Wireless automotive communications. Discovery **18**(53), 89–92 (2014)
20. Matsuzaki, M.: Reliability and stability of field wireless. Yokogawa Technical report English Edition, vol. 55:2, pp. 15–18 (2012)
21. ESnet: https://fasterdata.es.net/performance-testing/network-troubleshooting-tools/iperf-and-iperf3/
22. Lamping, U., Sharpe, R., Warnicke, E.: Wireshark user's guide (2008)

Real-Time Estimation of the Interference in Random Waypoint Mobile Networks

Luis Irio[1,2(✉)] and Rodolfo Oliveira[1,2]

[1] CTS, UNINOVA, Department of Electrical Engineering, Nova University of Lisbon (UNL),
Lisbon, Portugal
[2] IT, Instituto de Telecomunicações, Lisbon, Portugal
l.irio@campus.fct.unl.pt

Abstract. It is well known that the stochastic nature of the interference deeply impacts on the performance of emerging and future wireless communication systems. In this work we consider an ad hoc network where the mobile nodes adopt the Random Waypoint mobility model. Assuming a time-varying wireless channel due to slow and fast fading and, considering the dynamic path loss caused by the node's mobility, we start by characterizing the interference caused to a receiver by the moving nodes positioned in a ring. Based on the interference distribution, we evaluate two different methodologies to estimate the interference in real-time. The accuracy of the results achieved with the proposed methodologies in several simulations show that they may be used as an effective tool of interference estimation in future wireless communication systems, being the main contribution of this work.

Keywords: Interference estimation · Ad hoc networks · Mobility

1 Introduction

Interference is an important metric in the future generation of wireless communication systems because the traditional single transmitter and receiver model is being progressively replaced by a different approach, where multiple nodes may transmit simultaneously for a single or even multiple receivers.

The interference in wireless mobile networks, and particularly its characterization, is important for many applications. In most wireless mobile scenarios the characterization of the interference is a non-trivial task. While several authors model the interference in non-mobile networks [1], the assumption nodes' mobility introduces a novel degree related with the time-varying nature of nodes' positions. The works already published approaching a formal description of the interference in mobile networks are mainly focused on modeling. The use of statistics describing the level of mobility of the interferers in the modeling process was considered in [2–4]. [2] models the aggregate interference caused by static interferers, being considered that the nodes' mobility only causes a time-varying displacement with respect to the different non-mobile cells. [3] admits a mobile scenario where the nodes adopt the Random Direction mobility model.

© IFIP International Federation for Information Processing 2016
Published by Springer International Publishing Switzerland 2016. All Rights Reserved
L. Camarinha-Matos et al. (Eds.): DoCEIS 2016, IFIP AICT 470, pp. 302–311, 2016.
DOI: 10.1007/978-3-319-31165-4_30

[4] considers that the mobile nodes adopt the Random Waypoint mobility model (RWP), but only the interference power from the nearest interferer to the receiver is considered. [5] proposes an interference model for ad hoc mobile networks where the nodes move in accordance with the RWP and all the contributions of the nodes located within a defined region are considered.

This work starts by characterizing the distribution of the interference caused to a receiver by multiple moving nodes located in a ring, considering path loss and slow and fast fading. Based on the interference distribution, we evaluate two different methodologies to estimate the interference in real-time. The major contribution of this work is the identification of a method to estimate the aggregate interference in random waypoint mobility networks, leading to accurate results when used in real-time.

The next section describes the main contributions of this work. Section 3 presents the general assumptions. In Sect. 4 the distribution of the interference values obtained through simulation is approximated by known distributions in order to identify possible approximations. Section 5 describes two estimation methodologies as well as the real-time estimates obtained through simulation and finally the conclusions are present in Sect. 6.

2 Relationship to Cyber-Physical Systems

Recently, Cyber-physical Systems have attracted much attention from the academic community. These systems are mainly focused on the link between computation and physical processes in terms of their reciprocal interaction. Instead of considering stand-alone physical devices, Cyber-Physical Systems adopt an integrated network of multiple physical devices to enrich the interactions and cooperation between the devices and the virtual worlds available through computation.

Recent advances in wireless communications systems and distributed wireless networks have supported a plethora of innovation in Cyber-Physical Systems. Significant progresses have been observed in mobile ad hoc networks and wireless sensor networks. Examples of Cyber-Physical Systems include mobile robotics and mobile sensors or actuators.

Our work contributes to the development of mobile Cyber-Physical Systems, by studying interference phenomena in mobile wireless networks formed without a central coordinator. By characterizing the interference caused by multiple mobile nodes, the wireless communication process can be improved. Consequently, mobile Cyber-Physical Systems may benefit when higher throughput or reliability is needed.

Basically, we show the impact of the mobility, in terms of average velocity of the cyber-physical devices, in the interference caused to a central receiver. We characterize the interference power at the receiver taking into account the specifics associated with the propagation and mobility scenario. In this way, we contribute to the advance of Cyber-Physical Systems, by proposing an effective solution to estimate the interference, which may be used for different purposes ranging from wireless energy harvesting to the improvement of the wireless communication system.

3 System Description

3.1 Mobility Assumptions

This work considers that the nodes move in accordance with the RWP mobility model [6]. In a RWP model all nodes are firstly placed in a random position (x, y). (x, y) is sampled from an uniform distribution denoted by $x \in [0, X_{max}]$ and $y \in [0, Y_{max}]$. (x, y) denotes the starting point, and the following procedure is the definition of the ending point (x', y'), which is uniformly selected as the starting point (i.e. $x' \in [0, X_{max}]$ and $y' \in [0, Y_{max}]$). Afterwards a node samples the velocity $v \in [V_{min}, V_{max}]$ from an uniform distribution, which is adopted to travel from the starting point to the ending point.

After arriving at the ending point (x', y'), a node selects the duration of a pause (T_p) during which it remains stopped at the ending point. After the time T_p, a node selects another value for the velocity to travel to a different ending point. After arriving at the ending point, a node repeats the same procedure as many times as parameterized in the mobility simulations.

Considering that E[L] represents the expected distance between two random points and $E[V_{wp}]$ represents the expected velocity of the nodes without considering pause, the expected velocity of the nodes considering pause is given by

$$E[V] = \left(\frac{E[L]}{E\left[V_{wp}\right]^{-1} E[L] + E[T_p]} \right),$$ (1)

where $E[V_{wp}]$ and E[L] are defined in [7] as $E\left[V_{wp}\right] = \left(\frac{V_{max} - V_{min}}{\ln(V_{max}/V_{min})} \right)$, $E[L] \approx 521.405$ m, and $E[T_p]$ represents the expected value of the pause duration.

3.2 Network Scenario

The network scenario considered in this work assumes a RWP scenario where n nodes travel in a rectangular region with area $X_{max} \times Y_{max}$. The network model considered in this work assumes that a static node N_c is placed in the center of the considered scenario (located at $(X_{max}/2, Y_{max}/2)$). N_c is a receiver of the mobile transmitters. The objective of this paper is the study (in terms of statistics and estimation) of the interference caused to N_c by the transmitting nodes $\{N_1, N_2, \dots, N_k\}$ positioned within the interference region, i.e. the mobile transmitters located within the ring bounded by the smaller circle of radius R_i and the larger circle of radius R_o, represented in Fig. 1. Knowing that the interference depends on the distance between the transmitter and receiver, a circular model was chosen. The parameters describing the network and the mobility conditions are described in Table 1.

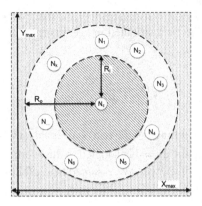

Fig. 1. Interference sensed by N_c due to the mobile interferers located in the annulus area $\pi \left(R_o^2 - R_i^2 \right)$.

Table 1. Parameters adopted in the simulations.

X_{max}	1000 m	n	100
Y_{max}	1000 m	T_p	0 s; 300 s
Simulation time	3000 s	R_i	20 m
V_{min}	5 m/s	R_o	120 m
V_{max}	20 m/s		

3.3 Radio Propagation Assumptions

This subsection describes the radio propagation scenario considered in this work.

The aggregate interference power received by the node N_c located in the centre is expressed by

$$I_{agg} = \sum_{i=1}^{n_{AR_iR_o}} I_i, \qquad (2)$$

where I_i is the interference caused by the node i, and $n_{AR_iR_o}$ is the number of transmitters positioned in the interference area $\pi \left(R_o^2 - R_i^2 \right)$. The interference power I_i is given by

$$I_i = P_{Tx}\psi_i r_i^{-\alpha}, \qquad (3)$$

where P_{Tx} is the transmitted power of the node i ($P_{Tx} = 10^3$ mW). ψ_i denotes the fading occurring in the channel between the node N_c and a transmitter i and r_i is the distance between the transmitter i and the receiver. Finally, α denotes the path-loss coefficient. In this works we consider that the transmitters do not adjust its transmitting power (i.e. no power control is considered).

The fading ψ_i includes the small-scale fading and shadowing effects. The small-scale fading effect is assumed to be distributed in accordance with a Rayleigh distribution, which is represented by

$$f_\zeta(x) = \frac{x}{\sigma_\zeta^2} e^{\frac{-x^2}{2\sigma_\zeta^2}},$$ (4)

where x is the envelope amplitude of the received signal, and $2\sigma_\zeta^2$ is the mean power of the multipath received signal. $\sigma_\zeta = 1$ is adopted in this work.

Regarding the fading effect, we have assumed that it follows a Lognormal distribution

$$f_\xi(x) = \frac{1}{\sqrt{2\pi}\sigma_\xi x} e^{\frac{-(\ln(x)-\mu)^2}{2\sigma_\xi^2}},$$ (5)

where σ_ξ is the shadow standard deviation when $\mu = 0$. The standard deviation is usually expressed in decibels and is given by $\sigma_{\xi dB} = 10\sigma_\xi / \ln(10)$. For $\sigma_\xi \to 0$, no shadowing results. Although (5) appears to be a simple expression, it is often inconvenient when further analyses are required. Consequently, [8] has shown that the log-normal distribution can be accurately approximated by a gamma distribution, defined by

$$f_\xi(x) = \frac{1}{\Gamma(\vartheta)} \left(\frac{\vartheta}{\omega_s}\right)^\vartheta x^{\vartheta-1} e^{-x\frac{\vartheta}{\omega_s}},$$ (6)

where ϑ is equal to $1/\left(e^{\sigma_\xi^2} - 1\right)$ and ω_s is equal to $e^\mu \sqrt{(\vartheta+1)/\vartheta}$. $\Gamma(.)$ represents the Gamma function. The probability distribution function of the fading ψ is thus represented by

$$f_\psi(x) = \frac{2}{\Gamma(\vartheta)} \left(\frac{\vartheta}{\omega_s}\right)^{\frac{\vartheta+1}{2}} x^{\frac{\vartheta-1}{2}} K_{\vartheta-1}\left(\sqrt{\frac{4\vartheta x}{\omega_s}}\right),$$ (7)

which is the Generalized-K distribution, where $K_{\vartheta-1}(.)$ is the modified Bessel function of the second kind.

4   Characterization of the Interference Distribution

Following the assumptions considered in the previous section, several simulations were performed considering two different mobility scenarios:

- Mobility scenario 1 - $V_{min} = 5$ m/s, $V_{max} = 20$ m/s, and $T_p = 0$ s, representing an average node's velocity $E[V] = 10.82$ m/s;
- Mobility scenario 2 - $V_{min} = 5$ m/s, $V_{max} = 20$ m/s, and $T_p = 300$ s, representing an average node's velocity $E[V] = 1.50$ m/s.

Regarding the propagation conditions, we have considered the following scenario:

- Radio scenario – $\alpha = 2$ and $\sigma_{\xi dB} = 3$ dB.

During the simulations, the interference power sensed by the node N_c was sampled every second in order to compute its Cumulative Distribution Function (CDF). The samples acquired in 1000 different simulations, totaling a sample set of $l = 3 \times 10^6$ samples, were also used to determine the parameters of a set of different probability density functions (PDFs) using a maximum-likelihood (ML) fitting methodology. For each one of the considered PDF f, an average logarithm likelihood was defined as follows

$$\hat{g} = \frac{1}{l} \sum_{k=1}^{l} \ln f(x_k | \Theta), \tag{8}$$

where Θ represents the parameters of the PDF and x_k represents each individual sample. ML was used to maximize the likelihood in order to determine Θ, which is described as follows

$$\hat{\Theta}_{MLE} = \underset{\Theta}{\mathrm{argmax}} \, \hat{g}(\Theta; x_1, \dots, x_l). \tag{9}$$

Figure 2 represents the CDFs computed with the parameters obtained in (9) for the Generalized Extreme Value (GEV) and Gamma distributions. As illustrated, the fitting obtained with the GEV distribution presents a better approximation for the two mobility scenarios. Because of this observation, the estimation methods proposed in the next section assume that the interference distribution follows a GEV distribution.

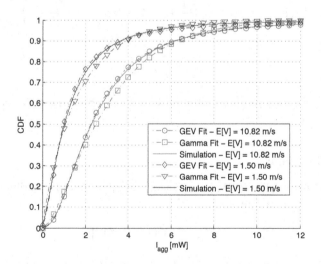

Fig. 2. Interference CDF.

5   Interference Estimation

This section assumes that the aggregate interference can be approximated by a GEV (Generalized Extreme Value) distribution, being its PDF represented by

$$f(x;\sigma,\gamma,\mu) = \frac{1}{\sigma}t(x)^{\gamma+1}e^{-t(x)}, \tag{10}$$

where

$$t(x) = \begin{cases} \left(1 + \gamma\frac{x-\mu}{\sigma}\right)^{-1/\gamma}, & \gamma \neq 0 \\ e^{-(x-\mu)/\sigma}, & \gamma = 0 \end{cases} \tag{11}$$

A Maximum Log-likelihood estimator (MLE) and a Probability Weighted Moments (PWM) estimator are introduced in the next subsections, in order to be used in real-time to estimate the aggregate interference. Hereafter, we denote the elements of an interference sample set by $\chi = X_1, X_2, \ldots, X_m$. We also consider the ordered sample set, which is denoted by $X_{1,m} \leq \ldots \leq X_{m,m}$.

5.1   Log-Likelihood Estimator

The log-likelihood function for a sample set $\chi = \{X_1, \ldots, X_m\}$ of i.i.d GEV random variables is given by

$$\begin{aligned} \log L(\sigma,\gamma,\mu) = & -m\,\log\sigma - \left(\frac{1}{\gamma}+1\right)\Sigma_{i=1}^m \log\left(1 + \gamma\frac{X_i-\mu}{\sigma}\right) \\ & -\Sigma_{i=1}^m \log\left(1 + \gamma\frac{X_i-\mu}{\sigma}\right)^{-1/\gamma} \end{aligned}, \tag{12}$$

under the condition $1 + \gamma\frac{X_i-\mu}{\sigma} > 0$. The MLE estimator $(\hat{\sigma}, \hat{\gamma}, \hat{\mu})$ for (σ, γ, μ) is obtained by maximizing (12).

5.2   PWM Estimator

As described in [9], the PWM of a random variable X with distribution function $F(X) = P(X \leq x)$ are the quantities

$$M_{p,r,s} = E\left[X^p (F(X))^r (1 - F(X))^s\right], \tag{13}$$

for real p, r and s values. For the GEV distribution, [10] shows that $E\left[X(F(X))^r\right]$ can be written as

$$M_{1,r,0} = \frac{1}{r+1} \left\{ \mu - \frac{\sigma}{\gamma} \left[1 - (r+1)^\gamma \, \Gamma(1-\gamma) \right] \right\}, \tag{14}$$

with $\gamma < 1$ and $\gamma \neq 0$. The PMW estimators $(\hat{\sigma}, \hat{\gamma}, \hat{\mu})$ of the Generalized Extreme Value distribution parameters (σ, γ, μ) may be computed through the following system of equations

$$\begin{cases} M_{1,0,0} = \mu - \frac{\sigma}{\gamma}(1 - \Gamma(1-\gamma)) \\ 2M_{1,1,0} - M_{1,0,0} = \frac{\sigma}{\gamma}\Gamma(1-\gamma)(2^\gamma - 1) \, , \\ \frac{3M_{1,2,0} - M_{1,0,0}}{2M_{1,1,0} - M_{1,0,0}} = \frac{3^\gamma - 1}{2^\gamma - 1} \end{cases} \tag{15}$$

in which $M_{1,r,0}$ is replaced by the unbiased estimator proposed in [11]

$$\hat{M}_{1,r,0} = \frac{1}{m} \sum_{j=1}^{m} \left(\prod_{l=1}^{r} \frac{j-l}{m-l} \right) X_{j,m}. \tag{16}$$

5.3 Simulation Results

Figure 3 presents the simulation results obtained for the same scenario adopted in Fig. 2. The "Simulation" curve represents the CDF obtained with the entire set of samples (3×10^6 samples). To apply the MLE and the PWM estimators in real-time we have considered a sample set χ of length $m = 100$ samples. The estimators were computed for 20 different sets of samples, thus 20 different CDFs were computed (one per set). The CDF presented in Fig. 3 is the average of the 20 CDFs computed for each sample set. All of the algorithms to find the estimators were solved by using the MATLAB.

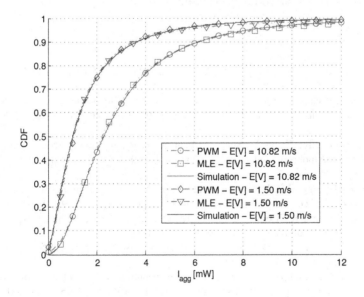

Fig. 3. Simulation results obtained with the MLE and PWM estimators.

Regarding the accuracy of the proposed estimators, both MLE and PWM present high accuracy. As a final remark, the results presented in Fig. 3 validate the proposed estimation methodologies, being the PWM estimator more adequate for the real-time estimation due to its higher accuracy. Finally, we highlight that approximate results were observed for smaller sample set sizes using the PWM estimator, and similar results may be achieved using only $m = 10$ samples per sample set, which is a remarkable low number of samples. The results are not show in the paper due to lack of space.

6   Conclusions

In this work we consider an ad hoc mobile network where the nodes move in accordance with the Random Waypoint mobility model. Assuming a time-varying wireless channel due to slow and fast fading and, considering the dynamic path loss due to the mobility of the nodes, we start by characterizing the interference distribution caused to a receiver by the mobile interferers located in a ring. The simulation results confirmed that the distribution of the aggregated interference may be accurately approximated by a Generalized Extreme Value distribution. Based on the interference distribution, two different methodologies (MLE and PWM) were assessed to estimate the interference in real-time. The accuracy of the results achieved with the proposed methodologies show that they may used as an effective tool of interference estimation in future wireless communication systems. Moreover, the low number of required samples constitutes one of the advantages of the proposed PWM estimator, even when the samples are highly correlated.

Acknowledgments. The authors gratefully acknowledge financial support from the Portuguese Science and Technology Foundation (FCT/MEC) through the project ADIN-PTDC/EEITEL/ 2990/2012 and the grant SFRH/BD/108525/2015.

References

1. Win, M.Z., Pinto, P.C., Shepp, L.A.: A mathematical theory of network interference and its applications. Proc. IEEE **97**, 205–230 (2009)
2. Yarkan, S., Maaref, A., Teo, K., Arslan, H.: Impact of mobility on the behavior of interference in cellular wireless networks. In: Proceedings of IEEE BECOM 2008, pp. 1–5 November 2008
3. Zhang, X., Wu, L., Zhang, Y., Sung, D.: Interference dynamics in MANETs with a random direction node mobility model. In: Proceedings of IEEE WCNC 2013, pp. 3788–3793 April 2013
4. Gong, Z., Haenggi, M.: Interference and outage in mobile random networks: expectation, distribution, and correlation. IEEE Trans. Mob. Comput. **13**, 337–349 (2014)
5. Irio, L., Oliveira, R., Bernardo, L.: Aggregate interference in random waypoint mobile networks. IEEE Commun. Lett. **19**, 1021–1024 (2015)
6. Johnson, D., Maltz, D.: Dynamic Source Routing in Ad Hoc Wireless Networks. Kluwer Academic Publishers, Boston (1996)
7. Bettstetter, C., Resta, G., Santi, P.: The node distribution of the random waypoint mobility model for wireless ad hoc networks. IEEE Trans. Mob. Comput. **2**, 257–269 (2003)

8. Abdi, A., Kaveh, M.: On the utility of gamma PDF in modeling shadow fading (slow fading). In: Proceedings of IEEE VTC 1999, pp. 2308–2312 (May 1999)
9. Greenwood, J., Landwehr, J., Matalas, N., Wallis, J.: Probability weighted moments: definition and relation to parameters of several distributions expressable in inverse form. Water Resour. Res. **15**, 1049–1054 (1979)
10. Hosking, J., Wallis, J., Wood, E.: Estimation of the generalized extreme-value distribution by the method of probability-weighted moments. Technometrics **27**, 251–261 (1985)
11. Landwehr, J., Matalas, N., Wallis, J.: Probability weighted moments compared with some traditional techniques in estimating Gumbel parameters and quantiles. Water Resour. Res. **15**, 1055–1064 (1979)

Energy - Smart Grids

Adaptive Multi-agent System for Smart Grid Regulation with Norms and Incentives

Thiago R.P.M. Rúbio[✉], Henrique Lopes Cardoso, and Eugénio Oliveira

LIACC/DEI, Faculdade de Engenharia, Universidade do Porto,
Rua Dr. Roberto Frias, 4200-465 Porto, Portugal
{reis.thiago,hlc,eco}@fe.up.pt

Abstract. Regulatory policies applied to traditional energy systems are not sufficient when considering smart grids' new requirements such as distributed and decentralised coordination. New management techniques are needed in order to shape consumers conducts by prohibiting, sanctioning or incentivising specific behaviours leading to more efficient utilisation of resources. This paper discusses the possibility of regulating demand in smart grids by the application of processes reflecting the utilisation of norms and incentives in order to better adjust supply to demand. Thus we consider this *soft-control* policy based on the business layer rather than traditional control for cyber-physical systems. The Business Process Modelling (BPM) approach will make easier the design, management and observation of the norms and incentives as flexible artefacts, highlighting decision and coordination processes.

Keywords: Smart grids · Multi-agent systems · Regulation · Coordination · Norms · Incentives · Business process modelling

1 Introduction

The introduction of renewable power sources, storage appliances, and domestic production developed in the last years introduced complexities that exceed current systems management capabilities. From one hand, the natural evolution to a smart grid paradigm becomes very important since current structures are not efficient to control loads in a dynamic and decentralised fashion.

From the other hand, the possibility of direct and timing interaction (e.g. control/actuation) among the various actors in the system brings changes also in the structure of the electricity market. Currently, operators keep generation at peak demand level, leading to large load wastes of energy and monetary losses. Typical actors in market are generator companies (GenCo) and brokers (retailer companies) representing the energy *supply*, and consumers, such as big industrial facilities, households and electric vehicles representing the *demand*. In the smart grid concept consumers can have also production capabilities; this is the case of *prosumers*. The regulation of the demand means the application of control policies in order to lead the consumers to adopt specific conducts and dynamic adapting strategies will allow efficient supply-demand adjustments.

© IFIP International Federation for Information Processing 2016
Published by Springer International Publishing Switzerland 2016. All Rights Reserved
L. Camarinha-Matos et al. (Eds.): DoCEIS 2016, IFIP AICT 470, pp. 315–322, 2016.
DOI: 10.1007/978-3-319-31165-4_31

Spontaneously some research questions are rising about the way we can gain insights about the smart grid's ecosystem properties and processes. In which way is it possible to influence entities' behaviours and create those adaptive mechanisms that lead the system into an efficient coordination? And last, how could these strategies support managers in real-world's decision processes? The current work results from a first effort to tackle the aforementioned issues. During our research we will seek to address all the related issues that will come up.

Given the domain's inherent characteristics, such as distributed control through autonomous entities, decentralised approaches as Multi-Agent Systems (MAS) are being pointed as the most appropriate [1]. We intend to produce a MAS to represent a smart grid and the connected energy market. In this context, acting as smart grid actors, agents are connected by energy contracts and the regulatory policies are expressed through energy tariffs. Each agent in the system can represent either a supply-side organisation (e.g. GenCo, brokers) or demand-side individual entity (e.g. household, electric vehicles, or industrial facilities).

Our MAS benefits from Business Process Modelling (BPM) enabling the design of more flexible regulation mechanisms composed by norms and incentives represented as process artefacts.

This paper is structured as follows: Sect. 2 identifies the relationship of this work to cyber-physical systems; Sect. 3 gives an overview of related literature; Sect. 4 discusses the contributions and novelty aspects of the work; and Sect. 5 overviews the critical analysis of this project and forward research directions. Finally, Sect. 6 describes some conclusions and future work.

2 Relationship to Cyber-Physical Systems

Smart grids are characterised by the combination of physical power system with information processing and communication capabilities, and business layers. This synergy creates a large-scale distributed system that reflects the Cyber-Physical System (CPS) features [2]. Our contributions are projected to CPS concept in the way it comprises control and coordination of human-agents with autonomous artificial artefacts (with different behaviours and intentions), energy and information. Although CPS are mostly oriented in controlling physical entities through the actuation on hardware agents applying traditional control techniques, our attempts here comprise a type of control based on a behavioural-shift fashion. The controller does not try to regulate the distribution load, but instead tries to modulate the demand by shifting the load peaks. This is achieved by coordinating the consumption (of not necessarily cooperative agents) in order to avoid herding phenomena. Such *soft-control* [3] is obtained by using price signals, as they derived from the energy market evolution. Load usage will ensure the feedback for regulation. Many issues in this domain are still open and the full development of the smart grid depends on solving them.

The importance of adjusting demand and supply and the mobility of energy, linking multiple grids are discussed in [4, 5]. Concerns exist also regarding political gridlocks, market inefficiency, and short-sighted planning as some of the existing bottlenecks.

In this sense, BPM is presented as a systematic approach to support the understanding of the interaction between business and control layers to achieve compliance. Moreover, regulatory mechanisms and incentives for small consumers are requirements to achieve success with smart grids [6].

3 State of the Art

According to the European Commission, "A smart grid is an electricity network that can cost efficiently integrate behaviours and actions of all users connected to it - generators, consumers and those that do both - in order to ensure economically efficient, sustainable power system with low losses and high levels of quality and security of supply and safety" [7].

Regulation in the current grid is almost exclusively lead by governmental regulatory policies that control supply and demand. In a fully operational smart grid, new regulatory policies will address the necessary coordination for efficient management. Some examples of regulation practises are: legal restrictions, contract regulations, self-regulations, or social regulations [8].

Norms represent means to control interactions between individuals to achieve desired behaviours and responses [9]. Normative MAS can be created for environments where agents can decide whether to follow or not the explicit norms [10]. Incentives, on the other hand are the instrument that can push agents in a desired way [11, 12, 13]. Some characteristics of the MAS paradigm [1, 14] make this an appropriate paradigm to model smart grids. Recent works using MAS in smart grids address coalition formation, distributed control, and market-driven competition [15].

BPM allows to decouple the business logic layer from agent's inner reasoning and behaviours, centring the focus on the functional aspects of the system management [16, 17]. Few research relates BPM and MAS, although some works on translating from one to another can be found in [18] and [19]. Efforts on the combination of normative systems with incentives mechanisms still lack implementation and analysis, as seen in [20]. Besides, new models might be created to help solving challenging problems as the participation of Electric Vehicles (EVs) [21], tariff composition and efficient appliance scheduling [13].

4 Contributions and Innovative Aspects

Smart grid managers may benefit from regulation since agents should be able to adapt their strategies given system's responses and their own goals [22]. Customers can also benefit from agents automatic negotiation by improving their ability to buy and sell energy [4]. However, selfish behaviours are present and managers will try to minimise their impact for the sake of social welfare [21]. Recent works [9, 12] show some efforts in decentralised approaches to regulate dynamic environments. Due to the inherent complexity of smart grids, big challenges are still open [4].

Research on MAS has, in the last years, devoted a strong attention to demand regulation. Although other existing proposals concern the use of norms and incentives to regulate multi-agent systems separately, appropriate models that combine norms and incentives in order to promote cooperative behaviours still lack flexibility and realistic characteristics. The challenge of using regulation as coordination mechanism is among the objectives of our research effort. Figure 1 show the four cornerstones in our smart grid representation. Customers (end-users), market agents (managers), operators and transmission are the main parts of the environment. Regulation is a pervasive element that coordinates all the actions: managers establish contracts that govern customer's consumption and production, constrained by energy availability and prices from both physical (transmission) and operational (economics) parts.

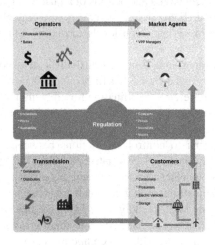

Fig. 1. Regulation in a smart grid architecture

4.1 Incentives and Norms as Processes Artefacts

In this work we discuss the use of BPM aspects that could help users (either acting on supply, or in demand side) to define and authorise agent to act on their behalf and trigger the proper processes. BPM systems consist on the systematic characterisation of the actor's actions when executing a process. Figure 2 shows a typical process, represented with BPMN. The process carry all the logic flow that correspond to the control tasks in a specific scenario, without defining the lower-level actions. Our approach intends to merge processes' tasks into agent's actions and benefit from such flexibility.

This approach offer a dynamic way to analyse and design managers' goals and user's actions, coordinating their interactions. Connecting the business logic layer to physical and economical aspects that constrain agent's actions allows to create a modular configuration for the smart grid scenario.

Describing the business processes in the setting of current and future grids will help to understand and develop novel regulation strategies. Figure 3 shows that processes will develop a major role on organising the interactions of the agents. We propose that the soft-control should occur in business-logic level activities.

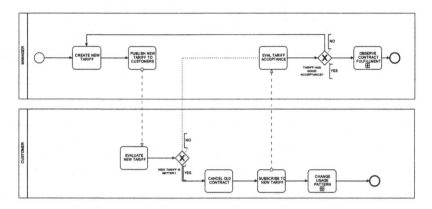

Fig. 2. A typical BPMN process

Managers can observe and act accordingly to enforce system's states and achieve their goals. To do so, we will study how to design norms and incentives as artefacts, such as sub-processes that can be triggered and respond to system's changes on-line. On the other hand, customers will use processes as the linking element that can respond to the environment. Manipulating preferences and contract restrictions, customers can act changing behaviour or facing consequences. Automatic scheduling from agent's decisions will allow customers to optimise demand, decrease costs and improve budgeting.

Following a BPM approach, our models will comprise the process design cycle, describing regulation in current grids **as-is** and new processes proposed for coordination in the smart grid context **to-be**. The problems tackled will focus the creation and adaptation of tariffs in order to ensure management efficiency (e.g. demand peak avoidance). Therefore, the test bed and validation of the current work is being considered in this context.

Fig. 3. Acting on the business process layer will help to achieve the desired states

5   Results and Critical View

The insights of this work are to create a vision and to formalise concepts necessary to future lines of research. Regarding the regulatory tools on smart grids, default control is related to the physical transmission lines and the electrical system. The other way, used on this work, is a system management based on tariff actuation.

Representing the contract between the broker (manager) and the customer, tariffs can wrap financial tools enabling regulation.

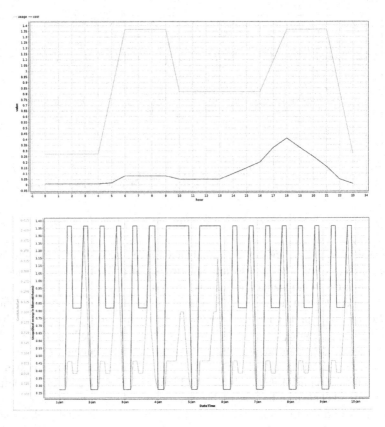

Fig. 4. Consumption vs price in a typical day (left) and consumption vs average daily price during a week (right)

Our preliminary work on this thread considers the PowerTAC[1] framework, a suitable simulation focused on the tariff composition problem. We have created and validated a new model for tariff creation, based on fuzzy logic that uses energy imbalances as a feature for incentivising specific prosumer behaviours [23]. The model creates new tariffs aiming for controlling broker's portfolio and improving its participation on the market. Results showed great potential in our fuzzy strategy. The broker agent is been refined in order to be able to compete in PowerTAC homonym official international competition.

Changing tariff features, such as the energy price, sign-up or withdraw fees can lead to customers' behaviour changes [24]. The first step in order to characterise relationship between the consumption and tariffs consists on analysing real consumption data-sets. Using smart metering data from the U.S. Department of Energy's Green Button initiative[2] we observe that there are some consumption changes accordingly to energy

---

[1] www.powertac.org.

[2] Green Button initiative - http://energy.gov/data/green-button.

prices. Figure 4 (left) shows the patterns for a typical day on the data-set. Market prices goes higher as consumption increases. In the same way, Fig. 4 (right) shows similar patterns for both consumption and prices on working days. Weekend days are evidenced on days 4th and 5th. It appears that exists a correlation between consumption behaviour and energy prices. This is in line with our intentions to apply control mechanisms for shaping customers' behaviour to desired states.

Finally, our best efforts will consider norms and incentives as process artefacts.

6 Conclusions and Future Work

The adoption of smart grids brings new challenges in energy management and thus in the society [4]. The development of an enhanced business layer infrastructure will allow the analysis and design of more complex and distributed strategies for demand coordination. Adopting *soft-control* mechanisms allow some kind of social management, based on incentivising specific behaviours while avoiding others.

Here, a flexible process-oriented approach for regulation in multi-agent systems is proposed, describing how management on the business layer can include norms and incentives as flexible process artefacts. For this purpose the BPM approach is discussed. Following, some preliminary studies are considered. First we have used an existing electricity market simulation framework to address the tariff creation problem in order to incentivise specific customer behaviours. Results indicate the feasibility of applying demand-side soft control based on tariff actuation.

Next, we have analysed the relationship between consumption and energy price from real data-sets. This opens the possibility for exploring the correlation between consumption behaviour and energy prices.

Future work will deploy data-mining techniques for prediction and creation of synthetic load profiles. We will also study how the combination of norms and incentives can suppress or support the emergence of behaviours that are not benevolent to the system.

Acknowledgments. This work is funded through an IBRASIL Grant. IBRASIL is a Full Doctorate programme selected under Erasmus Mundus, Action 2 – STRAND 1, Lot 16 and coordinated by University of Lille.

References

1. Wooldridge, M.: An introduction to multiagent systems. Wiley, Hoboken (2009)
2. Aman, S., Simmhan, Y., Prasanna, V.: Energy management systems: state of the art and emerging trends. Commun. Mag. IEEE **51**(1), 114–119 (2013)
3. Han, J., Li, M., Guo, L.: Soft control on collective behavior of a group of autonomous agents by a shill agent. J. Syst. Sci. Complex. **19**(1), 54–62 (2006)
4. Ramchurn, S.D., Vytelingum, P., Rogers, A., Jennings, N.R.: Putting the 'smarts' into the smart grid: a grand challenge for artificial intelligence. Commun. ACM **55**(4), 86–97 (2012)
5. Ramchurn, S.D., Vytelingum, P., Rogers, A., Jennings, N.: Agent-based control for decentralised demand side management in the grid. In: AAMAS, pp. 5–12 (2011)

6. Caron, S., Kesidis, G.: Incentive-based energy consumption scheduling algorithms for the smart grid. In: 2010 First IEEE International Conference on Smart Grid Communications (SmartGridComm), pp. 391–396. IEEE (2010)
7. The European Commission: Definition, expected services, functionalities and benefits of Smart Grids. The European Commission (2011)
8. Levi-Faur, D.: Regulation and regulatory governance. In: Handbook on the Politics of Regulation, pp. 1–25 (2011)
9. Lopes Cardoso, H., Oliveira, E.C.: Social control in a normative framework: an adaptive deterrence approach. Web Intell. Agent Syst. 9(4), 363–375 (2011)
10. Boella, G., Van Der Torre, L.: Norm negotiation in multiagent systems. Int. J. Coop. Inf. Syst. 16(01), 97–122 (2007)
11. Hermoso, R., Lopes Cardoso, H., Fasli, M.: From roles to standards: a dynamic maintenance approach using incentives. Inf. Syst. Front. 17(4), 763–778 (2015)
12. Centeno Sánchez, R.: Mecanismos incentivos para la regulación de sistemas multiagente abiertos basados en organizaciones (2012)
13. Logenthiran, T., Srinivasan, D., Shun, T.Z.: Demand side management in smart grid using heuristic optimization. In: Smart Grid, pp. 1244–1252. IEEE (2012)
14. Jennings, N.R.: On agent-based software engineering. In: AI, pp. 277–296 (2000)
15. Ketter, W., Collins, J., Reddy, P.: Power tac: a competitive economic simulation of the smart grid. Energy Econ. 39, 262–270 (2013)
16. Scholz-Reiter, B., Stickel, E.: Business Process Modelling. Springer, Berlin (2012)
17. Chinosi, M., Trombetta, A.: BPMN: an introduction to the standard. Comput. Stand. Interfaces 34(1), 124–134 (2012)
18. Küster, T., Lützenberger, M., Heßler, A., Hirsch, B.: Integrating process modelling into multi-agent system engineering. Multiagent Grid Syst. 8(1), 105–124 (2012)
19. Coria, J.A.G., Castellanos-Garzón, J.A., Corchado, J.M.: Intelligent business processes composition based on multi-agent systems. Expert Syst. Appl. 41(4), 1189–1205 (2014)
20. Ossowski, S., Vasirani, M.: Agent-based applications for the smart grid–a playground for agreement technologies. In: Highlights on Practical Applications of Agents and Multi-agent Systems, p. 13 (2014)
21. Marques, V., Bento, N., Costa, P.M.: The "smart paradox": stimulate the deployment of smart grids with effective regulatory instruments. Energy 69, 96–103 (2014)
22. Keshtkar, A., Arzanpour, S., Keshtkar, F., Ahmadi, P.: Smart residential load reduction via fuzzy logic, wireless sensors, and smart grid incentives. Energy Build. 104, 165–180 (2015)
23. Rúbio, T.R.P.M., Lopes Cardoso, H., Oliveira, E.: Tugatac broker: A fuzzy logic adaptive reasoning agent for energy trading. In: Agreement Technologies: Proceedings of the Third International Conference (AT 2015)
24. De Weerdt, M.M., Ketter, W., Collins, J.: Pricing mechanism for real-time balancing in regional electricity markets. In: Proceedings of the 2011 Workshop on Trading Agent Design and Analysis (TADA), Barcelona, Spain, 17 July 2011

Computational Models Development and Demand Response Application for Smart Grids

Rita Pereira[1,2(✉)], João Figueiredo[1,3], and José Carlos Quadrado[4]

[1] CEM/IDMEC, University of Évora, Évora, Portugal
rpereira@deea.isel.pt
[2] Lisbon Superior Engineering Institute, Lisbon, Portugal
[3] IDMEC, Instituto Superior Técnico, Technical University of Lisbon, Lisbon, Portugal
[4] Porto Superior Engineering Institute, Porto, Portugal

Abstract. This paper focuses on computational models development and its applications on demand response, within smart grid scope. A prosumer model is presented and the corresponding economic dispatch problem solution is analyzed. The prosumer solar radiation production and energy consumption are forecasted by artificial neural networks. The existing demand response models are studied and a computational tool based on fuzzy clustering algorithm is developed and the results discussed. Consumer energy management applications within the InovGrid pilot project are presented. Computation systems are developed for the acquisition, monitoring, control and supervision of consumption data provided by smart meters, allowing the incorporation of consumer actions on their electrical energy management. An energy management system with integration of smart meters for energy consumers in a smart grid is developed.

Keywords: Smart grids · Prosumers · Demand response · Energy management applications

1 Introduction

Traditional power grid was designed to operate according to a vertical structure defined by generation, transmission and distribution, supported by several control devices which guarantee the power grid stability, reliability and efficiency [1]. Nowadays the traditional power grid is a system supported by obsolete technology [2] and at the same time it has to deal with new challenges such as increasing consumption, more inaccessible and costly fossil fuels, penetration of renewable source generation, energy markets and several power grid stakeholders. Allowing an active participation of energy consumers, reducing greenhouse gas emissions and minimizing the new implantation of traditional power plants, are other challenges that should be considered [1–3]. In order to provide an answer to these challenges, the smart grid appears as a key element for future power grid design. Mainly because smart grids allow bidirectional power flow and data communication, also because they are based on digital technology and permit to offer new services to consumers supported by smart metering, digital control technologies and by the increasing consumption awareness.

© IFIP International Federation for Information Processing 2016
Published by Springer International Publishing Switzerland 2016. All Rights Reserved
L. Camarinha-Matos et al. (Eds.): DoCEIS 2016, IFIP AICT 470, pp. 323–339, 2016.
DOI: 10.1007/978-3-319-31165-4_32

Smart grids encompasses a panoply of themes, methodologies and technologies, however in this paper highlight is given to smart consumption where three main contributions are introduced: (i) to model a prosumer, showing the economic dispatch problem solution based on generation and consumption forecast given by neural networks; (ii) to develop a demand response (DR) computational tool to support consumers decisions based on fuzzy clustering and (iii) to design smart applications for demand side. These main contributions allow giving support to those challenges derived from the incorporation of distributed generation into the power grid. In addition, the proposed DR model is developed with the purpose of "giving intelligence" to consumers, allowing them to take advantage of smart grids implementation, in what regards to consumers active participation in power grid management and also contributing for energy efficiency increase. The smart applications provide the interface between computational tools, developed models and consumers. The smart meters, when integrated in a control structure, allow the implementation of advanced mathematical models.

2 Relation to Cyber-Physical Systems

Considering Cyber-Physical Systems (CPS) as the interaction between computers and physical devices is an important aggregation element in smart grids development and operation. The information exchange between consumers and power grid operator through smart meters is one of the several prevailing CPS in smart grid environment. The mobile communication used to perform demand response actions, allowing household devices management, helps supporting the desired consumers' active role in grid management. This feature also depends on CPS.

In this paper the relation to CPS is present, in a straightforward way and in an indirect way. An indirect relation to CPS can be found in the prosumer modelling. The prosumer model accuracy depends on consumption and solar radiation forecast data. For the forecast implementation radiation data derived from the Alcáçovas weather station, is used. A processing system is needed to handle the data provided by the sensors and signal acquisition hardware. In addition, a communication system is also involved in this process. Several straightforward relations to CPS can be found in the demand response model deployment. Demand response actions depend on information provided by smart meters. In this paper this provided information is used to define a computation tool that supports consumers in energy management. Through the knowledge of energy price evolution given by the grid operator, available power information and power consumption history data, a fuzzy clustering based software is developed to define which loads can be connected by a domestic consumer, taking into consideration his consumption profiles and operation modes, as described in Sect. 4. These preferences can be related with energy price or lifestyle, namely by defining which loads are eligible to be managed (connected or disconnected), or by defining which hours are more adequate to connect or disconnect loads. Also it is important to notice that the smart applications development and implementation described in Sect. 5 are intrinsically related to CPS. The consumer owned smart meter development incorporates a sensor unit, processor unit, transmitter and display. The hardware is based on Arduino platform. The distributor owned smart meter is the Energy Box [4], which consists on a CPS itself. Moreover, the implementation of smart application presented is based on a strategy that follows an advanced control strategy. This advanced control strategy is designed to perform temperature control actions. The advanced control actions are designed and implemented in Matlab software. The Matlab inputs are references provided by users and the outputs are transmitted to a SCADA system through OPC protocol, in order to perform consumers' tasks, resorting to a programmable logic controller network (PLC). The advanced control is developed in Matlab, because the required computation is not possible on a SCADA system. However the SCADA system is the interface between Matlab and the PLC network. The PLC network is the system actuator, because it receives the information provided by the SCADA system and executes the control action. Computers are used to monitor and to carry out control actions while physical systems, such as sensors and actuators, are the bridge between the cyber system and the end energy consumer.

3   Prosumer Modeling

Several prosumer models are described in literature [5–12]. There are models based in intelligent systems [5, 7, 8], in stochastic correlation [6], in predictive control [10], in graphs [11] in multi-objective methodology [12] and in the aggregation of consumer and producer models [9]. In this paper the analyzed prosumer is a domestic consumer, which is simultaneously a solar photovoltaic energy producer. Because distributed generation has intrinsically intermittent characteristics, the prosumer´s production forecast and consumption forecast is included in economic dispatch problem solution. The prosumer model (1–7) is based in [9] and considers energy buying prices given by energy market, obtained from OMEL's data base [13]. In addition constant selling energy price is considered. The considered generation and consumption power result from forecast models.

$$Max_{P_b,P_l,P_s} U\left(P_l\right) \tag{1}$$

Subjected to the following constraints:

$$\lambda_b P_b - \lambda_s P_s \leq \overline{W} \tag{2}$$

$$\lambda_b - \lambda_s \geq 0 \tag{3}$$

$$P_b + \overline{P_g} = P_l + P_s \tag{4}$$

$$P_{l_{min}} \leq P_l \leq P_{l_{max}} \tag{5}$$

$$P_s - \overline{P_g} \leq 0. \tag{6}$$

Where P_b, P_l and P_s are, buying power, load power and selling power, respectively, $U\left(P_l\right)$ is the utility function given by (7). λ_b and λ_s are the buying and selling price, respectively. $\overline{W}$ is the consumer budget and $\overline{P_g}$ is generation power. P_{lmin} and P_{lmax} are minimum and maximum load power, respectively.

$$U\left(P_l\right) = \sum_{t=1}^{n} \gamma\left(t\right) P_l\left(t\right) = [\gamma\left(1\right) \ldots \gamma\left(n\right)] \begin{bmatrix} P_l(1) \\ \vdots \\ P_l(n) \end{bmatrix}. \tag{7}$$

In (7), $\gamma\left(t\right)$ is the consumption preference factor vector given by (8–9), which depends on consumption time and is settled accordingly to the Portuguese winter and summer hourly cycles definition [14].

$$\gamma\left(t\right) = \gamma_{winter}\left(t\right) = [3\ 3\ 3\ 3\ 3\ 3\ 3\ 3\ 2\ 1\ 1\ 2\ 2\ 2\ 2\ 2\ 2\ 2\ 2\ 1\ 1\ 2\ 2\ 3\ 3] \tag{8}$$

$$\gamma\left(t\right) = \gamma_{summer}\left(t\right) = [3\ 3\ 3\ 3\ 3\ 3\ 3\ 3\ 2\ 2\ 1\ 1\ 1\ 2\ 2\ 2\ 2\ 2\ 2\ 1\ 1\ 2\ 2\ 3\ 3] \tag{9}$$

The consumption preference factor values are shown in Table 1.

Table 1. Consumption preference factor values

Consumption period	$\gamma\left(t\right)$
Peak hour	1
Shoulder hour	2
Off-peak hour	3

The consumer budget is 1.00 €/day, which under some consumption circumstances, can imply the occurrence of consumers load management in order to assure that this budget limit is not exceeded.

The mathematical model (1–7) incorporates prosumer preferences on selling or buying energy without considering prosumer energy storage capability and its cost. Energy selling price is 0.142 €/kWh, which was chosen accordingly to the Portuguese current selling price values.

Solar photovoltaic energy production is higher on summer; however, despite that fact, energy buying prices are superior to winter energy prices, which can be justified by the summer increased average consumption when compared to the average winter consumption.

In this paper, consumption and production are forecasted resorting to artificial neural networks (NN). For the consumption forecast, the forecasted radiation is converted into power generation, resorting photovoltaic solar system PV curves [13] and considers a 5 kW prosumer's power generation capacity. For radiation forecast a feedforward multi-layer perceptron NN with 19 inputs, 43 units in the hidden layers and 1 output, is considered. A sigmoid activation function is used in the hidden layer and an identity active function on the output layer is used. The training process is carried out by Laven-berg Marquard backpropagation algorithm, using the gradient descendent method and the root mean square error is the chosen performance. The stopping criterion is carried out by cross validation method. The NN is trained with radiation and temperature hourly mean values from 4 years (2005, 2007 to 2009). For the consumption forecast, a feed-forward multi-layer perceptron NN with 19 inputs, 49 units in the hidden layer and 1 output is considered. The activation functions used in the hidden and output layers are the same considered in the radiation forecast NN, as well as the training process, performance indicator and stopping criterion. The NN is trained using consumption and temperature hourly mean values available from year 2012. Both consumption and production forecast NN's are tested for a winter and a summer month, using data not provided to the NN during the learning process. Because input data is mainly time series, Pearson correlation is used in order to define inputs' pattern. The implemented consumption NN system is shown in Fig. 1. The last two inputs showed in Fig. 1 give the consumption pattern for a 24 h period [15].

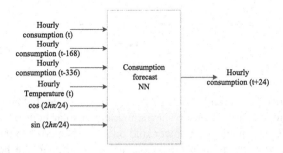

Fig. 1. Consumption forecast NN.

Three scenarios and six case studies are described for the prosumer model analysis, considering daily and monthly time horizons: (a) self-consumption and energy bought in energy market; (b) total produced energy sold and total consumed energy bought in free energy market, and (c) total produced energy sold and total consumed energy bought in energy market considering load shifting.

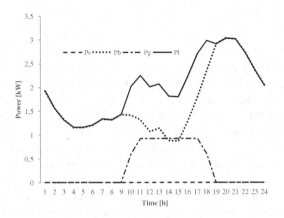

Fig. 2. Daily January EDPS.

The economical dispatch problem solution (EDPS) for scenario (a) considering a daily time horizon is shown in Figs. 2 and 3.

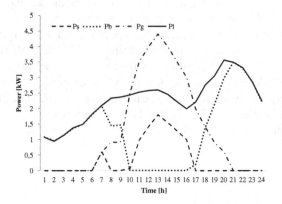

Fig. 3. Daily July EDPS.

From comparison between Figs. 2 and 3 it can be observed that generated power on July is superior to January because solar radiation and solar hour duration is superior during summer. Consequently the power sold on July is superior to the power sold on January and therefore the bought power on July is inferior to the bought power on January. In scenario a) a daily time horizon is used to show in detail the EDPS result and the balance between generation and demand, regarding the possibilities of self-consumption or selling the generated power. The comparison of EDPS between winter and summer is also shown with more detail when daily time horizon is considered. The EDPS results for this scenario are shown in Table 2.

Table 2. Daily EDPS results

Scenario (a)	Energy cost (€)	Utility value
January	0.30	110.01
July	−1.13	112.30

In Table 2, the utility value is adimensional and behaves like a satisfaction indicator, because it gives information about the appreciation of using prosumer's production.

From Fig. 2 and Table 2 analysis considering January, a prosumer's profit would have been achieved if the generated energy was sold instead of consumed. From Fig. 3 and Table 2 analysis considering July, generated power was enough to meet consumption and the remaining generated power was sold, resulting on prosumer's profit.

The EDPS for scenario (b) considering a monthly time horizon is shown in Figs. 4 and 5.

On Figs. 4, 5, 6 and 7, the load power matches with buying power and generated power matches with the power sold. In order to contribute to representation clarity, only the first and the last 24 h of each month are detailed, i.e., 0–24 h and 720–744 h. For the

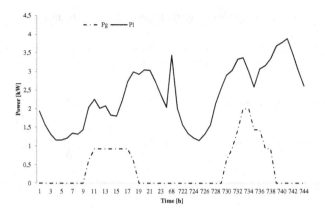

Fig. 4. Monthly January EDPS for scenario (b).

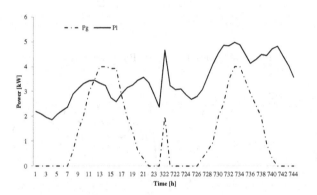

Fig. 5. Monthly July EDPS for scenario (b).

generated power 9 solar hours are considered during winter and 15 solar hours during summer. In this scenario the necessary budget to accomplish the desired consumption without performing load shifting or load shedding actions, is superior to 1 €/day. The EDPS results for this scenario are shown in Table 3.

Table 3. Monthly EDPS results for scenario (b)

Scenario (b)	Energy cost (€)	Utility value
January	52.13	3596.53
July	54.28	2666.73

The lack of prosumer flexibility in performing consumption adjustments resulted into the predefined budget limit violation and therefore the energy cost has increased, despite selling all generated energy. The utility values are compatible with the previous scenario, showing a slight increase on January and a slight decrease on July. For this compatibility analysis the conversion of monthly utility values into average daily utility values is considered.

The EDPS for scenario (c) considering a monthly time horizon is shown in Figs. 6 and 7.

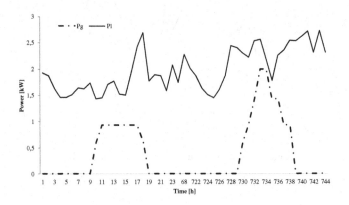

Fig. 6. Monthly January EDPS for scenario (c).

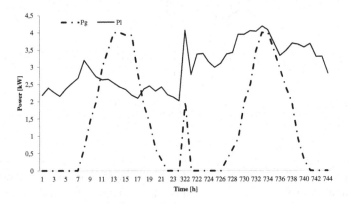

Fig. 7. Monthly July EDPS for scenario (c).

The EDPS results for this scenario (c) are shown in Table 4.

Table 4. Monthly EDPS results for scenario (c)

Scenario (c)	Energy cost (€)	Utility value
January	30	3065.45
July	30	2338.12

In this scenario, the compliance of budget ceiling led to a load shifting and resulted on the utility value reduction, on both months. The referred load shifting resulted from

EDPS outcome as a conjunction of budget ceiling compliance and the P_b, P_l, P_s and P_g values.

In the six case studies the generated power on summer is superior to the one observed during winter as well as the load power. This is justified, not only by the increased solar radiation during summer, but also with geographical characteristics that show higher thermal amplitude during summer when compared to the one verified during winter. The prosumer model drives to a lower energy expenditure when the load shifting occurs, however if consumption needs increases the load shedding is mandatory, in order to accomplish the budget limit [16].

4 Demand Response Model

Usually, DR programs are classified into two main types: Time-Based Program (TBP) [17–20], which is also described as Price-Based Programs [17, 18] or Time-Based Rate program [21] and Incentive-Based Program. The proposed model [22] is TBP type based on fuzzy subtractive clustering algorithm and intends to give consumers flexibility in order to take advantage of economic benefits allowing the load management that best fits consumer's profiles or life-styles. Consumers load management is possible through methods of load scheduling and load shedding. The loads selected to be under consumer management actions are named as controllable load. Consumers' profiles and operation modes were obtained from analysis of consumption behaviour that allowed a consumption pattern definition. Three consumer profiles were set to ensure the coverage of DR generalization on the model and help consumers cope with price changes over one day period. Two operation modes are set for each consumer profile. An example of a priority list is shown in Table 5 [22].

Table 5. Controllable load priority list for cleaning mode.

Economic profile	Moderate profile	Extravagant profile
	Dishwasher Washing machine Dryer machine Air conditioner Thermo ventilator	
Low price	Low or medium price	Low, medium or high price

The priority list shown in Table 5 is the same for all consumer profiles because it assures that consumer choice and preferences are kept, despite a possible adjustment on energy price through a different profile setting. I.e., considering the settled operation mode, all consumption profiles can be chosen accordingly with energy price that consumers' are willing to pay for. Therefore, for any consumption profile, the consumers' preferences are obeyed because CL priority list is common to the 3 profiles. Table 5 shows the relation between consumer profiles and energy prices, where economic profile is only related with low energy price, moderate profile is related with low energy price and also allows a medium energy price and finally the extravagant profile is related to all levels of energy prices, allowing a consumption which is energy price independent. The necessity of pattern recognition associated with a control which supports consumers' decisions for DR model design is fulfilled using fuzzy clustering method. For the control implementation, an off-line fuzzy clustering technique is used because it is intended to determine a DR behaviour pattern and to design a controller that performs the adequate adjustments between the inputs and controller parameters, in order to guarantee an appropriate DR model behaviour. The controller is implemented in the Matlab-Simulink® software resorting to the Fuzzy Logic Toolbox.

The demand response model is shown in Fig. 8 [22].

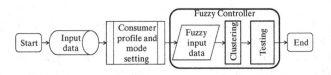

Fig. 8. Demand response model scheme.

The subtractive clustering technique applied in this paper and all DR model considerations are described in [22]. For DR model analysis Figs. 9 and 10 are used as comparison base. The consumer load diagram without DR model implementation is shown in Fig. 9.

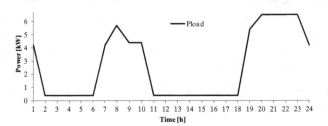

Fig. 9. Load diagram without DR model.

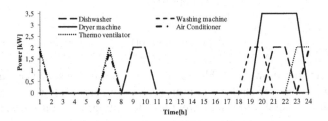

Fig. 10. Controllable loads evolution stage without DR model.

The correspondence between Fig. 9 and the controllable load power diagram is shown in Fig. 10.

The available power, P_a, the base power consumption, P_c, and the energy price (Ep) with DR model implementation are shown in Fig. 11. The available power is assumed to be given by the electric power grid and cannot be exceeded by load power. The base power consumption corresponds to the non-controllable load power consumption.

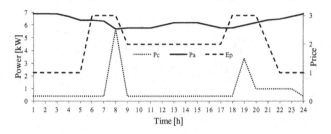

Fig. 11. Power and energy price evolution with DR model.

Considering that an economic profile is selected, the total power consumption, P_t, the base power consumption and available power are shown in Fig. 12. Where the total power consumption diagram is the controllable load power consumption added by the controllable load power consumption.

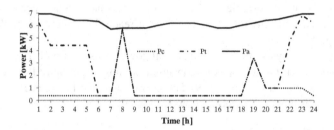

Fig. 12. Power evolution for economic profile.

The correspondence between Fig. 12 and the controllable loads power diagram is shown in Fig. 13. The load power diagram corresponds to the load scheduling resulted from the developed fuzzy subtractive clustering algorithm detailed in [22].

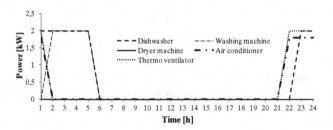

Fig. 13. Controllable loads evolution stage for economic profile.

From comparison between Figs. 11 and 13, it is shown that controllable loads are only connected when energy price is low. In addition, DR model assures that consumption is never superior to available power.

Considering that an ideal profile is selected, the total power consumption, the base power consumption and available power are shown in Fig. 14.

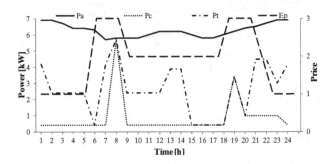

Fig. 14. Power and energy price evolution with DR model for ideal profile.

The ideal profile allows consumers to connect the same controllable loads used without DR model implementation, i.e., is a combination of the 3 profiles in order to allow the same consumption flexibility that consumers have in absence of DR model. As a result, even with DR model consumer can reach the same consumption profile shown in Fig. 9.

From Fig. 14 analysis, it is shown that total power consumption is never superior to available power and that total power consumption diagram is more flattened than the load power diagram shown in Fig. 9. The correspondence between Fig. 14 and the controllable loads power diagram is shown in Fig. 15.

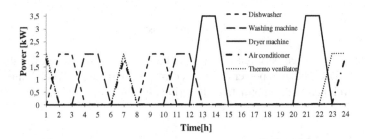

Fig. 15. Controllable loads evolution stage for ideal profile.

For the DR model implementation in the physical system, smart applications development is needed, in order to allow incorporation of the advanced mathematical models into the smart metering and monitoring system.

5 Smart Applications Development and Implementation

The developed strategy is based on the incorporation of two Smart Meters (SM) types in the smart consumer environment (consumer owned SM and distributor owned SM). The consumer owned SM runs over a wireless platform and the distributor owned SM employs the wired environment [4]. The consumer owned SM is a set of simple devices, essentially constituted by a sensor unit and a mobile display, and usually run over a wireless network. Therefore, this system is appropriate to domestic purposes.

The main intention of consumer owned SM is to supply simple data in order to support consumption patterns of consumers. The consumer owned SM is developed with double interface between Supervisory Control And Data Acquisition (SCADA) systems and mobile displays [4]. For the implementation, ZigBee protocol is used, which is suitable for small distances and commonly applied for domestic environments. The developed hardware is based on Arduino platform. Part of this system implementation is shown in Fig. 16. This is an original implementation and consists of 3 sub-systems: the sensor and processor unit, the mobile display and centralized unit, which is the SCADA system.

Fig. 16. Consumer owned SM sensor unit and mobile display [4].

The distributor owned SM is composed by sheltered devices with closed communication protocols, which are constituted by a sensor unit that measures the consumed electricity and remotely informs the local distributer. The consumption information exchange is the main distributed owned SM purpose, because it allows the electricity local distributor to know about the clients' consumption, providing a significant reduction of operational costs. In addition, this data provides important information to the distributor, concerning the consumer patterns, the optimization of energy selling prices, the electric grid management and consumption trends [4]. The presented strategy pursues the advanced control structure shown in Fig. 17 which is composed by 2 levels: the Operational level (SCADA system) and the Interactive level that optimizes the consumers' preferences in relation to control references [4]. The interactive level allows the assumption of advanced control actions. The considered controller development consists on a room control temperature, with distributed interfaces that allows receiving inputs accordingly to room users' temperature preferences. The considered room is connected to the master actuator unit which receives commands from SCADA system in order to perform control temperature regarding restrictions of consumption minimization. On its turn, the master actuator unit commands the HVAC actuator which performs room temperature adjustments. The control algorithm topology is a model-based predictive controller, which is implemented on Matlab platform. Matlab is connected to the SCADA system through OPC protocol. Details of controller design and achieved results can be found on [4]. The operational level consists on a hierarchical cascade control shown in Fig. 18.

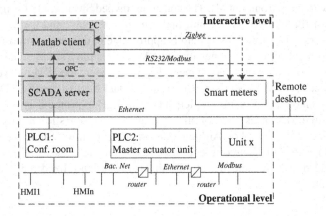

Fig. 17. Two-level supervisory control architecture [4].

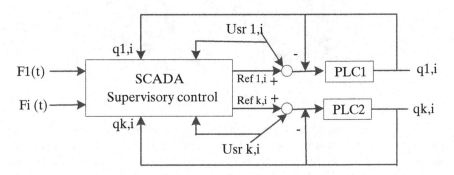

Fig. 18. Operational level controller [4].

The first control loop is managed by local programmable logic controllers and the second control loop is controlled by a SCADA system. The SCADA system inputs referred by (F1(t) .. Fi(t)) are named as comfort laws, because they allow to accommodate users decisions about room temperature, luminosity and consider energy consumption supplied by the SM, as described in [4]. These comfort laws derived from the interactive level shown in Fig. 17. The SCADA system output gives the necessary data to the interactive level. The operational level also incorporates and manages sensors and actuators, namely temperature, luminosity and HVAC. The resulted data is fed as input to the SCADA system.

6 Conclusions

The prosumer economical dispatch problem solution is obtained considering a time dependant cost function that incorporates consumption preferences. Because consumption preferences increase on off-peak and shoulder hours, this prosumer model is in line with DR actions. In addition, the NN incorporation supports grid operator and enhances the grid management efficiency by providing forecast information about production and consumption. In this paper the implemented NN for winter and summer periods gave satisfactory results. For the scenarios considered in this paper, it can be concluded that self consumption is not always the best decision for decreasing the prosumer energy bill. Mainly because prosumer's profit depends on energy selling and buying prices, which are dependant of wholesale energy market. In addition, it can be concluded that the budget limit compliance in association with consumption preferences led to a load shifting.

The DR model is obtained resorting to subtractive fuzzy clustering techniques and intends to be an efficient domestic consumer's supporting tool on load management. Three consumption profiles and two operation modes are considered, in order to give consumers' flexibility to perform DR actions accordingly to their consumption preferences. From the considered case studies, it is visible that, the proposed DR model assures that controllable loads priority list is obeyed and that controllable loads are connected in accordance to the settled consumption profile. Also, the case studies analysis shows that the proposed DR model allows consumer to connect the same controllable loads which were also connected considering the absence of DR model, because the model allows consumers to take advantage of the offered consumption profiles and operation modes. For the analyzed case studies, the DR model guarantees that total consumed power is never higher than the available power. Moreover, it can be concluded that power valley filling can be achieved with the DR model resulting on reshaped consumption diagrams that are mainly instigated by energy price information.

This paper shows an energy management system development with SM for electricity consumers in a smart grid context. The integration of two types of SM is considered; the consumer owned SM and the distributor owned SM. The SM are connected to a SCADA system that supervises a PLC network. The developed control strategy is based on a hierarchical cascade controller. A complete new platform connecting the SCADA supervisory system, the Matlab software, and the two existing main topologies of electricity smart meters (distributor owned and customer owned) is developed. This methodology contributes to provide SCADA systems with the ability to handle advanced control techniques for consumer energy management systems.

References

1. Momoh, J.: Smart grid: fundamentals of design and analysis. In: El-Hawary, M.E. (ed.) IEEE Press, pp. 1–232. Wiley (2012)
2. Collier, S.E.: Ten steps to a smart grid. In: IEEE Industry Applications Magazine (2010)
3. Lo, C.H., Ansari, N.: The progressive smart grid system from both power and communications aspects. IEEE Commun. Surv. Tutorials 14(3), 799–821 (2011)

4. Pereira, R., Figueiredo, J., Melicio, R., Mendes, V.M.F., Martins, J., Quadrado, J.C.: Consumer energy management system with integration of smart meters. Energy Rep. **1**, 22–29 (2015)

5. Vale, Z.A., Morais, H., Khodr, H., Canizes, B., Soares, J.: Technical and economic resources management in smart grids using heuristic optimization methods. In: Proceedings of IEEE Power and Energy Society General Meeting, pp. 1–7, Minneapolis (2010)

6. Shi, Y., Xiong, J.: Contingency constrained economic dispatch in smart grids with correlated demands. In: IEEE International Conference on Smart Grid Communications, pp. 333–338, Brussels (2011)

7. Arif, A., Javed, F., Arshad, N.: Integrating renewables economic dispatch with demand side management in micro-grids: a genetic algorithm-based approach. Energy Effi. **7**(2), 271–284 (2013)

8. Lazzerini, B., Pistolesi, F.: Neural network-based objectives prioritization for multi-objective economic dispatch in microgrids. In: Proceedings of IEEE/SICE International Symposium on System Integration, pp. 665–671, Tokyo (2014)

9. Sun, Q., Beach, A., Cotterell, M.E., Wu, Z., Grijalva, S.: An economic model for distributed energy prosumers. In: Proceedings of 46th Hawaii International Conference on System Sciences (HICSS), pp. 2103–2112, Wailea (2013)

10. del Real, A.J., Arce, A., Bordons, C.: Combined environmental and economic dispatch of smart grids using distributed model predictive control. Electr. Power Energy Syst. **54**, 65–76 (2014)

11. Kellerer, E., Steinke, F.: Scalable economic dispatch for smart distribution networks. IEEE Trans. Power Syst. **30**(4), 1739–1746 (2015)

12. Sousa, T., Morais, H., Vale, Z., Castro, R.: A multi-objective optimization of the active and reactive resource scheduling at a distribution level in a smart grid context. Energy **85**, 236–250 (2015)

13. Silva, A.: Sistema de conversão de energia solar fotovoltaica para interligação à rede doméstica de energia elétrica, 230 V, 50 Hz. In: Master thesis in Electric Engineering and Computers, FEUP, (2008) (in Portuguese)

14. Energy Services Regulatory Authority. http://www.erse.pt/consumidor/electricidade (in Portuguese)

15. Ramezani, M., Falaghi, H., Haghifam, M., Shahryari, G.A.: Short-term electric load forecasting using neural networks. In: The International Conference on Computer as a Tool, EUROCON, pp. 1525–1528, Belgrade (2005)

16. Pereira, R., Pereira, D., Figueiredo, J., Quadrado, J.C., Martins, J., Melício, R., Mendes, V.M.F.: Prosumers Economic Dispatch Model and DC Power Flow Analysis. Energy, Elsevier (2015) (submitted and under review)

17. Songa, M., Alvehaga, K., Widénb, J., Parisio, A.: Estimating the impacts of demand response by simulating household behaviours under price and CO2 signals. Electr. Power Syst. Res. **111**, 103–114 (2014)

18. Albadi, M.H., El-Saadany, E.F.: A summary of demand response in electricity markets. Electr. Power Syst. Res. **78**(11), 1898–1996 (2008)

19. Aalami, H., Yousefi, G.R., Moghadam, M.P.: Demand response model considering EDRP and TOU programs. In: Proceedings of IEEE/PES Transmission Distribution Conference Exposition, pp. 1–6, Chicago (2008)

20. Han, J., Piette, M.A.: Solutions for summer electric power shortages: demand response and its applications in air conditioning and refrigeration systems. J. Refrig. Air Conditioning Electr. Power Mach. **29**(1), 1–4 (2008)

21. Wang, J., Kennedy, S., Kirtley, J.: Optimization of time-based rates in forward energy markets. In: Proceedings of IEEE International Conference on the European Energy Market, pp. 1–7, Madrid (2010)
22. Pereira, R., Fagundes, A., Melício, R., Mendes, V.M.F., Figueiredo, J., Martins, J., Quadrado, J.C.: Demand response analysis in smart grids resorting to fuzzy clustering model. In: Camarinha-Matos, L.M., Tomic, S., Graça, P. (eds.) Contribution to Technological Innovation, pp. 403–412. Springer, Heidelberg (2013)

Load Forecasting in Electrical Distribution Grid of Medium Voltage

Svetlana Chemetova[1(✉)], Paulo Santos[1], and Mário Ventim-Neves[2]

[1] Department of Electrical Engineering ESTSetúbal, Polytechnic Institute of Setúbal,
Rua Vale de Chaves Estefanilha, 2910-761 Setúbal, Portugal
{svetlana.chemetova,paulo.santos}@estsetubal.ips.pt
[2] Department of Electrical Engineering, Universidade Nova de Lisboa,
Faculty of Sciences and Technology Quinta da Torre, 2829-516 Caparica, Portugal
ventim@uninova.pt

Abstract. The importance of forecasting has become more evident with the appearance of the open electricity market and the restructuring of the national energy sector. This paper presents a new approach to load forecasting in the medium voltage distribution network in Portugal. The forecast horizon is short term, from 24 h up to a week. The forecast method is based on the combined use of a regression model and artificial neural networks (ANN). The study was done with the time series of telemetry data of the DSO (EDP Distribution) and climatic records from IPMA (Portuguese Institute of Sea and Atmosphere), applied for the urban area of Évora - one of the first Smart Cities in Portugal. The performance of the proposed methodology is illustrated by graphical results and evaluated with statistical indicators. The error (MAPE) was lower than 5 %, meaning that chosen methodology clearly validate the feasibility of the test.

Keywords: Electric power systems · Load forecasting · Smart-grids · Distribution systems · Electric substations · Artificial Neural Networks

1 Introduction

The recognition of future electricity consumption patterns is an important part of the planning, operation and exploration of Electrical Power System. The quality control of Power Systems and effectiveness of its operation are very sensitive to forecasting errors. Thus, estimating the future energy consumption correctly is a mandatory prerogative in production management, transport and distribution of that energy.

The importance of load forecasting has grown with the emergence of the open electricity market, in the 90 s of the last century. The vertical structure (which was traditional) of the electricity sector was restructured. Relations between production companies, transportation, distribution and marketing in this new scenario became more complex both from a technical and from a commercial point of view.

© IFIP International Federation for Information Processing 2016
Published by Springer International Publishing Switzerland 2016. All Rights Reserved
L. Camarinha-Matos et al. (Eds.): DoCEIS 2016, IFIP AICT 470, pp. 340–349, 2016.
DOI: 10.1007/978-3-319-31165-4_33

In order to improve the efficiency and reliability of the distribution sector, Distribution Management System (DMS) applications are developed, that monitor and control the electricity distribution system [1]. Those applications are illustrated in Fig. 1, in the form of a diagram.

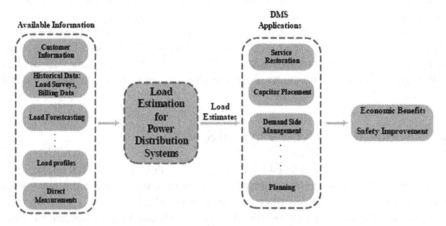

Fig. 1. The main functions of the Distribution Management Systems (taken from [1]).

Load Forecasting is considered in this context as one of the most important functions for the Power System operators. Among the applications, the most expected development is the Demand Side Management (DSM).

As an example of electric demand Fig. 2 represents two load curves, real and forecasted, made by the System Operator (SO) on December 13, 2012 [3]. This load refers to the Portuguese national transmission grid operator REN, without consideration of the interconnection with the European grid.

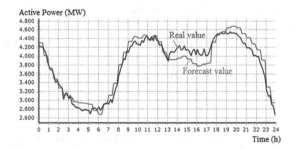

Fig. 2. Real and Forecast values collected in to the REN.

These load curves were obtained for all REN busbars including all the tree levels of voltage (400 kV, 220 kV and 150 kV). The data were acquired in the most important substations using data acquisition system SCADA.

Electrical power is delivered to the Distribution Network, operated by EDP Distribution, which next performs the 60/30 kV, 60/15 kV and 60/10 kV HV/MV transformation in the substations located in the main consumption centers.

Short-term load forecast allows the System Operator to answer to issues such as: network reconfiguration, voltage control, scheduling of maintenance actions and power factor correction, amongst others.

In Portugal there are few studies of load forecasting in distribution networks in medium voltage [5, 6]. Our work will try to contribute to fill this gap. This paper deals with a methodological approach, based on ANN, to forecast the next-hour load, applied to the distribution grid of medium voltage (15 kV), located in the urban area of the city of Évora.

2   Relationship to Cyber-Physical Systems

To test and evaluate the developed load forecasting model, the choice fell on the urban area of the city of Évora, in the context of development of smart grids in this city, a pilot project carried out by EDP Distribution, called InovGrid [2].

The energy sector is facing several challenges: capacity to meet demand for electricity, security of supply, energy efficiency and reduced environmental impact. Évora has been giving priority to smart, sustainable and inclusive projects that contribute to achieving the objectives of Strategy Europe 2020 [4].

The European Smart Grid Task Force defines smart grids as energy networks that can efficiently integrate the behaviour of all users connected to them in order to ensure an economically efficient, sustainable power system with low losses and high quality and security of supply and safety:

The InovGrid project aims to move towards a system of intelligent electrical distribution, focused on energy telemanagement, which revolutionized the networks and their style of consumers/producers interaction.

This proposed change was implemented by a technological renovation and organizational adequacy of the distribution system operation and relationship with other stakeholders, based on a infrastructure that aims to respond to the needs arising from energy efficiency, remote management, distributed generation and microgeneration, and assume active control of the intelligent network (Fig. 3).

This change is associated with the installation of measuring equipment for metering. By the end of 2010, about 30 thousand low voltage customers (domestic, small trade and industry), covering the entire city of Évora, were linked to this integrated, intelligent electrical system.

The time series of the electrical energy consumption obtained by InovGrid telemetry system will be an essential database for the development of load forecasting applications. Load forecasting models, object of this study, will use these data.

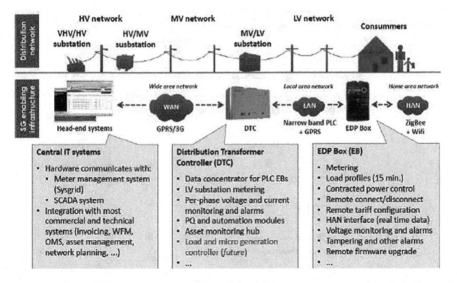

Fig. 3. Supervision and control of the distribution network, InovGrid Project (taken from [7]).

3 Data Analysis

The urban area of the city of Évora is fed with 15 kV medium voltage, by two primary substations (Évora and Caeira) connected to REN - ÉVORA transmission substation (Fig. 4).

Fig. 4. HV and MV distribution networks around Évora.

In the substations, HV/MV transformers are installed: two 60/15 kV in Évora substation and two 60/30/15 kV in Caeira substation (Fig. 5).

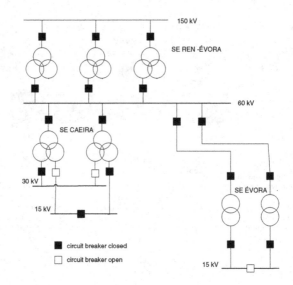

Fig. 5. One line diagram of HV substations in city of Évora.

The Évora substation has two transformers of 31.5 MVA each. This substation is located in an urban area. The substation of Caeira also has two power transformers of 31.5 MVA each and is situated on the border of the city, in the industrial zone. Among the Évora City consuming sectors prevail domestic, services and industry.

The data that are part of the "historic" set of endogenous variables (load) were collected at 15 kV windings of each transformer, through the energy metering system during the years 2002–2013.

Among the social factors that affect the load diagram are the holidays, and other atypical days such as the football world championships etc.

The climatic factors that influence energy consumption are temperature, relative humidity, wind speed, rainfall and luminosity (collected from IPMA).

4   Forecast Methodology

Forecasting methodologies have registered an evolution, over the past four to five decades. They started with the regressive approaches based on time series of active power [8]. The evolution of the electric market increased the complexity of the electrical energy systems and therefore the models based on regressions approaches became more and more complex. Methodologies based on analysis of uncertainty deal better with the information absence, reflecting sometimes more clearly the registered changes in the electrical power systems [9].

In resume, the development of methodologies for forecasting experienced a significant advance. In the sixties years of the XX century, methodologies mainly based on the regressive approach emerged; in the eighties and early nineties of the last century, methodologies emerged, that are based on knowledge and fuzzy techniques [10, 11], artificial neural networks [12, 13], hybrid systems [14] and genetic algorithms [15].

The methods for load prediction used in the energy industry in last years are mainly based on Artificial Intelligence theories because they allow the best way to deal with uncertainty, as well as non-linear functions. There are several scientific publications that prove the quality and robustness of the predictions based on these theories, specifically in artificial neural networks (ANN) [13, 16, 17].

In this research a standard feedforward backpropagation (BP) ANN is used. The neural network is trained on input data as well as the associated target values. The trained neural network can then make predictions based on the relationships learning during training. The data employed for training and validation the neural network were obtained from EDP Distribution (see Part 3 of this paper) for the year 2013. For the simulation data is used the year of 2014.

To define the structure and composition of the input vector to the model based on ANN it is necessary to follow a sequence of tasks [16]. Initially, data pre-processing is performed, mainly in order to fill the gaps caused by failures occurred in the data acquisition system operation. Next, a correlation analysis is made, between the endogenous and exogenous variables, to identify relevant influences of independent random factors on consumption dynamics. Endogenous variables are active and reactive power, exogenous variables are meteorological (among exogenous factors temperature is the most significant). Then, active power time series is subjected to an entropy analysis in order to determine the length of the series of short-term memory. This is to identify the maximum length of the relevant data sequence to be included in the input vector in the ANN algorithm. In a last phase, an auto-correlation analysis is carried out backwards in the time series of power, looking for the best correlation coefficients in such a way as to identify short sequences of data representing the presumable trend of consumption evolution. This is based on the behavior of load in homologous periods of past weeks [17].

After making the analysis described above, we concluded that practically there is no correlation between the load and the climatic variables, and we defined the following input vector, consisting of 13 endogenous variables: [$P(t - 1)$, $P(t - 2)$, $P(t - 3)$, $P(t - 23)$, $P(t - 24)$, $P(t - 25)$, $P(t - 167)$, $P(t - 168)$, $P(t - 169)$, Δ_1, Δ_2, Δ_3 e Δ_4]. Figure 6 shows the predicted variable of the load and two contiguous variables in example of daily load diagram of Évora. The variables $P(t - 1)$, $P(t - 2)$ ... $P(t-n)$ are load values used in the construction of the input vector. Δ_i gradient is calculated for the adjacent load values $P(t - k)$, $P(t - k + 1)$ where k is the "backward displacement".

For building the forecast model it was used a standard backpropagation feedforward ANN having a fully connected architecture with a single hidden layer. The number of neurons in the single hidden layer was half of the one of the input layer [17]. The hyperbolic tangent function was chosen for the middle layer. A linear function was used for the output layer. The input vector was normalized between −1 and 1. This is a well-proven arrangement, adequate when, as in the present case, the relations between the variables at stake have a strong non-linear behavior. In many short-term forecast models this type of ANN structure is widely used [18].

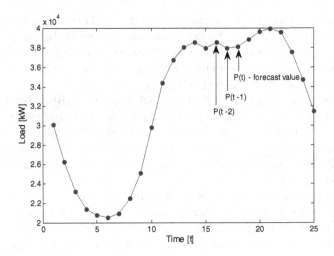

Fig. 6. Representation of contiguous endogenous variables.

5 Results

The results obtained from simulation with the trained neural network are presented below in graphical form (Figs. 7 and 8). Each graph shows a plot of both the "real" (obtained from telemetric reading) and "forecast" data.

The performance of the forecast model has to be evaluated; for this purpose the most current statistical indicators were obtained:

Forecast error: $\qquad e_t = P_t - \hat{P}_t$

Mean error: $\qquad ME = \sum_{t=1}^{n} \frac{e_t}{n}$

Percentage error: $\qquad PE_t = \frac{(P_t - \hat{P}_t)}{\hat{P}_t} 100$

Mean absolute percentage error: $\quad MAPE = \sum_{t=1}^{n} \frac{|PE_t|}{n}$

The forecast error e_t is the difference between the load value recorded in the data acquisition process and the numerical result of our forecasting model. The maximum percentage error (PE_t) calculated value is 6.26 %, the same for the first day and the first week, and is 7.64 % for the first two months of the year 2014 (at this moment no more data referring to consecutive 2014 days is available). The mean absolute percentage error (MAPE) is the most significant indicator in these studies. MAPE calculated value is 1.7 %, 1.3 % and 1.07 % for the same periods of time for 2014, respectively. These are satisfactory results, considering that the maximum MAPE value in such models should not exceed 5 %.

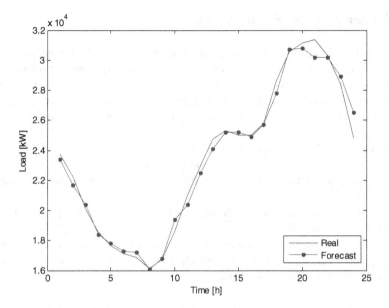

Fig. 7. Load diagram of January 1st, 2014.

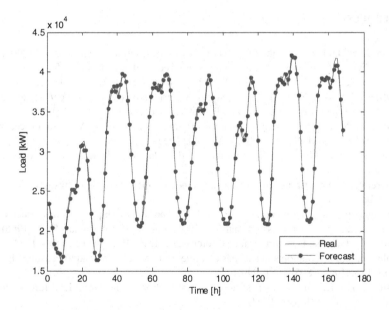

Fig. 8. Load diagram of first week of January, 2014.

6   Conclusions and Future Work

Short-term load forecasting assumes great importance in distribution networks, especially considering the smart grid environment. In fact, the distributors or retailers activities closely related to network management or energy purchase decisions might largely benefit from load forecast models in short time range. The time series of the electrical energy consumption obtained by InovGrid telemetry system used in conjunction with the values acquired by the SCADA system will constitute an essential database for implementation of load forecasting models.

The innovative nature of the forecasting methodology in this work consists in the search for the best technique to achieve a better prediction, for the electrical medium voltage distribution grid.

Forecasting model based on ANN suggested in this work shows good results with very satisfactory error values.

As a future step we will develop a methodology to predict the maximum power daily load diagram (load peak forecasting), which will serve as a support application for operators of Electrical Power Systems.

Acknowledgments.   The authors would like to thank EDP Distribuição and IPMA.

References

1. Wan, J.: Nodal load estimation for electric power distribution systems. Drexel (2003)
2. Smart Energy Grid. http://www.inovgrid.pt/en. Accessed November 2015
3. National Energetic Networks REN, 2012. http://www.ren.pt/. Accessed December 2012
4. GRID INNOVATION online, http://www.gridinnovation-on-line.eu/Articles/Library/InovGrid-Project—EDP-Distribuicao-Portugal.kl. Accessed November 2015
5. Fidalgo, J.N.: Load curve estimation for distribution systems using ANN. In: Proceedings of 7th WSEAS—CIMMACD 2008, pp. 197–202 (2008)
6. Guimarães, A.: Forecast of medium term load evolution in distribution networks. In: IST (2008)
7. Loureiro, M.: Smart grids: an integrated perspective on efficiency, from supply to demand. Coimbra (2014)
8. Heinemann, G.T., Nordman, D.A., Plant, E.C.: The relationship between summer weather and summer loads - a regression analysis. IEEE Trans. Power Apparatus Syst. (1966)
9. Gross, G., Galiana, F.: Short term load forecasting. IEEE Proc. **75**(12), 1558–1573 (1987)
10. Rahman, S.: Formulation and analysis of a rule-based short-term load forecasting algorithm. IEEE Proc. **78**(5), 805–816 (1990)
11. Hsu, Y., Ho, K.: Fuzzy expert systems: an application to short term load forecasting. IEE Proc-C. **139**(6), 471–477 (1992)
12. Park, D., El-Sharkawi, M., Marks, I.R.: Electric load forecasting using an artificial neural network. IEEE Trans. Power Syst. **6**(2), 442–449 (1991)
13. Marin, F., Garcia-Lagos, F., Joya, G., Sandoval, F.: Global model for short-term load forecasting using artificial neural networks. IEEE Proc. Gener. Transm. Distrib. (2), 121–125 (2002)

14. Sfetsos, A.: Short-term load forecasting with a hybrid clustering algorithm. IEEE Proc. Gener. Transm. Distrib. **150**(3), 257–262 (2003)
15. Maifeld, T., Sheblé, G.: Short-term load forecasting by a neural network and refined genetic algorithm. Electr. Power Syst. Res. **31**, 147–152 (1994)
16. Hippert, H., Pereira, C., Souza, R.: Neural networks for short-term load forecasting: a review and evaluation. IEEE Trans. Power Syst. **16**(1), 44–55 (2001)
17. Santos, P., Martins, A., Pires, A.: Designing the input vector to ANN-based models for short-term load forecast in electricity distribution systems. ELSEVIER Electr. Power Energy Syst. **29**, 338–347 (2007)
18. Lu, C., Wu, H., Vemuri, S.: Neural network based on short term load forecasting. IEEE Trans. Power Syst. **8**(1), 336–342 (1993)

Renewable Energy

Control and Supervision of Wind Energy Conversion Systems

Carla Viveiros[1,2,3], R. Melício[1,2(✉)], José M. Igreja[3], and Victor M.F. Mendes[2,3]

[1] IDMEC/LAETA, Instituto Superior Técnico, Universidade de Lisboa, Lisbon, Portugal
ruimelicio@gmail.com
[2] Departamento de Física, Escola de Ciências e tecnologia, Universidade de Évora, Évora, Portugal
[3] Instituto Superior de Engenharia de Lisboa, Lisbon, Portugal

Abstract. This paper is about a PhD thesis and includes the study and analysis of the performance of an onshore wind energy conversion system. First, mathematical models of a variable speed wind turbine with pitch control are studied, followed by the study of different controller types such as integer-order controllers, fractional-order controllers, fuzzy logic controllers, adaptive controllers and predictive controllers and the study of a supervisor based on finite state machines is also studied. The controllers are included in the lower level of a hierarchical structure composed by two levels whose objective is to control the electric output power around the rated power. The supervisor included at the higher level is based on finite state machines whose objective is to analyze the operational states according to the wind speed. The studied mathematical models are integrated into computer simulations for the wind energy conversion system and the obtained numerical results allow for the performance assessment of the system connected to the electric grid. The wind energy conversion system is composed by a variable speed wind turbine, a mechanical transmission system described by a two mass drive train, a gearbox, a doubly fed induction generator rotor and by a two level converter.

Keywords: Modelling · Simulation · Wind energy · Controllers · Supervision · Performance assessment

1 Introduction

The energy crisis of 1973, when it was increased six-fold the price of oil and the blockage of oil-producing countries to Denmark, the Netherlands, Portugal, South Africa and the United States, provided conditions for the resurgence of renewable energies [1]. This crisis has highlighted political consequences that have materialized in actions whose aim is to ensure diversity and security of energy supply. Thus, the motivation and the interest for renewable energies have emerged, and research and development activities in wind energy, as an alternative source of electricity, were stepped up significantly, particularly in Europe and USA.

© IFIP International Federation for Information Processing 2016
Published by Springer International Publishing Switzerland 2016. All Rights Reserved
L. Camarinha-Matos et al. (Eds.): DoCEIS 2016, IFIP AICT 470, pp. 353–368, 2016.
DOI: 10.1007/978-3-319-31165-4_34

With the growing need for electricity production from renewable energy sources, wind turbines are an effective response. Wind turbines are the most common form to describe the wind energy conversion systems (WEnCS) in the form of electricity [2]. After the energy crisis of 1973, in the 80 s, the first wind turbines had rotor diameters between 10 m to 20 m and output power ranging from 25 kW to 100 kW. Research yielded the technological development which allowed the growth conditions favorable for mass production, making possible the development of construction techniques for more robust wind turbines and allowing the increase of the installed power [3].

In the period 2011–2014, there was an average annual growth in installed wind power worldwide by 13 %, reaching a value of approximately 336 GW in mid 2014 [4]. This production technology is considered one of the technologies with the largest and fastest growing worldwide due to the level of penetration and maturity. The wind power installed worldwide in the period 2011–2014 is presented in Fig. 1.

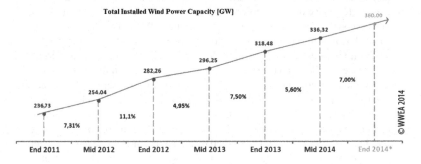

Fig. 1. Installed wind power 2011–2014 [4].

In 2014, India, China, USA, Spain and Germany, are the countries with the largest installed wind power capacity and represent a global market share of 72 % of the world-wide installed wind power [4]. Portugal is positioned in 11th place with an installed wind power capacity of 4829 MW, reached in the first half of 2014, and is considered the second largest electricity generation source in Portugal, reaching 11.8 TWh [5].

In the overview of this paper, Sect. 2 presents the connection of the PhD work to cyber-physical systems. Section 3 presents the modeling and structure of WEnCS connected to the electric grid and the operating regions according to the wind speed are also defined. Control and supervision of WEnCS is presented in Sect. 4. WEnCS control is achieved using different types of controllers such as integer-order controllers, e.g. classical proportional integral (PI), fractional-order controllers e.g. fractional order proportional integral (FOPI), fuzzy logic controllers (fuzzy PI), adaptive controllers, e.g. linear quadratic Gaussian (LQG) or predictive, e.g. model predictive control (MPC). The supervision system is based on FSM. Section 5 presents the simulation results as well the performance of WEnCS with and without supervisor. Conclusions remarks are given in Sect. 6.

1.1 Motivation

Since 2000, wind technology has seen a continuous growth in Portugal, motivated by a political strategy, at European and National levels, in endogenous and renewable resources with the objective of diversifying sources, improving the security on supply, decreasing of energetic dependency and reducing the environmental impact of electro production system. The promotion of renewable energies, particularly wind energy, is particularly important in this international and community context taking into account the objectives and goals of which the country is committed to progressively decrease external energy dependence and reduce the carbon intensity of its economy.

With recent technological advances and reduction of electronic converters and position actuators costs, most WEnCS are equipped with electronic power converters and servomechanisms that control the blade pitch angle. The widespread use of electronic converters power and position servo allows flexibility and controllability of WEnCS behavior in terms of energy harvesting but increases the level of complexity of the control loop system. Hence, WEnCS should be carefully designed to meet the international specifications of the power quality injected in the power grid. The WEnCS control involving electrical and mechanical subsystems is considered complex, thus challenging.

The motivation to address the issue of control and supervision of WEnCS results from the need to respond to the challenge related to the control of the various subsystems constituting the WEnCS. This paper deal with an interesting, current and important research topic which includes appropriate mathematical models describing WEnCS dynamics, the study of different types of controllers and the study of a supervisor based on FSM. Considering the mentioned above, this work aims providing solutions to answer the following research question:

Q.1 how can the electric power of a complex system meet the international specifications of its quality injected in the electric grid?

The adopted work hypothesis to address the research questions is defined below:

Using a stratified structure with two distinct levels, a FSM supervisor in the supervision level, and five different control approaches in the execution level, to improve the performance of the WEnCS.

1.2 Original Contributions

This paper presents original contributions to the development of hierarchical structures with supervision and control applied to WEnCS with special emphasis on its performance when using different types of controllers with and without supervisor. In particular: the study on the performance of five different controllers, PI, FOPI, Fuzzy PI, LQG and MPC applied to WEnCS [6, 7]; the development of a hierarchical structure with two operating levels: execution and supervision level. The supervision level, based on FSM's which represent the operating regions, determines the operational states. The execution level receives the information from the operational states and acts accordingly using one of the controllers [8, 9]; the comparative study of the performance assessment of the developed hierarchical structure, for five different types of controllers in the absence and presence of supervisor [9, 10].

2   Relationship to Cyber-Physical Systems

In the past few years, control and computer science researchers have come together in the development of dominant engineering methods and tools, namely in the areas of system identification, robust control, optimization and stochastic control. In the meantime, researchers also made significant breakthroughs in embedded architectures and systems software, on programming languages, real-time techniques, and innovative approaches ensuring computer system reliability, fault tolerance and cyber security.

The significant breakthroughs in science and technology allow the opportunity to link cyber space and physical components. This bridge leads to Cyber-Physical Systems (CyPS) [11]. CyPS researches are constantly developing strategies in order to combine knowledge and engineering principles crosswise the computational and engineering areas such as mechanical, electrical, networking, control, learning theory to overcome the difficulties existing in CyPS supporting technology and science.

The conception of CyPS is to use communication, computing, and control to build autonomous and intelligent systems. Autonomous controls are widely used in industrial process with control loops and the production process is monitored using sensors placed along the production line and acts with the industrial process using actuators. CyPS focuses on bridging control network architectures using sensors and actuators in complex processes [12].

Currently, CyPS is too involved in the area of energy distribution and management. The research developments in cyber-physical energy systems (CyPES) are concentrated on demand side management, consumption and distribution of technologies, such as smart grids and energy managements of buildings and physical structures. For the generation phase, the innovative CyPES should be focused on renewable sources such as solar and wind energy. Regarding wind farms, its construction and planning implicates a problematical decision because it depends on economic, technical, social and environmental aspects. Following a detailed investigation of WEnCS, it is possible to formulate physical and cyber layers with diverse technical and non-technical layers. One can consider collaborative wind farms, wind turbine (WiT) and integration with smart grids.

The interaction with social context and coordination of cyber layers and physical components of WEnCS require handling with complex challenges regarding competitiveness and practicability of upcoming WEnCS [13].

The success of the integration of wind power into the electric power grid depends on interconnected, distributed CyPS, which is evolving in the engineering industry. The models of CyPS components are composed by state variables, usually discrete values that are updated at discrete events. FSM's are systems with finite states, inputs and outputs assuming values from discrete sets that are updated at discrete transitions triggered by its inputs. The FSM can be used to describe CyPS.

3   WEnCS Modeling

In this chapter it is developed a mathematical model that represents an appropriate dynamic of a WEnCS. The model should be thorough enough to be used as a simulation model. The development of the mathematical model is based on the standard model developed in [14], whose electric power output is 4.8 MW.

3.1   WEnCS

WEnCS are designed in order to convert wind kinetic energy into electrical energy. The wind kinetic energy is captured by the blades causing the rotation of the blades; this rotation transforms the wind kinetic energy into mechanical energy increasing the speed of an electric generator thus obtaining electrical energy. WEnCS operation can be divided into functional subsystems that describe the WEnCS overall operation. The subsystems of the benchmark structure with a supervisor are shown in Fig. 2.

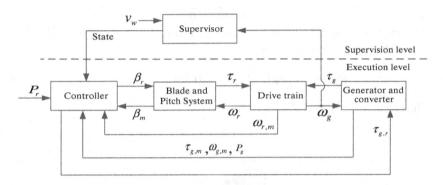

Fig. 2.   Benchmark model structure [14].

Figure 2 shows the relationship between the functional subsystems and the variables involved. The variables are: v_w is the wind speed; τ_r is the turbine rotor torque; ω_r is the turbine rotor speed; τ_g is the generator rotor torque; ω_g is the generator rotor speed; β_r is the rated pitch angle; P_r is the turbine rated power and P_g is the generator output power. The m subscripts designate measured values.

Blade and Pitch Subsystem: This subsystem combines the aerodynamic with blade and pitch models. The torque acting on the blades can be determined by the aerodynamics of the WiT. The aerodynamic torque is expressed as:

$$\tau_r(t) = \frac{\rho \pi R^3 C_p \left(\lambda(t), \beta(t)\right) v_w^2(t)}{2\lambda}. \tag{1}$$

where ρ is the air density, $C_p\left(\lambda(t), \beta(t)\right)$ is the power coefficient, which is, respectively, a function of the tip speed ratio and the pitch angle and R is the radius of the blades. The tip speed ratio is given by:

$$\lambda(t) = \frac{\omega_r(t)R}{v_w(t)}. \tag{2}$$

Observing (2), the variation of the wind speed can lead to two consequences: if the mechanical speed is constant, then $\lambda(t)$ will change, leading to a consequent change in C_p hence in the power capturing; if the mechanical speed is suitably adjusted, then $\lambda(t)$ can be held at a reference point and as a result C_p can be kept at a desired value. The power coefficient of a WiT using pitch control [15] is expressed as:

$$C_p(\lambda,\ \beta) = 0.73\ (\ \frac{151}{\lambda_i} - 0.58\ \beta - 0.002\beta^{2.14} - 13.2\ )\ e^{\frac{-18.4}{\lambda_i}}.$$

(3)

where λ_i is expressed as:

$$\lambda_i = \frac{1}{\dfrac{1}{(\lambda-0.02\ \beta)} - \dfrac{0.003}{(\beta^3+1)}}.$$

(4)

The pitch angle is defined by three hydraulic actuators and can be modeled as a second order system [14] expressed as:

$$\ddot{\beta}(t) = -2\xi\omega_n(t)\dot{\beta}(t) - \omega_n^2\beta(t) + \omega_n^2\beta_r(t).$$

(5)

The C_p curve is shown in Fig. 3.

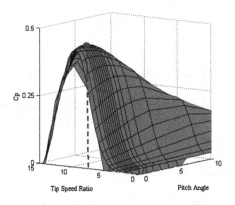

Fig. 3. Power coefficient curve.

Drive Train Subsystem. This subsystem is configured by a two-mass model has a first mass J_r to concentrate the inertia of the turbine blades, hub and low-speed shaft inertia and a second mass J_g to concentrate the generator inertia and high-speed shaft. The low-speed and high-speed shafts are connected by a gear box ratio N_g, with torsion shaft stiffness K_{dt} and torsion shaft damping B_{dt}. This results in the angular deviation $\theta_\Delta(t)$ due to the damping and stiffness coefficients between turbine and generator. The linearized model for this subsystem is expressed as:

$$J_r\dot{\omega}_r(t) = \tau_r(t) + \frac{B_{dt}}{N_g}\omega_g(t) - K_{dt}\theta_\Delta(t) - \left(B_{dt} + B_r\right)\omega_r(t).$$

(6)

$$J_g \dot{\omega}_g(t) = \frac{K_{dt}}{N_g} \theta_\Delta(t) + \frac{B_{dt}}{N_g} \omega_r(t) - (\frac{B_{dt}}{N_g^2} + B_g) \omega_g(t) - \tau_g(t). \tag{7}$$

$$\dot{\theta}_\Delta(t) = \omega_r(t) - \frac{1}{N_g} \omega_g(t). \tag{8}$$

Generator and Converter Subsystem. This subsystem dynamics is described by a first order system. The generator and the power converter model as well as the generator output power are expressed as:

$$\dot{\tau}_g(t) = \alpha_{gc} \left(\tau_{g,r}(t) - \tau_g(t) \right). \tag{9}$$

$$P_g(t) = \eta_g \omega_g(t) \tau_g(t). \tag{10}$$

where α_{gc} is a first order time constant and η_g denotes the generator efficiency.

3.2 Operating Regions

The overall goal of WEnCS control is to optimize the power supplied to the electric grid within a certain range of wind speed, and to minimize energy production and maintenance costs [16]. These costs depend on the conditions in which the WiT is subjected while converting the wind energy captured. Hence, four operating regions are considered, according to the wind variation [17], shown in Fig. 4.

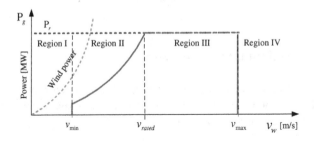

Fig. 4. Operating regions by wind speed.

The maximum electrical power associated with energy supplied to the electric grid is also known as rated or nominal power. The rated wind speed, v_{rated}, is the wind speed at which the rated power is reached.

When the wind speed is too slow, i.e., below 5 m/s [14] the operating region is classified as Region I, and the wind turbine is stopped in this region. When the wind speed is between 5 m/s and 13 m/s the operating region is classified as Region II [14]. The control objective is to capture all available wind power by forcing the pitch angle equal to zero degrees within safety conditions.

Above the nominal wind speed, i.e., 14 m/s the operating region is classified as Region III [14], and the wind turbine operates at the rated power of the generator.

The control objective is to operate the WiT at the rated power. When the wind speed is above 25 m/s [14], the operating region is classified as Region IV and the wind turbine is shut down for its own protection.

For control purpose, only Region II and Region III are considered. For those regions, the implemented controllers provide the pitch angle and the generator torque reference.

For power optimization region, the reference for pitch angle is equal to zero degrees and the electric torque reference is given by:

$$\tau_{g,r}(k) = K_{opt} \left(\frac{\omega_g(k)}{N_g} \right)^2 . \tag{11}$$

$$K_{opt} = \frac{1}{2} \rho A R^3 \frac{C_{pmax}}{\lambda_{opt}^3}. \tag{12}$$

where λ_{opt} is the optimal point in C_p and A is the area swept by the blades. The optimal solution can be seen in Fig. 4 and is given by:

$$\begin{cases} C_{pmax}(\lambda_{opt}(0), 0) = 0.4554 \\ \lambda_{opt}(0) = 6.743 \end{cases} . \tag{13}$$

For rated power region, the pitch and generator torque reference should be tuned at the same time and the latter is expressed as:

$$\tau_{g,r}(k) = \frac{P_r(k)}{\eta_g \omega_g(k)}. \tag{14}$$

4 WEnCS Control and Supervision

The control of WEnCS is achieved using different types of controllers such as integer and fractional-order controllers, fuzzy logic controllers, adaptive and predictive controllers. The supervision system is based on FSM.

4.1 Control Strategies

The control strategies used in this paper include the switching of the control mode from Region II to Region III if $P_g(k) > P_r(k)$ or $\omega_g(k) > \omega_{nom}(k)$ and switching back from Region III to Region II if $\omega_g(k) < \omega_{nom}(k) - \omega_\Delta$, where ω_Δ is a small offset used to avoid numerous switches between control modes. The mathematical equations that describe the dynamics of the controllers are described thoroughly in [18], thus, in the current section it will only be presented the final control action equation regarding each controller.

Integer-Order Proportional Integral Controller. The integral proportional integral control action is given by:

$$\begin{cases} u(k) = u(k-1) + K_p e(k) + \left(K_i T_s - k_p\right) e(k-1) \\ e(k) = \omega_g(k) - \omega_{nom}(k) \end{cases} . \tag{15}$$

where ω_{nom} is the nominal WiT speed, K_p is the proportional gain and K_i is the integral gain.

Fractional-Order Proportional Integral Controller. This controller is based on power series expansion of the trapezoidal rule [19], the controller is expressed as:

$$\begin{cases} G(s) = K_p + K_i \, s^{-\mu} \\ s^\mu \approx \left[\dfrac{2}{T_s} \dfrac{1-z^{-1}}{1+z^{-1}}\right]^\mu \end{cases} . \tag{16}$$

where μ is the integral fractional-order satisfying $0 < \mu < 1$ and T_s is sampling time. The discrete PI^μ control parameters were obtained using a MATLAB function [20].

Fuzzy Proportional Integral Controller. This controller is expressed as:

$$\begin{cases} u(k) = u(k-1) + k_{\Delta u} f_{NL}(e(k), k_e, \Delta e(k), k_{\Delta e}) \\ e(k) = \omega_g(k) - \omega_{nom}(k) \end{cases} . \tag{17}$$

where $r(k)$ is the input, $y(k)$ is the output, f_{NL} is a non linear function which represents the inference fuzzy system with scaling factors $k_e, k_{\Delta e}, k_{\Delta u}$.

Linear Quadratic Gaussian Controller. This controller is expressed as:

$$u(k) = \frac{\hat{b}_0}{\hat{b}_0^2 + \rho^2} \left[\left(\frac{\rho^2 - \hat{b}_1 \hat{b}_0}{\hat{b}_0}\right) u(k-1) + \hat{a}_1 y(k) + \hat{a}_2 y(k-1) + \omega_{nom}(k) \right] . \tag{18}$$

where $\hat{\theta}(k) = [\hat{a}_1 \ \hat{a}_2 \ \hat{b}_0 \ \hat{b}_1]$ are estimated using recursive least squares (RLS) algorithm.

Model Predictive Controller. The minimization of cost function in order to determine the optimal control action is given by:

$$u * (k) = \min_{u(k) \, \cdots \, \hat{u}(k+Np-1)} J(k) = \overbrace{\sum_{j=1}^{Np} \left[e(k+j)|k\right]^T Q(j) \left[e(k+j)|k\right]}^{\text{quadratic error}} + \cdots$$

$$+ \underbrace{\sum_{j=0}^{Np-1} \left[u(k+j)|k\right]^T R(j) \left[u(k+j)|k\right]}_{\text{control effort}} . \tag{19}$$

4.2   Supervisor

The WEnCS operational state is determined by the event-based supervisor in the supervision level. To model the event-based controller, the following operational states, shown in Fig. 5, are used. It is considered the following operational states: park, start-up, generating and brake.

In the park state, Region I, the WiT should be stopped and the generator should not be connected to the grid.

In the start-up state, Region II, the wind must be above a minimum speed, hence, the WiT should be rotating in order to capture all available power. The generator is connected to the electric grid, but not necessarily at rated power the majority of the time.

The power production state or generating state, Region III, is the region where the turbine speed is in the rated wind range, hence, the generator is connected to the electric grid at rated power all the time.

The brake state is based on a number of conditions which will allow the supervisor to exit the generating state, enter into the start-up state or enter into the park state.

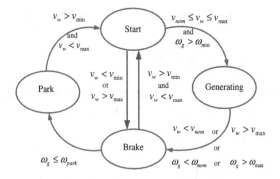

Fig. 5.   Representation of the operational states and conditions.

4.3   Performance Assessment of the Controllers

The metrics used in the evaluation of the performance of the controllers are the integral of time multiplied by the absolute value of the error (ITAE) given by:

$$ITAE = \int_0^{t_f} t. |e(t)| dt. \tag{20}$$

and the integral of the square value (ISV) of the control input given by:

$$ISV = \int_0^{t_f} u^2(t) dt. \tag{21}$$

where ITAE is used as numerical measure of tracking performance for the entire error curve and ISV is used as numerical measure of the control effort.

5 Results

The numerical results and the conclusions about the performance of the WEnCS, using computer simulations are presented. The performance of WEnCS is studied using the PI controllers, FOPI, Fuzzy PI, LQG or MPC with or without the inclusion of the supervisor. The mathematical model for the WEnCS and the simulations with the two-level power converter topology are implemented in Matlab/Simulink. The time horizon considered in the simulations is of 4500 s, and the sampling time $T_s = 0.01$ s.

The wind speed considered in the simulations has a profile in the range of 7.5 m/s to 22.5 m/s (between Region II and Region III) and white noise is added to the wind speed to make it more realistic. The wind speed with white noise is shown in Fig. 6.

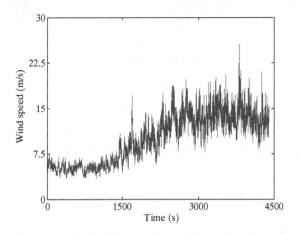

Fig. 6. Wind speed sequence with white noise.

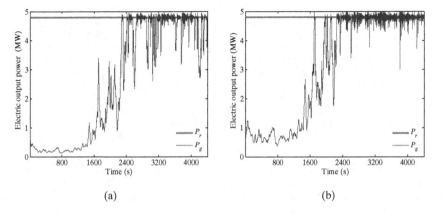

(a)   (b)

Fig. 7. Generated and reference power (PI): (a) without supervisor, (b) with supervisor.

5.1   Generated and Reference Power Simulation

All the generated and reference power without the supervisor are shown in Fig. 7(a) and with the supervisor are shown in Fig. 7(b).

PI Controller. With the PI controller, one can see that the electric output power, with or without the supervisor, follows the reference power presenting higher levels of oscillation.

FOPI Controller. The generated and rated power without the supervisor is shown in Fig. 8(a) and with the supervisor is shown in Fig. 8(b).

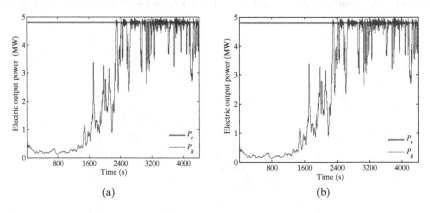

Fig. 8. Generated and reference power (FOPI): (a) without supervisor, (b) with supervisor.

With the FOPI controller, it can be seen that the electric output power still presents higher level of oscillation around the reference power, with or without the presence of the supervisor and also presents frequent decreases in electric output power due to wind variations.

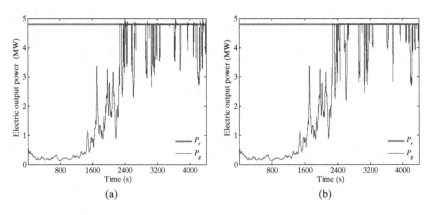

Fig. 9. Generated and reference power (Fuzzy PI): (a) without supervisor, (b) with supervisor.

Fuzzy PI Controller. The generated and rated power without the supervisor is shown in Fig. 9(a) and with the supervisor is shown in Fig. 9(b).

With the Fuzzy PI controller, in both situations, the electric output power follows the reference power with a smoother response around the reference power having some decreases in the electric output power due to sudden wind variations.

LQG Controller. The generated and rated power without the supervisor is shown in Fig. 10(a) and with the supervisor is shown in Fig. 10(b).

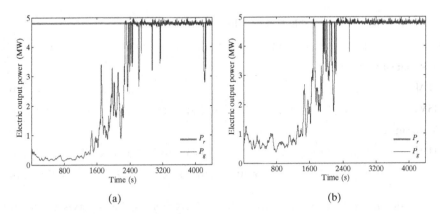

(a)   (b)

Fig. 10. Generated and reference power (LQG): (a) without supervisor, (b) with supervisor.

With the LQG controller, in both situations, the electric output power follows the rated power with few oscillations around the reference power having some decreases in the output power without the supervisor.

MPC Controller. The generated and rated power without the supervisor is shown in Fig. 11(a) and with the supervisor is shown in Fig. 11(b).

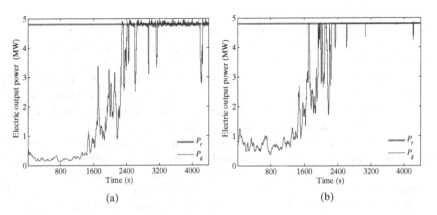

(a)   (b)

Fig. 11. Generated and reference power (MPC): (a) without supervisor, (b) with supervisor.

With the MPC controller, without the supervisor, the electric output power follows the rated power with few oscillations around the reference power having some decreases in the output power. With the supervisor, the electric output power follows the rated power with a smoother response around the reference power.

5.2 Performance Assessment

Table 1 summarizes the controller performance results.

Table 1. Controller performance results.

Controller	PI	FOPI	Fuzzy PI	LQG	MPC
Without supervisor					
ITAE	1.2103×10^{15}	1.2073×10^{15}	1.1752×10^{15}	1.0792×10^{15}	1.0886×10^{15}
ISV	6.054×10^{6}	5.7895×10^{6}	6.4604×10^{6}	1.0770×10^{7}	1.4791×10^{7}
With supervisor					
ITAE	1.2048×10^{15}	1.2087×10^{15}	1.1643×10^{15}	7.0328×10^{14}	7.0250×10^{14}
ISV	5.7652×10^{6}	5.8518×10^{6}	6.1677×10^{6}	2.7171×10^{5}	1.7276×10^{7}

Considering the values obtained without the action of the supervisor, the error between electric output and reference power is smaller with LQG controller, meaning that the electric output power follows the rated power more accurately. From a control effort point of view, FOPI consumes less energy given the narrower variation of the pitch angle. Considering the values obtained under the action of the supervisor, the error between electric output and reference power is smaller with LQG and MPC controllers. Regarding energy consumption, LQG presents a superior performance.

6   Conclusions

A hierarchical structured is presented composed by a FSM supervisor in the top level and five distinct controllers: PI, FOPI, Fuzzy PI, LQG and MPC in the lower level. The WEnCS operational state is determined by the operating conditions, which are analyzed by the event-based supervisor. The implemented controllers in the lower level are intended to process the operational state information provided by the supervisor. The controllers in the lower level act in order to sustain the output power near the region of the nominal value by acting on the pitch angle of the blades.

Comparisons between the implemented controllers regarding closed loop response, without supervisor, unveil the fact that the LQG controller outperforms the remaining controllers at expense of higher control effort. Regarding control effort, FOPI controller presents better performance in what regards the expenditure effort, i.e., the controller presents lower control effort but at expense of an oscillatory closed loop response.

Comparisons between the implemented controllers regarding closed loop response, with supervisor, show that LQG and MPC controllers follow the reference power with a smoother response outperforming the remaining controllers. Regarding control effort, LQG controller presents better performance while MPC continues to have high values of control effort.

Overall, the simulation results with supervisor presented better performance, where LQG controller stood out among the other controllers.

Acknowledgments. This work is funded by Portuguese Funds through the Foundation for Science and Technology-FCT under the project LAETA 2015-2020, reference UID/EMS/50022/2013.

References

1. Toffler, A.: La tercera ola. Plaza y Janes Editores, Barcelona, España (1982)
2. Resende, F.: Evolução tecnológica dos sistemas de conversão de energia eólica para ligação à rede. E-LP Eng. Technol. J. **2**, 22–36 (2011)
3. Melicio, R., Mendes, V.M.F., Catalão, J.P.S.: Modeling and simulation of wind energy systems with matrix and multilevel power converters. IEEE Latin Am. Trans. **7**(1), 78–84 (2009)
4. World Wind Energy Association. Half-year report 2014, October 2014
5. da Costa, A.S.: Estratégia para o crescimento verde: A eletricidade renovável em Portugal. In: Ciclo de Conferências Engenharia em Movimento, pp. 1–29 (2015)
6. Viveiros, C., Melicio, R., Igreja, J.M., Mendes, V.M.F.: Application of a discrete adaptive LQG and Fuzzy control design to a wind turbine benchmark model. In: International Conference on Renewable Energy Research and Applications, pp. 488–493, Madrid, Spain (2013)
7. Viveiros, C., Melicio, R., Igreja, J.M., Mendes, V.M.F.: Fuzzy, integer and fractional order control: application on a wind turbine benchmark model. In: 19th International Conference on Methods and Models in Automation and Robotics, pp. 252–257, Międzyzdroje, Poland (2014)
8. Viveiros, C., Melicio, R., Igreja, J.M., Mendes, V.M.F.: Fractional order control on a wind turbine benchmark. In: 18th International Conference on System Theory, Control and Computing, pp. 76–81, Sinaia, Romania (2014)
9. Viveiros, C., Melicio, R., Igreja, J.M., Mendes, V.M.F.: Performance assessment of a wind energy conversion system using a hierarchical controller structure. Energy Convers. Manage. **93**, 40–48 (2015)
10. Viveiros, C., Melicio, R., Igreja, J.M., Mendes, V.M.F.: Performance assessment of a wind turbine using benchmark model: fuzzy controllers and discrete adaptive LQG. Procedia Technol. **17**, 487–494 (2014)
11. Rawat, D., Rodrigues, J., Stojmenovic, I.: Cyber-physical systems: from theory to practice. CRC Press, Boca Raton (2016)
12. Samad, T., Annaswamy, A.M.: The impact of control technology. In: IEEE Control System Society (2011)
13. Moness, M., Moustafa, A.M.: A survey of cyber-physical advances and challenges of wind energy conversion systems: prospects for internet of energy. IEEE Internet Things J. **99**, 1 (2015)
14. Odgaard, P.F., Stroustrup, J., Kinnaert, M.: Fault tolerant control of wind turbines: a benchmark model. IEEE Trans. Control Syst. Technol. **21**(4), 1168–1182 (2013)

368   C. Viveiros et al.

15. Melício, R., Mendes, V.M.F., Catalão, J.P.S.: Wind turbines equipped with fractional-order controllers: stress on the mechanical drive train due to a converter control malfunction. Wind Energy **14**, 13–25 (2010)
16. Munteanu, I., Bratcu, A.I., Cutululis, N.A., Caenga, E.: Optimal Control of Wind Energy Systems Towards a Global Approach. Springer, London (2008)
17. Bianchi, F.D., de Battista, H., Mantz, R.J.: Wind Turbine Control Systems. Springer, London (2007)
18. Viveiros, C.: Controlo e supervisão em sistemas de conversão de energia eólica. Ph.D. thesis, Universidade de Évora (2015)
19. Petráš, I.: The fractional-order controllers: methods for their synthesis and application. J. Electr. Eng. **50**, 284–288 (2015)
20. Chen, Y.Q., Petráš, I., Xue, D.: Fractional order control–a tutorial. In: Proceedings of the American Control Conference, pp. 1397–1411, St. Louis, USA (2009)

Review of Novel Topologies for PV Applications

Elena Makovenko[1,2(✉)], Oleksandr Husev[1,3], Carlos Roncero-Clemente[2],
and Enrique Romero-Cadaval[2]

[1] Department of Electrical Engineering, Tallinn University of Technology,
Ehitajate tee 5, 19086 Tallinn, Estonia
elena-makovenko@yandex.ru
[2] Power Electrical & Electronic System (PE&ES), University of Extremadura, Campus
Universitario, Avda. de Elvas s/n, Escuela de Ingenierías Industriales, Lab. C2.7,
06006 Badajoz, Spain
croncero@peandes.unex.es
[3] Department of Biomedical Radioelectronic Apparatus and Systems,
Chernihiv State Technological University, Shevchenko Street 95, Chernihiv 14027, Ukraine
oleksandr.husev@ieee.org

Abstract. Renewable energy capacity has been growing rapidly, exceeding
140 GW of installed power in solar Photovoltaic (PV) power generation. Along
with PV installations, the variety of applied power electronics topologies has also
increased, resulting in a key point of future Smart Grids, as long as they allow
new operation possibilities. This paper reviews the emerging topologies for PV
applications that could be used in the generation of new smart inverters. A particular focus is on impedance-source converters and naturally clamped solutions.
Pros and cons along with areas of application are summarized.

Keywords: Soft-switching inverters · Z-source inverters · Smart inverters · PV
power generation

1   Introduction

Today's development is towards alternative and Renewable Energy Sources (RES) such
as hydro, geothermal, wind and solar energy. These types of energy are of prime importance because of limited natural resources and the increasing concern about environmental pollution. By use of RESs, consumer dependence on the main grid is lower and
they will not suffer from different types of perturbations such as voltage spikes, sags or
swells; at the same time, reliability is improved by an alternative supply in case of a
blackout in the main grid.

RES that are based on emerging integrated concepts, such as smart grids, smart meters
and smart buildings, are playing an important role in the modern world. They can be also
associated with Fuel Cell Vehicles (FCVs), which are transportation generation systems
with zero emission that reduce negative influence on the environment [1]. RESs are integrated as Distributed Generation (DG), essential in a back-up electric power generating
unit. The presence of DG in industrial or public infrastructures (hospitals, shopping

© IFIP International Federation for Information Processing 2016
Published by Springer International Publishing Switzerland 2016. All Rights Reserved
L. Camarinha-Matos et al. (Eds.): DoCEIS 2016, IFIP AICT 470, pp. 369–377, 2016.
DOI: 10.1007/978-3-319-31165-4_35

centers) is very important, because when a grid blackout happens, DG will provide emergency power during that time. Also, home's DG can contribute to avoid or mitigate different negative effects caused by voltage sags or voltage swells. If some Energy Storage System (ESS) is present, it could support extra energy if the energy from RES is insufficient to supply the consumption. It can also store surplus energy if RES is producing more energy than the consumer requires.

Figure 1a shows the general functional structure of back-up DG, which contains the following blocks: RES (PV, fuel cell or wind turbine), bidirectional and unidirectional dc-dc converters to connect the RES and ESS to the constant dc-link voltage bus, the dc-ac converters to connect to the main grid for energy interchanges, and loads connected directly to the dc bus.

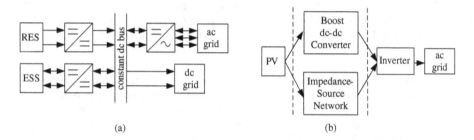

(a)   (b)

Fig. 1. (a) Functional structure of general back-up DG [1]; (b) grid-connected solar converter.

In case of solar systems the proper design of converters and controllers is essential because the voltage and current characteristics of the PV modules are non-linear and depend on solar irradiance and temperature. The Maximum Power Point Tracking (MPPT) is not an easy task with typical PV modules and the Maximum Power Point (MPP) voltage in the range from 20 V to 50 V. The most simple is conventional Voltage Source Inverters (VSI) or Current Source Inverters (CSI), but they cannot provide very wide input voltage regulation range.

In case of wide input voltage regulation range is demanded the intermediate voltage boost dc-dc converters can be used. Figure 1b shows a generalized concept of the buck-boost dc-ac converter for PV application. The first one is two stage energy conversion that is based on the voltage boost dc-dc converter along with VSI [1]. These converters can use a High Frequency (HF) transformer, as they allow achieving high step-up ratio and provide galvanic isolation between the PV modules and the grid. Second is based on an intermediate Impedance-Source (IS) network. IS based converters are also being considered as good candidates since they are able to obtain an additional voltage boost by applying the shoot-through states.

The aim of this paper is to review the emerging converter topologies for PV applications. This overview is meant as a starting point of enabling technologies; further research will focus on the design of an advanced control algorithm for a selected converter, specifically on distorted, unbalance and unstable grid-connected applications, which would cover scenarios in future Smart Grids where distributed and RES and Cyber-Physical System (CPS) integrated within power electronic systems will be key factors.

2   Technological Innovation for Cyber-Physical Systems

New power grids need to migrate to a new Smart Grid where Information and Communication Technologies (ICT) and CPSs will be of higher importance. These new ICT facilities make it easier to move from centralized to distributed generation, allowing implementation of new benefits in a more optimized electric grid management, dividing the distribution grid in micro or even nano-grids, and giving back-up support. The success of this new grid paradigm will also be related to the correct use of a distributed energy resource, principally based on renewable resources, which should be controlled by ICT, thus the result will be a Cyber-Physical System (CPS).

The PV inverter is the key element in the grid integration of PV systems because it enables an efficient and flexible interconnection of different elements to the electric power system. The electronic control platforms integrated in these inverters usually have communication interface to send measurements and state variables and to receive operation set-points. At the same time, these control platforms have sufficiently free resources to be employed in the CPS used to optimize the whole electric system operation. For a long time, the function of the PV inverters was merely to inject power into the main grid with a unitary power factor as the control reference [2]; however, under new trends and policies, PV plants are integrated as active and smart devices. In this way, PV inverters are able to contribute to the local voltage support, to improve the power quality and to give rise to flexibility and supply reliability. Some of those new demands for inverters are power flow controls [3], voltage level restoration at the Point of Common Coupling (PCC) [3–5], active filtering capabilities [5], integration with energy storage systems [6], and communications compatibilities [7].

These new features of PV inverters require sensors, monitoring and data exchanging between different agents: electrical companies, producers and end-customers to make its integration in the Smart Grid possible. Concurrently, due to the high cost of solar energy (the cost of PV panels is \$2.75/W in 2014 [8]), industry and researchers are aiming at cost reductions and enhanced performance of the energy conversion process, with a focus on the inverter topologies.

The first step here is to analyze the enabling technologies that are mainly focused on new advances topologies, to determine the control platform that can be used (in terms of sampling period related to the required switching frequency, number of magnitudes to be measured and number of switches to be controlled) and how many resources will be available for CPS implementation later.

3   Novel Impedance-Source Converters for PV Application

Today different types of the IS converters are available that can be applied to RES. Many papers cover this topic [9] and some authors provide good reviews of the state of the art [10–13]. All these types of converters have a common feature, they use an additional shoot-through (ST) state, which allows boosting the output voltage and enhancing the converter reliability. Hence, IS based inverters can buck-boost the output voltage in a single energy conversion stage, minimizing the number of components, increasing the efficiency and reducing the cost [15]. IS network provides a second-order filter in their input side, which makes it more effective to suppress input voltage and current ripples. After the first publications of the Z-source inverter [9, 16], many other topologies have been proposed to overcome its drawbacks [15] (limited boost capability, discontinuous input current and high inrush current). The main goal of further research and development regarding IS networks is to reduce the inrush current, improve the input current profile, lower the components stress, and improve the boost capability using modulation techniques variations.

IS inverters can work in two operation modes according to the input current is continuous or discontinuous. They can be associated with transformers in order to boost the output voltage by proper selection of the turns-ratio. Figure 2 presents only three recently proposed IS networks that can be most suitable for PV applications: Trans-qZ-source inverter [17]; LCCT-Z-source Inverters [18]; L-Z-source inverters with two inductors [19].

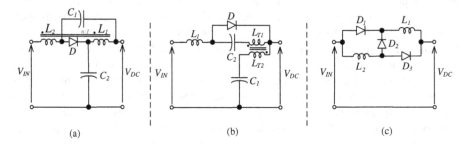

Fig. 2. (a) Trans-qZ-source inverter; (b) LCCT-qZ-source inverters; (c) L-Z-source inverters.

Table 1 summarizes the boost factor and ac voltage amplitude obtained with the discussed topologies. All these topologies operate in continuous mode, improving the Electromagnetic Inteferences (EMI) and allowing reduction of the size of the input filter and the stress of PV modules and an increase in their life-time.

Table 1. Boost factor and ac output voltage amplitude

Parameter	Trans-qZS	LCCT-ZSI	L-ZSI
Boost factor	$B = \dfrac{1}{1-(n+1)D_0}$	$B = \dfrac{1}{1-(1+\frac{n_{T1}}{n_{T2}})D_0}$	$B = \dfrac{1}{1-(2+\frac{1}{n_1-1})D_0}$
Ac voltage	$\hat{v}_{ac} = 0.5MV_{dc}$	$\hat{v}_{ac} = 0.5MV_{dc}$	$\hat{v}_{ac} = 0.5MV_{dc}$

Trans-qZ-source inverter can be used to obtain higher boost factors maintaining the voltage stress on the semiconductor devices. To reach better dc-link voltage utilization the higher turns-ratio should be used, which in turns requires higher isolation level between the windings, and producing an increase of the leakage inductance. This leakage inductance is connected in series with the inverter bridge, without any snubber circuit, so *di/dt* caused by the current switching of the windings produces large switch-voltage spikes. Due to the tight coupling between the primary and secondary inductors, the voltage overshoot has an insignificant value. The effect on dv/dt is lower because of parasitic capacitance in the tightly coupled windings.

In LCCT-Z-source inverter topology, the transformer core does not reach the saturation due to the capacitors. They can operate in three different modes during active inverter states: in the first mode, all input currents are continuous; in the second mode, the first inductor current is continuous and that of the second inductor is discontinuous; and in the third mode, all input currents are discontinuous. The first mode is preferable because of better utilization of power devices that will reduce the conduction losses and the input current ripple. There is a higher turn ratio, higher boost factor and voltage stress on the capacitors. Nevertheless, if the turn ratio is more than one, the shoot-through duty cycle can be reduced to obtain the same output voltage. This topology guarantees stable inverter operation even at no load [18].

In the first and second IS networks, there are two capacitors that cause inrush current and voltage overshoot at startup. L-Z-source inverter with two inductors enables avoidance of the inrush current at startup and the resonance of Z-source between capacitors and inductors, providing also a ground path between the dc source and the inverter [19]. The increasing number of inductors raises the boost factor as well as the number of diodes, thus producing more losses. This topology does not suffer from voltage stress on the capacitor (as it has no snubber circuit) but requires that when working in discontinuous mode. 4 Novel soft-switching techniques for PV application. Soft-switching techniques are important in power electronics. These techniques allow reducing the losses in semiconductors and as a result, the switching frequency can be raised or the heat sink may shrink in size. They also help to reduce switching stresses (high voltage and current spikes) on the semiconductors during the turn-on and turn-off, increasing the Safe Operating Area (SOA) and reducing di/dt and dv/dt values responsible for EMI.

There are several ways to achieve soft-switching, such as a resonant circuit and active-clamping circuit. The converter should maintain the soft-switching capability in a wide range of input and output voltages. Also in IS converters where the shoot-through states are included, the soft-switching algorithm can be implemented as it is shown in paper [20].

3.1 Novel Naturally Clamped Topologies

Different types of topologies with naturally clamped commutated algorithm were proposed in the literature [21–24]. Several examples are shown in Fig. 3. Such topologies can be used as intermediate step-up dc-dc converters.

These novel naturally clamped topologies have the following features: Zero-Current Switching (ZCS) and Natural Voltage Clamping (NVC) that eliminate the need for active-clamping circuits or passive snubbers required to absorb surge voltage in conventional current-fed topologies. Switching losses are reduced significantly due to the ZCS of primary-side devices and the zero-voltage switching (ZVS) of secondary-side devices. Turn-on switching losses are also negligible in primary devices. Soft-switching and NVC are inherent and load independent. The voltage across primary-side devices is independent of duty cycle with varying input voltage and output power and clamped at rather low reflected output voltage, enabling the use of semiconductor devices of low voltage rating [23]. They use a secondary-modulation technique that naturally clamps the voltage across the primary-side devices with zero-current commutation, avoiding the necessity of an active-clamping circuit or passive snubbers (as mentioned above). All of these topologies have a HF transformer with a high step-up ratio that also provides galvanic isolation.

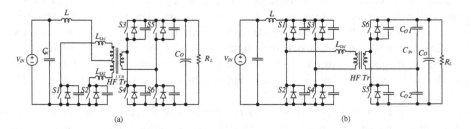

(a)   (b)

Fig. 3. (a) ZCS current-fed push-pull dc-dc converter [22]; (b) Soft-switching snubberless naturally clamped current-fed full-bridge front-end converter [23].

It should be mentioned that hard-switching of secondary switches leads to higher switching losses, which, however, are not so significant because this hard-switching occurs only four times by period. The losses are mainly due to the HF transformer, boost inductor, secondary diodes and switches [23]. These topologies were tested under different situations and the results showed that the stresses on the devices are increasing proportionally with the increasing output power [22–24]. All of these topologies exhibit high efficiency [21], but the value of the first one differs significantly from that of others, because the voltage across the primary switches is twice the value in other topologies. It can be concluded that naturally clamped topologies may be a suitable solution for use as part of a PV microinverter or a string single-phase PV inverter. At the same time, full soft-switching operation is not the case in these topologies, since the transistors on the secondary side have hard turn-on and off.

3.2   qZS dc-dc Converter with a Novel ZVS and ZCS Technique

Recently isolated qZS dc-dc converter with a novel ZVS and ZCS switching technique has been presented in [20]. Figure 4 shows this solution. The idea lies in the boundary conduction mode in the qZS network along with the snubber capacitors in the two out of four transistors and a special control algorithm implementation.

As a result, full soft-switching of the FB transistors without any auxiliary circuits is achieved. The main aim of the proposed idea lies in the elimination of the switching losses and resulting rise of the switching frequency, which turns in higher switching frequency with decreased value of the passive components and, consequently, also in a more compact design of the converter.

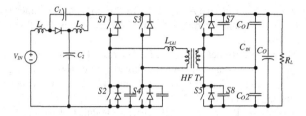

Fig. 4. Isolated dc-dc converter with full soft-switching based on the qZS network.

4 Conclusions and Further Work

This paper has reviewed novel topologies suitable for PV applications that have appeared recently. The main benefits of the discussed converters lie in the improved input voltage range regulation, possible decrease in the cost and size due to their benefits. The IS inverters possess several qualities that are also very interesting for PV applications, such as single-stage buck-boost operation capable of operation with a wide range of input voltage. In advance, the soft-switching algorithm can be implemented in the IS converters. Naturally clamped commutated converters have an ability to decrease switching losses, maintaining soft-switching in a wide range of the input and output voltage. Also, the topologies discussed are suitable for the Smart-Grid concept due to the possibility of the implementation of the advanced functions that can be controlled by using ICT. As a result, such kinds of converters are integrated to the CPS. Regarding the resources required for the control platform (and hence the free resources left for integrating these CPS functions), the advantage of the IS networks is that they need no dead-time control in one inverter branch as far as they can operate in ST states. The soft-switching topologies require a more complex modulation that consumes large amounts of control resources.

At the same time, it should be emphasized that none of the discussed topologies have been commercialized yet and they are under research and development. In the future work, these solutions should be tested to provide data for a comparative analysis between the topologies under development and those on the market.

References

1. Xuewei, P., Rathore, A.K.: Naturally commutated and clamped soft-switching current-fed push-pull voltage doubler based solar PV inverter. In: IEEE 23rd International Symposium on Industrial Electronics (ISIE), pp. 2631–2636, June 2014
2. Romero-Cadaval, E., Spagnuolo, G., Garcia Franquelo, L., Ramos-Paja, C.A., Suntio, T., Xiao, W.M.: Grid-connected photovoltaic generation plants: components and operation. IEEE Ind. Electron. Mag. **7**(3), 6–20 (2013)
3. Miñambres-Marcos, V., Guerrero-Martínez, M.A., Romero-Cadaval, E., González-Castrillo, P.: Grid-connected photovoltaic power plants for helping node voltage regulation. IET Renew. Power Gener. **9**(3), 236–244 (2015)

4. Roldán-Pérez, J., García-Cerrada, A., Zamora-Macho, J.L., Ochoa-Giménez, M.: Helping all generations of photo-voltaic inverters ride-through voltage sags. IET Power Electron. **7**(10), 2555–2563 (2014)
5. Miñambres-Marcos, V., Romero-Cadaval, E., Guerrero-Martinez, M.A., Milanés-Montero, M.I.: Three-phase single stage photovoltaic inverter with active filtering capabilities. In: Proceeding of 38th Annual Conference on IEEE Industrial Electronics Society, IECON 2012, pp. 5253–5258, 25–28 October 2012
6. Beltran, H., Bilbao, E., Belenguer, E., Etxeberria-Otadui, I., Rodriguez, P.: Evaluation of storage energy requirements for constant production in PV power plants. IEEE Trans. Ind. Electron. **60**(3), 1225–1234 (2013)
7. Navas-Matos, F.M., Romero-Cadaval, E., Milanes-Montero, M., Minambres-Marcos, V.: Distributed smart metering by using power electronics systems. In: IEEE 23rd International Symposium on Industrial Electronics (ISIE), pp. 2008–2013, 1–4 June 2014
8. https://www.ihs.com/Info/0115/top-solar-power-industry-trends-for-2015.html
9. Peng, F.Z.: Z-source inverter. In: Proceedings of 37th IAS, vol. 2, pp. 775–781, Pittsburgh, PA, USA, 13–18 October 2002
10. Yushan, L., Abu-Rub, H., Baoming, G.: Z-source/quasi-Z-source inverters: derived networks, modulations, controls, and emerging applications to photovoltaic conversion. IEEE Ind. Electron. Mag. **8**(4), 32–44 (2014)
11. Siwakoti, Y.P., Peng, F., Blaabjerg, F., Loh, P., Town, G.E.: Impedance source networks for electric power conversion Part-I: a topological review. IEEE Trans. Power Electron. **30**(2), 699–716 (2015)
12. Siwakoti, Y.P., Peng, F., Blaabjerg, F., Loh, P., Town, G.E.: Impedance source networks for electric power conversion Part-II: review of control method and modulation techniques. IEEE Trans. Power Electron. **30**(4), 1887–1906 (2015)
13. Chub, A., Vinnikov, D., Blaabjerg, F., Peng, F.Z.: A review of galvanically isolated impedance-source DC–DC converters. IEEE Trans. Power Electron. **31**(4), 2808–2828 (2016)
14. Shults, T., Husev, O., Zakis, J.: Overview of impedance source networks for voltage source inverters. In: EDM 2015, pp. 514–519 (2015)
15. Cao, D., Jiang, S., Yu, X., Peng, F.Z.: Low-cost semi-Z-source inverter for single-phase photovoltaic systems. IEEE Trans. Power Electron. **26**(12), 3514–3523 (2011)
16. Peng, F.Z.: Trans-Z-source inverters. IEEE Trans. Ind. Appl. **39**, 504–510 (2003)
17. Qian, W., Peng, F.Z., Cha, H.: Trans-Z-source inverters. IEEE Trans. Power Electron. **26**(12), 3453–3463 (2011)
18. Adamowicz, M.: LCCT-Z-source inverters. In: 2011 10th International Conference on Environment and Electrical Engineering (EEEIC), pp. 1–6 (2011). doi:10.1109/EEEIC.2011.5874799
19. Pan, L.: L-Z-source inverter. IEEE Trans. Power Electron. **29**(12), 6534–6543 (2014)
20. Husev, O., Liivik, L., Blaabjerg, F., Chub, A., Vinnikov, D., Roasto, I.: Galvanically isolated quasi-Z-source DC-DC converter with a novel ZVS and ZCS technique. IEEE Trans. Ind. Electron. **62**(12), 7547–7556 (2015)
21. Rathore, A.K., Bhat, A.K.S., Oruganti, R.: Wide range ZVS active-clamped L-L type current-fed DC-DC converter for fuel cells to utility interface: analysis, design and experimental results. In: Energy Conversion Congress and Exposition, pp. 1153–1160. IEEE (2009)
22. Xuewei, P., Rathore, A.K.: Naturally clamped zero-current commutated soft-switching current-fed push–pull DC/DC converter: analysis, design, and experimental results. IEEE Trans. Ind. Electron. **30**(3), 1318–1327 (2015)

23. Xuewei, P., Rathore, A.K., Prasanna, U.: Novel soft-switching snubberless naturally clamped current-fed full-bridge front-end converter based bidirectional inverter for, microgrid and UPS application. IEEE Trans. Ind. Electron. 2729–2736 (2013). doi:10.1109/ECCE. 2013.6647054

24. Prasanna, U.R., Rathore, A.K.: Analysis, design, and experimental results of a novel soft-switching snubberless current-fed half-bridge front-end converter-based PV inverter. IEEE Trans. Power Electron. 28(7), 3219–3230 (2013)

Contributions to the Design of a Water Pumped Storage System in an Isolated Power System with High Penetration of Wind Energy

Antonio Setas Lopes[1(✉)], Rui Castro[2], and Carlos Silva[3]

[1] IST, University of Lisbon, Lisbon, Portugal
as.l@netcabo.pt
[2] INESC-ID/IST, University of Lisbon, Lisbon, Portugal
rcastro@tecnico.ulisboa.pt
[3] IN+/IST, University of Lisbon, Lisbon, Portugal
carlos.santos.silva@tecnico.ulisboa.pt

Abstract. The increasing penetration of renewable energies in the electrical systems, particularly in small and isolated systems, like the case studied in this paper, Terceira Island in Azores, Portugal, creates challenges in the dispatch related to the variability and the difficulty to forecast the renewable resources. One way to deal with such issues is to use Water Pumping Storage Systems (WPSS) to regulate the system electricity production by storing energy surplus in low load periods and returning it back in high load periods, reducing at the same time the need to curtail wind. This paper describes and compares a deterministic and a metaheuristic methodology, using Particle Swarm Optimization (PSO), used to determine the best configuration of the WPSS in terms of number and unit power of pumps and turbines, and upper and lower reservoir capacity that lead to the best economic value, determined by the Net Present Value (NPV) of the investment.

Keywords: Renewables integration · Water Pumped Storage Systems (WPSS) · Particle Swarm Optimization (PSO) · Computational algorithms · Cyber-Physical Systems

1 Introduction

The integration of renewable energies in the electrical systems has brought issues related with the variability of the primary sources, such as network instability, with possible negative impacts, namely in small and isolated systems.

Wind curtailment strategies combined with high reserve margin values for the thermal base generating units are the usual measures to cope with this issue. This has the consequence of increasing system costs, and wasting renewable resources. One way to deal with the problem is to use WPSS, designed to absorb excess energy in low load periods and later return it in peak periods, providing a balance to the system. However, it is necessary to verify the feasibility of such systems, as compared to the above mentioned solutions.

© IFIP International Federation for Information Processing 2016
Published by Springer International Publishing Switzerland 2016. All Rights Reserved
L. Camarinha-Matos et al. (Eds.): DoCEIS 2016, IFIP AICT 470, pp. 378–386, 2016.
DOI: 10.1007/978-3-319-31165-4_36

The models used to simulate the energy balance of such systems generate the need to develop tools that are able to deal with large amounts of data. There are several tools available as described in [1], where an extended list of energy system analysis tools is covered, for a wide range of applications, in terms of time scale and application area; as well as several optimization methods described in [2], where a review on several studies approaching the use of optimization methods supporting the integration of renewables is done, concluding that this type of tools is becoming increasingly important.

The study presented in this paper intends to verify if the use of such optimization methods, namely Particle Swarm Optimization (PSO), to the design of a WPSS, leads to the same results for the configuration of the WPSS, with better performance in terms of computational times, than the deterministic model.

PSO was used, because there are several examples in the available literature showing that it presents good approximations of the correct results, the convergence is fast and the involved computational time is reasonable.

The decision to develop a dedicated algorithm for Terceira case, instead of using available tools, was based on the fact that none of the tools seems to cover, in an integrated way, the optimization of a WPSS, considering both the energy balance and investment analysis models.

All the algorithms were implemented using MATLAB language. MATLAB was chosen because it is a well-known language with a wide network of developers, providing a large set of toolboxes that can be used in this type of problems.

2   Benefits and Contribution from Cyber-Physical Systems

In this study several computational algorithms implemented in MATLAB language, using both deterministic and metaheuristics models were used to solve the problem of determining the configuration of a WPSS, leading to the best Net Present Value (NPV).

Using an integrated approach that includes the electrical system energy balance, the system design, and the economic analysis, leads to a complex problem that involves large amounts of data; the use of computational calculus is therefore mandatory.

Besides that, the use of metaheuristics methods, namely PSO, shows that with the support of computational optimization algorithms is possible to reach optimal results in solving complex problems, using less CPU time than the deterministic methods.

The research presented in this paper intends to be a contribution to the optimization of the electrical energy production in an islanded power system. This can be regarded also as a contribution to the development of the Smart Grid concept, in which renewable energies are expected to be fully integrated under the Smart Environment paradigm.

Given the nature and dimension of the problem, it would not be practical to build a real model, even at laboratory scale, where several WPSS configurations could be tested in order to find the one that leads to the best results. Nevertheless, it is the authors purpose to compare the results obtained using the presented model with the results of the real system, after its construction and commercial operation starts. Another issue that we would like to point out is that economic aspects are much relevant and have to be assessed. As so, a theoretical economic model seemed to be the best way to accomplish this task.

With the increasing complexity of Energy Systems (ES), Cyber-Physical Systems (CPS) will play an important role in the management of ES. As so, the development of simulation tools that are able to recreate the conditions found in the operation of ES is also of importance, as mentioned in [3].

We believe that a possible contribution of our work is to provide a simple simulation tool of the energy balance between load, generation and storage. Although the generation and storage is dealt in the model in an aggregated form, it would be possible to further refine it including unit commitment modules for optimum scheduling of the thermal units. Furthermore, the data supplied by this model could then be used to feed the models that deal with the dynamic behavior of the ES components.

3   State of the Art

Several papers on the utilization of computational optimization methods to find the best configurations of WPSS are available, in [4] the sizing of a WPSS is done with the support of Genetic Algorithms (GA), on the investor's perspective, by looking at the tariffs and applicable energy effects in the investment, and on the system perspective, by looking at the reduction of the Levelized Cost of Electricity (LCOE) and the increase of renewable energy sources penetration. [5] explores the application of a PSO approach to the design optimization of a pump storage system, combined with a photovoltaic system. In [6] the authors compare different approaches using several optimization methods, GA, PSO and Simulated Annealing, concluding that the suitable approach may depend on the type of application and user requirements, nevertheless, promoting the applicability of renewable energy systems (RES).

4   Research Contribution and Innovation

4.1   Energy Balance and Economic Models

Characterization of Terceira Power Generation System. This system, as planned for 2020, comprises the power plants of: Central Térmica de Belo Jardim (CTBJ), a thermal power plant using as prime movers internal combustion engines, with total installed power of 47.6 MW divided by 4 units of 5.9 MW each and 2 units of 12 MW each; Parque Eólico da Serra do Cume (PESC), a wind farm with 10 units, each of 0.900 MW, plus 4 equal additional units to be installed; 3 small-hydro adding to 1,432 MW; A geothermal power plant of 3 MW; A waste to energy power plant of 1,7 MW. The voltage and frequency control are assigned to the CTBJ power plant, by choice of the system operator.

The base data set consists of the electricity demand; hydro and wind based electricity production records, as well as wind speed, in periods of 30 min, for the year of 2012, geothermal and waste to energy were considered to run continuously at power plant rated power.

The used methodology aims at determining, for each analyzed period of 30 min, the excess of electric energy available after the demand is supplied by the existing power plants, using as much as possible the electricity produced by wind energy conversion systems. This surplus energy is stored, subject to operational restrictions and limits of the electricity production and water storage systems. The stored energy in a WPSS is dispatched for each period, in order to minimize the thermal based electricity.

Calculation of Excess Energy Available For Storage. The excess energy Eeei, for each period i is calculated by:

$$Eee_i = (Ectbj_i + Eh_i + Epesc_i + Egth_i + Ewte_i - Eload_i)\eta_{pump}. \tag{1}$$

where: $Eload_i$, is the total demand for period i; $Ectbj_i$, is the thermal power production for period i; Eh_i, is the hydro power production for period i; $Epesc_i$, is the wind power production for period i; $Egth_i$, is the total geothermal power production for period i, assumed equal to continuous rating; $Ewte_i$, is the total waste to energy power production for period i, assumed equal to continuous rating; η_{pump}, is the pumping efficiency.

Constraints and Limits. Constraints: Load satisfaction; Spinning Reserve; Technical operational minimum and maximum load of the thermal units; Feasible states of operation of the thermal power plant; Upper and lower reservoirs maximum capacity; Minimum number of consecutive operation periods for thermal units.

Limits: Pump efficiency and its minimum load; Turbine efficiency and its minimum load.

Load Satisfaction. As in any electrical power system, the load must be fulfilled independently of existing storage or not, and this is translated by:

$$Ectbj_i + Eh_i + Epesc_i + Egth_i + Ewte_i - Eee_{i-1}\eta_{turbine} = Eload_i. \tag{2}$$

Where, Eee_{i-1}, is the total excess energy stored up to the end of period i − 1 and $\eta_{turbine}$, is the turbine efficiency.

Spinning Reserve. SR is provided by the thermal units plus the stored energy (turbine units), and is established by the power system operator as a function of the average wind speed at PESC.

Technical Operational Minimum and Maximum Load of Thermal Units. A minimum of two thermal units operating and at no less than 6 MW.

Feasible States of Operation of the Thermal Units. Determined by simple enumeration, considering that a minimum of two units has to be operating, running on 50 % minimum load. Note that for the units with 5.9 MW rated power, the minimum power was set to 3 MW, in order to satisfy the operator's demand of 6 MW minimum.

Upper and Lower Reservoirs Maximum Capacity. The upper and lower reservoir maximum capacity was obtained by calculating the volume of water corresponding to the potential energy equal to the maximum excess energy available, for each 24 h period, affected by the pump efficiency.

Minimum Number of Consecutive Operation Periods for Thermal Units. Established as a minimum of 2 h of continuous operation for the same configuration of units, except where transition to a higher power state is needed in order to satisfy demand.

Pump and Turbine Operating Limits. For units 1–5 MW, based on existing references [7] efficiencies of 80 %, for both the pump and turbine, were considered.

The considered minimum operating loads are 70 % of nominal rated power for pumps and 15 % for the Pelton turbines [8].

Economic Model. The economic assessment is based on an NPV model, for a 30 years period, of the cash flows resulting from the difference between the WPSS investment and operation costs and the savings obtained by not running the thermal units.

The WPSS investment costs are obtained by estimates of the civil works, including reservoir, electromechanical and engineering costs; based on the cost functions determined in [9] and cost breakdown explained in [10], as: Civil costs: 60 % of total investment; Electromechanical costs: 33 %; Engineering costs: 7 %.

Electromechanical equipment costs were determined using the cost function for single Pelton units [9]:

$$Em_{cost1} = aP^{b-1}H^c, a = 17.693, b = 0.635275, c = -0.281735. \tag{3}$$

Adapted to multiple units using the cost formulas determined in [11], resulting in the following cost formulas:

$$\text{Two units} \quad Em_{cost2} = a_2P^{b-1}H^c, a_2 = 27.070. \tag{3a}$$

$$\text{Three units} \quad Em_{cost3} = a_3P^{b-1}H^c, a_3 = 35.209. \tag{3b}$$

$$\text{Four units} \quad Em_{cost4} = a_4P^{b-1}H^c, a_4 = 42.109. \tag{3c}$$

Where: Em_{cost}, is the cost of the electromechanical equipment, in €/kW; P, is the rated power, in kW; H, is the net head in meters, in this case 300 m.

The costs for the pump units, including the electrical drive, were estimated as equal to the turbine/generator groups.

Calculation Process: Excess energy, storage, wind curtailment and NPV, A flow chart describing the process is depicted in Fig. 1.

4.2   Optimization Approaches

For both approaches, the same models that simulate the energy balance and that make the investment economic analysis, are used.

Fast Deterministic Method. To reduce computational time, the applied method uses a stepped approach:

#1 considers combinations of one turbine and one pump, starting with 100 kW for each and increasing in steps of 100 kW, up to the maximum of 5000 kW; the combination leading to the best NPV is then identified. #2 a new range of values is defined, by adding and subtracting 1000 kW to the previous values of the best combination obtained; these values are then divided in 1 to 4 units each, for pump and turbine, and run again for NPV calculation with steps of 10 kW; #3 is similar to #2, but with steps of 1 kW and a tighter variation band for the range of power values, the combination with best NPV results is then chosen as the best one. With this approach is possible to reach to the pump and turbine combination values for the WPSS configuration, without the need of looking through all possible combinations, which would take too long time to be considered, for instance, considering that each iteration takes about 0.1 s, all the combinations of 4 pumps plus all the combination of 4 turbines, would take approximately 6.7e+20 days. In the stepped approach it took only 31697 s.

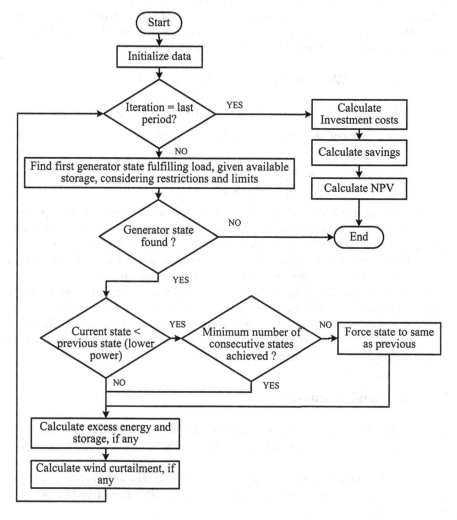

Fig. 1. Excess energy, storage, wind curtailment and NPV calculation

Particle Swarm Optimization Method. The PSO method is a meta-heuristic optimization model initially developed by James Kennedy and Russell Eberhart [12].

The method used here results from several evolutions of the first method [13–15] and is based in the work of [16]. The method starts by randomly initializing a group of possible configuration of pump and turbines, the so called particle positions, however testing for limits in the number and unit power of each unit. For each particle position the corresponding best NPV is calculated and stored (called Local Best), the best NPV of all particles is also stored (called Global Best). Then the next particle position is calculated based on its previous position and its velocity. The velocity is calculated from the difference between its current position and the Local Best and Global Best. The procedure is: Initialize particle positions Pij (pnij, ppij, tnij, tpij) and velocity vij; For each particle i position and current iteration j, calculate its NPV value and determine its Local Best (LBi); For all particles determine the best NPV, which becomes the Global Best (GB); Update next iteration (j + 1) particle position, by adding its updated velocity to its current position, according to following equations:

$$P_{ij+1}\left(pn_{ij+1}, pp_{ij+1}, tn_{ij+1}, tp_{ij+1}\right) = P_{ij}\left(pn_{ij}, pp_{ij}, tn_{ij}, tp_{ij}\right) + v_{ij+1} \tag{4}$$

$$v_{ij+1} = K(v_{ij+1} + \varphi_1 \text{rand}\left(\text{Pbest}_i - P_{ij}\left(pn_{ij}, pp_{ij}, tn_{ij}, tp_{ij}\right)\right)$$
$$+ \varphi_2 \text{rand}\left(\text{Gbest} - P_{ij}\left(pn_{ij}, pp_{ij}, tn_{ij}, tp_{ij}\right)\right)) \tag{5}$$

$$K = \frac{2}{\left|2 - \phi - \sqrt{\phi^2 - 4\phi}\right|} \tag{6}$$

With $\varphi = \varphi1 + \varphi2$. The values of $\varphi1$ and $\varphi2$ are chosen with basis on experience from several runs of the algorithm, as the ones that lead to convergence and best results. Several literature [13, 16, 17] advise the use of $\varphi1 = 2.8$; $\varphi2 = 1.3$; in any case, $\varphi > 4$.

Pbesti, is the position of the particle that corresponds to the best result of all particles i, up to the current iteration, therefore corresponding to the value LBi.

Gbest, is the position of the particle that corresponds to the best result of all the particles, up to the current iteration, therefore corresponding to the value GB.

The stopping criteria is determined by a minimum number of consecutive iterations, where the position of a significant number of particle remains unchanged. In any case, the PSO stops at the maximum number of iterations.

5   Results and Discussion

With the Fast Deterministic method, values reached for the optimal configuration are: 2 pumps of 853 kW each and 1 turbine of 3843 kW; NPV: 2,574,857 €; Computation time: 31,697 s.

With the PSO method, applied for several different combinations of parameters, each combination run 50 times, the results leading to the best NPV in each combination, are displayed in Table 1.

Table 1. PSO results

NPV (M€)	pn	pp (kW)	tn	tp (kW)	φ_1	φ_2	ptn	nmx	Stopping criteria	TCT (s)	ANI	f (%)
2574900	2	853	1	3843	2.3	1.8	20	3000	15 of 50	95,577	2564	22
2574900	2	853	1	3843	2.3	1.8	10	1000	8 of 20	16,105	818	8
2574900	2	853	1	3843	2.3	1.8	10	2000	8 of 50	24,183	1291	16
2574900	2	853	1	3843	2.3	1.8	15	2000	11 of 50	57,238	1512	8
2574900	2	853	1	3843	2.1	2.0	10	2000	8 of 50	23,607	1280	8
2574900	2	853	1	3843	2.1	2.0	15	2000	11 of 50	38,152	1363	11
2574900	2	853	1	3843	2.05	2.05	10	2000	8 of 50	12,519	250	10
2574900	2	853	1	3843	2.05	2.05	15	2000	11 of 50	25,069	501	6
2574900	2	853	1	3843	2.05	2.05	20	3000	15 of 50	31,973	640	10

The variables shown in Table 1 are NPV: Net Present Value; pn: number of pumps; pp: unit power of pumps; tn: number of turbines; tp: unit power of turbines; ptn: number of particles; nmx: maximum number of iterations for each PSO run; TCT: total calculation time for the 50 runs of the PSO; ANI: average number of iterations; f: frequency of occurrence of the best solution in 50 runs of the PSO; Stopping criteria is number of particles with unaltered position in the last number of iterations.

With the PSO method, the obtained results for the combination that leads to the best NPV, are the same as the ones obtained with the fast deterministic method, Furthermore, the best PSO result in terms of CPU time has a total CPU time of 12,519 s, which is less than half of the FD result. This shows a clear advantage for the PSO method.

Important to note that, although the frequencies of occurrence of the optimal solution are generally low, the sub-optimal solutions, not shown here, have NPV values close to the optimal NPV.

6   Conclusions

This study shows that metaheuristics methods can be used with good results to solve complex and large data problems in the field of renewable energy integration, namely in the study of economically feasibility of WPSS to support the integration of renewable energy sources.

The main advantage of using PSO is the reduction in computational time as compared to the fast deterministic approach. With PSO, it is possible to reach very good approximate results for the optimal WPSS configuration, in less than half of the computational time required for the fast deterministic approach. In fact, the results are the same, due to the precision used (1 kW for the pumps and turbines capacity).

Relaxing the number of iterations of the PSO method, as well as the number of particles, leads to higher dispersion of results, however resulting in faster total calculation times, as compared with the FD method.

Future work could be done in comparing other metaheuristic methods, like genetic algorithms, and compare its results with the findings of this work.

Acknowledgements. This work was supported by national funds through Fundação para a Ciência e a Tecnologia (FCT) with reference UID/CEC/50021/2013. Alstom Portugal, S.A. is deeply acknowledged for supporting the execution of this work. The authors are intensely grateful to EDA – Eletricidade dos Açores (Azores System Operator) for providing all the necessary data to perform this work.

References

1. Connolly, D., Lund, H., Mathiesen, B.V., Leahy, M.: A review of computer tools for analysing the integration of renewable energy into various energy systems. Appl. Energy **87**(4), 1059–1082 (2010)
2. Baños, R., Manzano-Agugliaro, F., Montoya, F.G., Gil, C., Alcayde, A., Gómez, J.: Optimization methods applied to renewable and sustainable energy: a review. Renew. Sustain. Energy Rev. **15**(4), 1753–1766 (2011)
3. Ilic, M.D., Xie, L., Khan, U.A., Moura, J.M.F.: Modeling of future cyber–physical energy systems for distributed sensing and control. IEEE Trans. Syst. Man Cybern. Part A Syst. Hum. **40**(4), 825–838 (2010)
4. Papaefthymiou, S.V., Papathanassiou, S.A.: Optimum sizing of wind-pumped-storage hybrid power stations in island systems. Renew. Energy **64**, 187–196 (2014)
5. Stoppato, A., Cavazzini, G., Ardizzon, G., Rossetti, A.: A PSO (particle swarm optimization)-based model for the optimal management of a small PV(Photovoltaic)-pump hydro energy storage in a rural dry area. Energy **76**, 168–174 (2014)
6. Erdinc, O., Uzunoglu, M.: Optimum design of hybrid renewable energy systems: overview of different approaches. Renew. Sustain. Energy Rev. **16**(3), 1412–1425 (2012)
7. Katsaprakakis, D.A., Christakis, D.G., Pavlopoylos, K., Stamataki, S., Dimitrelou, I., Stefanakis, I., Spanos, P.: Introduction of a wind powered pumped storage system in the isolated insular power system of Karpathos–Kasos. Appl. Energy **97**, 38–48 (2012)
8. Ardizzon, G., Cavazzini, G., Pavesi, G.: A new generation of small hydro and pumped-hydro power plants: advances and future challenges. Renew. Sustain. Energy Rev. **31**, 746–761 (2014)
9. Ogayar, B., Vidal, P.G.: Cost determination of the electro-mechanical equipment of a small hydro-power plant. Renew. Energy **34**(1), 6–13 (2009)
10. Zhang, Q.F., Smith, B., Zhang, W.: Small Hydropower Cost Reference Model. Oak Ridge National Laboratory, October 2012
11. Singal, S.K., Saini, R.P.: Cost analysis of low-head dam-toe small hydropower plants based on number of generating units. Energy Sustain. Dev. **12**(3), 55–60 (2008)
12. Kennedy, J., Eberhart, R.: Particle swarm optimization. In: Proceedings of the IEEE International Conference on Neural Networks, vol. 4, pp. 1942–1948 (1995)
13. Tuppadung, Y., Kurutach, W.: Comparing nonlinear inertia weights and constriction factors in particle swarm optimization. Int. J. Knowl.-Based Intell. Eng. Syst. **15**(2), 65–70 (2011)
14. Shi, Y., Eberhart, R.: Empirical study of particle swarm optimization. In: Proceedings of the 1999 Congress on Evolutionary Computation, pp. 1945–1950 (1999)
15. Clerc, M.: The swarm and the queen: towards a deterministic and adaptive particle swarm optimization. In: Proceedings of the 1999 Congress on Evolutionary Computation, CEC 1999, vol. 3, pp. 1951–1957 (1999)
16. Carlisle, A., Dozier, G.: An off-the-shelf PSO. In: Proceedings of the Workshop on Particle Swarm Optimization 2001, Indianapolis (2001)
17. Clerc, M., Kennedy, J.: The particle swarm - explosion, stability, and convergence in a multidimensional complex space. IEEE Trans. Evol. Comput. **6**(1), 58–73 (2002)

Offshore Wind Energy Conversion System Connected to the Electric Grid: Modeling and Simulation

Mafalda Seixas[1,2,3], Rui Melício[1,2(✉)], and Victor M.F. Mendes[2,3]

[1] IDMEC/LAETA, Instituto Superior Técnico, Universidade de Lisboa, Lisbon, Portugal
ruimelicio@gmail.com
[2] Universidade de Évora, Évora, Portugal
[3] Instituto Superior de Engenharia de Lisboa, Lisboa, Portugal

Abstract. This paper is on modeling and simulation for an offshore wind system equipped with a semi-submersible floating platform, a wind turbine, a permanent magnet synchronous generator, a multiple point clamped four level or five level full-power converter, a submarine cable and a second order filter. The drive train is modeled by three mass model considering the resistant stiffness torque, structure and tower in deep water due to the moving surface elevation. The system control uses PMW by space vector modulation associated with sliding mode and proportional integral controllers. The electric energy is injected into the electric grid either by an alternated current link or by a direct current link. The model is intend to be a useful tool for unveil the behavior and performance of the offshore wind system, especially for the multiple point clamped full-power converter, under normal operation or under malfunctions.

Keywords: Modeling · Simulation · Offshore wind energy conversion · Power converters · Energy transmission · Harmonic distortion

1   Introduction

The focus of this paper is on the modeling and simulation of offshore wind energy conversion systems (OWECS) connected to the electric grid either by an alternated current link or by a direct current link in a view of timeliness developments on electricity sector restructuring and integrating the relevant dynamics [1]. Besides the electricity transmission, the system consists of a semi-submersible floating platform; a variable speed wind turbine; a mechanical transmission system described respectively by one, two, three, or five masses; a synchronous generator with excitation provided by means of permanent magnets; an electronic power converter, respectively described by a two level converter or by a multilevel converter in a multiple point clamped topology of three, four, five, or p levels [1–3].

The control of the system employs pulse width modulation (PWM) by space vector modulation (SVM) associated with sliding mode (SM) and proportional integral (PI) controllers. The behaviors resulting from the fact that wind energy is a variable, intermittently source of energy, as well as, due to eventual malfunctions of devices controlling the systems are studied using computer simulations.

© IFIP International Federation for Information Processing 2016
Published by Springer International Publishing Switzerland 2016. All Rights Reserved
L. Camarinha-Matos et al. (Eds.): DoCEIS 2016, IFIP AICT 470, pp. 387–403, 2016.
DOI: 10.1007/978-3-319-31165-4_37

The research question is about the model of the offshore wind system who is intend to be a useful tool for unveil the behavior and performance of the offshore wind system, especially for the multiple point clamped full-power converter, under normal operation or under malfunctions.

The novelty of the contribution of the research work are the modeling allowing a simulation of OWECS integrating the fundamental components and considering both AC or DC power transmission; the integration of the modeling of the multilevel converter in a multi point clamped diode configuration of three, four, five levels or p levels; the mitigation of the imbalance in the multilevel converter capacitor banks voltages [1–3].

2   Relationship to Cyber-Physical Systems

Nowadays there is a global integration increase of renewable energy systems in the existing electric grid, i.e., power system. The majority of renewable technologies are reliant on weather conditions and demanding regarding the integration into the electric grid. The answer to this demand, that needs to be fully resolved, entails advances in smart grid systems. This type of systems possibly will include micro-grids, energy storage facilities and might enable the coexistence of power systems with transportation and heating/cooling, representing in this manner complex energy systems [4].

Assuring sustainable development is a huge challenge for power systems [5]. Traditionally, power systems were designed to supply electrical energy observing security conditions. Due to the opening of the electricity markets the need of an adequate response to severe faults is even more important [6].

Wind power systems manifest variations in the output power due to the variations of wind speed or the marine waves, thus introducing a new factor of uncertainty and risk on the electrical grid posing challenges in terms of power system security, power system stability, power quality and malfunctions [2].

Power systems are usually pointed out as a reference example of cyber-physical systems (CyPS) [6], intending for example for construction of smart grids, for attaining a blackout-free electricity generation and distribution system or for optimization of energy consumption [5].

CyPS can be described as smart systems that include software and hardware, namely components for sensing, monitoring, gathering, actuating, computing, communication and controlling physical infrastructures, completely integrated and directly interacting to sense the alterations in the state of the surrounding environment [7].

The integration OWECS into a electric grid, raises many concerns regarding operation and reliability, not only because of the weather conditions, as well as the severity of the environment location and the eventual difficulty to access the physical installation. Thus, the use of smart devices such as CyPS, to assist monitoring, control and operating OWECS in a smart grid context [2] are fundamental pieces in the success of these systems.

However, integrating the CyPS engineering and technology in the already existing electric grid and other utility systems is a challenge. A major challenge for CyPS is the conception and the implementation of a power system infrastructure that is capable of providing blackout free electricity generation and distribution, that is flexible enough to allow mixed energy supply to or withdrawal from the grid, and that is resistant to accidental or intentional manipulations [7]. So, the successful integration of the cyber and the physical system components will obligate to an understanding of the multi-scale, multi-physics models and abstractions that will be required to allow the coexistence of software, communications, and interacting physical subsystems [7]. In this sense, the successfully integration of an OWECS into an electric grid, needs models more accurate and nearer to the reality. The model presented in this paper for an OWECS is intended to provide a useful toll in the analysis of a multi-disciplinary physical system in smart grid context.

3 Modeling

The modeling of the OWECS takes into account the mechanical, the electrical and the electronic fundamental components, integrating those inner systems in order to unveil the behavior and performance of the OWECS.

Figure 1 shows the layout of the OWES, considering AC energy transmission with a five level converter.

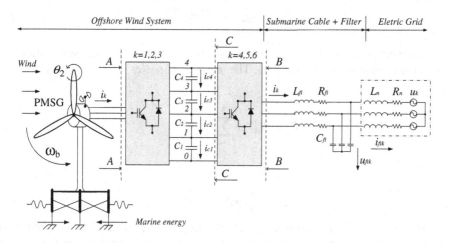

Fig. 1. Layout of the OWES, AC transmission, five level converter.

Figure 2 shows the layout of the OWES, considering DC energy transmission with a four level converter.

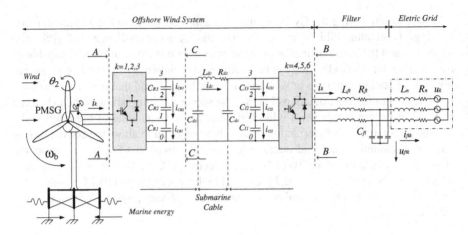

Fig. 2. Layout of the OWES, DC transmission, four level converter.

While the wind speed has a stochastic nature and generally varies considerably, for the purpose of the simulation in this paper the wind speed is modeled by a finite sum of harmonic terms, in the range $0.1 - 10\,\mathrm{Hz}$ and take three influences of the dynamic related with the action excited by wind on all physical structure as reported in [8]. The wind speed subject to disturbance is given by:

$$u = u_0[1 + \sum_n A_n \sin(\omega_n t)] \tag{1}$$

The mechanical power of the wind turbine has a model taking in consideration the three perturbations mentioned in (1). The mechanical power of the wind turbine P_t is given by:

$$P_t = P_{tt}[1 + \sum_{n=1}^{3} I_n(t)] \tag{2}$$

where

$$P_{tt} = \frac{1}{2}\rho\pi R^2 u^3 c_p \tag{3}$$

The power coefficient cp is a function of the pitch angle θ and the tip speed ratio λ_R. Normally, numerical approximations are advised, as the one developed in [9] and followed in this paper. The mechanical power in (2) is computed by the mechanical power captured from the wind turbine without dynamic perturbations [8], in (3) associated with three perturbations, namely the asymmetry in the turbine, the vortex tower interaction and the eigenswings in the blades. The perturbations are computed by the formula given by:

$$I_n(t) = A_n \left(\sum_{m=1}^{2} a_{nm} \, g_{nm}(t) \right) h_n(t) \tag{4}$$

where

$$g_{nm}(t) = \sin \left(\int_0^t m\omega_n(t')dt' + \varphi_{nm} \right) \tag{5}$$

The dynamic associated with the asymmetry in the turbine has the following data:

$A_1 = 0.01, a_{11} = 4/5, a_{12} = 1/5, \omega_1(t) = \omega_t(t), \varphi_{11} = 0, \varphi_{12} = \pi/2.$

The dynamic associated with the vortex tower interaction has the following data:

$A_2 = 0.08, a_{21} = 1/2, a_{22} = 1/2, \omega_2(t) = 3\omega_t(t) \, \varphi_{21} = 0, \varphi_{22} = \pi/2.$

The dynamic associated with the eigenswings in the blades has the following data:

$A_3 = 0.15, a_{31} = 1, \omega_3(t) = 1/2 [g_{11}(t) + g_{21}(t)], \phi_{31} = 0.$

The marine wave model [10] acting on the drive train is given by:

$$\eta(x, y, t) = \sum_{j=1}^{n} \eta_a(j) \cos[\vartheta(j)t + \varepsilon(j) - \phi(j)(x \cos(\psi(j)) + y \sin(\psi(j)))] \tag{6}$$

The elastic behavior of the tower and platform due to the surface motion action by marine waves in deep water, causes a resistant torque on the drive train [1] given by:

$$T_{hs} = k_{hs}\omega_w \tag{7}$$

The wind turbines augmentation in size involves that the blades are more flexible and tend to bend. The blade bending occurs at a significant range of the joint between the blades and the hub. The modeling of the blades considers a splitting in two parts: a rigid one and a flexible one. Hence, the drive train is modeled by a discrete three mass model. The model for the drive train illustrated as shown in Fig. 3.

The first mass represents the concentration of the inertia of the flexible part of the blades; the second mass represents the concentration of the rigid part of the blades, hub, tower and platform, discarding the displacement between the different elements, but including the floating motion influence as a whole; the third mass represents the concentration of the inertia of the generator. The connection between the three masses is made through elastic couplings [11]. This three mass model has been proven as a good option [1] to study the system behavior in response to heavy disturbance. The model assumes a radius r of 2.5 m for the rigid part of the blades. The flexible part is in the range 2.5–45 m. Figure 4 shows a comparison between both parts.

The state equations for modeling the mechanical drive train are based in the torsional version of the second law of Newton. The state equations for the rotational speeds of the three masses modeling shown in Fig. 3 are given by:

$$\frac{d\omega_b}{dt} = \frac{1}{J_b}(T_b - T_{db} - T_{bh}) \tag{8}$$

$$\frac{d\omega_h}{dt} = \frac{1}{J_h}(T_{bh} + T_{rb} + T_{hs} - T_{dh} - T_{hg}) \tag{9}$$

$$\frac{d\omega_g}{dt} = \frac{1}{J_g}(T_{hg} - T_{dg} - T_g) \tag{10}$$

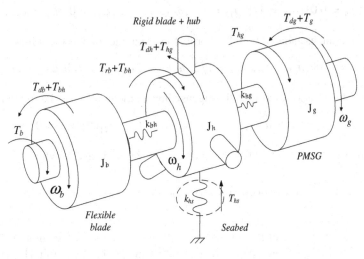

Fig. 3. Three mass drive train model.

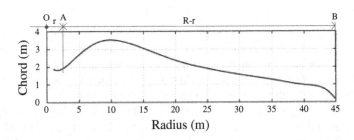

Fig. 4. Blade profile: rigid part OA, flexible part AB.

The equations that model the permanent magnet synchronous generator (PMSG) can be found in various texts [12]. In order to prevent demagnetization of the permanent magnet is imposed a null reference to the direct component of the stator current $i_{sd}^* = 0$ [1].

An equivalent three-phase active symmetrical circuit given by a series of a resistance and an inductance models the electric grid. Thus, the electric current injected into the electric grid is given by:

$$\frac{di_{ftk}}{dt} = \frac{1}{L_n}(u_{ftk} - R_n i_{ftk} - u_k) \quad k \in \{4,5,6\} \tag{11}$$

In transient simulations this is the model that is normally used and corresponds to the model of an infinite grid linked by the equivalent series impedance.

3.1 AC Energy Transmission

The AC-DC-AC five level power converter is implemented by forty eight unidirectional commanded insulated gate bipolar transistors (IGBTs), identified by S_{ik}. Both the functionalities of rectifier and of inverter are implemented with twenty four [3], IGBTs each connected to a diode in an anti-parallel configuration. Following the PMSG the rectifier is linked before the voltage divider, which consists of four capacitor banks, C_j with $j \in \{1, \dots (p-1)\}$. Following the capacitor banks the inverter is linked before a second order filter, which is linked to an electric grid, see Fig. 1. A group of height IGBTs connected to the same phase composes the arm k of the converter. For the five level power converter there is $p = 5$ voltage levels. On each phase the voltage level is related with the switching variable n_k with $n_k \in \{0, \dots, (p-1)\}$. The switching variable identifies the conduction or blockage state of the IGBT S_{ik} with $i \in \{1, \dots 8\}$ of the converter k arm defining the switching function on the IGBTs. The IGBTs state combinations for the converter k arm, establish the level variable δ_{jnk}. The level variable is related with the charging state of the capacitor banks C_j. The switching variable n_k [1] is given by:

$$n_k = \begin{cases} 4, & (S_{5k}, S_{6k}, S_{7k} \text{ e } S_{8k}) = 1 \text{ e } (S_{1k}, S_{2k}, S_{3k} \text{ e } S_{4k}) = 0 \\ 3, & (S_{4k}, S_{5k}, S_{6k} \text{ e } S_{7k}) = 1 \text{ e } (S_{1k}, S_{2k}, S_{3k} \text{ e } S_{8k}) = 0 \\ 2, & (S_{3k}, S_{4k}, S_{5k} \text{ e } S_{6k}) = 1 \text{ e } (S_{1k}, S_{2k}, S_{7k} \text{ e } S_{8k}) = 0 \quad k \in \{1,\dots,6\} \\ 1, & (S_{2k}, S_{3k}, S_{4k} \text{ e } S_{5k}) = 1 \text{ e } (S_{1k}, S_{6k}, S_{7k} \text{ e } S_{8k}) = 0 \\ 0, & (S_{1k}, S_{2k}, S_{3k} \text{ e } S_{4k}) = 1 \text{ e } (S_{5k}, S_{6k}, S_{7k} \text{ e } S_{8k}) = 0 \end{cases} \tag{12}$$

The level variable [1] is given by:

$$\delta_{jn_k} = \begin{cases} 0 & j > n_k \\ 1 & j \le n_k \end{cases} \tag{13}$$

The output voltage of the rectifier [3] is given by:

$$u_{sk} = \frac{1}{3} \sum_{j=1}^{p-1} (2\delta_{jnk} - \sum_{\substack{l=1 \\ l \ne k}}^{3} \delta_{jnk}) U_{Cj} \quad k \in \{1,2,3\} \tag{14}$$

The input voltage of the inverter is given by a similar function to (14), but with $k \in \{4, 5, 6\}$.

The capacitor bank current i_{Cj} [5] as a function of δ_{jnk} is given by:

$$i_{Cj} = \sum_{k=1}^{3} \delta_{nk} i_k - \sum_{k=4}^{6} \delta_{nk} i_k \quad k \in \{1, \ldots, 6\} \tag{15}$$

The sum voltage at the capacitor banks of the voltage divider is the voltage U_{dc} and is given by:

$$\frac{dU_{dc}}{dt} = \sum_{j=1}^{p-1} \frac{1}{C_j} i_{Cj} \tag{16}$$

The submarine cable model is represented by an inductance L_{cable} and a resistance R_{cable} in series connected between the inverter and the second order filter. The model for the second-order filter is represented by an inductance L_{filter}, a resistance R_{filter}, and a capacitor bank with a capacity C_{ft}. The resistances and the inductances of the cable and filter are associated in equivalents resistance and inductance [1], respectively given by:

$$\begin{cases} L_{ft} = L_{cable} + L_{filter} \\ R_{ft} = R_{cable} + R_{filter} \end{cases} \tag{17}$$

3.2   DC Energy Transmission

The AC-DC-AC four level power converter is implemented with thirty six unidirectional commanded IGBTs. Both rectifier and inverter functionalities are implemented with eighteen IGBTs, each connected to a diode in an anti-parallel configuration. Following the PMSG the rectifier is linked before a first voltage divider, which consists of three capacitor banks C_{Rj}, located in the offshore semi-submersible platform. The inverter is connected between a second voltage divider formed by three capacitor banks C_{Ij} and the second order filter connected before the electric grid, located in an electric substation on onshore. The submarine cable is connected between the first voltage divider downstream of the rectifier and the second voltage divider upstream of the inverter, see Fig. 2. For the four level power converter there is $p = 4$ voltage levels. On each phase the voltage level is related with the switching variable n_k, given by:

$$n_k = \begin{cases} 3, & (S_{1k}, S_{2k} \text{ and } S_{3k}) = 1 \quad \text{and} \quad (S_{4y}, S_{5k} \text{ and } S_{6y}) = 0 \\ 2, & (S_{2k}, S_{3k} \text{ and } S_{4k}) = 1 \quad \text{and} \quad (S_{1y}, S_{5y} \text{ and } S_{6y}) = 0 \\ 1, & (S_{3k}, S_{4k} \text{ and } S_{5k}) = 1 \quad \text{and} \quad (S_{1y}, S_{2y} \text{ and } S_{6y}) = 0 \\ 0, & (S_{4k}, S_{5k} \text{ and } S_{6k}) = 1 \quad \text{and} \quad (S_{1y}, S_{2y} \text{ and } S_{3y}) = 0 \end{cases} \tag{18}$$

The output voltage of the rectifier is given by:

$$u_{sk} = \frac{1}{3} \sum_{j=1}^{p-1} (2\delta_{jnk} - \sum_{\substack{l=1 \\ l \neq k}}^{3} \delta_{jnk}) U_{CRj} \quad k \in \{1, 2, 3\} \tag{19}$$

The input voltage of the inverter is given by:

$$u_{sk} = \frac{1}{3} \sum_{j=1}^{p-1} (2\delta_{jnk} - \sum_{\substack{l=4 \\ l \neq k}}^{6} \delta_{jnk}) U_{CIj} \quad k \in \{4, 5, 6\} \tag{20}$$

The current on each capacitor bank of the first voltage divider i_{CRj} [2] is given by:

$$i_{CRj} = \sum_{k=1}^{3} \delta_{jn_k} i_k - i_{dc} \quad k \in \{1, 2, 3\} \tag{21}$$

The current on each capacitor bank of the second voltage divider i_{CIj} [2] is given by:

$$i_{CIj} = i_{dc} - \sum_{k=4}^{6} \delta_{nk} i_k \quad k \in \{4, 5, 6\} \tag{22}$$

The voltage U_{dcR} as a sum of the capacitor banks voltages of the first voltage divider is given by

$$\frac{dU_{dcR}}{dt} = \sum_{j=1}^{3} \frac{1}{C_{Rj} + C_{dc}} i_{CRj} \tag{23}$$

The voltage U_{dcI} as a sum of the capacitor banks voltages of the second voltage divider is given by:

$$\frac{dU_{dcI}}{dt} = \sum_{j=1}^{3} \frac{1}{C_{Ij} + C_{dc}} i_{CIj} \tag{24}$$

A π equivalent circuit models the submarine cable [2]. The submarine cable current i_{dc} state equation is given by:

$$\frac{di_{dc}}{dt} = \frac{1}{L_{dc}} (U_{dcR} - R_{dc} i_{dc} - U_{dcI}) \tag{25}$$

4   Control Method

The OWECS employs PI controllers and the converters employ PWM by SVM associated with SM. Power converters have variable structure behavior, because of the IGBTs switching states, conduction or blockage. Also, the variance in wind speed introduces model uncertainty. SM is known as a good choice for the control method for this type of structure. Switching frequency finite values 2 kHz, 5 kHz or 10 kHz are usually reported.

The physical limitation of finite switch frequency pf the IGBTs implies an expected error $e_{\alpha\beta}$ between the control value and the reference value. So, to guarantee that the system does not drifts away from the sliding surface $S(e_{\alpha\beta}, t)$ is proved that has to hold a stability condition [13, 14, 15] given by:

$$S(e_{\alpha\beta}, t)\frac{dS(e_{\alpha\beta}, t)}{dt} < 0 \qquad (26)$$

In order to establish a reference for the generator stator currents the error between the reference voltage and the capacitor voltage U_{dc} is used as the input of the rectifier *PI* controller. Meanwhile, the error between the reference stator current and the stator current is applied in the selection of the output space voltage vector in the (α, β) space, which in turn copes with the triggering of the rectifier IGBTs. The error between the reference voltage and the electric grid voltage is treated by the *PI* controller of the inverter intending to establish a reference for the inverter currents. The triggering of the inverter IGBTs is carried out using the error between the electric grid current and the inverter controller reference current.

In practical implementations is allowed a limited error $\varepsilon > 0$ for $S(e_{\alpha\beta}, t)$, consequently is feasible to accept an error window, being this error window over time the sliding surface. So, the switching strategy is given by:

$$-\varepsilon < S(e_{\alpha\beta}, t) < +\varepsilon \qquad (27)$$

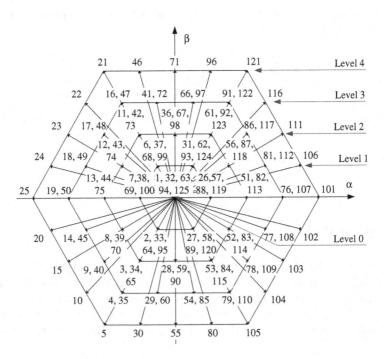

Fig. 5. Output space voltage vectors, in (α, β) plane.

The integer voltage variables values σ_α and σ_β are in a domain set, limited by the converter voltage levels p, i.e., comply with the relation given by:

$$\sigma_\alpha, \sigma_\beta \in \{-(p-1), \dots 1, \dots, (p-1)\} \tag{28}$$

The integer variables σ_α and σ_β permits to choose the most suitable vector from a set of 125 space vectors for the five level converter or from a set of 64 space vectors for the four level converter.

From the entire set of allowed space vectors there are redundant vector that identify different voltage levels. The control strategy processes a lessening on the unbalanced voltage of the capacitors banks through the consideration of p vector tables [1], which consider the charging state of each capacitor bank and the voltage level. Figure 5 shows the output vectors for levels 0 to $(p-1)$ in the $\alpha\beta$ plane for the five level and four level converter.

5   Case Studies

The mathematical model for the OWECS with the five level converter with AC transmission or with the four level converter with DC transmission is executed in Matlab/Simulink. The nominal power is 2 MW. The switching frequency for the IGBTs is 10 kHz. The mechanical mathematical model is represented by a three mass model. Figure 6 shows the average wind speed profile with perturbations.

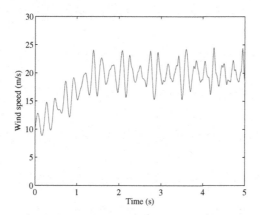

Fig. 6. Wind speed profile.

Figure 7 shows the marine height elevation.

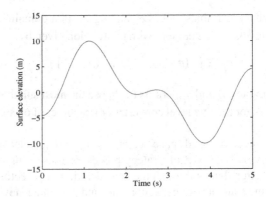

Fig. 7. Marine height elevation.

5.1   AC Energy Transmission, Five Level Power Converter

Figure 8 shows the DC voltage at the capacitor banks without unbalancing and the reference voltage.

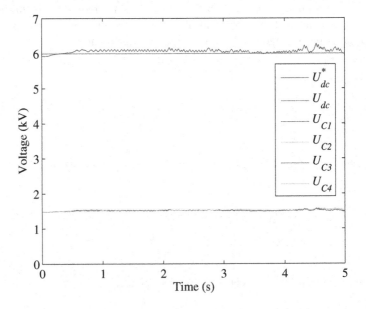

Fig. 8. Capacitors banks DC voltages without unbalancing.

Figure 9 shows the instantaneous current injected into the electric grid.

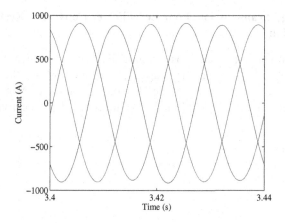

Fig. 9. Current injected in the electric grid.

The Discrete Fourier Transform applied to process the total harmonic distortion (THD) is given by:

$$\text{THD (\%)} = 100 \; \frac{\sqrt{\sum_{H=2}^{50} X_H^2}}{X_F} \tag{29}$$

where X_H is the harmonic H root mean square value and X_F is the fundamental component root mean square value.

Figure 10 shows the THD of the current injected into the electric grid.

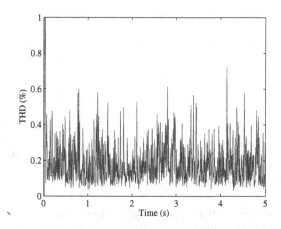

Fig. 10. THD of the current injected into the electric grid.

5.2　DC Energy Transmission, Four Level Power Converter, Rectifier Voltage Malfunction

The rectifier malfunction at one phase on an IGBT occurs between 2.45 s and 2.80 s, imposing $u_{s2} = 87.5\%$ of the normal voltage value. Figure 11 shows the submarine cable DC current. Figure 12a and b shows the DC voltages on the capacitor banks and the reference voltage at the rectifier side and at the inverter side, respectively.

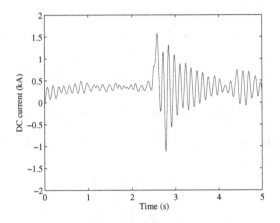

Fig. 11.　Submarine cable DC current.

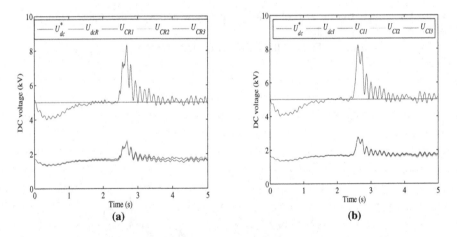

Fig. 12.　Capacitor banks DC voltages; a) rectifier side; b) inverter side.

Figure 13 shows the rectifier input voltage malfunction.

Figure 14 shows the DC current at the capacitor banks at the rectifier side and at the inverter side, respectively.

Figure 15 shows the THD of the current injected into the electric grid.

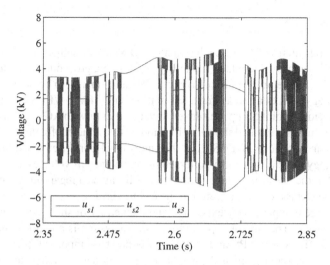

Fig. 13. Rectifier input voltage malfunction.

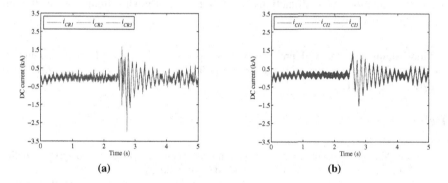

Fig. 14. Capacitor banks DC current; a) rectifier side; b) inverter side.

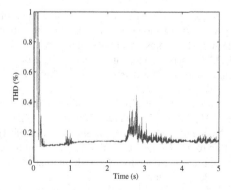

Fig. 15. THD of the current injected into the electric grid.

6   Conclusions

The increasing interest for the deployment of OWECS requires further research for modeling, allowing a value-added capability and suitable procedures not only at design phase, but also at operation, particularly on CyPS and smart grid context.

This paper presents a model for a simulation study for OWECS equipped with a PMSG with a multi point clamp multilevel power converter with AC or DC transmission.

The case studies presented show that the model is a useful tool for unveil the behavior and performance of the OWECS, that the multi-level converter associated with the control strategy unveils the ability to lessen the unbalance of voltages at the capacitors bank, and that the use of multilevel converters is in favor in regarding the effect on the quality of the electric current injected in the electric grid.

OWECS are complex, very expensive and located far of shore where maintenance is very expensive. The suitable operation of such a system is a major issue. The integration of OWECS on CyPS and smart grid context is important because real-time monitoring is allowed as well as the use of smart equipment with the capability to pre-detect malfunctions providing means to take the best decision on the capturing of power and acting measures on malfunctions through the use of model simulations as the one projected in this paper.

Acknowledgments. The work presented in this paper is funded through Portuguese Funds by the Foundation for Science and Technology-FCT for project scope LAETA 2015-2020, UID/EMS/50022/2013.

References

1. Seixas, M.: Offshore wind energy conversion system connected to the electric grid: modeling and simulation. Ph.D. Thesis, Universidade de Évora (2015)
2. Seixas, M., Melício, R., Mendes, V.M.F.: Simulation of rectifier voltage malfunction on OWECS, four-level converter, HVDC light link: smart grid context tool. Energy Convers. Manage. **97**, 140–153 (2015)
3. Seixas, M., Melício, R., Mendes, V.M.F., Couto, C.: Simulation of OWES with five-level converter linked to the grid: harmonic assessment. In: 9th International Conference on Compatibility and Power Electronics – CPE 2015, pp. 1–6, Lisbon, Portugal (2015)
4. Blaabjerg, F., Ionel, D.M.: Energy devices and systems – state-of-the art technology, research and development, challenges and future trends. Electric Power Compon. Syst. **43**(12), 1319–1328 (2015)
5. Ramos, C., Vale, Z., Faria, L.: Cyber-physical intelligence in the context of power systems. In: Kim, T.-H., Adeli, H., Slezak, D., Sandnes, F.E., Song, X., Chung, K.-I., Arnett, K.P. (eds.) FGIT 2011. LNCS, vol. 7105, pp. 19–29. Springer, Heidelberg (2011)
6. Faria, L, Silva, A., Ramos, C., Gomez, L., Vale, Z.: Intelligent behavior in a cyber-ambient training system for control center operators. In: 16th International Conference on Intelligent System Application to Power Systems, pp. 1–6 (2011)
7. Foundations for innovation in cyber-physical systems - workshop summary report. National Institute of Standards and Technology (2013)

8. Akhmatov, V., Knudsen, H., Nielsen, A.H.: Advanced simulation of windmills in the electric power supply. Int. J. Electr. Power Energy Syst. **22**, 421–434 (2000)
9. Slootweg, J.G., Polinder, H., Kling, W.L.: Representing wind turbine electrical generating systems in fundamental frequency simulations. IEEE Trans. Power Syst. **18**, 516–524 (2003)
10. Eikeland, F.N.: Compensation of wave-induced motion for marine crane operations. Msc. thesis, Norwegian University of Science, pp. 16–26 (2008)
11. Ramtharan, G., Jenkins, N.: Influence of rotor structural dynamics representations on the electrical transient performance of DFIG wind turbines. Wind Energy **10**, 293–401 (2007)
12. Ong, C.-M.: Dynamic simulation of electric machinery: using Matlab/Simulink, pp. 259–350. Prentice-Hall, New Jersey (1998)
13. Melicio, R., Mendes, V.M.F.: Simulation of power converters for wind energy systems. Información Tecnológica **18**(4), 25–34 (2007)
14. Barros, J.-D., Silva, J.F.: Optimal predictive control of three-phase NPC multilevel converter for power quality applications. IEEE Trans. Industr. Electron. **55**, 3670–3681 (2008)
15. Melicio, R., Mendes, V.M.F., Catalão, J.P.S.: Document Modeling and simulation of wind energy systems with matrix and multilevel power converters. IEEE Lat. Am. Trans. **7**, 78–84 (2009)

Energy Systems

Independent Energy Storage Power Limitations for Secured Power System Operation

Hussein H. Abdeltawab$^{(\boxtimes)}$ and Yasser Abdel-Rady I. Mohamed

Electrical, Computer Engineering, University of Alberta, Edmonton, T6G 1H9, Canada
abdeltaw@ualberta.ca

Abstract. This paper presents a tool to robustly allocate the allowable operating zones of active and reactive power trading margins for multi energy storage systems (ESSs) without violating typical distribution system constraints. This tool helps the distribution network operator (DNO) to facilitate ESS safe participation in day-ahead active and reactive power markets. It estimates the required ESS reactive power support to keep safe voltage margins. In order to avoid conservative results, an uncertainty budget designed by a fuzzy expert is imposed on the uncertainty domain. Case studies on one hundred different uncertainty scenarios are conducted on a real 41-bus Canadian system. Simulation results have shown that the proposed algorithm provides robust operating zones for ESSs with less conservatism.

Keywords: Energy storage system (ESS) · Particle swarm optimization (PSO) · Uncertainty budget · Wind power uncertainty · Worst-case power flow

1 Introduction

Day after day, the world witnesses an increasing penetration level of renewable energy resources (RESs) in power systems. As RESs are indispatchable, combining energy storage systems (ESSs) with RESs is a necessity for enhancing profitability and ensuring system stability. In addition to renewable integrations, ESSs have many other power applications [1, 2], and it participates in ancillary services markets [3], e.g. voltage support in weak grids [4].

Unfortunately, the lack of regulatory rules and grid codes for ESSs in different applications is one of the main challenges facing effective integration of ESSs in grid systems [1, 5]. While ESS acts as an electrical load or generator, the DNO needs to define the safe dispatchability zones of each ESS in case of charge or discharge modes. Within these zones (named hereinafter as robust operating zone (ROZ)), the DNO should guarantee that system operational limits are respected under renewable generation and load uncertainties, and possible contingencies. On the other hand, as each ESS has a different stakeholder with different profit portfolios and dispatching agendas (e.g. energy arbitrage or renewable integration), the DNO should not interfere in ESS commitment or impose a certain dispatching strategy on other assets. In addition to the aforementioned challenges, load and RES uncertainty makes ROZ identification for ESS a complicated problem.

© IFIP International Federation for Information Processing 2016
Published by Springer International Publishing Switzerland 2016. All Rights Reserved
L. Camarinha-Matos et al. (Eds.): DoCEIS 2016, IFIP AICT 470, pp. 407–415, 2016.
DOI: 10.1007/978-3-319-31165-4_38

Energy management system (EMS) of ESSs has different strategies related to the industrial application's nature. For instance, in [6], an intelligent EMS for a battery and a fuel cell in a high frequency microgrid is proposed, while in [6] EMS dispatches the Battery for cost minimization, meanwhile the authors of [7] uses a fuzzy expert to optimize both environmental and economic cost functions. EMS is especially more important with hybrid vehicles (V2G) research [8, 9]. However, none of the aforementioned works considered the AC power flow constraints with the EMS which is the main motive for this PhD research. This paper is arranged as follows. Section 2 explains the main contributions of this work while Sect. 3 briefly describes the problem formulation, whereas Sect. 4 explains the different stages for ROZ generation. Section 5 presents a case study on a radial feeder for results validation. Finally, the conclusions are drawn in Sect. 6.

2   Contribution to Cyber-Physical Systems

This work proposes a framework to facilitate ESS participation in day-ahead markets under distribution system uncertainty taking the power flow constraints into consideration. This is possible via defining ROZ for each ESS. The proposed framework has the following contributions to the research field in general:

1. It combines the merits of stochastic programming and robust optimization (RO) as it is applicable with or without uncertainty distribution availability.
2. For reducing RO conservatism, a fuzzy expert is designed to calculate the uncertainty budget limits according to the uncertainty probability and its risk level.
3. Developing the ROZ concept that draws the charge and discharge margins for each ESS taking into account RES, load uncertainties and possible system contingency.

The main contribution to Cyber-physical system (CPS) is that ROZ can facilitate a distributed trading framework between ESSs and renewable energy resources using Multiagent systems. This is possible since ROZ decentralizes the optimal power flow problem into a distributed economic one and a centralized technical problem (finding ROZ). When each ESS entity knows his feasible operating zone (ROZ), a decentralized trading system for services between ESS and renewable energy owners can be conducted easily via internet or an available local communication network.

3   Problem Formulation

In distribution system, we have two sides that control ESSs dispatch. Firstly, the ESS owner aims at maximizing his profit from market. Secondly, the DNO is a non-profit organization that keeps system reliability and guarantees its secure operation. As ESSs participate in various services, the DNO must guarantee that ESS dispatching decisions will not impair the system operation. Further, the DNO may require reactive power support from different ESSs in case of RES or load uncertainty. The ESS owner is paid for this reactive power support in the reactive power market [10, 11].

To sum up, the ESSs dispatch decision must fulfills the following techno-economic objectives:

1. The DNO must guarantee sound system operation and security for all dispatchability levels of ESS.
2. The ESS provides adequate reactive power support to overcome any uncertainties for a fair compensation from the local reactive power market.
3. The ESS owner has the freedom of dispatching his assets to participate in different electricity markets (given that conditions 1 and 2 are satisfied).

Conditions 1 and 2 are the DNO conditions in each ESS dispatch decision taken by the ESS owner. These conditions are easily embedded as a constraints (limits) on the dispatch power (charge and discharge) at each time, known here as ROZ. As depicted in Fig. 1, firstly, the DNO calculates the day-ahead ROZ and sends it to all ESS stations. Secondly, each ESS owner dispatches his own assets in different markets such that his profit is maximized. As given in (1), each owner dispatches his ESS to maximize his profit, given the network constraints embedded in the ROZ and other dynamical operational constraints for the ESS (e.g., state-of-charge, number of charging cycles, etc.).

$$\underset{p_s}{\text{Max}} \left(profit_s \right) S.t.$$
$$p_s \epsilon ROZ_s, \quad ESS \text{ Operational constraints} \tag{1}$$

The ROZ is defined as the allowable margins of charging and discharging for each ESS to participate in different active power markets, such that the grid system constraints are respected and the ESS can provide adequate reactive power support for compensating uncertainties.

$$ROZ_{sk} = \left[\underline{p_{sk}}, \overline{p_{sk}} \right] \forall s \epsilon \mathcal{N}_s, \forall k \epsilon \mathcal{N}_k \tag{2}$$

$$ROZ_s = \bigcup_k ROZ_{sk}, ROZ_s \in \mathbb{R}^{2 \times n_k} \tag{3}$$

$$ROZ = \bigcup_s ROZ_s, ROZ \in \mathbb{R}^{2 \times n_k \times n_s} \tag{4}$$

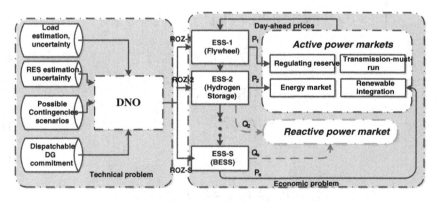

Fig. 1. Proposed ESS dispatch framework.

As given in (2), the ROZ is defined by the allowable upper and lower limits for each ESS s over the time horizon k. for a set of time horizon n_k, and a set of ESSs n_s, ROZ is defined by the sets given in (3) and (4). Given the uncertainty domain $\mathcal{D}$, the possible contingencies, the dispatchable units' commitment and RES and load expected values, two optimization problems define the ROZ. First, the ROZ upper limit $\overline{psk}$ is defined via (5). The objective function (5) aims at maximizing the per-unit active power participation of all ESSs. The resulting ESS active power represents the ROZ upper limits $p_{sk-ref} = \overline{p_{sk}}$. The reason for dividing each ESS active power on the apparent power (C_s) is the fair participations for reactive power support from the different ESSs as they have different sizes; otherwise, the large ESS units will have higher weights in the objective function. However, it should be noted that reactive power support is also location-based power; we can't guarantee equal VAR participation from different ESSs. Constraint (6), (7) represent the power flow models set $PF_k^m(\tilde{d})$ for active and, reactive powers, respectively. These are uncertain models as function in uncertainty set $\mathcal{D}$. While in (8), each branch current (I_{tk}) is calculated and the power loss (p_{loss_k}) is estimated in (9), constraints (10)–(13) represent the DNO different technical constraints. First, the bus voltage limits are defined in (10), while each branch ampacity constraint is limited by (11). In (12), the maximum power loss is limited (e.g., 5 %). Equation (13) limits the exchange power with the grid to avoid un-allowed reversal power flow or extra loading on the main grid. On the other hand; (14) and (15) define the apparent and discharge powers margins. Finally, (16) represents the discharging power losses in each storage in order to consider the ESS efficiency [12], where

$$\eta_{dcs} \geq 1$$

$$\max_{p_{sk-ref}, q_{sk}} \left(\sum_{k=1}^{n_k} \sum_{s=1}^{n_s} \frac{p_{sk}}{C_s} \right) s.t. \tag{5}$$

$$PF_k^m(\tilde{d}) \begin{cases} \hat{p}_{ik} + \tilde{p}_{ik} = \sum_{j=1}^{n_b} |v_{ik}^m| \, |v_{jk}^m| \left(G_{ij}^m \cos\left(\delta_{ijk}^m\right) \right) + \left(B_{ij}^m \sin\left(\delta_{ijk}^m\right) \right) & (6) \\[2mm] \hat{q}_{ik} + \tilde{q}_{ik} = \sum_{j=1}^{n_b} |v_{ik}^m| \, |v_{jk}^m| \left(G_{ij}^m \sin\left(\delta_{ijk}^m\right) \right) - \left(B_{ij}^m \cos\left(\delta_{ijk}^m\right) \right) & (7) \\[2mm] I_{tk}^m = |v_{ik}^m - v_{jk}^m| \, |G_{ij}^m + jB_{ij}^m| & (8) \end{cases}$$

$$p_{loss_k}^m = \sum_{t=1}^{n_T} I_{tk}^{m2} Z_t^m \tag{9}$$

$$v_{min} \leq v_{ik}^m \leq v_{max} \tag{10}$$

$$I_{tk}^m \leq \bar{I}_t \tag{11}$$

$$p_{loss_k}^m \leq \overline{P}_{loss_k} \tag{12}$$

$$\underline{p}_{grid} \leq p_{grid_k}^m \leq \overline{P}_{grid} \tag{13}$$

$$p^2_{sk-ref} + q^2_{sk} \leq C^2_s \tag{14}$$

$$0 \leq p_{sk-ref} \leq C_s \tag{15}$$

$$p_{sk} = \eta_{dcs}p_{sk-ref} \tag{16}$$

$$\forall i,j \in \mathcal{N}_b, k \in \mathcal{N}_k, t \in \mathcal{N}_t, s \in \mathcal{N}_s, m \in \mathcal{N}_c, [\tilde{P}_{ik}, \tilde{q}_{ik}] = \tilde{d}_{ik} \in D$$

Second, the charge power maximization is driven by (17).

$$\min_{p_{sk-ref}, q_{sk}} \left(\sum_{k=1}^{n_k} \sum_{s=1}^{n_s} \frac{p_{sk}}{c_s} \right) \tag{17}$$

$$s.t. \ (6) - (14)$$

$$-C_s \leq p_{sk-ref} \leq 0 \tag{18}$$

$$p_{sk} = \eta_{chs}p_{sk-ref} \tag{19}$$

Unlike (5), in (17), ESSs act as a load and the objective is to minimize their active power (maximizing charge power). The resulting ESS power in this case represents the lower ROZ limit ($p_{sk-ref} = p_{sk}$). This problem has the same constraints as (5) plus constraints (18), (19) and the first constraint guarantees that all ESSs act as load, whereas the second constraint represents the charging losses in each ESS-s, where $\eta_{chs} \leq 1$. Since the problem (5) aims at maximizing the discharge power of all ESS,

4   Roz Generation Framework

We assume a power flow model where all ESSs participate with their maximum discharge power and a zero reactive power support; thus, the PF linearization is conducted around the ESS power coordinates $[p_{sk}, q_{sk}] = [c_s, 0] \ \forall s \in \mathcal{N}_s, k \in \mathcal{N}_k$. The voltage sensitivity matrix $\Lambda_{ip}, \Lambda_{iq}$ is derived for this special power flow to obtain this linearized model; however, the problem still uncertain; thus, a possible framework is summarized in these three steps:

1. Relax the uncertainty domain from D into D_f.
2. Search for the worst case uncertainty (WCU) within the domain D_f to detect d* and pick up the worst contingency power flow structure.
3. Solve optimization problems (5) using semi-definite programming given the WCU-power flow. The ROZ generation is formalized in a three stages framework as explained next.

The proposed framework finds the ROZ for each ESS on three stages: Stage (A), given the uncertainty range for both RES and load D, it is relaxed using an uncertainty budget to get a new set D_f.

This work proposes a new method for uncertainty budget determination based on the uncertainty risk and associated probability (if available). In Stage (B), PSO detects

WCU by scanning $\boldsymbol{D}_f$. As a result, Stage (A) draws the search domain for PSO. Finally, in Stage (C), first, the full active power participation of all ESS in both full charge and full discharge scenarios with the WCU are tested. In case of any grid code violation, a semi-definite programming technique calculates the required reactive power support from each ESS such that no system violation occurs at the WCU. Figure 2 shows the ROZ generation process.

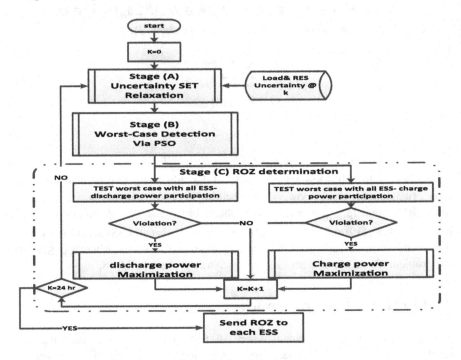

Fig. 2. Flowchart for ROZ generation.

5   Case Study

This study tests the proposed framework on a 41-bus real radial feeder in Ontario, Canada [13]. This medium-voltage (20 MVA, 16 kV) radial feeder is depicted in Fig. 3. The load represents real case residential profiles as stated in [14]. The feeder has five RESs with a gross penetration of 30 % (6 MVA). Further, it has two BESS represented in a PEV parking lot (BESS1) and a storage station (BESS2). The total storage penetration is (15 %) and aims at providing enough reactive power support if needed. The RES and storage allocation is optimized in [13] in order to minimize power losses, while RES profiles are historical data from Alberta system operator (AESO). Simulations study the effect of uncertainty domain choice D on the ROZ and voltage violation. Four cases with different conservatism degrees are compared here:

- **Deterministic case (D1):** the RES and load uncertainty are unconsidered.
- **Uncertainty budget (D2):** in this case, the fuzzy expert defines the uncertainty budget and uses 95 % probability confidence level.
- **No-budget case (D3):** same as D2 without uncertainty budget.
- **Six-sigma case (D4):** most conservative but highest robustness case considering 99.99 % of the uncertainty domain under the famous six-sigma rule. No uncertainty budget is assumed here. This emulates the robust optimization.

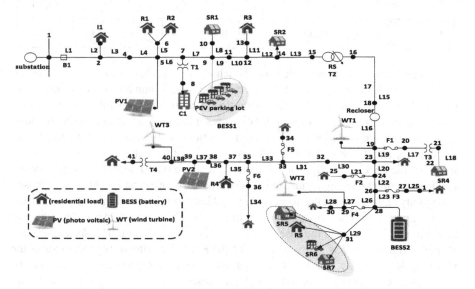

Fig. 3. 41-bus radial feeder.

A comparison is conducted between the four cases by testing the 100 differentuncertainty scenarios generated from the probabilistic distribution for all renewable resources and load. Further, the comparison is repeated at different RES penetration levels (20, 30 and 40 %). No contingency is assumed in all simulations. The results are investigated from two perspectives. First, the DNO cares about the network sound operation represented here by buses' voltage violations. This is represented here by the count of [over-voltage (OV) or under-voltage (UV)] during the 100 scenarios. No ampacity or loss violations occur; thus they are not included in the comparison. On the other hand, The BESS-owner cares about the ROZ size, as a higher ROZ-size means a higher operation margin and profit. Comparison between the gross violations (OV, UV) over the 100 different scenarios is shown in Table 1. It compares the four ROZ cases corresponding to the different uncertainties.

Table 1. ROZ size and voltage violations in 100 scenarios.

	20%-RES			30%-RES			40%-RES		
	OV	UV	ROZ$_2$	OV	UV	ROZ$_2$	OV	UV	ROZ$_2$
D1	133	47	98.2%	88	45	98%	43	41	97.6%
D2	7	3	95.7%	7	1	95.7%	6	1	95.3%
D3	6	2	95%	6	0	95.3%	5	1	94.9%
D4	0	0	92.9%	0	0	92.9%	0	0	92.1%

As expected, the deterministic case D1 achieves the highest ROZ size (best for the owner as less VAR support needed) but with the highest number of violations on all penetrations (all apparent power is committed as active power while reactive power support is not enough to overcome violations resulting from uncertainty). On the other hand, the six-sigma case D4 results in zero violations with the lowest ROZ size for all BESS, As a result, the ESS owner achieves less profit from the active power market; however, the ESS has always had the required reactive power support to overcome the uncertainty effect during the hundred different power flow scenarios. For the proposed uncertainty budget D2 and the no-budget case D3, the number of violations is very small (a maximum of ten times violations occur during the 100 scenarios in a 41-bus system); however, the ROZ size in the case of D2 is always greater than in D3. As a result, the proposed framework has managed to boost the ROZ size without big sacrifices in the power system security. On the other hand, the RES penetration effect is clear on the ROZ-size, for a higher RES penetration, higher reactive power support is required from the BESS. As a result, the active power limits decreases (ROZ size diminishes with RES penetration).

6   Conclusions

This paper presents a framework to define robust operating zones (ROZ) for independent energy storage units in power systems without imposing unit commitment on the ESSs owners. The technical constraints include permissible voltage level, ampacity, power losses, and reverse power flow limits. Further, RES and load uncertainty are taken into consideration when generating the ROZ. Furthermore, a security constrained-ROZ is possible for any contingences combinations. This tool provides the safe operating zone for each ESS which facilitates a decentralized co-operation between a high number of storages and other renewable facilities. The proposed framework is tested on a real 41-bus radial feeder using 100 different uncertainty scenarios. Four uncertainty sets with different conservatism levels are tested. The results have shown that the designed uncertainty budget managed to boost the ROZ size with a very low violation probability.

References

1. G.C. Jim Eyer, "Energy Storage for the Electricity Grid: Benefits and Market Potential Assessment Guide," SANDIA, Albuquerque, New Mexico, 2010
2. J.A. Rahul Walawalkar, "Market Analysis of Emerging Electric Energy Storage Systems," NETL, 2008
3. Rebours, Y., Kirschen, D., Trotignon, M., Rossignol, S.: A Survey of Frequency and Voltage Control Ancillary Services—Part I: Technical Features. IEEE Trans. Power Syst. **22**(1), 250–357 (2007)
4. Zhong, J., Bhattacharya, K.: Toward a competitive market for reactive power. IEEE Trans. Power Syst. **17**(4), 1206–1215 (2002)
5. M. G. Jenny Chen, "Energy Storage Initiative Issue Identification," AESO, Edmonton, 2013
6. Chakraborty, S., Weiss, M., Simoes, M.: Distributed Intelligent Energy Management System for a Single-Phase High-Frequency AC Microgrid. IEEE Trans. Industr. Electron. **54**(1), 97–109 (2007)
7. Chaouachi, A., Kamel, R., Andoulsi, R., Nagasaka, K.: Multiobjective Intelligent Energy Management for a Microgrid. IEEE Trans. Industr. Electron. **60**(4), 1688–1699 (2013)
8. Jiang, W., Fahimi, B.: Active Current Sharing and Source Management in Fuel Cell-Battery Hybrid Power System. IEEE Trans. Industr. Electron. **57**(2), 752–761 (2010)
9. Ma, T., Mohammed, O.: Economic Analysis of Real-Time Large-Scale PEVs Network Power Flow Control Algorithm With the Consideration of V2G Services. IEEE Trans. Ind. Appl. **50**(6), 4272–4280 (2014)
10. El-samahy, I.: Secure Provision of Reactive Power Ancillary Services in competitive Electricity Markets. University of waterloo, Waterloo (2008)
11. Rebours, Y., Kirschen, D., Trotignon, M., Rossignol, S.: A Survey of Frequency and Voltage Control Ancillary Services—Part II: Economic Features. IEEE Trans. Power Syst. **22**(1), 358–366 (2007)
12. Thatte, A., Xie, L., Viassolo, D., Singh, S.: Risk Measure Based Robust Bidding Strategy for Arbitrage Using a Wind Farm and Energy Storage. IEEE Transactions on Smart Grid **4**(4), 2191–2199 (2013)
13. Atwa, Y.M.: Distribution System Planning and Reliability Assessment under High DG Penetration. University of Waterloo, Waterloo, Ontario, Canada (2010)
14. Lopez, E., Opazo, H., Garcia, L., Bastard, P.: Online reconfiguration considering variability demand: applications to real networks. IEEE Trans. Power Syst. **19**(1), 549–553 (2004)

Greenhouse with Sustainable Energy for IoT

Filipe T. Oliveira[1], Ségio A. Leitão[2], Adelino S. Nabais[1],
Rita M. Ascenso[1,3], and João R. Galvão[1,4(✉)]

[1] Leiria Polytechnic Institute, 4163 I 2411-901 Apartado, Leiria, Portugal
adelino.nabais1@gmail.com,
{ftadeu,rita.ascenso,jrgalvao}@ipleiria.pt
[2] INESC-TEC, University of Trás-os-Montes and Alto Douro, 5001-801 Vila Real, Portugal
sleitao@utad.pt
[3] Computer Science and Communications Research Center, Polytechnic Institute of Leiria, Leiria,
Portugal
[4] INESC Coimbra - Institute of Systems Engineering and Computers at Coimbra R&D Unit,
Coimbra, Portugal

Abstract. In order to support the intensive development of agricultural crops
and, in particular the floricultural inside a greenhouse, with the perspective of a
quick distribution in the market, increasing the economic benefits and supported
on efficient and intelligent management systems energy, it is mandatory to
conceive a model based on Cyber-Physical Systems (CPS) This implies, accordingly, increases in renewable primary energy sources utilization coupled with
sensing technologies, include developments on Internet of Things (IoT) and
Cloud Computing (CC), supported with Information and Communication Technologies (ICT) that will lead to new architectural approach applied to a proposed
energy system, based on a sustainable and more engineering autonomous process.
This work comes up with a new energy model that retrofits the system of a greenhouse supported with multiple sensors in one grid, to expand into CPS concept
to manage sensors and controllers that will improve a profitable energy system.

Keywords: CPS · Energy management · Efficiency · Energy model · IoT ·
Computing

1 Introduction

The Census Bureau's latest projections imply that population growth will continue into
the 21st century, although more slowly [1]. To avoid starvation, the production of agriculture became massive and more productive in greenhouses [2, 3]. In 21st century,
greenhouses are used for food production and, among others, floriculture.

Increase the process of plant growth was one of the goals to be achieved with this
research, performing a characterization of the system, from plant conditions of growth,
to environmental conditions and electrical and thermal energy deep analysis to propose
an energetic and environmental sustainable model. The increasing of flower productivity
should take into account no increase in the greenhouse gases (GHG) emissions and

© IFIP International Federation for Information Processing 2016
Published by Springer International Publishing Switzerland 2016. All Rights Reserved
L. Camarinha-Matos et al. (Eds.): DoCEIS 2016, IFIP AICT 470, pp. 416–424, 2016.
DOI: 10.1007/978-3-319-31165-4_39

contribute to the fulfillment of international commitments set by 20/20/20 European Union Plan adopted by Portugal, based on renewable energy use as primary source, contributing to sustainable human activities and achieve eco-certification recognition [4]. The case study is a greenhouse with intensive flowers production where was performed an energy audit, comprising a survey on energy consumption by type: electrical and thermal and several simulations to improve energy efficiency. This research coupled with CPS concept could increase the energy performance to develop intense agricultural crops production and in particular the flori-cultural inside a greenhouse, with the perspective of a rapid response in placing flowers in the distribution market by increasing the economic benefits, but supported by an efficient and intelligent management energy model, in the process of plant growth [5].

The innovative proposed model consists of a mix of energy production processes based on photovoltaic panels and biomass boilers supported by CPS, where the amount of data acquired should be process with the computing power and communication interfaces between different networked systems, becoming more user friendly and less expensive. This concept rest on the interactions between multiple areas and promotes the emergence of design and process science applied to floriculture [6].

2   Benefits from Cyber-Physical Systems

CPS represents a class of control systems that consist of tightly linked computational and physical components [7]. The primary advantages of CPS when compared with others include increased autonomy, efficiency, functionality, reliability and safety level. These advantages allow CPS to be utilized in a wide range of technical fields including, power management, vehicle control and safety, advanced robotics, image analysis and IoT [8, 9]. The sensing data and accumulated information from sensors and internet knowledge is analyzed by intelligent data mining processing and the final extracted information is provided to the managers, through a diversity of interface applications, including mobile. A permanent and constant information storage in cloud computing and data mining allows real-time and newer decision-making through a human–machine interface [10]. Beyond the data collection with sensors and monitoring, the advantage is also on control the physical processes, usually with feedback loops where physical processes affect computations and vice versa. Lee mentions that CPS could significantly improve energy efficiency and demand variability, reducing our dependence on fossil fuels and our greenhouse gas emissions [7].

Energy in the real world represents the physical part of action, but on the other hand, the concept that models how systems decide on, manage, and control their actions is the information [11]. This study is mainly focused on energy audit and energy efficiency, proposing an energy model. This proposed architecture was designed to achieve, firstly energy savings in the floricultural area, increasing floricultural and agricultural in greenhouse management with major economic incomes and sustainability, as other models of energy management [12]. The objective is similarly to obtain large and quality productions due to the integrated management system, saving water resources by careful monitoring the crop vegetation needs regarding the hydric stress. Finally, was taken into account provide a low cost solution for the greenhouse owner that has limited financial possibilities, enabling its' implementation.

Rad emphasizes CPS playing an important role in the field of precision agriculture and proposed a model of agricultural management integrated system architecture based on CPS design technology [5].

The floriculture greenhouse case study presented in this paper, beyond the proposal of a new energy model that retrofits the system supported with multiple sensors in one grid, is expected to expand into CPS concept, including IoT and cloud computing to improve, even more, a profitable energy system, based on sustainable and autonomous process. The application of CPS complements the existing sensors installed to monitor the conditions inside and outside the greenhouse, which already exist. Cyberization would allow integrating the data and saving in cloud data warehouses. The data collected associated with weather forecast freely available on the internet, would improve the capacity of prediction to act in order to save energy and benefit productivity. For energy saving, the work performed included the utilization of a different biomass fuel and produce electricity, but this savings could be even more raised if coupled with controlling systems to avoid waste. So, CPS concept undertake could get information from the data collected, trough computing, to ground the management of energy spent in lightning, heating and boiler supply. All this permit to fundament decision making, that can be performed automatically through local or cloud controlling systems or by workers by remote web based or mobile devices (Fig. 1).

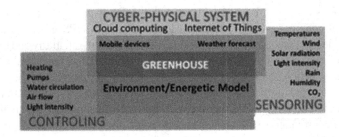

Fig. 1. CPS applied to energy efficiency and environment sustainability model.

The physical system and the energetic and environmental model already performed, improves efficiency and sustainability, and has advantages and synergies if included in such wider inclusive CPS model.

3   Greenhouse Case Study

This study is based on an energy and environmental audit and identifies critical areas of a physical complex system, in order to propose a new energy model, which intends to contribute to energy and environmental sustainability of the case study, a greenhouse. Its realization required technical knowledge in terms of electricity, thermal and mechanical energy, environment, as well as the processes associated with the species to produce and also construction parameters, materials, processes, environmental conditions, finally the local climate. The information referred to study "the agro-system" is related with the energy consumption disaggregation, to determine and assess the energy and environmental efficiency of each of the available energy sources, whether fossil or renewable sources.

The greenhouses where the study was done are located in the Alpiarça municipality at an altitude of 56 m. The greenhouses are the curved roof multi-tunnel, with a total 27 tunnels, with varying lengths (75–145 m) and width (8–9.6 m). The height is on average 5.6 m ridge and 4 m in height (growth zone) with infrastructures total area of 24,104 m^2 (Fig. 2a). The local minimum temperature in the winter months is about 2°C and the maximum temperature in summer averages about 30°C. The structure is made of galvanized steel, the cover material (ceiling and side walls) is composed of a three layer plastic film with a thickness of 200 μm, anti-UV treatment, and light transmission of 92 %, light diffusion of 14 % and thermicity of 86 % [13]. All tunnels have zenith windows for circulation/air renovation, with opening 1/2 arch, located in the ridge for a more effective output of hot air. There are no side windows, nor any forced ventilation system, but there are zenith windows. For control of brightness and temperature, the greenhouses are equipped with thermal curtains, opaque, XLS obscure placed horizontally and laterally.

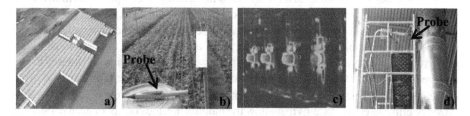

Fig. 2. Data collection in greenhouse: (a) Greenhouses overview; (b) Temperature, humidity, CO_2; (c) Thermographic images of hot water pumps; (d) Gaseous emission probe.

Chrysanthemum plant is the produced species being used as cut flower, through a variety of species and sizes. It's kind of the family *Asteraceae* sp., *Dendranthema* genus [13]. Production of the chrysanthemum is made in beds, with an approximate width of 1.35 m and 0.30 m spaced for workers passage. The optimal ambient temperature for a good air conditioning plant in the winter period should be between 12°C–18°C nighttime, 19°C–24°C daytime and soil temperature with identical values. The greenhouse for plant production involves a building with the boiler, the fuel, water accumulation tanks, pumps for water circulation and biomass storage tanks. The heating indoor the greenhouse is performed using four hot water circuits, divided by tunnels. The distribution of heated water is made at a temperature of about 40°C in round ≈30°C in return. A similar system of hot water circulation of used to maintain the soil temperature. The heating fluid used is obtained from the power station, using a boiler with a rated output of 2,000 kW thermal, whose primary energy is the solid biomass. The system is complemented by two accumulation buffer tanks with a unit capacity of 30,000 litres to address the point of consumption. To perform water circulation there are several electrical pumps to pump the hot water along the circuit that crosses all the greenhouses rooms and soil.

4   Data Collection and Results Analysis

In order to perform a deep characterization of energy consumption and environmental impact it was collected data on the energy sources used, namely electrical and thermal, using portable analyzers. The environment inside the greenhouse was quantified according to CO_2, humidity, air and soil temperature, using portable analyzers, although there is a system to measure these parameters, still without recording or storing the data, Fig. 2b. Also the combustion and gaseous emissions due to boiler operation, its performance and power calorific value were determined (Methods ASTM D240) and analyzed, based on the biomass used, namely pine wood, olive wood and olive-pomace pellets. In the study, it was also considered the calorific value analysis of each of the used biomass fuels and its moisture, as well as analysis of the ash from the burn. It was calculated the energy requirements for heating and associated costs, as well as determined the footprint carbon resulting from this activity. To fulfill the electricity needs was proposed a viable photovoltaic (PV) system. As mentioned the audit included an analysis of the environmental data, once it is fundamental to the development of crops as intense as these. This analysis was performed using several scattered multifunction equipment as indicated in the areas: indoor greenhouse and thermal power plant.

The results obtained for most parameters were within the reference values for the species production, above mentioned, except for air temperature that, for some nighttime, went down to levels below the minimum set for this species. Likewise, it was detected low values for CO_2 in some periods of the day and the maximum does not exceed 626 ppm. These values out of the optimal conditions for growth should be minimized or eliminated by systems that allow the monitoring and control of each parameter. The thermal energy comes from renewable sources (solid biomass) mainly to heat water to maintain the greenhouse internal temperature, through pipe hot water circulation.

The thermal energy losses should be avoided to ensure the temperature conditions for plants growth, and achieve best energy efficiency performance Fig. 2c. There were significant losses detected in several infrastructure components. The red color represents the highest temperature (higher losses) and blue color at lower temperature (lower losses). These heat losses reach resulting mainly from the bad/poor insulation of the components of hot water circulation.

Evaluations to combustion and exhaust emissions were held on different days because of equipment unavailability. The results for the combustion boiler as well as the exhaust emissions are interrelated, since a boiler malfunction involves exposure limit value in misfit emissions. The evaluation of higher calorific value and lower calorific value for biomass used pine wood (17.17 MJ/Kg, 16.33 MJ/Kg), olive wood (16.75 MJ/Kg, 16.33 MJ/Kg) and olive-pomace pellets (18 MJ/Kg, 18 MJ/Kg) identified the olive-pomace pellets as best biomass, allowing the system automation. Thermal energy is so improved, increasing the calorific energy yield, but also reducing losses, once the loss at 105 °C determined was 29 % for pine wood, 18 % for olive wood and 10 % for olive-pomace pellets and reducing toxic emissions, Fig. 2d.

The electricity used in greenhouses is only from the public power grid (normal low voltage tri-hourly tariff of 41.40 kVA) and the significant energy consumptions were in the engine room (50 %), cold chambers (31 %) and in lighting, sockets and engines for opening the zenith windows (19 %). To analyze the electricity consumption was used a portable analyzer during one week, in January, as shown in Fig. 3a, and the analysis was performed according to consumption of a whole year through itemized billing values.

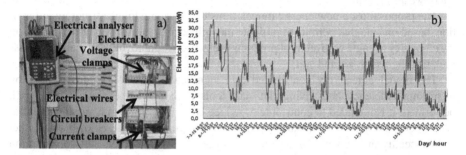

Fig. 3. Collection of electrical consumption: (a) Data analyzer collecting data in electrical box; (b) Electrical power recorded in the sampling week.

During the period analyzed the peak power was 33.35 kW electrical corresponding to the largest energy needs recorded at night due to space heating requirements with the operation of circulator pumps and boiler. The results obtained show a wide range of power requirements over 24 h, from 0.49 kW electrical up to 33.35 kW electrical, mentioned, Fig. 3b. The average power in a full day in the analyzed period corresponds to 13.81 kW electrical.

After determining the average hourly power consumption of monthly cycles and daily schedules, using the PVsyst software [14], introduced the monitoring data of the installation in order to scale the PV system, which is justified by the existence of abundant solar radiation in the area. The PV system to install should have 90 modules with a total power of 23.4 kWp photovoltaic field, with groups of five panels strings connected in parallel, and 18 panels connected in series. Economic analysis support the system viability because net present value is 110,725.30€ and internal rate return is 18 % and 5.4 years of break even, for an initial investment of 38,672.00€ in 20 years. Also, it contributes by 668.7 MWh of self-consumption, 54.9 MWh sold to the grid and only with 340,09 KgCO$_2$e.

The carbon footprint of 88,381.12 KgCO$_2$e per year along the greenhouse life cycle, using the PAS 2050, which is a cross tool of environmental sustainability providing analysis, information on the performance of the most important environmental indicators of GHG emissions (KgCO$_2$e) on energy consumption, namely Biomass (zero), fuel (34,300.17), electricity (45,973.52), and on fungicide (135.52), insecticide (135.70), herbicides (135,74), fertilizers (941.59) and other [15].

5   Proposal of a New Energy and Environment Model with CPS

Energy audit found weaknesses in the use of energy, both electricity and thermal, namely in boiler supply and biomass storage. These deficiencies stem largely behavioral actions by lack or absence of proactive in real time decisions. In the structure there are many employees with different and varied functions, whose responsibility for ensuring a good energy and environmental performance will be difficult to achieve, so, the installation of automatic controllers, commanded by applications that control the pumps, zenith windows and curtains.

The proposal for a new energy model, aims to contribute to infrastructure automation, based on data that permits to mitigate the influence of human behavior, increasing the efficiency triangle with areas of production electricity (photovoltaic panels), biomass fuel, heating water, air and soil,

The model objectively contributed to a significant reduction of costs either by increasing efficiency and giving the staff tools to improve the yield during work performed. It is intended to contribute to reduce GHG emissions with significant influence to environmental sustainability, namely using controlled source of biomass fuel and especially in terms of implementation of the photovoltaic (PV) self-consumption system. Also, related to thermal energy, several advices on storage and use of biomass and in hot water storage, circulation and thermal insulation.

In this regard, among the various components that make up the new model, the implementation of CPS technologies, including progress towards the IoT, CC, Infrastructure-as-a-Service and Platform-as-a-Service that largely determine the extent of the success of this energy and environmental model (Fig. 4).

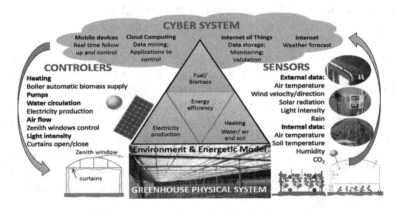

Fig. 4. Proposal of a new energy and environment model with CPS.

The greenhouse system in study has already installed an AGRITEC C800 system with sensors to climate conditions (wind direction and velocity, rain, temperature, humidity), and internal greenhouse conditions sensoring of temperature, humidity and CO_2, with the possibility to record this data. Although the software was not installed yet to transfer and save data. The actual systems of internet communication and cloud computing storage, monitoring and validating data, would be useful to install a computing system connected to the actual AGRITEC C800. Also, this system could allow to control the hot water tanks and circulation, as also the zenith windows and curtains, if a mechatronics system is installed, these could be controlled and monitored by software, acting automatic and/or validated by the owner in real time through mobile devices or web based applications.

6 Conclusions

This research was performed to improve energy efficiency and environmental sustainability based on a floriculture greenhouse study. Beyond the proposal of a new energy model that retrofits the system of a greenhouse supported with multiple sensors in one grid, this work intends to expand into CPS concept, including IoT and CC to improve, even more, a profitable energy system, based on sustainable and autonomous process. The energetic and environmental model of this physical system was characterized in order to improve efficiency and sustainability, having advantages and synergies if included in a wider inclusive CPS model (Fig. 4).

As future perspective, it is intended to promote the cyberization of the greenhouse, this is a challenge of integrating computing using already existing sensing equipment and implementing mechatronics, which has been recognized as beneficial in precision agriculture.

References

1. U.S. Census Bureau. https://www.census.gov/population/international/data/idb/worldpop graph.php. Accessed September 2014
2. Nelson, P.V.: Greenhouse Operation and Management. Prentice Hall, NJ (1995)
3. Jensen, M., Malter, A.: Protected Horticulture: a Global Review. World Bank, USA (1991)
4. Horizon2020 EU http://ec.europa.eu/programmes/horizon2020/en/area/energy. Accessed November 2015
5. Rad, C.R., Hancu, O., Takacs, I.A., Olteanu, G.: Smart monitoring of potato crop: a cyber-physical system architecture model in the field of precision agriculture. A. A. Sci. Proc. **6**, 73–79 (2005). doi:10.1016/j.aaspro.2015.08.041
6. Nie, J., Sun, R., Li, X.: A precision agriculture architecture with cyber-physical systems design technology. Appl. Mech. Mater. **543–547**, 1567–1570 (2014)
7. Edward A. Lee: CPS foundations. In: Proceedings of 47th Design Automation Conference, pp. 737–742. ACM, June 2010
8. Mišiū, V., Mišiū, J.: Machine-to-Machine Communications Architectures, Technology, Standards, and Applications, pp. 1–30. CRC Press, Taylor & Francis Group, UK and USA (2014)
9. Tung, S., Liu, Y., Wejinwa, U.: Special issue on control and automation in cyber-physical systems. Trans. Inst. Meas. Control **36**(7), 867 (2014)
10. ICRI: Cyber-Physical Systems. http://www.cities.io/project/cps/. Accessed October 2015
11. Moreira, L., Leitão, S., Vale, Z., Galvão, J., Marques, P.: Analysis of power quality disturbances in industry in the centre region of Portugal. In: Camarinha-Matos, L.M., Barrento, N.S., Mendonça, R. (eds.) DoCEIS 2014. IFIP AICT, vol. 423, pp. 435–442. Springer, Heidelberg (2014)
12. Galvão, J.R., Jesus, C.D., Ascenso, R.M.T.: Sustainable energy model in high rate hotel. In: International Conference on Energy for Sustainability (EfS) Proceedings, Coimbra (2015)

13. Kessler, J.R.: Chrysanthemum-commercial greenhouse production. Auburn University (2015). http://www.ag.auburn.edu/hort/landscape/Potmum.htm. Accessed October 2015
14. PVsyst: photovoltaic software, January 2015. http://www.pvsyst.com/
15. Carbon Trust and Defra: Guide to PAS 2050 How to assess the carbon footprint of goods and services. BSI, London (2008)

Decentralised Coordination of Intelligent Autonomous Batteries

Evgeny Nefedov[1(✉)] and Valeriy Vyatkin[1,2]

[1] Department of Electrical Engineering and Automation, Aalto University, Helsinki, Finland
evgeny.nefedov@aalto.fi
[2] Department of Computer Science, Electrical and Space Engineering,
Luleå University of Technology, Luleå, Sweden
vyatkin@ieee.org

Abstract. This paper proposes enabling intelligence for cyber-physical system of intelligent collaborating energy storages. Two intelligent batteries coordinate their behaviour in a dynamic electricity price scenario, accumulating the energy when the electricity price is low, and replacing the grid when the price is high. Both batteries directly exchange their state information with each other without any centralized processing agent, following the coordination algorithm developed in this paper. This simplifies their integration and enables achieving more optimal behaviour with regards to state of their charge. When one battery is depleted, the other one immediately compensates the losses by a higher discharge rate. Such a distributed coordination approach enables plug-and-play formation of system of batteries, demonstrates the efficiency of such formation and allows for reduction of costs due to longer discharge time of the batteries.

Keywords: Intelligent battery · Energy management · Battery monitoring · Plug-and-play · Multi-agent coordination

1   Introduction

Battery energy storage is expected to play a major role in Smart Grid systems, providing more flexibility and efficiency to the grid. Storing the energy, it is possible to use efficiently energy generated by renewable sources, reducing costs and decreasing the load on the electrical grid [1]. One possible source of energy storage capacity is parking lots of hybrid and electric vehicles. However, such systems are highly dynamic in their structure: cars are joining and leaving parking lots regularly. Besides, such systems of batteries are highly heterogeneous in their technical parameters, state of charge, schedules and intentions of their owners.

There exist approaches of forming energy storages from such batteries [2], but a typical coordination concept is that every battery coordinates its activity with a management centre, which has access to the utility and market data. In the present paper, another, peer-to-peer coordination mechanism between system components as embedded agents is proposed. The agents that represent individual batteries take into account own

L. Camarinha-Matos et al. (Eds.): DoCEIS 2016, IFIP AICT 470, pp. 425–433, 2016.
DOI: 10.1007/978-3-319-31165-4_40

schedule and conditions. One possible implementation of this concept is that each battery has its own embedded control system, which interfaces a barebones battery, and communicates (e.g. via the Internet) with other batteries, utility and market data. The research question, driving this study is whether decentralised decision-making in such distributed energy resource systems can bring substantial economic benefits and how to measure them. Removing the central aggregator is the feature that leads to more independent role of each battery in the system, which improves flexibility and adaptability of the energy storage. Another application of the same approach could be in seamless integration of the household battery storages.

2 Contribution to Cyber-Physical Systems

The intelligence and methods of collaboration for networking systems in automation are developing for several years and can be found in literature, for example in [3]. As one of the practical applications, communication between power system components for rapid supply restoration was presented in [4]. These works, among others, refer to the Internet of Energy as the future energy distribution environment, enabled by the Internet communication technologies and cyber-physical systems theory. Still, the effective use of renewable energy resources is impossible without efficient energy storages. The current battery technologies are still quite costly to make the renewable energy a dominant source. To achieve affordability of energy storages, this paper proposes to use available batteries, e.g. of electric vehicles, or standard batteries with low residual cost. Enhanced by the appropriate sensing, control and communication infrastructure, large sets of such devices can represent reliable energy storages of substantial capacity. Such sets will be dynamically changing their structure; therefore, self-configuration algorithms must be developed to support dynamic reconfiguration of such distributed storage systems. Each of the batteries is equipped with embedded control device that coordinates its activity with the other such devices of the peers, forming a cyber-physical energy storage system.

The paper investigates application scenarios of the proposed cyber-physical energy storage in the Internet of Energy. The considered system is relatively simple and controls mainly the State of Charge (SOC) parameter; nevertheless, it first demonstrates the operability of suggested peer-to-peer communication between batteries as physical systems. The future development assumes monitoring of more parameters with the consequent use of more complex algorithms.

3 Related Works

In recent years, various researches have been carried out at building the intelligent service for using batteries. Plug-In Electric Vehicles technology is the perspective example, where intelligent services and scheduling may have a large impact on the efficiency of such vehicles [5]. A role of energy storage was investigated also for large systems with other energy sources, like photovoltaic panels [6] or wind farm [7], where batteries balance the load smoothing out an intermittent energy profile.

The work of battery management system (BMS) has been studied for several years. It can be presented, for example, as a layered structure for battery state, monitoring and management with mutual data flow between those layers [8]. Some commercial BMS products for electric and hybrid vehicles were compared and summarized by Xing et al. [9]. Waag et al. [10] did the essential review concerning battery monitoring. Estimation of battery parameters, as SOC is critical to control the work of the battery, methods for such estimation were observed in the same paper. The battery monitoring algorithm has to satisfy certain requirements, such as consider battery characteristics, limitations due to operation conditions, hardware characteristics, influences of load profile and common requirements on automotive software.

The challenge of battery management is to operate the batteries in a most effective way, considering the above-mentioned constrains. Control approaches in energy management have been studied for many years. Vega et al. [11] reviewed the existing home energy management models, with various coordination approaches between participants. For example, algorithms based on stochastic game theory [12], or service-oriented architecture [13] were observed. Home energy management mechanism considering primarily the state of involved batteries was also studied [14]. The related works, however, do not follow the Internet of Things architecture with batteries communicating to each other; this paper attempts to bridge this gap.

4   Problem Formulation and Case Study

The mechanism of composing energy storage is investigated for a simple household application. Each participating battery is assumed to have own controller, which plays role of a communication agent in addition to the charge control. The battery controller has to monitor battery's SOC through current-voltage measurements. The SOC value updates frequently and is used as an input data for the BMS. For brevity, further in this article by interaction of two batteries is meant interaction of their BMSs.

The main structure of the test system is shown in Fig. 1. Each battery gets utility data from the grid and demand data from the consumer, and, in addition, receives the state (remaining capacity) data from the other battery. BMS determines own charge and discharge mode based on all these data. Since the main objective of the paper is first to demonstrate working mechanism of battery coordination, some simplifications are used. Thus, only two batteries are considered. The model also neglects the state of health estimation and therefore costs related to battery aging.

A two-level Matlab Simulink model is proposed. The low-level component is an electric scheme of battery connection. The load, represented as a resistor, must be supplied from the battery during discharge, or otherwise directly from the grid, which is assumed to be a DC voltage source. In the latter case, battery also consumes energy for recharging. Current controllers drive the current to supply the load and to recharge the battery, and work as switches, holding and controlling the circuit in the correct state. Those controllers are regulated by external signals from BMS.

State of the battery is monitored via the voltage measurements, which enable estimation of the battery's depth of discharge, and accordingly the SOC. The voltage V

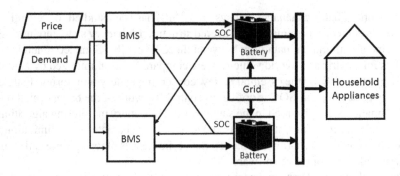

Fig. 1. The layout of the proposed model.

changes with the SOC, and is not a constant value. Therefore, it must be measured to control the current I, which is related to the electric power P with the relationship $I = P/V$. The value of power provided to the load is determined based on consumer's demand. The charge current to the batteries is constant; however, it automatically decreases when the SOC is close to a higher threshold.

The high-level component is the embedded BMS, determining the operation mode of the battery. The BMS is modelled in Simulink Stateflow mode, and represents a hierarchical state chart directly derived from the operating principle. To satisfy the energy demand from the customer E_{dem}, the following equality must hold all the time:

$$E_{dem} = E_{gtl} + E_{disch}, \tag{1}$$

where E_{gtl} is the energy taken from the grid, hereinafter grid-to-load. E_{disch} is the energy to be extracted from the batteries and consists of two components, as follows:

$$E_{disch} = E_{B1} + E_{B2}, \tag{2}$$

where E_{B1} is the energy taken from the current battery; E_{B2} is the energy from the other one. As seen from (1), the customer's load should be supplied either from batteries or directly from the grid. Factors determining a source of power to the load are the utility price p and states of charge of both batteries: SOC_1 means the state of the current battery; SOC_2 is the state of the second battery.

Despite the fact that BMS gets the data from all batteries in the model, each of them is responsible to control only the own battery. Thus, the operating principles further are shown for the single BMS, although they also contain the parameters and symbols related to another battery. The assumption made, that in normal condition (when all the batteries are in the working SOC range) the load is distributed evenly on them, so that $E_{B1} = E_{B2}$. In other words, all BMSs generate commands E_{ex} to use a certain amount of energy from the battery or from the grid, in normal SOC range $E_{ex} = E_{dem}/2$, but all the time the total energy should be equal to E_{dem}.

The following hierarchy of modes is used:

(1) Normal or high SOC_2 (when $SOC_2 > 0.2$)
    (a) Normal SOC_1 (when $0.2 < SOC_1 <= 0.9$)

    i. If $p > 3$: demand is satisfied from the battery ($E_{ex} = E_{Bl} = E_{dem}/2$)

    ii. If $p <= 3$: demand is satisfied directly from the grid ($E_{ex} = E_{gtl}/2$); and battery is being charged

  (b) Low SOC_1 (when $SOC_1 <= 0.2$)

    i. If $p > 3$: no command; all demand should be satisfied from the other battery ($E_{ex} = 0$)

    ii. If $p <= 3$: demand is satisfied directly from the grid ($E_{ex} = E_{gtl}/2$); and battery is being charged

  (c) High SOC_1 (when $SOC_1 > 0.9$)

    i. If $p > 3$: demand is satisfied from the battery ($E_{ex} = E_{Bl} = E_{dem}/2$)

    ii. If $p <= 3$: demand is satisfied directly from the grid ($E_{ex} = E_{gtl}/2$)

(2) Low SOC_2 (when $SOC_2 <= 0.2$)

  (a) Normal SOC_1 (when $0.2 < SOC_1 <= 0.9$)

    i. If $p > 3$: all demand is satisfied from the battery ($E_{ex} = E_{Bl} = E_{dem}$)

    ii. If $p <= 3$: demand is satisfied directly from the grid ($E_{ex} = E_{gtl}/2$); and battery is being charged

  (b) Low SOC_1 (when $SOC_1 <= 0.2$)

    i. If $p > 3$: demand is satisfied directly from the grid ($E_{ex} = E_{gtl}/2$)

    ii. If $p <= 3$: demand is satisfied directly from the grid ($E_{ex} = E_{gtl}/2$); and battery is being charged

  (c) High SOC_1 (when $SOC_1 > 0.9$)

    i. If $p > 3$: all demand is satisfied from the battery ($E_{ex} = E_{Bl} = E_{dem}$)

    ii. If $p <= 3$: demand is satisfied directly from the grid ($E_{ex} = E_{gtl}/2$)

A case study of a household with quite standard power demand and connected batteries is presented in this paper. The electricity price values were chosen based on MISO (Midcontinent Independent System Operator, Inc.) hourly price data [15], and vary between 1.9 and 4.9 cents for kWh. The customer's demand values are taken from hourly energy consumption data of a sample household in [16]. Price and demand values repeat every 24 h and their variations are shown in Fig. 2.

A system of two identical NiMH batteries was modelled. Both are fully charged initially, i.e. $SOC = 1$. In normal SOC range, battery charges during low-price periods, and discharges when the price is high. If SOC is lower than 0.2, discharging process stops, and conversely, if SOC value is higher than 0.9 the battery should stop charging to avoid the negative effects related to overcharge [17]. To evaluate the interchangeability of the batteries, one of them switches off periodically. This emulates the case when the user needs to take the battery out for any other purposes. When such an extraction takes place, BMS assigns the SOC value to zero and no charge-discharge operations are possible until next switching-on. It is assumed that the real SOC stays unchanged during this time. To evaluate the results, two different scenarios studied. In scenario 1, batteries work independently, and each of them is responsible to supply its own half of demand. In scenario 2, those batteries use the coordination algorithm above to replace each other when needed.

5   Results and Discussion

Figure 3 shows two plots for the SOC of the batteries: left for scenario 1, and right for scenario 2, for 72 h period. SOC decreases when the price value from Fig. 2 is greater than 3 cents/kWh, and increases when the price is getting equal or lower than 3 cents/kWh, which represents battery's charge-discharge operations. At some point, when the first battery becomes inactive, it can no longer supply the load until reconnected. Such transition can be seen in Fig. 3 (left) at time moment 40 h, for example. The second battery is still able to provide energy, and it continues to discharge. In scenario 1, when only one battery is working, the demand is satisfied by taking a half from the second battery, and a half from the grid.

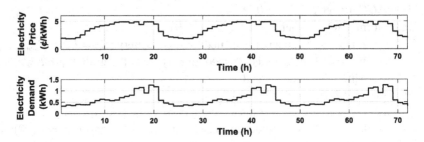

Fig. 2.   Price and customer's demand profiles, used in simulations.

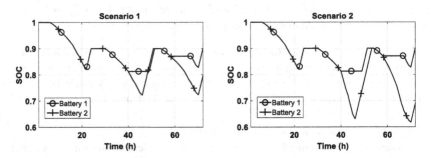

Fig. 3.   SOC of the batteries in scenario 1 *(left)* and scenario 2 *(right)*.

In Scenario 2, when such extraction of the first battery occurs, the second one reacts with producing more energy. It continues to provide its previous "normal" amount of energy, but now its task is also to give energy for another inactive battery. This is reflected in Fig. 3 (right): at the same time moment 40 h, SOC of the second battery starts to decrease faster, and this difference with scenario 1 is the result of the work for two batteries. When price value becomes lower than 3 ¢/kWh, the connected batteries use this opportunity for recharging, and their SOC values are getting higher. In the considered case, both batteries are able to restore the full SOC (90 % threshold) before next discharge (i.e. next period of high-price).

Figure 4 shows energy consumption from the grid for two scenarios. The grid-to-load line repeats demand profile given in Fig. 2 when both batteries are inactive, and becomes zero when they are being discharged. When batteries work on the scenario 1, a half of the required energy is also taken from the grid when the first battery switches off, regardless of the price. However, in Scenario 2 grid-to-load line repeats demand profile only when the price is less than 3 ¢/kWh. Thereby, regardless of the demand, all duration of high-price periods is covered by the energy stored in batteries. Only when the second battery also becomes not available, due to either extraction or depletion, the grid will start to supply all the demand. It can happen with higher energy consumption or using the batteries with less capacity.

The economical profit of using batteries is presented on Fig. 5. It captured three cost cases. The first one is the cost if no battery is used, i.e. the integrated cost of the product of the price and consumption if the demand is supplied directly from the grid. The other two cases are the costs for the considered scenarios, i.e. the consumption in the form of grid-to-load plus recharging of the batteries. In Scenario 1, the profit from the use of batteries amounted to 5.56 $ for 30 days period. It is clear that the longer the battery can be discharged during high-price periods, the greater the total costs difference. In scenario 2, avoiding the necessity of buying electricity by higher price for a longer time, the effect of the presence of the batteries becomes even greater than before. The results demonstrated an additional saving of 1.75$ for a month period comparing with scenario 1, or about 13 %; and total profit of 7.31$ with respect to the case when the batteries are not used at all. However, since the model neglects all kinds of losses related to the energy flowing in and out of the batteries, the real savings will be a bit less. The profit may be greater or lower depending on battery capacity, electricity prices and demand value.

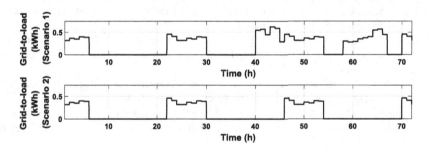

Fig. 4. Grid-to-load profiles comparison between scenario 1 and scenario 2.

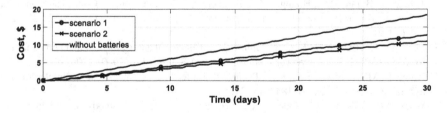

Fig. 5. Costs comparison between scenario 1 and scenario 2.

6 Conclusion

A new energy storage control principle is proposed, based on plug-and-play integration of multiple intelligent batteries with behaviour emerging from the collaboration of components. Such coordination was illustrated on a simulated example with two batteries and compared with the case when batteries work independently. The results demonstrate the efficiency of the approach to the direct interaction between batteries. Such decentralized concept could be implemented, for example, using embedded microcontrollers with network connectivity. Mutual interchangeability allows increasing total working time of the batteries, and therefore a significant cost reduction due to maximizing of battery usage as an energy source during periods with high electricity prices.

Future related research is planned in such directions as achieving scalability, functional completeness of such battery constellations, theoretical estimation of limitations and parameters of their performance. The ongoing work also includes real validation of the proposed algorithm. Practically it is planned to develop a prototype laboratory implementation of an intelligent battery and demonstrate the scenarios of their coalition formation in particular Internet of Energy applications.

References

1. Wade, N.S., Taylor, P.C., Lang, P.D., Jones, P.R.: Evaluating the benefits of an electrical energy storage system in a future smart grid. Energy Policy **38**, 7180–7188 (2010)
2. Lukic, S.M., Cao, J., Bansal, R.C., Rodriguez, F., Emadi, A.: Energy storage systems for automotive applications. IEEE Trans. Ind. Electron. **55**(6), 2258–2267 (2008)
3. Colombo, A., Karnouskos, S., Bangemann, T.: A system of systems view on collaborative industrial automation. In: 2013 IEEE International Conference on Industrial Technology (ICIT), pp. 1968–1975 (2013)
4. Zhabelova, G., Vyatkin, V., Dubinin, V.: Towards industrially usable agent technology for smart grid automation. IEEE Trans. Ind. Electron. **62**(4), 2629–2641 (2015)
5. Honarmand, M., Zakariazadeh, A., Jadid, S.: Optimal scheduling of electric vehicles in an intelligent parking lot considering vehicle-to-grid concept and battery condition. Energy **65**, 572–579 (2014)
6. Delfanti, M., Falabretti, D., Merlo, M.: Energy storage for PV power plant dispatching. Renewable Energy **80**, 61–72 (2015)
7. Teleke, S., Baran, M.E., Huang, A.Q., Bhattacharya, S., Anderson, L.: Control strategies for battery energy storage for wind farm dispatching. IEEE Trans. Energy Convers. **24**(3), 725–732 (2009)
8. Meissner, E., Richter, G.: Battery monitoring and electrical energy management precondition for future vehicle electric power systems. J. Power Sources **116**, 79–98 (2003)
9. Xing, Y., Ma, E.W.M., Tsui, K.L., Pecht, M.: Battery management systems in electric and hybrid vehicles. Energies **4**, 1840–1857 (2011)
10. Waag, W., Fleischer, C., Sauer, D.U.: Critical review of the methods for monitoring of lithium-ion batteries in electric and hybrid vehicles. J. Power Sources **258**, 321–339 (2014)
11. Vega, A.M., Santamaria, F., Rivas, E.: Modeling for home electric energy management: a review. Renew. Sustain. Energy Rev. **52**, 948–959 (2015)
12. Das, S.K., Roy, N., Roy, A.: Context-aware resource management in multi-inhabitant smart homes: a framework based on Nash H-learning. Pervasive Mobile Comput. **2**, 372–404 (2006)

13. Chehri, A., Mouftah, H.T.: Service-oriented architecture for smart building energy management. In: Proceedings of the IEEE ICC 2013 - Selected Areas in Communications Symposium, Budapest, pp. 4099–4103 (2013)
14. Boynuegri, R., Yagcitekin, B., Bays, M., Karakas, A., Uzunoglu, M.: Energy management algorithm for smart home with renewable energy sources. In: Proceedings of the 4th International Conference on Power Engineering, Energy and Electrical Drives, Istanbul, pp. 1753–1758 (2013)
15. Power Smart Prices, Ameren Illinois Co., May 2015. http://www.powersmartpricing.org/prices/?date=20150506
16. Ardakanian, O., Koochakzadeh, N., Singh, R.P., Golab, L., Keshav, S.: Computing electricity consumption profiles from household smart meter data. In: Proceedings of EDBT/ICDT Workshops (EnDM), pp. 140–147 (2014)
17. Zhu, W.H., Zhu, Y., Tatarchuk, B.J.: Self-discharge characteristics and performance degradation of Ni-MH batteries for storage applications. Int. J. Hydrogen Energy **39**, 19789–19798 (2014)

Energy Management

Impacts of Energy Market Prices Variation in Aggregator's Portfolio

Eduardo Eusébio[1(✉)], Jorge A. M. Sousa[1,2], and Mário Ventim Neves[3]

[1] ISEL, R. Conselheiro Emídio Navarro no. 1, 1959-007 Lisbon, Portugal
eaeusebio@deea.isel.ipl.pt
[2] INESC-ID, R. Alves Redol no. 9, 1000-029 Lisbon, Portugal
jsousa@deea.isel.ipl.pt
[3] FCT/UNL, Campus Caparica, 2829-516 Lisbon, Portugal
ventim@uninova.pt

Abstract. After liberalization of the electric sector and due to the expansion of distributed generation with the appearing of new kinds of producers and consumers, a new power player emerged taking an major role in the commercialization of electricity - the commercial agent or aggregator. The aggregator thus enables small and medium clients to access to market prices that were impossible to obtain by themselves, since scale is an important factor in the electric energy market. This paper is focused on analyzing how the variation of prices in the energy market affects the aggregator's customer portfolio energy sold and its total profits. The weekly market prices considered showed different levels of volatility. The effect of market price variation, both in terms of average value and variance, was analysed for a typical clients' portfolio in terms of profitability and risk. As it was expected the highest levels of profitability were attained in the weeks of lowest average prices that also correspond to the highest price volatilities increasing also the risk of the aggregator.

Keywords: Aggregation · Portfolio optimization · Load diagrams · Energy prices · Risk and profitability

1 Introduction

The electricity system is facing new challenges with respect to its development and future shape [1]. These challenges are the result of the growing emergence of an increasingly large number of small and average energy producers, they are already today a very reasonable size group for which it is necessary to develop new forms of participation and integration into the power grid and consequently the energy market, to achieve these goals it is necessary to develop new inclusion strategies [2].

This new type of electricity producers based on renewable sources has different characteristics from those traditionally found in the electric power system, such as, for example, the intermittency of production and, sometimes reduced volume due to these constraints, so an association that aggregates and represents them will bring benefits to both [3].

© IFIP International Federation for Information Processing 2016
Published by Springer International Publishing Switzerland 2016. All Rights Reserved
L. Camarinha-Matos et al. (Eds.): DoCEIS 2016, IFIP AICT 470, pp. 437–445, 2016.
DOI: 10.1007/978-3-319-31165-4_41

1.1 Purpose of the Paper

Nowadays after the liberalization of the electric sector and due to the expansion of distributed generation with the appearing of new kinds of producers and consumers, a new entity emerged "aggregator".

In order to develop this study one simulator specially built for this purpose was used: the "Commercial Agents in the Electricity Market (CAEM)" and the optimization module (PA). This simulator will analyze each aggregator costumer's portfolio, showing both risks and profitability associated [2].

The Pareto frontier establishes the relationship between profitability and risk expected relating to a diversified portfolio of assets, in which each asset has a nominal weight. Graphically draws a curve of optimum places on risk profitability plan. The best portfolios are located on the Pareto curve, i.e. those that offer either the highest rentability for a given risk level or the lowest risk for a given rentability

The Pareto frontier allows the aggregator to characterize the optimal directions in the profitability/risk plan that it should follow when introducing new customers in the portfolio according to its risk aversion [4].

1.2 Outline of the Paper

This study is divided into six sections, in the first section an introduction to the development of power systems is done stressing the importance of the inclusion of small consumers and producers in the Electricity Market. The second section presents the aggregation phenomenon contribution to the electrical market for the "cyber-physical systems. In the third section the study of investment portfolios and how to aggregate to the electric market is done, particularly in regard to the grouping of different types of energy assets and the advantages of grouping. The fourth section illustrates the adaptation of the economic theory of Harry Markowitz to portfolios of the electricity market, all the necessary equations for the preparation of simulations are presented and in the fifth section, the simulations are performed. The variation in profitability and risk of a portfolio of electricity market assets, given the different electricity price profiles observed in the electricity market is studied. The work ends with the conclusions over the results of prepared simulations.

2 Contribution to Cyber-Physical Systems

Nowadays it is not possible for small producers and consumers to reach the global market, therefore it should exist an entity that interacts and establishes the needed contracts. This entity, called Aggregator, will settle a contract with producers and consumers allowing them to establish connections in the global market [5]. Ultimately this will lead to the creation of a Network since there's real interaction between producers/consumers and aggregators, the outcome gives them access to the global market (Fig. 1).

There are some methods that helps the aggregator obtain a better efficiency such as smart grids. Smarts grids have the ability to enable an active participation in Demand

Dispatch (DD), which ultimately represents more penetration in Distributed Energy Resources (DER), new storage capacity, assets optimization and new and better market opportunities [1]. A smart grid combined with smart meters is a synonym of knowledge and allows the aggregator to have real time feed-back regarding consumption levels.

There are two words in this theme, related to our Cyber-Physical System that must be emphasized; these words are accuracy and efficiency. Accuracy is directly related with smart meters, allowing a better control and supervision of energy consumption. This enables the aggregator to be more precise in contracts decisions with consumers and producers. With this in mind, there will be a better efficiency in energy and contractual terms optimizing the relation between producers and consumers allowing them to reduce energy consumption on full peak hours and increase it in the valley periods.

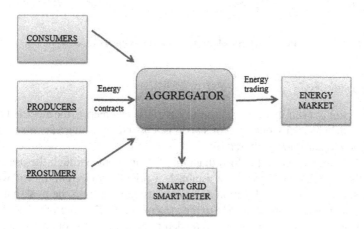

Fig. 1. Cyber-physical system aggregator network.

3 Related Literature

3.1 Economic Theory Overview

The Portfolio Theory has its origins on the scientific article "Portfolio Selection of Nobel Economy" by Harry Markowitz. In this article he addresses the choice of financial assets portfolios of uncertain future value by following completely new criteria at the time: the expected return criteria variance [6].

Harry Markowitz sensed that different investor's choices and diversification (Fig. 2) were based on two distinct issues: the trade-off between risk and rentability of investments and the correlation between the returns of different assets [6].

The risk suggested was measured by Harry Markowitz as dispersion return. Thus, Harry Markowitz provided an analytical foundation for the portfolios diversification on financial markets and established the optimal and most rational composition methodology for assets portfolios whose future value is uncertain [7, 8].

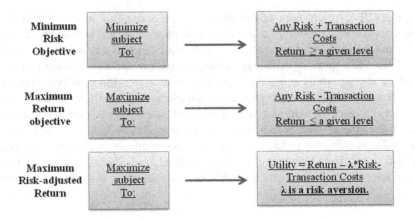

Fig. 2. Harry Markowitz, portfolio theory cases.

3.2 Aggregators

The aggregator is an entity that eases the access of its customers to the electricity market, establishes the electricity purchase or sale to its customers, with great versatility in the kind of contractual clauses offered [9].

Based on knowledge of the load diagram of their clients through the use of smart-meters, the aggregating agent may establish a pricing policy contracted with their customers, promoting the relocation of some consumption from the peak hours to the valley hours or other more favorable to the system operator's point of view [9, 10].

This kind of aggregating agent is intended to combine the economic interests of a large and diverse number of small and medium customers in terms of volume of traded energy, thus providing critical mass to enable them to trade on the energy market [11].

4   Methodology

The economic behavior analysis, considering the risk and profitability of the customers aggregator portfolio from the electricity company, is based on the market traded energy, as well as the client's loads forecast, and was conducted with the help of the tool developed in MATLAB (Fig. 3):

It shows the main inputs and outputs and how the simulators integration is to be used in the case studies, for a client's portfolio, with several consumers, producers and "prosumers", with aggregator contract for the supply and energy purchase. The first main inputs are the historical data series of the customers.

Also the contracted prices for sale and/or energy purchase predefined in the contract terms are indispensable: all this information available (input) allows, with appropriate treatment, the simulators to obtain the desirable output [2, 4].

The profitability and risk of the portfolio for one entire week, 168 periods (hours), are computed, according to Eq. 1, and for its calculation a set of different scenarios of energy and prices are considered.

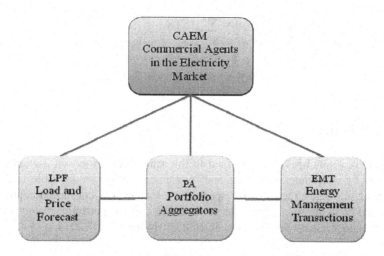

Fig. 3. Simulator CAEM and the computation modules.

$$E(R_P) = \sum_{i=1}^{n} E(R_i) \times w_i \tag{1}$$

Where $E(Rp)$ profitability of portfolio P, $E(Ri)$ profitability of client i, in the set of scenarios, and w_i the weight of client in portfolio P, n is the number of clients in portfolio P, finally, i is the type of clients in portfolio.

$$\sigma_P = \sqrt{\sigma_P^2} \tag{2}$$

Where σ_P is the risk the portfolio P.

$$\sigma_P^2 = \sum_{i=1}^{n} w_i^2 \, \sigma_i^2 + \sum_{i=1}^{n-1} \sum_{j=i+1}^{n} w_i \, w_j \sigma_{i,j} \tag{3}$$

Where σ_p^2 is the variance of the portfolio P, σ_i^2 is the variance of client i, w_i and w_j are the weights of the clients in the portfolio and $\sigma_{i,j}$ the covariance between the clients i and j, and n the total number of different type of clients. The Pareto curve of the portfolio is computed according to the Eqs. 4 and 5.

$$
\begin{aligned}
\min \sigma_P^2 &= \sum_{i=1}^{n} w_i^2 \, \sigma_i^2 + \sum_{i=1}^{n-1} \sum_{j=i+1}^{n} w_i \, w_j \sigma_{i,j} \\
s.a \quad &\sum_{i=1}^{n} E\left(R_i\right) w_i \geq E(R_c) \\
&\sum_{i=1}^{n} w_i = 1 \\
0 \leq w_i \leq 1 \quad &i = 1, 2, 3, \ldots \ldots, n
\end{aligned}
\tag{4}
$$

$$\max E\,(Rp) = \sum_{i=1}^{n} E\left(R_i\right) w_i$$

$$s.a \quad \sum_{i=1}^{n} w_i^2\,\sigma_i^2 + \sum_{i=1}^{n-1} \sum_{j=i+1}^{n} w_i w_j \sigma_{i,j} \leq \sigma_v^2 \tag{5}$$

$$\sum_{i=1}^{n} w_i = 1$$

$$0 \leq w_i \leq 1 \qquad i = 1, 2, 3, \ldots \ldots, n$$

5   Case Study

The simulations and results performed in this study were based on a clients group, composed by consumers, producers and "prosumers" as shown on Table 1. The time horizon of this study requires one specific week of energy consumption/production levels, and applying to compute the profit and risk of the portfolio, twenty weeks of different energy market prices, between 10/03/2014 until 21/07/2014.

Table 1. Customers' portfolio characterization

Type profile	Nature	Type source	Power (kVA)	No	Energy (kVA/week)	Weight (%)	Risk, P1 (σ_i)
A	Consumer	--------	3.45	6000	203	9.24	0,84
B	Consumer	--------	6.90	4000	755	22.93	0,88
C	Consumer	--------	13.80	2500	1234	23.45	0,93
L	Producer	Wind	350.0	200	20665	31.4	8,19
M	Producer	Photovol-taic	4.00	600	195	0.89	7,26
O	Producer	Photovol-taic	12.00	600	569	2.59	6,43
Q	Prosumer	Photovol-taic	90/350	40	25683	7.81	4,84
R	Prosumer	Cogenera-tion	130/160	10	22234	1.69	1,02

In Fig. 4, the profiles of clients B, L, O and Q, are shown, all the profiles are presented in a 168 h time period.

For the computation of rentability and risk variation of the portfolio composition, several weeks of energy market prices are considered, four of them are displayed in Fig. 5.

The simulations are done on the same portfolio composition considering to all of them, the same consumption profiles and production levels. The results up to four weeks simulations are showed on Table 2.

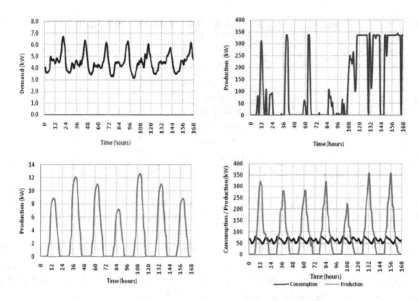

Fig. 4. Load profiles: (a) B - residential consumer; (b) L - wind producer; (c) O - photovoltaic producer; (d) Q - consumer and photovoltaic producer, prosumer.

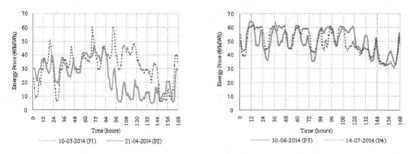

Fig. 5. Energy market load profiles: (a) weeks 10/03/2014 (P1) and 21/04/2014 (P2); (b) weeks 30/06/2014 (P3) and 14/07/2014 (P4).

Table 2. Profitability and risk for portfolio, P, considering different energy price market

Week	13-01-2014 (P1)	10-02-2014 (P2)	21-04-2014 (P3)	12-05-2014 (P4)
Profitability (R_P)	−14.22	−57.63	0.38	6.7
Risk (σ_P)	2.52	8.09	1.22	1.2
E. price (average)	31.9	23.7	49.5	49.6
E. price (Std)	12.3	12.0	8.7	9.1

Computation of profitability and risk are made with the PA simulator, Fig. 6, presents an example of the simulation, and the results for the week 10/03/2014 (P1). .

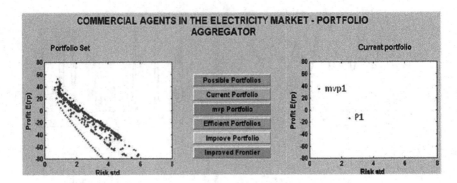

Fig. 6. Efficient frontier design, location of the R_P, σ_P to the current portfolio and minimum variance point. Results of PA simulator using energy week market prices 10/03/2014 (P1).

Figure 7 shows the values of R_P, σ_P of the portfolio, P, with the considered composition, according to Table 1, along with the set of price market scenarios in the study, with specific reference of some chosen weeks. Finally on Table 2 the results of the simulations for the week prices mentioned on Fig. 5.

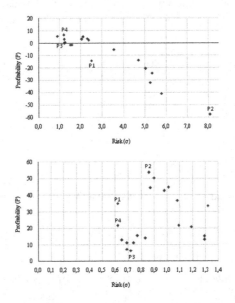

Fig. 7. Evaluation of profitability, RP and risk, σ_p: (a) portfolio P; (b) minimum variance point, mvp.

6   Conclusions

The profitability and risk computation of a portfolio aggregator under a certain composition was possible by using the developed simulator. In this case study, a composition portfolio with a significant percentage of energy producers with higher risk, namely the wind and photovoltaic, promotes a higher risk level on the portfolio. The simulation results show the high dependence of the energy price vector, such is noticeable through the analysis of his standard deviation.

The combination of the high risk level of some customers portfolio with the price vector, leads to high risk values in the portfolio. It is also possible to verify that the profitability of the portfolio depends on the energy price contracts established between aggregator and customers, but another decisive influence is the average value of the energy market price vector.

References

1. Hossain, M.R., Oo, A.M.T., Ali, A.B.M.S.: Evolution of smart grid and some pertinent issues. In: Universities Power Engineering Conference, AUPEC (2010)
2. Eusébio, E., Sousa, J., Ventin Neves, M.: Risk analysis and behavior of electricity portfolio aggregator. In: DoCeis15 - 6th Doctoral Conference on Computing, Electrical and Industrial Systems, Caparica, 13–15 April 2015
3. Lampopoulos, I., Vanalme, G.M.A., Kling, W.L.: A methodology for modeling the behavior of electricity prosumers within the smart grid. In: IEEE ISGT Europe (2010)
4. Eusébio, E., Sousa, J., Ventim Neves, M.: Commercial agent's portfolio optimization in electricity markets. In: International Conference European Electricity Markets – EEM12, Florence, April 2012
5. Stern, P.C.: Information, incentives, and proenvironmental consumer behavior. J. Consum. Policy **22**, 461–478 (1999)
6. Markovitz, H., Sharpe, W.F., Miller, M.: Founders of modern finance: their portfolio selection. J. Finance **7**(1) (1952)
7. Markovitz, H.: Portfolio Efficient: Efficient Diversification of Investments. Wiley, New York (1959)
8. Wu, L., Shahidehpour, M.: Financial risk evaluation in sthochastic PBUC. IEEE Trans. Power Syst. **24**(4), 4 (2009)
9. Lambert, Q.: Business models for an aggregator - is an aggregator economically sustainable on Gotland? M.Sc thesis, XR – EE – ICS 2012:003, Sthockholm (2012)
10. Bollen, M.: Adapting electricity networks to a sustainable energy system. Energy Markets Inspectorate, EI R2011:03 (2011)
11. Eurelectric: flexibility and aggregation, requirements for their interaction in the market. Union of the Electricity Industry - EURELECTRIC aisbl (2014)

Impact of Self-consumption and Storage in Low Voltage Distribution Networks: An Economic Outlook

Fernando M. Camilo[1(✉)], Rui Castro[1,2], M. E. Almeida[1,2], and V. Fernão Pires[2,3]

[1] IST, University of Lisbon, 1049-001 Lisbon, Portugal
{fernando.camilo,rcastro,d2527}@tecnico.ulisboa.pt
[2] INESC-ID/IST, University of Lisbon, Lisbon, Portugal
[3] EstSetúbal, Polythecnics of Setúbal, Setúbal, Portugal
vitor.pires@estsetubal.ips.pt

Abstract. A paradigm shift is taking place in Low Voltage (LV) distribution networks, motivated by progressive implementation of renewable micro-generation (μG), mainly Photovoltaic (PV), near household consumers. The concept of self-consumption linked to battery storage is emerging as a way to enhance the quality of electrical network. Smart-Grid (SG) environment comes close to this approach and may have a crucial relevance on management of intelligent power distribution networks, in the framework of a Smart Environment. This paper proposes an additional contribution on the subject by investigating the economic profitability of PV battery systems being analyzed with respect to its impact and economic feasibility, taking into consideration their initial investment and operation costs. The purpose is to verify if prosumer's investment is financially more interesting than purchasing all electricity needed for consumption from the LV grid. The results of the performed economic analysis show that self-consumption with storage is a potential solution.

Keywords: Micro-generation · Self-consumption · Battery storage · Economic analysis

1   Introduction

Nowadays, a paradigm change is ongoing in LV distribution networks driven by the progressive implementation of renewable μG next to the consumers, mainly from small PVs type. This change can bring benefits, mainly in rural areas, as it allows to improve the quality of electrical energy supply [1]. This novel operational model allows to highlight the concept of prosumers (consumers that both produce and consume electricity), considering that this strategy may be interesting both from a technical and economical point of view [2, 3]. Self-consumption and storage of electricity surplus in batteries during daylight, to be used later when there is no sun power, is being recently adopted as a way to facilitate the integration of more small renewable energy sources in LV distribution networks. The advantages resulting from the implementation of prosumers in the distribution network, as well as the possibilities for integration of energy storage

© IFIP International Federation for Information Processing 2016
Published by Springer International Publishing Switzerland 2016. All Rights Reserved
L. Camarinha-Matos et al. (Eds.): DoCEIS 2016, IFIP AICT 470, pp. 446–454, 2016.
DOI: 10.1007/978-3-319-31165-4_42

technologies, can be further enhanced by combining the intelligent power distribution networks or SG [4, 5].

The SG concept comes very close to this paradigm change. It is nowadays well-known that SG can bring benefits in terms of safety, economy and efficiency of the network [6]. Moreover, the evolution of communication and information technologies, is supporting the development of SG and appears as an assimilated vision for the future power systems [7, 8].

Nowadays, PV feed-in tariffs incentives are progressively being removed. This can be perceived as a prelude for the promotion of self-consumption. Intense research on economic viability of PV systems have been recently presented. These works intend to validate the viability of PV systems from different perspectives: net-metering and bill savings [9, 10], feed-in tariffs assessment [11, 12], energy policies to support the promotion of PV systems [13, 14]. As a result, it is now commonly accepted that it is essential to have investment and operating costs information, in order to enable timely planning and on-line buying and selling of electrical energy, bearing in mind the net-metering context [15].

This paper proposes a supplementary contribution on the subject by investigating the economic profitability of PV battery systems. Battery Energy Storage (BES) aims at reducing the electricity consumption from the distribution networks, because residential clients consume and store the energy they produce on their own. PV battery systems are analyzed with respect to its impact and economic feasibility, taking into consideration, their initial investment and operation costs, where the purpose is to verify if the prosumer's investment, is financially more interesting than purchasing all electricity needed for consumption from the LV grid.

2 Contribution to Cyber-Physical Systems

Cyber-Physical Systems (CPS) are the new generation of engineering systems that offer close interaction between cyber and physical components [16]. CPS are at the center of current discussions about the next generation of intelligent services, that power areas as diverse as medicine, aerospace, electrical networks, consumer appliances [17]. This approach may be the new vision of the future in several areas of society. The potential investment gains of CPS are immense, inasmuch as the traditional systems can fail in solving problems of physical and computational nature. Continuous advances in science and engineering improve the bond between the computational and physical elements through smart mechanisms, significantly increasing the adaptability, autonomy, efficiency, reliability, security and functionality of cyber-physical systems [18].

The SG concept is connected to the evolution of communication and information technologies, where CPS may be the answer to bring this concept to another level. Some entities, such as the NIST Engineering Laboratory [19], are dedicated to the development of SG study in a CPS context. This approach, from SG perspective, can bring many benefits in terms of safety, economy, efficiency and reliability [20, 21].

Classical SG example are the smart-meters, in which is possible through CPS, monitor and manage the power of the smart house. On another scale, in theory, the aim of CPS is not only managing and monitor the smart house but manage and monitor the entire electrical network as a whole. In sequence of this management through CPS, the net-metering system can be applied. Net-metering is based on the concept that the surplus production and injected into the network is recorded as credits, that can be compensated for consumption, in greater time period, using the grid as a long-term storage [13]. An alternative would be to manage the power grid with the possibility of using the batteries of residential prosumers, if required, by the power grid. The CPS, in this case, could manage and use energy from these batteries, bridging any gaps of the network. In this case, there would be a compensation to the prosumer, credited for the energy used in this process, similar to net-metering, where CPS could be an excellent contribution, allowing the optimization of management of the entire electrical system.

Is essential to have a knowledge of the investment and operating costs, in order to enable timely planning and, considering the net-metering context, the on-line buying and selling of electricity. In future, the storage system can be fundamental, in electricity sales perspective by credits, in which CPS may reveal great importance on the management of this process. Consequently, this paper aims to make an investigation related to this issue, by verifying the economic viability and profitability of PV battery systems, in several scenarios.

3   Research Contribution

3.1   Methodology

Commonly, the purpose of an economic analysis is to make an economic assessment of the viability, stability and profitability of project investment, wherein this analysis helps to reach a decision about that investment. There are some key-indicators of economic parameters that should be considered [15], as follows.

The net present value NPV is defined as the sum of present values of incoming and outgoing cash flows over a period of time of the project.

$$NPV = \sum_{j=1}^{n} \frac{R_{Lj}}{(1+a)^j} - \sum_{j=0}^{n-1} \frac{I_j}{(1+a)^j}.$$

(1)

Where n is the project period time, a the discount rate, I_j is the investment for year j, R_{Lj} is the net revenue obtained for year j calculated from the difference between gross revenues R_j and maintenance and operation costs d_{omj}, I_t relates to the total investment.

$$R_{Lj} = R_j - d_{om_j} I_t.$$

(2)

The internal rate of return IRR is a rate of return used in a capital expenditure budget to measure and compare the profitability of the investment.

$$\sum_{j=1}^{n} \frac{R_{Lj}}{(1+IRR)^j} - \sum_{j=0}^{n-1} \frac{I_j}{(1+IRR)^j} = 0.$$

(3)

The discounted payback period DPP is the length of time required to recover the initial investment from the present value of the expected future cash flows.

$$\sum_{j=1}^{DPP} \frac{R_{Lj}}{(1+a)^j} = I_t \Leftrightarrow DPP = \frac{\ln\left(\frac{R_L}{R_L - aI_t}\right)}{\ln(1+a)}. \tag{4}$$

3.2   Case Study Definition

According to the current Portuguese legislation (Decree-Law 153/2014), the production of electricity for self-consumption is allowed with the possibility of injection of surplus to the network. Another important aspect, is that individual capacity of μG must be less than or equal to 100 % of prosumer contracted power. These legal specifications have been taken into consideration in this work.

Load and μG data, for a typical prosumer, was estimated based on the annual profiles of consumption and μG made available by the Portuguese Energy Services Regulatory Authority (ERSE), based on average 15 min intervals. On the other hand, it was assumed, based on consumption data profile, that prosumer's contracted power is 4.6 kVA, where bi-hourly tariff was selected. The average annual energy consumed per consumer is 2293 kWh.

For purposes of economic analysis of PV systems, considering the concept of self-consumption by prosumers, eight scenarios were proposed: (i) **PV of 0.5 kW**, where all the energy produced by PV system is used only for self-consumption; (ii) **PV of 1 kW + net-metering**, the surplus is injected in network, on net-metering format; (iii) **PV of 1 kW + batteries**, a part of surplus is used to charge the batteries; (iv) **PV of 1.5 kW + grid injection**, the surplus is injected in network; (v) **PV of 1.5 kW + net-metering**, the surplus is injected in network, on net-metering format; (vi) **PV of 1.5 kW + batteries**, a part of surplus is used to charge the batteries; (vii) **PV of 1.5 kW + grid injection + batteries**, the surplus is used to charge the batteries and to be injected in network; (viii) **PV of 1.5 kW + net-metering + batteries**, the surplus is used to charge the batteries and to be injected in network, on net-metering format.

In scenarios (iv) and (vii), is considered that the prosumers sell the surplus of energy to the network. In this context, the prosumers have to install and pay a bidirectional counter, pay a fee and get a certificate of exploitation. The remuneration R_m of surplus (Decree-Law 153/2014) is calculated by

$$R_m = E_m OMIE_m 90\,\%. \tag{5}$$

R_m [€] corresponds to the remuneration of surplus energy, E_m [kWh] is the provided energy to grid, $OMIE_m$ [€/kWh] is the average Iberian market value, where m is the month referred to electricity counting. Considering the net-metering format, the prosumers in scenarios (ii), (v) and (viii) are paid by credits.

The battery bank was sized to support only 7 h of consumption per day, resulting in a set of 18 OPzV gel battery units of 2 V – 220 Ah, at scenarios ((iii), (vi), (vii), (viii)).

Batteries sizing was made bearing in mind the depth of discharge (DOD) and the efficiency of the energy-conversion process. Accordingly to the manufacturer, these type of battery has an approximate lifetime of 7 years, considering 2500 cycles at 35 % DOD.

In the economic analysis, for the eight proposed scenarios, was considered: a period of 20 years, a discount rate of 4 %, a maintenance and operation costs of 1 % of total investment of the project (based on [22, 23]) and also a depreciation factor D_f of 0.75 % per year has been considered [24]. The D_f factor accounts for the decrease in the efficiency of the PV panel, over its lifetime. Table 1 shows the total investment for the proposed scenarios.

Table 1. Total investment (I_t) of PV system to eight scenarios proposed for economic analysis.

Scenario	I_t [€]	Scenario	I_t [€]	Scenario	I_t [€]
(i)	1135.5	(iv)	3234.4	(vii)	14014.2
(ii)	2449.4	(v)	3234.4	(viii)	14014.2
(iii)	13027.0	(vi)	13477.2		

PV system values include: PV panels; micro-inverters ((i), (ii), (iv), (v)); battery bank and inverter ((iii), (vi), (vii), (viii)), where the inverter is replaced after 11 years; connecting (DC/AC) cables structure; bidirectional meter (scenarios (ii), (iv), (v), (vii), (viii)) and the value of the installation. A tax of 23 % is included on these values.

4   Simulation Results and Discussion

Figure 1 represents the average curves of the load and PV μG profiles of a residential prosumer, on a summer typical day. The radiance and temperature were taken into account on PV μG profile. Self-consumption with PV 0.5 kW and PV 1.5 kW is about

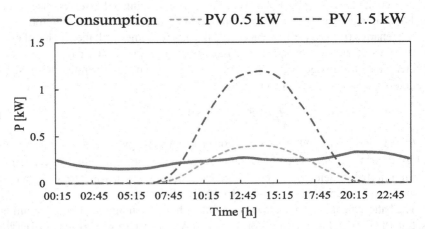

Fig. 1. Load and PV *μG* profiles of a residential prosumer on a summer typical day of the year.

78.8 % and 32.1 %, respectively. The energy produced by PV 0.5 kW and PV 1.5 kW is approximately 800 kWh and 2400 kWh, per year.

Figure 2 presents the savings resulting from self-consumption, at first year of operation. The electricity bill was calculated using the present bi-hourly tariff for Portuguese electricity consumers. The first bar represents the annual electricity bill, considering that no PV system is installed. The subsequent blue bars represent the annual electricity bill considering that different PV systems are installed. The orange bars display the savings as compared to the case in which no PV system is installed. As it can be seen, each type of PV system allows a reduction of the electricity bill from network consumption. These calculations refer to the net electricity bill, no investment costs in the PV systems being considering. As is possible to perceive, PV 0.5 kW allows a annual reduction of 30.2 %, PV 1.5 kW of 40.4 %, and PV 1.5 kW + Batteries of 63.3 %.

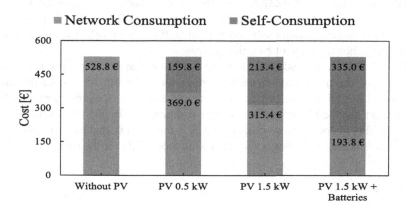

Fig. 2. The resulting savings from self-consumption, accordingly with PV system type.

Table 2 presents the economical analysis results for the eight scenarios proposed.

Table 2. The results of economic analysis.

Scen.	NPV [€]	IRR [%]	DPP [Years]	Scen.	NPV [€]	IRR [%]	DPP [Years]
(i)	807.9	11.2	9.3	(v)	2491.7	11.8	8.7
(ii)	1544.8	10.5	9.6	(vi)	−7215.9	−16.5	N/A
(iii)	−6877.9	−16.1	N/A[a]	(vii)	−7255.8	−16.0	N/A
(iv)	267.7	4.9	16.8	(viii)	−6065.7	−12.7	N/A

[a]Not Applicable.

The results presented in Table 2 allow to conclude that scenarios (iii), (vi), (vii) and (viii) have a payback superior to 20 years. Contrariwise, the remaining scenarios are feasible for investment.

Currently, a PV system for self-consumption, sized to the minimum peak consumption throughout the day, avoiding the injection in network as much as possible, presents a reasonable profitable solution for the prosumer. Scenario (i), as presented in Fig. 1, comes close to this requirement. Although feasible, (iv) has a payback on a long period of time, making it unattractive as investment. This happens because, currently, the sale of electricity to the network is not economically feasible. On the other hand, the scenarios (ii) and (v) present an investment with an interesting *IRR*. It should be mentioned that was not considered any percentage penalty, relatively, to the net-metering approach.

Ideally, the PV with storage allows to increase the amount of self-consumption by prosumer, as presented in Fig. 2, in comparison with a conventional PV system, without batteries. However, results of all scenarios with batteries, presented in Table 2, demonstrate that these scenarios are not profitable. This can be explained by the fact that the batteries, at present, are still expensive, although its technology has evolved. Currently, there are batteries with a long life-cycle reaching up to 5000 cycles, as lithium technology for instance. Consequently, investment on PV system with storage may become interesting, when the price of batteries decrease.

It is possible to conclude, by analyzing the results that *IRR* improves, if increasing the *Rm*, thus making it economically, more viable and profitable for the prosumer's investment. In addition, the energy traders should encourage the prosumers with financial support, for the PV system and storage equipments. Despite the net-metering not be yet a reality in Portugal, this study revealed, in this context, that the self-consumption with storage is feasible, being a very interesting possibility for the future.

5   Conclusions

An economic analysis of profitability and feasibility of PV battery systems were proposed. The main contribution of this work is to investigate, if the prosumer's investment is attractive on an economic perspective. Is essential to have a knowledge about the investment and operating costs, in order to enable timely planning and, considering the net-metering context, the on-line buying and selling of electricity where CPS can be an excellent contribution in this process.

Eight scenarios were proposed for economic analysis. The results reveal that payback of prosumer's investment can improve: if increase the *Rm*, or if is on a net-metering format. On the other hand, the batteries are still expensive, despite its technology has evolved.

In future, is expected, with the prices reduction of batteries, that most of the residential PV systems will be equipped with storage. On a net-metering context, self-consumption with storage promises to be a very feasible solution, with high potential, on an economic outlook.

Acknowledgements. This work was supported by national funds through Fundação para a Ciência e a Tecnologia (FCT) with reference UID/CEC/50021/2013.

References

1. Luque, A., Hegedus, S.: Handbook of Photovoltaic Science and Engineering. John Wiley & Sons, Hoboken (2011)
2. Rathnayaka, A.J.D., Potdar, V.M., Kuruppu, S.J.: An innovative approach to manage prosumers in smart grid. In: 2011 Sustainable Technologies (WCST), pp. 141–146 (2011)
3. Sun, Q., Beach, A., Cotterell, M.E., Wu, Z., Grijalva, S.: An economic model for distributed energy prosumers. In: 2013 46th Hawaii International Conference on System Sciences (HICSS), pp. 2103–2112 (2013)
4. Lampropoulos, I., Vanalme, G.M.A., Kling, W.L.: A methodology for modeling the behavior of electricity prosumers within the smart grid. In: Innovative Smart Grid Technologies Conference Europe (ISGT Europe), 2010 IEEE PES, pp. 1–8 (2010)
5. Pagani, G.A., Aiello, M.: Power grid complex network evolutions for the smart grid. Phys. A Stat. Mech. Appl. **396**, 248–266 (2014)
6. Putrus, G.A., Bentley, E., Binns, R., Jiang, T., Johnston, D.: Smart grids: energising the future. Int. J. Environ. Stud. **70**, 691–701 (2013)
7. Miceli, R., Favuzza, S., Genduso, F.: A perspective on the future of distribution: Smart grids, state of the art, benefits and research plans. Energy Power Eng. **5**, 36 (2013)
8. Borlase, S.: Smart Grids: Infrastructure, Technology, and Solutions. CRC Press, Boca Raton (2012)
9. Darghouth, N.R., Barbose, G., Wiser, R.: The impact of rate design and net metering on the bill savings from distributed PV for residential customers in California. Energy Policy. **39**, 5243–5253 (2011)
10. Darghouth, N.R., Barbose, G., Wiser, R.H.: Customer-economics of residential photovoltaic systems (Part 1): The impact of high renewable energy penetrations on electricity bill savings with net metering. Energy Policy **67**, 290–300 (2014)
11. Campoccia, A., Dusonchet, L., Telaretti, E., Zizzo, G.: An analysis of feed'in tariffs for solar PV in six representative countries of the European Union. Sol. Energy **107**, 530–542 (2014)
12. Forbes, I., Pearsall, N., Georgitsioti, T.: Simplified levelised cost of the domestic photovoltaic energy in the UK: the importance of the feed-in tariff scheme. IET Renew. Power Gener. **8**, 451–458 (2014)
13. Eid, C., Guillén, J.R., Marín, P.F., Hakvoort, R.: The economic effect of electricity net-metering with solar PV: Consequences for network cost recovery, cross subsidies and policy objectives. Energy Policy **75**, 244–254 (2014)
14. Dusonchet, L., Telaretti, E.: Comparative economic analysis of support policies for solar PV in the most representative EU countries. Renew. Sustain. Energy Rev. **42**, 986–998 (2015)
15. Baker, H.K., Powell, G.: Understanding Financial Management: A Practical Guide. Blackwell Publishing, Hoboken (2009)
16. Abad, F.A.T., Caccamo, M., Robbins, B.: A fault resilient architecture for distributed cyber-physical systems. In: 2012 IEEE International Conference on Embedded and Real-Time Computing Systems and Applications, pp. 222–231. IEEE (2012)
17. Khaitan, S.K., McCalley, J.D.: Design techniques and applications of cyberphysical systems: a survey. IEEE Syst. J. **9**, 350–365 (2015)
18. Alippi, C.: Intelligence for Embedded Systems. Springer, Berlin (2014)
19. NIST: National Institute of Standards and Technology. http://www.nist.gov
20. Karnouskos, S.: Cyber-physical systems in the smartgrid. In: 9th International Conference on Industrial Informatics (INDIN), pp. 20–23. IEEE (2011)
21. Hu, F.: Cyber-Physical Systems: Integrated Computing and Engineering Design. CRC Press, Boca Raton (2013)

22. Peters, M., Schmidt, T.S., Wiederkehr, D., Schneider, M.: Shedding light on solar technologies—a techno-economic assessment and its policy implications. Energy Policy **39**, 6422–6439 (2011)
23. NREL - National Renewable Energy Laboratory: Best Practices in PV System Operations and Maintenance (2015)
24. Jordan, D., Kurtz, S.: Photovoltaic degradation rates—an analytical review. Prog. Photovoltaics Res. Appl. **21**(1), 12–29 (2013)

Demand Side Management Energy Management System for Distributed Networks

Filipe A. Barata[1(✉)], José M. Igreja[1,2], and Rui Neves-Silva[3]

[1] Instituto Superior de Engenharia de Lisboa (ISEL),
R. Conselheiro Emídio Navarro 1, 1959-007 Lisbon, Portugal
{fbarata,jigreja}@deea.isel.ipl.pt
[2] INESC-ID, Rua Alves Redol, 9, 1000-029 Lisbon, Portugal
[3] Universidade Nova de Lisboa,
Monte da Caparica, 2829-516 Caparica, Portugal
rns@fct.unl.pt

Abstract. This paper is focused on the development of a demand side management control method in a distributed network, aiming the creation of greater flexibility in demand and better ease the integration of renewable technologies. In particular, this work presents a novel multi-agent model-based predictive control method to manage distributed energy systems from the demand side, in presence of limited energy sources with fluctuating output and with energy storage in house-hold or car batteries. Specifically, here is presented a solution for thermal comfort which manages a limited shared energy resource via a demand side management perspective, using an integrated approach that includes an auction and a shifting load strategy. The control is applied individually to a set of Thermal Control Areas, demand units, where the objective is to minimize the energy usage and not exceed the limited and shared energy resource, while simultaneously indoor temperatures are maintained within a comfort frame. The developed solution is explained and applied to different scenarios wherein the results illustrate the benefits of the proposed approach.

Keywords: DMPC · Intermittent energy resource · DSM · Energy auction · Shifting loads · Energy efficiency · Limited energy resource

1 Introduction

Nowadays buildings sector is responsible for 40 % of the world's energy consumption and almost 50 % greenhouse gas emissions. Buildings emit more gases than transports and industry sector, estimated in 31 % and 28 % respectively. By analyzing the energy profile of buildings, it is clear that most of the consumption is to heat/cool the spaces to provide indoor comfort [1]. The energy consumption in buildings sector has increased along the last 20 years. Consequently, data from 2009 [2], showed that in residences the space heating accounted for about 70 % of the 68 % of total final energy consumed in buildings.

This increasing energy consumption is mainly to fulfil the demand for thermal comfort, being presently the HVAC systems the principal energy end use in buildings [3].

© IFIP International Federation for Information Processing 2016
Published by Springer International Publishing Switzerland 2016. All Rights Reserved
L. Camarinha-Matos et al. (Eds.): DoCEIS 2016, IFIP AICT 470, pp. 455–471, 2016.
DOI: 10.1007/978-3-319-31165-4_43

By these facts, it is socially, environmentally and economically imperative to decrease the energy consumption by increasing the buildings efficiency. A viable choice to achieve the reduction of energy consumption in the building sector is the application of demand response (DR) mechanisms. Demand Response program, is an efficient load management strategy for customer side is a load allocation scheme from the demand side, that it is nowadays mostly used to encourage users to shift their energy usage to periods with low demand and, consequently, lower prices [4]. The DR potential it is not sufficiently explored, being the two key challenges to work with diverse heterogeneous loads and with the distributed nature of renewable sources. New technology advances in communication are providing solutions to overcome the current electricity demand requirements. This new development has accelerated devising various industrial programs for scheduling utilization of residential appliances [5]. This DR mechanism combined with Demand Side Management (DSM) methodologies will be relevant in the future distributed Smart Grids (SGs) [6], and provide solutions that allow buildings to be fully integrated and prepared to efficiently coexist in a dynamic and inconstant environment typically supported by renewable resources [7].

Currently, the grid balance is mainly made with the generation following the electricity demand. Nevertheless, being renewable resources mainly weather dependent, it is vital to provide flexibility to the grid, in order to respond to the resource variability. In future SGs, it is expected that the production will control the energy consumption. Therefore, buildings settings must be controlled and adaptable to the clean resources intermittency. With this approach, users will have a more active position instead of being merely spectators in the electricity grid systems. Several solutions are emerging to deal with the variability, flexibility and poor controllability of the *green* sources and consequently on the ability to maintain the balance between demand and supply. Remark that, DSM strategies must have into account the control of many kinds of appliances, for instance, HVAC systems, lighting or electric vehicles charging.

The methodology here presented seeks a solution to respond to this variability, implementing, with the technological and advanced environment that SGs will provide [8], pursuing new technologies and solutions that will allow simple home appliances to be entirely controllable. This active DSM [9] will manage the loads to obtain harmony in demand supply ratio. DSM can include a mixture of several approaches, load control manipulation models, pricing, with distinguished electricity tariffs along the day to encourage load management and other approaches that promote energy efficiency and conservation [10, 11]. Therefore, a novel integrated control solution with DSM automated response is proposed based on Model Based Predictive Control (MPC) techniques. The MPC technique, in comparison with traditional HVAC systems used in the domestic sector, is able to save 16–41 % of energy, and also it adds robustness, adjustment and flexibility to overall system [12]. Thus, due to its features, MPC is suitable for energy savings, energy management and optimization, in particular to control temperature set-points [13], see [14] for a review on MPC in HVAC control systems.

The MPC control has also evolved as a solution to act in distinguished distributed environments [15]. Distributed Model Predictive Control (DMPC) algorithms, offers the same features of MPC but have the advantage of supporting the distribution of sensing and control using local controllers/agents that cooperate by exchanging information to decide their control actions [16]. This is the reason why it was the chosen method to deal with this kind of system. These DMPC infrastructures are suitable to use in a Multi Agent System (MAS) framework where, in a distributed environment, several agents employing individually a MPC control strategy are able to interact and receive influence from neighbour subsystems, exchange predictions on their future state and incorporate this information into their local MPC problems [17]. An agent can be defined as a complex software entity or intelligent entities with three main characteristics: Pro-activeness (they react to external events and they are driven to their objectives), social ability (they can cooperate or compete between them) and autonomy (they can decide in order to archive their objectives) [18]. Remark that the SG vision can be based on this concept, a framework where distinguished identities cooperate to obtain a collective perception. Thus, the developed system considers each agent as an autonomous entity. Designated by Thermal Control Area (TCA) the entity is embedded in a distributed environment, where several TCA's are working individually to achieve their own goal but sharing their information among them to maintain some kind of coordination in face of a global objective.

The outline of the rest of the chapter is as follows. In Sect. 1.1 is presented the research question and hypothesis and Sect. 1.2 describe the innovative contributions. In Sect. 2 is presented the contribution of our work to the main theme of the conference "Technological Innovation for Cyber-Physical Systems", in Sect. 3 is made the scenario description and the developed dynamical models. In Sect. 4, the developed MPC control scheme is described, in Sect. 5 results from numerical simulations are depicted and comment, and finally Sect. 6 offers the concluding statements.

1.1 Research Question

With SGs, domestic customers will have an important role in the electric grid system. Final consumers will no longer be merely spectators; their contribution will be relevant due the new technological advances that allow fully manageable households appliances. Shifting their electricity consumption in time or, by changing their work conditions, these devices can be controllable, adjusting the demand to the desired intermittent source without decreasing the comfort of the residents. In distributed networks, aggregated to an intermittent source, an unlimited number of this kind of loads can exist, representing a control problem to achieve the network efficiencies, involving stakeholder's satisfaction.

Consequently, new distributed, coordinative and cooperative strategies are necessary to ensure that the control decisions of all identities present in the system, contribute for the global objective, and also as referred, is desirable any or negligible comfort impact on end users/occupiers.

Considering the mentioned above, this work aims providing solutions to respond to the next questions:

Q.1 How, in a distributed network, can the demand be adjusted to an intermittent source to maximize the energy efficiency?

Q.2 How to improve energy efficiency using the domestic potential in a distributed network?

Q.3 Which control methods should be applied in a distributed network with demand side management to obtain all the existing energy potential from intermittent energy sources?

The adopted work hypothesis to address the research question is defined below:

Using an integrated approach, that in a distributed environment, considers multi-agent control scheme and an optimization MPC multi-objective approach with anticipative effect, capable to deal in a DSM perspective, with fluctuating energy sources, smart load control, thermal comfort and real-time price negotiation.

1.2   Research Contribution and Innovation

The novel proposed sequential multi-agent DMPC scheme for thermal house comfort and energy savings provides robustness, adjustment and flexibility to the global system. The sequence is built based on an energy bid where the highest biddings are placed first in the access order. After consume, each agent predicts its consumption profile and pass through the next, the information about the predicted available renewable energy. At each hour, the sequence is established and the energy price depends from the offered bid, amount and type (renewable or grid) of energy consumed. Through this DSM energy usage optimization scheme, the consumer has the flexibility to choose hourly between comfort or energy savings. The anticipated knowledge of the energy source value (by forecasts in renewable case or by quantifying in fuel case) allows the system to decide how to split and when and in what quantity the energy is spend to respect all the power and comfort constraints. Thus, the innovative cost function optimization scheme is also suitable to apply in environments where at the same time, the energy sources renewable or not, have strong limitations and must be rationed and the comfort issues must be also taken in to consideration, as for example, in remote areas or in boats/cruises where the rooms must be acclimatized and the fuel source is limited.

A shifting load algorithm for loads allocation is also established. The customer sets features of flexible loads and the algorithm fits them in the most favorable time interval gap.

2   Relationship to Cyber-Physical Systems

Cyber Physical Systems (CPS) are organized systems that are connected and related between them and have the capability to act, react and cooperate with multiple technological and biological agents (devices and humans). To allow this kind of features, CPS incorporates sensors, processors and actuators that provide safety and interoperable actions between components in real-time applications.

Advances in CPS are providing new solutions to control and manage buildings. Thus, buildings may have a fully integrated and embedded network of devices in their infrastructure to provide superior management solutions in the energy usage, namely in heating, cooling, lighting and elevators/transportation. Operating with several sensors and actuators planned to provide comfort, security, operational efficiency and intelligent performability, new buildings may be viewed as the CPS stereotype with a high level of interrelated systems that sense and process the data in the net. As mentioned, in future SGs household appliances are network connected and perfectly manageable in real time.

The developed framework considers a set of buildings, compose each one with several floors and each floor by several rooms, that consist in many distinct appliances with the capability to feel the surroundings (e.g., temperature, energy consumption) and actuators (e.g., air conditioning equipment) to influence the environment. Thus, building is perceived entirely as a cyber-physical energy system, where distinguished identities cooperate, sense, act and react to obtain a collective perception. Therefore, each agent is observed an autonomous entity, TCA that physically is a CPS that is embedded/integrated in a distributed environment, where several CPS are working individually to achieve their own goal but sharing their information among them to maintain some kind of coordination in face of a global objective.

3 Dynamical Models and Scenarios Description

The scenario here described intends to be a realistic solution to take advantage of SG's features. Smart meters, or commonly named advanced metering infrastructure (AMI), are applied in SG to allow the equipment monitoring and control. Because renewable energies are nowadays significantly expanded, electricity is being fed into both the medium- voltage and low-voltage grids, depending on changing external conditions (e.g., weather, time of day, etc.). These fluctuating energy resources can severely impair the stability of the distribution grid. One of the key challenges of a smart grid is therefore to quick balance out the energy supply and energy consumption in the distribution grid. Thus, the scenario here presented has in consideration this challenge, intend to provide a solution to it based on, and taking advantage of all the AMI and communication technologies that SG provide.

The conceptual scenario involves a set of buildings with an electricity source provided by their own renewable energy park and energy storage as presented in Fig. 1. Henceforward, the term house will be applied to classify any type of structure for habitation (houses), office buildings or other kind of analogous constructions. The set $W = \{w_1, w_2, \ldots, w_{N_S}\}$ identifies the group of houses in consideration and the different spaces or divisions in each house are specified by N_S sets given by $D_i = \{d_{i1}, d_{i2}, \ldots, d_{iNd_i}\}$ where $i = 1, \ldots N_S$ and Nd_i is the number of divisions for house i.

Due to the existing division diversity in houses, each area may vary in: construction materials, sun exposure, occupancy and indoor temperature set-points, thus, each division has its own energy needs to weatherize the space, and for this reason a TCA is considered. As mentioned, one TCA represents an autonomous thermal control entity within an environment where the actions and reactions are made in order to achieve a common goal. Therefore, depending from the desired infrastructure intent to be implemented, a set of buildings or a simple division may represent a TCA. Each TCA receives several external inputs; the outdoor temperature, available renewable power forecasts, the Market Operator (MO) kWh price, access order to the renewable resource

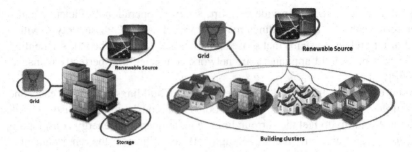

Fig. 1. Implemented scheme.

and, when applied, the neighbour's indoor temperature forecasts. Figure 2 presents an example of the implemented TCA framework. Thus, in a scenario that privileges the renewable energy usage, the used demand side management approach allows the management of distributed loads, aiming the adjustment of the demand to the supply, providing thermal comfort, lower energy costs and lowering CO_2 emissions. By using this active DSM control, the optimal control strategies for various appliances can be generated whilst maximum utilisation of energy supplied from intermittent systems is guaranteed.

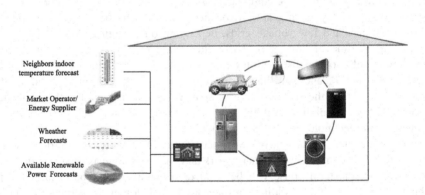

Fig. 2. Example of a TCA conceptual framework.

3.1 Dynamical Models

The electro thermal-electrical modular approach used to develop the house models is described in detail in [19]. So, in this work are only showed the main equations. Firstly, a first order linear energy balance model is used to define the dominant dynamics of a generic division,

$$\mathbf{x}(k+1) = \mathbf{A}\mathbf{x}(k) + \mathbf{B}\mathbf{u}(k) + \mathbf{v}(k), \tag{1}$$

where $\mathbf{x} \in \mathbb{R}^n$ is the state variable, indoor temperatures, vector containing all the divisions temperatures (°C), $\mathbf{u} \in \mathbb{R}^m$ is the input vector containing all the heating and cooling power sources (W) to weatherize each division, and $\mathbf{v} \in \mathbb{R}^n$ includes all the disturbances (W), including load generated by occupants, solar radiation, any other heat/cooling sources or doors and windows aperture to recycle the indoor air, k is an integer number that denotes discrete time and $\mathbf{A} \in \mathbb{R}^{n \times n}$, $\mathbf{B} \in \mathbb{R}^{n \times m}$, are matrices.

Generically, using Euler discretization with a sampling time of Δ, the discrete model space-state representation of (1) can be written, for the N_s TCAs and for each division (l) as,

$$
T_l^i(k+1) = A_{ll}^{ii} T_l^i(k) + B_l^i u_l^i(k) + \underbrace{\sum_{\substack{g=1 \\ (g \neq l)}}^{Nd_i} \left(A_{lg}^{ii} T_g^i(k) \right)}_{\substack{thermal contributions \\ from adjacent areas \\ inside the same house}} + \underbrace{\sum_{\substack{h=1 \\ (h \neq i)}}^{Ns} \sum_{m=1}^{Nd_h} \left(A_{lm}^{ih} T_m^h(k) \right)}_{\substack{thermal contributions \\ from adjacent areas \\ from other houses}} + v_l^i(k),
$$

(2)

where

$$
A_{ll}^{ii} = \left(1 - \frac{\Delta t}{R_{l_{eq}}^i C_{l_{eq}}^i}\right), B_l^i = \frac{\Delta t}{C_{l_{eq}}^i}
$$

$$
D_l^i = \sum_{\substack{g=1 \\ (g \neq l)}}^{Nd_i} \frac{T_g^i - T_l^i}{R_{l_{g_{eq}}}^i C_{l_{eq}}^i} \Delta t + \sum_{\substack{h=1 \\ (h \neq i)}}^{Ns} \sum_{m=1}^{Nd_h} \frac{T_m^h - T_l^i}{R_{l_{m_{eq}}}^i C_{l_{eq}}^i} \Delta t, v_l^i = \frac{P_{l_{Pd}}^i \Delta t}{C_{l_{eq}}^i} + \frac{T_{oa} \Delta t}{R_{l_{eq}}^i C_{l_{eq}}^i}.
$$

(3)

where Ns is number of TCA's, Nd_i is number of divisions inside subsystem (i), x_l^i is the indoor temperature in TCA/subsystem (i) inside division (l), u_l^i is the used power to provide comfort in TCA/subsystem (i) inside division (l), A_{lm}^{ih} is an element from the state matrix A that relates the state (indoor temperature) in division (m) from TCA/subsystem (h), with the state from division (l) in TCA (i) and v_l^i is the thermal disturbance in TCA/subsystem (i) inside division (l) e(g. load generated by occupants, direct sunlight, electrical devices or doors and windows aperture to recycle the indoor air), and T_{oa}, the temperature of outside air (°C). Remark that the number of states variables in general model (1) is $n = \sum_{i=1}^{Ns} Nd_i$. For a more complex scenario, Fig. 3 shows a distributed environment with five houses with different plans and with different zones that may thermally interact. Remark that as described in the sequel, adjacent areas can be doubly coupled, thermally and by the power constraint. In Fig. 3, u_i represents the input (heat/cooling power) and y_i is the output vector containing the temperatures/states inside the several divisions of house (i)

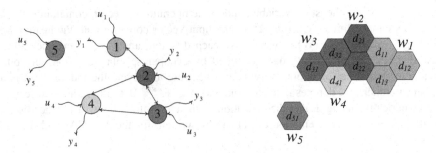

Fig. 3. Generalized house/TCA scheme example.

4   Model Predictive Control Cost Function

At each time step, each one of the agents must solve his MPC problem. The objectives are: minimize the energy consumption to heating and cooling; minimize the peak power consumption; maintain the zones within a desired temperature range and maintain the used power within the *green* available bounds. Feedback stability is provided by choosing a sufficiently long predictive horizon and proven by results presented in Sect. 5. Feasibility is achieved by the use of *soft* constraints in the optimization problem formulation as explained in the sequel. The generic linear convex optimization problem to be solved with a *Matlab* routine (*fmincon*) by each agent at each instant, assumes the following form:

$$
\min_{U,\bar{\varepsilon},\underline{\varepsilon},\bar{\gamma},\underline{\gamma}} J_i(k) \sum_{j=0}^{H_p-1}\left[\underbrace{\sum_{l=1}^{Nd_i} u_l^i(k+j)^T \varphi_l^i u_l^i(k+j)}_{Consumption}\right] + \underbrace{\sum_{l=1}^{Nd_i}\phi_i max\{u_l^{i2}(k),\ldots,u_l^{i2}(k+H_p-1)\}}_{Power\ peaks}
$$

$$
+ \sum_{j=1}^{H_p}\left[\underbrace{\varepsilon_l^i(k+j)^T \Xi_l^i \varepsilon_l^i(k+j)}_{Comfort\ violation} + \underbrace{\gamma_l^i(k+j)^T \Psi_l^i \gamma_l^i(k+j)}_{Power\ violation}\right],
$$

$$\tag{4}$$

$$
\min_{U_i,\varepsilon_i,\gamma_i} J_i(k) = \varepsilon_i^T(k)\Xi_i\varepsilon_i(k) + \gamma_i^T(k)\Psi_i\gamma_i(k) + U_i^T(k)R_iU_i(k)
$$

$$
+ \sum_{l=1}^{Nd_i}\phi_i max\{u_l^{i2}(k),\ldots,u_l^{i2}(k+H_p-1)\},
$$

$$\tag{5}$$

and subject to the following constraints,

$$x_l^i(k+j+1) = A_{ll}^{ii}x_l^i(k+j) + B_l^i u_l^i(k+j) + \underbrace{\sum_{\substack{g=1 \\ (g \neq l)}}^{Nd_i} \left(A_{lg}^{ii}\tilde{x}_l^i(k+j)\right)}_{\substack{\text{predicted temperatures} \\ \text{from adjacent areas} \\ \text{inside the same house}}}$$

$$+ \underbrace{\sum_{\substack{h=1 \\ (h \neq i)}}^{Ns} \sum_{m=1}^{Nd_h} \left(A_{lm}^{ih}\tilde{x}_m^h(k+j)\right)}_{\substack{\text{predicted temperatures} \\ \text{from adjacent areas} \\ \text{from other houses}}} + v_l^i(k+j), \quad j=1\ldots H_P \tag{6}$$

$$\underline{T}_l^i(k+j) - \underline{\varepsilon}_l^i(k+j) \leq x_l^i(k+j) \leq \bar{T}_l^i(k+j) + \bar{\varepsilon}_l^i(k+j), \tag{7}$$

$$\underline{U}_i(k+j-1) - \underline{\gamma}_i(k+j-1) \leq \sum_{l=1}^{Nd_i} u_l^i(k+j-1) \leq \bar{U}_i(k+j-1) + \bar{\gamma}_i(k+j-1), \tag{8}$$

$$\underline{\gamma}_i, \bar{\gamma}_i, \underline{\varepsilon}_l^i, \bar{\varepsilon}_l^i \geq 0. \tag{9}$$

In (4) Nd_i is the number of divisions of house (i), u_i^l represents the power control inputs from house (i) division (l), ϕ_i is the penalty on peak power consumption, Ξ_i is the penalty on the comfort constraint violation, Ψ_i the penalty on the power constraint violation and H_P is the length of the prediction horizon. In (7), $\bar{\varepsilon}_l^i$ and $\underline{\varepsilon}_l^i$ are the vectors of temperature violations that are above and below the desired comfort zone defined by $\bar{T}_l^i$ and $\underline{T}_l^i$. In (8), the coupled power constraint, $\bar{\gamma}_i$ and $\underline{\gamma}_i$ are the power violations that are above or lower the maximum, $\bar{U}_i$, and minimum, $\underline{U}_i$, available *green* power for heating/cooling the house, with $\underline{U}_i = -\bar{U}_i$. Remark that, in each TCA (i), the power sum in all divisions cannot exceed $\bar{U}_i$.

5 Results

The results are presented with two different approaches. In Sect. 5.1 the energy split performance is based on a fixed sequential order, A_1, A_2 and A_3, established from a previously done auction wherein the bids are daily made by each TCA, acting as demand side management agents and based on the energy daily price ([20] allows the bidding value vary hourly and consequently, the agents order to access to the clean energy also varies). In Sect. 5.2 the results show the developed shifting and loads allocation. All the presented results were obtained with an optimization *Matlab* routine.

5.1   DMPC for Thermal House Comfort with Sequential Access Auction

In the first approach three houses are considered, two of them thermally interacting (with a thermal resistance between them of R_{12} = 30 °C/kW and the third is isolated.

As mentioned, agents can also have distinct penalties on power and temperature constraints violations, they can hourly privilege comfort or cost according to consumer choice. Therefore, to explore the concept two scenarios are here presented.

In the first scenario (S1) the penalty values of the parameters related with consumption were increased. The comfort issues are less important, with the agents mainly concerned with lower consumptions and in satisfy the power constraint. With this variation, the *soft* power constraint was transformed in a *hard* constraint. The second scenario (S2) is focused in maintaining the indoor comfort, all agents want to respect the established temperature gap regardless the required consumption. To accomplish this goal, the temperature penalty was significantly increased, and the consumption parameters decreased (Tables 1 and 2).

Table 1. Scenario distributed parameters

Parameter	A1	A2	A3	Units
Req	50	50	75	°C/kW
Ceq	9.2×10^3	9.2×10^3	9.2×10^3	kJ/°C
Green Price (per kWh)	0.09	0.08	0.07	€
Red Price(per kWh)	0.18			€

Table 2. Penalty values

Parameter	A_1		A_2		A_3	
	S1	S2	S1	S2	S1	S2
Ξ	50	50000	50	50000	50	50000
ψ	100000	1	100000	1	30000	3
φ	20	0.2	20	0.2	20	0.2
φ	10	0.1	10	0.1	10	0.1

It is considered that all TCA's have the same outdoor temperature pictured in Fig. 12.

The thermal disturbances forecasts and the indoor temperature with its constraints, as well the power profile for the TCA, A_1, A_2 and A_3 are pictured above (Figs. 4, 5 and 6).

Due to the scarcity of *green* resource and the obligation to respect the power limits, A_3 was the one that presented lower consumptions, Fig. 10(a), and consequently minor costs, on the other hand, the indoor temperature was the most penalized with the highest deviation from the chosen comfort range. With this parameterization, the consumption profile always preserved inside bounds, for all TCA's, and can be seen that any or negligible *red* resource was consumed.

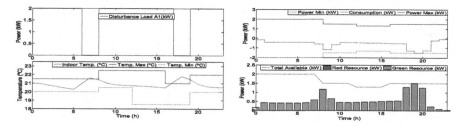

Fig. 4. Scenario 1,A_1. (a) Thermal disturbance and indoor temperature; (b) Power profile.

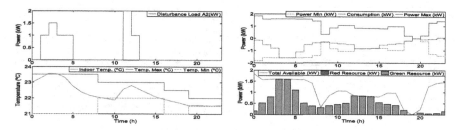

Fig. 5. Scenario 1, A_2. (a) Thermal disturbance and indoor temperature; (b) Power profile.

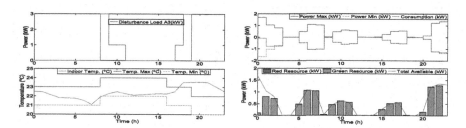

Fig. 6. Scenario 1, A_3. (a) Thermal disturbance and indoor temperature; (b) Power profile.

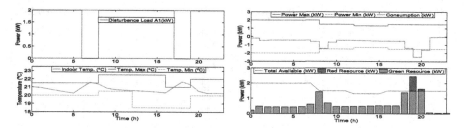

Fig. 7. Scenario 2, A_1. (a) Thermal disturbance and indoor temperature; (b) Power profile.

As can be seen in Figs. 7(a), 8(a) and 9(a) that all the indoor temperatures are mostly maintained inside the comfort gap. As consequence, the power constraints, Figs. 7(b), 8(b) and 9(b), are violated and the *red* resource consumption increased

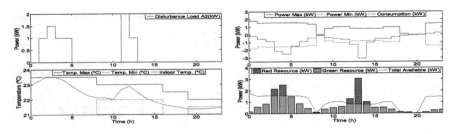

Fig. 8. Scenario 2, A$_2$. (a) Thermal disturbance and indoor temperature; (b) Power profile.

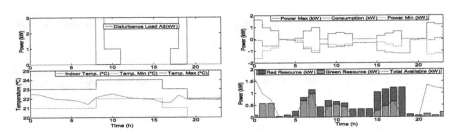

Fig. 9. Scenario 2, A$_3$. (a) Thermal disturbance and indoor temperature; (b) Power profile.

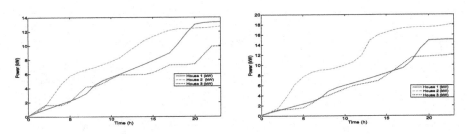

Fig. 10. (a) Scenario 1, total consumption; (b) Scenario 2, total consumption.

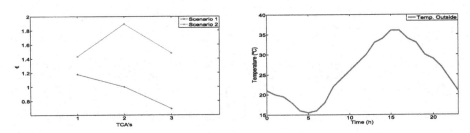

Fig. 11. Daily heating/cooling total cost.   **Fig. 12.** Outdoor temperature forecasting

significantly. Being now the comfort a priority it's quite clear that the controller tries to accomplished the pre-defined comfort range leading to a consumption and cost surplus (Fig. 11).

Comparing the economic differences between scenarios, the S2 has shown to be the most economical however, this decreased consumption led to a higher indoor temperature deviation from the established boundaries. On the other hand, S1 was the most expensive, all the necessary resources were consumed, leading to higher costs, in order to respect the comfort limits. The possibility of obtaining comfort in detriment of the cost may be important in various situations. For example: To acclimatize rooms with children or areas in laboratories and/or hospitals. Remark that, despite this comfort preference, the cost function (in due proportion given by the parameters) also minimizes all the other terms.

5.2   Implemented Shifting and Loads Allocation Scheme

In buildings, DSM is based on an effective reduction of the energy needs by changing the shape and amplitude consumer's load diagrams. So, DSM can involve a combination of several strategies; pricing, load management curves and energy conservation are implemented for a more energy efficient use. Load shifting is considered a common practice in the management of electricity supply and demand, where the peak energy use is shifted to less busy periods. Properly done, load shifting helps meeting the goals of improving energy efficiency and reducing emissions, smoothing the daily peaks and valleys of energy use and optimizing existing generation assets. With new technological advances, DR programs may shift loads by controlling the function of air conditioners, refrigerators, water heaters, heat pumps and other similar electric loads at maximum demand times. The work here presented is distinct because provides an integrative solution which is able to, in a distributed network with multiple TCAs, adjust the demand to an intermittent limited energy source, using load shift and maintaining the indoor comfort [21].

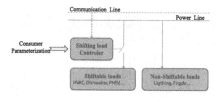

Fig. 13. Shifting load communication infrastructure.

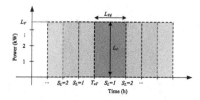

Fig. 14. Implemented shifting load scheme

Figure 13, exemplifies a shifting load communication infrastructure and Fig. 14 the shifting load characteristics. Each division selects the load value (L_V), the duration (L_{Vd}), the turned on time (T_{oT}) and the "sliding level" (S_L) of the "shifted loads". The S_L

indicates that the load can be turned on S_L hours before and after the chosen T_{oT}. With this data, all the possible loads schedule combinations ($PLSC_S$) are establish (see Fig. 19 *e.g.*). At each time step, it's verified if inside the predictive horizon, any $PLSC_S$ exceeds the maximum available *green* energy. The sequences that are at any instant above the limit are removed, and the remaining are the feasible load schedule combinations ($FLSC_S$). The $FLSC_S$ are tested in the minimization problem (4) as maximum available *green* resource for comfort (8). The hypothesis that provided less consumption is chosen. Once one sequence is started, all the others that are different until the current step time are eliminated until the final load sequence is chosen, *FLSeq*. The results here presented show only the shifting loads procedure for one house represented by one division with thermal disturbance. A distributed scenario, algorithm and details

Table 3. Thermal characteristics and cost function parameterization.

R_{eq} (°C/kW)	C_{eq} (kJ/°C)	Ξ	Ψ	ϕ	φ	Δt(h)	H_P	N_C	$T(0)$ (°C)
50	9.2×10^3	500	500	2	1	1	24	24	21

Table 4. Shifted loads characteristics.

Loads	L_V (kW)	L_{Vd} (h)	T_{oT} (h)	S_L (h)
Load 1	1.5	3	7	1
Load 2	2	2	19	3

about the implemented scheme can be seen in [21]. The house thermal characteristics and cost function parameterization are showed in Table 3, and the *Loads* that can be daily shifted have the characteristics present in Table 4.

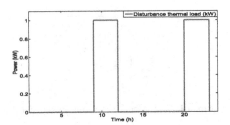

Fig. 15. Thermal disturbance forecasting

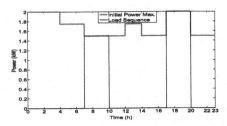

Fig. 16. Maximum available green energy and chosen sequence

In order to minimize the energy costs by consuming only *green* resource, the implemented algorithm chooses the gaps that fit properly in the maximum available *green* energy.

In Fig. 20 are the total energy costs of the $FLSC_S$ shown in Table 5, and can be seen that the chosen sequence, third hypothesis, is the less expensive (Figs. 15, 16 and 17).

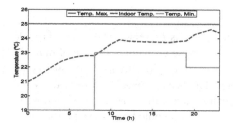

Fig. 17. Indoor temperature

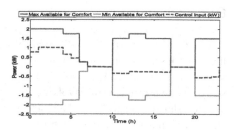

Fig. 18. Used power to heat/cool the space and the maximum green resource available for comfort

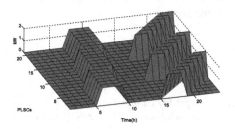

Fig. 19. Possible loads schedule

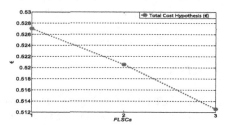

Fig. 20. Total energy costs of $FLSC_S$

Table 5. Feasible loads sequence combinations

FLSC	Time (h)											
	0	1	2	3	4	5	6	7	8	9	10	11
1	0	0	0	0	0	1,5	1,5	1,5	1,5	0	0	0
2	0	0	0	0	0	0	1,5	1,5	1,5	1,5	0	0
3	0	0	0	0	0	0	0	1,5	1,5	1,5	1,5	0
FLSC	Time (h)											
	12	13	14	15	16	17	18	19	20	21	22	23
1	0	0	0	0	0	2	2	2	0	0	0	0
2	0	0	0	0	0	2	2	2	0	0	0	0
3	0	0	0	0	0	2	2	2	0	0	0	0

The periods between 7–10 h and 17–20 h are extremely demanding, all *green* energy is consumed by the shifted loads, with no remaining one for comfort proposes. Although, Fig. 18 shows that in that periods the algorithm choose to not use the *red* resource and, taking advantage of the prediction horizon, pre-heat or pre-cool the spaces when only renewable resource is available.

6 Conclusions

In this paper, a distributed MPC control technique was presented in order to provide thermal house comfort. The solution obtained solves the problem of control of multiple subsystems dynamically coupled subject to a coupled constraint. Each subsystem solves its own problem by involving its own and adjacent rooms state predictions and also the shared constraints. Changing the penalty values, the consumer can choose in each division between indoor comfort and lower costs. It could be observed through the simulations and results analysis that suitable dynamic performances were obtained. Also, the approach shows that distributed predictive control is able to provide house comfort within a DSM policy, based in a price auction and the rescheduling of appliance loads. It is a valid methodology to achieve reduction in consumption and price. The method is more effective with wider periods where the loads are allowed to slide and consequently allocate the most favorable zone.

References

1. StorePET Project, November 2014. http://www.storepet-fp7.eu/project-overview
2. Nolte, I., Strong, D.: Europe's buildings under the microscope. Buildings Performance Institute Europe (2011). ISBN 9789491143014
3. Korolija, I., Marjanovic-Halburd, L., Zhang, Y., Hanby, V.I.: Influence of building parameters and HVAC systems coupling on building energy performance. Energy Build. **43** (6), 1247–1253 (2011)
4. Siano, P.: Demand response and smart grids-a survey. Renew. Sustain. Energy Rev. **30**, 461–478 (2014)
5. Wang, C., Zhou, Y., Jiao, B., Wang, Y., Liu, W., Wang, D.: Robust optimization for load scheduling of a smart home with photovoltaic system. Energy Conv. Manag., 1–11 (2015). doi:10.1016/j.enconman.2015.01.053
6. Mahmood, A., Ullah, M.N., Razzaq, S., Basit, A., Mustafa, U., Naeem, M., Javaid, N.: A new scheme for demand side management in future smart grid networks. Proc. Comput. Sci. **32**, 477–484 (2014). doi:10.1016/j.procs.2014.05.450
7. Figueiredo, J., Martins, J.: Energy production system management - renewable energy power supply integration with building automation system. Energy Convers. Manag. **51**(6), 1120–1126 (2010). doi:10.1016/j.enconman.2009.12.020
8. Paul, S., Rabbani, M.S., Kundu, R.K., Zaman, S.M.R.: A review of smart technology (Smart Grid) and its features. In: 1st International Conference on Non Conventional Energy (ICONCE 2014), pp. 200–203 (2014). doi:10.1109/ICONCE.2014.6808719
9. Gelazanskas, L., Gamage, K.A.A.: Demand side management in smart grid: a review and proposals for future direction. Sustain. Cities Soc. **11**, 22–30 (2014). doi:10.1016/j.scs.2013.11.001

10. Molderink, V., Bakker, V., Bosman, M., Hurink, J., Smith, G.: Management and control of domestic smart grid technology. IEEE Trans. Smart Grid **1**(2), 109–119 (2010). doi:10.1109/TSG.2010.2055904

11. Ullah, M.N., Javaid, N., Khan, I., Mahmood, A., Farooq, M.U.: Residential energy consumption controlling techniques to enable autonomous demand side management in future smart grid communications. In: 2013 Eighth International Conference on Broadband and Wireless Computing, Communication and Applications, pp. 545–550 (2013). doi:10.1109/BWCCA.2013.94

12. Maasoumy, M., Razmara, M., Shahbakhti, M., Sangiovanni Vincentelli, A.: Selecting building predictive control based on model uncertainty. In: American Control Conference, no. 1, pp. 404–411 (2014). doi:10.1109/ACC.2014.6858875

13. Bruni, G., Cordiner, S., Mulone, V., Rocco V., Spagnolo, F.: A study on the energy management in domestic micro-grids based on model predictive control strategies. In: Energy Conversion and Management, pp. 1–8 (2015). doi:10.1016/j.enconman.2015.01.067

14. Afram, A., Janabi-Sharifi, F.: Theory and applications of HVAC control systems - a review of model predictive control (MPC). Build. Environ. **72**, 343–355 (2014). doi:10.1016/j.buildenv.2013.11.016

15. De Souza, F.A., Camponogara, E., Junior, W.K.: Distributed MPC for urban traffic networks: a simulation-based performance analysis. J. Optimal Control Appl. Methods (2014). doi:10.1002/oca

16. Maestre, J., Negenborn, R.: Distributed Model Predictive Control Made Easy. Springer, Berlin (2014). ISBN 978-94-007-7006-5

17. Chandan, V., Alleyne, A.G.: Decentralized predictive thermal control for buildings. J. Process Control 1–16 (2014). doi:10.1016/j.jprocont.2014.02.015

18. Raza, S.M.A., Akbar, M., Kamran, F.: Use case model of genetic algorithms of agents for control of distributed power system networks. In: Proceedings of the IEEE Symposium on Emerging Technologies, pp. 405–411 (2005). doi:10.1109/ICET.2005.1558916

19. Barata, F.A., Neves-Silva, R.: Distributed MPC for thermal comfort in buildings with dynamically coupled zones and limited energy resources. In: Camarinha-Matos, L.M., Barrento, N.S., Mendonça, R. (eds.) DoCEIS 2014. IFIP AICT, vol. 423, pp. 305–312. Springer, Heidelberg (2014)

20. Barata, F.A., Silva, R.N.: Distributed model predictive control for housing with hourly auction of available energy. In: Camarinha-Matos, L.M., Tomic, S., Graça, P. (eds.) DoCEIS 2013. IFIP AICT, vol. 394, pp. 469–476. Springer, Heidelberg (2013)

21. Barata, F.A., Igreja, J.M., Neves-Silva, R.: Distributed MPC for thermal comfort and load allocation with energy auction. Int. J. Renew. Energy Res. (IJRER) **4**(2), 371–383 (2014). ISSN 1309-0127

Optimization in Energy Management

Optimal Wind Bidding Strategies
in Day-Ahead Markets

Isaias L.R. Gomes[2], Hugo M.I. Pousinho[1], Rui Melício[1,2($\boxtimes$)],
and Victor M.F. Mendes[2,3]

[1] IDMEC/LAETA, Instituto Superior Técnico,
Universidade de Lisboa, Lisbon, Portugal
ruimelicio@gmail.com
[2] Departamento de Física, Escola de Ciências
e Tecnologia Universidade de Évora, Évora, Portugal
[3] Instituto Superior de Engenharia de Lisboa, Lisbon, Portugal

Abstract. This paper presents a computer application (CoA) for wind energy (WEn) bidding strategies (BStr) in a pool-based electricity market (EMar) to better accommodate the variability of the renewable energy (ReEn) source. The CoA is based in a stochastic linear mathematical programming (SLPr) problem. The goal is to obtain the optimal wind bidding strategy (OWBS) so as to maximize the revenue (MRev). Electricity prices (EPr) and financial penalties (FiPen) for shortfall or surplus energy deliver are modeled. Finally, conclusions are addressed from a case study, using data from the pool-based EMar of the Iberian Peninsula.

Keywords: Bidding strategies · Wind power system · Stochastic linear programming · Day-ahead market

1 Introduction

A growing attention in ReEn sources has been followed over the last years due to its conversion into electric energy (EEn) free of pollutant emission and availability in all over the world [1]. E.U. countries are integrating ReEn into the electric grid (EGr) to fulfill Energy–2020 [2]. ReEn incentive has appeared and after 2012 the adoption by countries of one or even more incentives is a noteworthy fact [3]. Incentives include feed-in-tariff, guaranteed grid access, green certificates, investments incentives, tax credits and soft balancing (Bal) costs [4].

In restructured EMar, power producers (PPr) are entities owning power resources and taking part in the EMar with the objective of maximizing total profits. A PPr exploiting sources of WEn also face the uncertainty (Uncer) on the availability of these sources, meaning Uncer in complying with power contracts [5]. The closing of the EMar defines power trading and EPr. However, the remuneration depends on the conformity achieved on the level of the real deliver with the accepted value of the bid at the closing of the EMar. If no conformity is revealed, economic penalization for imbalances (Imb) is due to happen [6].

© IFIP International Federation for Information Processing 2016
Published by Springer International Publishing Switzerland 2016. All Rights Reserved
L. Camarinha-Matos et al. (Eds.): DoCEIS 2016, IFIP AICT 470, pp. 475–484, 2016.
DOI: 10.1007/978-3-319-31165-4_44

Uncers on sources of WEn can be accommodated to mitigate the final Uncer on the delivered energy (DEn). WEn cannot provide a continuous source of energy without Uncer due to the change of wind speed from one period to another [7]. Typically a WEn farm has more availability of WEn during the night and particular during the winter.

This paper is a research contribution to get the most out of WEn systems in order to conveniently accommodate BStr by the use of a CoA based in a SLPr approach.

2   Relationship to Cyber-Physical Systems

Power systems (PSy) face a shift from non-interactive, manually-controlled power grid (PGr) to the increased mixing of both cyber information and proper physical representations at all scales of the grid. A novel grid which features this cyber and physical grouping is the called Cyber-Physical Energy system (CyPES) [8]. ReEn reliance on weather conditions is a challenge in what regards integration into EGr [9]. A smart grid (SGr) provides flexibility needed to integrate ReEn. The SGr monitors energy flows, reducing cost and improving reliability [10]. Smart Systems (SSy) have been widely used in PSy [11]. The intelligent supervisory control and SCADA systems are pointed out as cyber-physical systems (CyPS) [12]. A CyPS is a combination of computation with physical processes [13]. CPS can be stated as a SSy, having software and hardware for sensing, monitoring, gathering, actuating, computing and controlling [14]. Cyber-Physical Intelligence (CyPI) is one topic to deal with SGrs, distributed generation and EMar [11]. The SGr needs CyPI to support tools for planning, EMar monitoring and risk management. Forecasting demand and EPr of EMar use Neural Network, Fuzzy Logic or both [15]. When big database about events is available, an ANN is a favorable option [15].

Multi-Agent System and Data Mining are considered for solving problems of EMar [15]. Intelligent systems are used for diagnosis and alarm processing at dispatch, at generation plant or substations. CyPES is estimated to improve the reliability, efficiency, flexibility, cost-effectiveness, and, very important, electric grid security [8] to avoid attacks [16].

3   State of the Art

Non-dispatchable ReEn sources, like WEn, are challenging for managing the bids in an EMar due to the Uncer in the unavailability of the ReEn, implying eventual FiPen for energy Imb [6]. So, the eventual deviations (Dev) and consequently penalty of Imb has to be considered. The technical literature presents methods for WEn BStr solved using different approaches. For example, in [17–21] PPr minimize the Dev losses through a portfolio approach, where the PPr can combine the WEn production with energy storage technologies so as to submit the optimal bids.

In [22], an approach is addressed so as to find out the OWBS, considering WEn production time series obtained from a Markov process. In [23] the Dutch EMar is analyzed, taking into account the robustness of the results obtained from point forecasts and probabilistic forecasts. In [17], the development of BStr is investigated for a WEn farm owner and a deterministic mixed-integer linear programming (MILP) approach for the optimal operation is proposed. In [17] is revealed an absence of treatment of WEn subjected to EMar price Uncer. In [24], linear programming is used for a WEn problem instead of mixed-integer nonlinear programming, concluding gains on robustness, simplicity and computational efficiency, implying adequacy to be integrated in SGrs in order to act in an enough small time processing, supporting decisions not only for helping bidding, but also for comparison in due time with eventual bilateral contracts. This linear programming used for the WEn problem is followed in this paper.

4 Description of the Problem

The wind energy traded (ETr) in a day-ahead electricity market (D-AEM) is subject to Uncers that must be considered in developing of offer strategies (OStr). The Uncers derive from the intermittence and volatility on the WEn and the EPr. These Uncers, if not adequately considered, can trigger losses of revenue (Rev) as a result of penalty incurred by the Imb. The Imb is set as a difference between the supplied energy (SpEn) and the offered energy (OfEn). If a surplus of SpEn in the PGr, then the Imb is positive (Posit); otherwise, the Imb is negative (Negat).

The system operator (SOp) minimizes the Imb value in a PGr using a methodology based on prices penalization for the Dev of the SpEn from the one offered by a producer and accepted in the closing D-AEM. In case of the Imb to be Negat, the SOp maintains the EPr for PPr with surplus of offers and pays a premium EPr produced above the offer. The EPr are as:

$$\vartheta_x^+ = \vartheta_x^D \tag{1}$$

$$\vartheta_x^- = \max(\vartheta_x^D, \vartheta_x^{UP}) \tag{2}$$

In (1) and (2), ϑ_x^+ and ϑ_x^-, are used in the complementary EMar to the energy Dev, ϑ_x^D is the D-AEM price and ϑ_x^{UP} is the EPr for the energy that desires to be added to the system. In the case of the Imb to be Posit, the EPr are as:

$$\vartheta_x^+ = \min(\vartheta_x^D, \vartheta_x^{DN}) \tag{3}$$

$$\vartheta_x^- = \vartheta_x^D \tag{4}$$

In (3), ϑ_x^{DN} is the EPr of offers in exceeds. The Uncer in the available of ReEn source results in differences between the OfEn and the actual SpEn. The Rev H_x for period x is as:

$$H_x = \vartheta_x^D I_x^D + P_x \tag{5}$$

In (5), I_x^D is the ETr in the D-AEM and P_x is the Imb income derived from the Bal penalty.

The total Dev for period x is as:

$$\Theta_x = I_x - I_x^D \tag{6}$$

Where I_x is the actual power for period x. P_x is set as:

$$P_x = \vartheta_x^+ \, \Theta_x, \; \Theta_x \geq 0 \tag{7}$$

$$P_x = \vartheta_x^- \, \Theta_x, \; \Theta_x < 0 \tag{8}$$

In (6), a Posit Dev happens (DevH) when the actual production is greater than the ETr in the D-AEM and a Negat DevH when the actual production lower than the ETr. In this manner, ϑ_x^+ is the EPr paid for surplus of production and ϑ_x^- the EPr to be charged for the scarcity. Suppose that ϑ_x^D is not null and let:

$$r_x^+ = \frac{\vartheta_x^+}{\vartheta_x^D}, \; r_x^+ \leq 1, \; h_x^+ = r_x^+ \tag{9}$$

$$r_x^- = \frac{\vartheta_x^-}{\vartheta_x^D}, \; r_x^- \geq 1, \; h_x^- = r_x^- \tag{10}$$

(9) and (10) are known respectively by the Posit and the Negat Imb EPr per unit of the closing EPr in the D-AEM. Substituting (9) and (10), respectively in (7) and (8), then:

$$P_x = \vartheta_x^D h_x^+ \, \Theta_x, \; \Theta_x \geq 0 \tag{11}$$

$$P_x = \vartheta_x^D h_x^- \Theta_x, \; \Theta_x < 0 \tag{12}$$

A producer that desires to rectify the energy Dev in a complementary EMar is subject to an opportunity cost, since D-AEMs have more competitive EPr.

Equation (5) can be rewritten to reveal the opportunity cost. Hence, in case of the energy Dev to be Posit, $\Theta_x > 0$, the Rev is set as:

$$H_x = \vartheta_x^D I_x^D + \vartheta_x^D h_x^+ \, \Theta_x \tag{13}$$

Using the total Dev stated in (6), the Rev is set as:

$$H_x = \vartheta_x^D I_x^D - \vartheta_x^D (1 - h_x^+) \Theta_x, \; \Theta_x \geq 0 \tag{14}$$

Also, in case of the energy Dev to be Negat, the Rev is set as:

$$H_x = \vartheta_x^D I_x^D + \vartheta_x^D (h_x^- - 1)\Theta_x, \ \Theta_x < 0 \tag{15}$$

Considering (14) and (15) can be expressed as:

$$H_x = \vartheta_x^D I_x^D - C_x \tag{16}$$

where:

$$Z_x = \vartheta_x^D (1 - h_x^+)\Theta_x, \ \Theta_x \geq 0 \tag{17}$$

$$Z_x = -\vartheta_x^D (h_x^- - 1)\Theta_x, \ \Theta_x < 0 \tag{18}$$

In (16) $\vartheta_x^D I_x^D$ is the MRev obtained from ETr when Uncer is disregarded.

With Uncer a scenarios set Φ is accounted for WEn and Imbs. The scenario φ is weighted with a probability of occurrence η. The η is the same for φ.

The Rev over a time horizon is the objective function for the problem and is given as:

$$\sum_{\varphi=1}^{N_\Phi} \sum_{x=1}^{N_x} \eta_\varphi \left(\vartheta_{x\varphi}^D I_x^D + \vartheta_{x\varphi}^D h_{x\varphi}^+ \Theta_{x\varphi}^+ - \vartheta_{x\varphi}^D h_{x\varphi}^- \Theta_{x\varphi}^- \right) \tag{19}$$

In (19) $\vartheta_{x\varphi}^D I_x^D$ is the Rev from the accepted energy at the EPr of EMar closing, the other terms are the penalty for Imb.

The OWBS is given by the maximization of (19) subjected to the constraints as:

$$0 \leq I_x^D \leq I^{\max}, \ \forall x \tag{20}$$

$$\Theta_{x\varphi} = \left(I_{x\varphi} - I_x^D \right), \ \forall x, \forall \varphi \tag{21}$$

$$\Theta_{x\varphi} = \Theta_{x\varphi}^+ - \Theta_{x\varphi}^-, \ \forall x, \forall \varphi \tag{22}$$

$$0 \leq \Theta_{x\varphi}^+ \leq I_{x\varphi} d_x, \ \forall x, \forall \varphi \tag{23}$$

In (20), the offers are limited by the technical range of operation, i.e., the installed maximum power in the WEn farm. In (21)–(23) is imposed that $\Theta_{t\varphi}^+ = 0$ when $\Theta_{t\varphi}^+$ is Negat, $I_{x\varphi} < I_x^D$, and is imposed that $\Theta_{x\varphi}^- = 0$ when $\Theta_{x\varphi}^-$ is Negat, $I_x^D < I_{x\varphi}$.

When the system balance (SBal) is Negat, the producer is penalized for the deficit of energy generated (EnGen) below the ETr in the D-AEM, so the term $\vartheta_{x\varphi}^D h_{x\varphi}^+ \Theta_{x\varphi}^+$ is null and the term $\vartheta_{x\varphi}^D h_{x\varphi}^- \Theta_{x\varphi}^-$ is subtracted from the Rev in the situation of no Dev, $\vartheta_{x\varphi}^D I_x^D$.

When the SBal is Posit, the producer is penalized for the EnGen above the ETr in the D-AEM, so the term $\vartheta_{x\varphi}^D h_{x\varphi}^- \Theta_{x\varphi}^-$ is null and the term $\vartheta_{x\varphi}^D h_{x\varphi}^+ \Theta_{x\varphi}^+$ is added to the Rev in the situation of no Dev.

Notice by the inequality in (3) that for the SBal Posit the added Rev is never greater than the Rev if no Posit Imb happens.

5   Results

The SLPr methodology is presented by a case study composed by two sets of data from the Iberian EMar [25].

The data set, i.e., price scenarios (PSc) includes 10 days of 2013-November and the other one 10 days of 2014-June shown in Fig. 1.

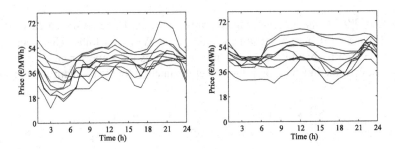

Fig. 1. PSc; (a) 2013-November, (b) 2014-June

Figure 1 shows hourly PSc at the closing of the D-AEM and each line is intended to represents a possible PSc.

The 2013 energy produced and 2014 energy produced are shown in Fig. 2.

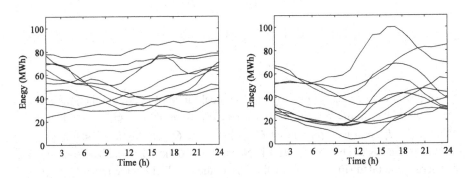

Fig. 2. WEn: energy (a) 2013-November, (b) 2014-June.

The EnGen is found through the total EnGen from the maximum power capacity of wind farm, $I^{\max} = 120\,\mathrm{MW}$.

The SOp matches the total EnGen to the system desires. This is reached by setting the price multipliers (PMul) r_t^+ and r_t^- specified by (9) and (10).

The r_t^+ and r_t^- in 2013 are shown in Fig. 3.

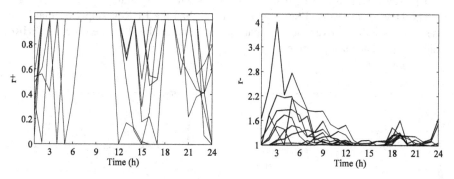

Fig. 3. Imb PMul 2013; (a) r_t^+, (b) r_t^-.

In 2014 the r_t^+ and r_t^- are shown in Fig. 4.

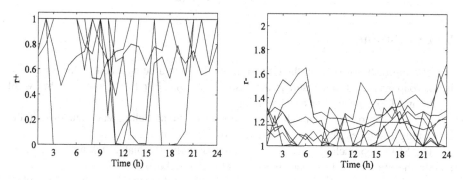

Fig. 4. Imb PMul 2014; (a) r_t^+, (b) r_t^-.

The optimal energy (OEn) offer that MRev estimated is determined by (19)–(23). The OEn offer and the estimated hourly revenue (EHRev) are shown in Figs. 5 and 6.

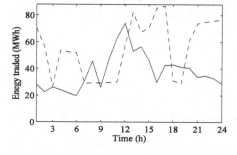

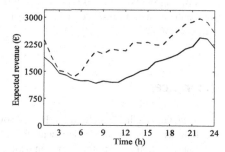

Fig. 5. ETr in D-AEM; 2013 dashed-line; 2014 solid line.

Fig. 6. EHRev; 2013 dashed-line; 2014 solid line

For the 24 h the forecasted Rev in year 2013 is 52,861 € and in year 2014 39,656 €. The Rev Dev, considering that the 24 periods will be equal in φ, is shown in Fig. 7. The Dev for the worst PSc (Fig. 7) is lower than the EHRev, implying that in the case of the condition over the day matches any one of the PSc, the outcome EHRev would be permanently positive.

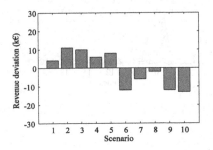 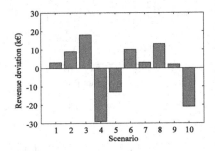

Fig. 7. Rev Dev; (a) 2013, (b) 2014.

6   Conclusion

A SLPr approach for solving the OStr of a wind PPr in a deregulated EMar is handled in this paper. The crutial result is the BStr for a wind PPr facing not only the wind power and EPr Uncers, but also the Imbs.

A SLPr approach is suitable to address OStr, in modeling via scenarios, if the computing resources needed are not excessive for use in a day basis.

The SLPr approach showed to be computationally acceptable, taking into consideration the number of PSc, and periods over the time horizon.

Additionality, the proposed approach is a useful tool for the PPr so as to be applied several times to reconfigured the bids according to short run changes of information.

Hence, the approach has adequacy to be integrated in SGrs in order to act in an enough small time processing, supporting decisions not only for helping bidding, but also for comparison with eventual bilateral contracts in due time.

Acknowledgments. The work presented in this paper is funded through Portuguese Funds by the Foundation for Science and Technology-FCT for project scope LAETA2015-2020, UID/EMS/50022/2013.

References

1. Tina, G., Gagliano, S., Raiti, S.: Hybrid solar/wind power system probabilistic modelling for long-term performance assessment. Sol. Energy **80**, 578–588 (2006)
2. Energy 2020 - a strategy for competitive, sustainable and secure energy (2015). http://europa.eu

3. Ritzenhofen, I., Birge, J.R., Spinler, S.: The structural impact of renewable portfolio standards and feed-in-tariffs on electricity markets (2015). http://papers.ssrn.com/sol3/papers.cfm?abstract_id=2418196

4. Wang, T., Gong, Y., Jiang, C.: A review on promoting share of renewable energy by green-trading mechanisms in power system. Renew. Sustain. Energy Rev. **40**, 923–929 (2014)

5. Shrestha, G.B., Kokharel, B.K., Lie, T.T., Fleten, S.-E.: Medium term power planning with bilateral contracts. IEEE Trans. Power Syst. **20**(2), 627–633 (2005)

6. Giannitrapani, A., Paoletti, S., Vicino, A., Zarrilli, D.: Bidding strategies for renewable energy generation with non stationary statistics. In: Proceedings of 19th International Federation of Automatic Control World Congress (2014)

7. Notton, G., Diaf, S., Stoyanov, L.: Hybrid photovoltaic/wind energy systems for remote locations. Energy Procedia **6**, 666–677 (2011)

8. Al Faruque, M.A.: A model-based design of cyber-physical energy systems. In: Proceedings of 19th Asia and South Pacific Design Automation Conference (ASP-DAC), pp. 97–104 (2014)

9. Blaabjerg, F., Ionel, D.M.: Renewable energy devices and systems – state-of-the-art technology, research and development, challenges and future trends. Electr. Power Compon. Syst. **43**(12), 1319–1328 (2015)

10. IEEE: Smart grid: reinventing the electric power system. In: IEEE Power and Energy Magazine for Electric Power Professionals. IEEE Power and Energy Society, USA (2011)

11. Ramos, C., Vale, Z., Faria, L.: Cyber-physical intelligence in the context of power systems. In: Kim, T.-h., Adeli, H., Slezak, D., Sandnes, F.E., Song, X., Chung, K.-i., Arnett, K. P. (eds.) FGIT 2011. LNCS, vol. 7105, pp. 19–29. Springer, Heidelberg (2011)

12. Faria, L, Silva, A., Ramos, C., Gomez, L., Vale, Z.: Intelligent behavior in a cyber-ambient training system for control center operators. In: Proceedings of 16th International Conference on Intelligent System Application to Power Systems, pp. 1–6 (2011)

13. Lee, E.A., Seshia, S.A.: Introduction to embedded systems – A cyber-physical systems approach. 2nd edn (2015). http://leeseshia.org/releases/LeeSeshia_DigitalV2_0.pdf

14. Foundations for Innovation in Cyber-Physical Systems - Workshop Summary Report. National Institute of Standards and Technology (2013)

15. Vale, Z.: Intelligent Power Systems. Wiley Encyclopedia of Computer Science and Engineering, Wiley, New York (2008)

16. Xie, L., Mo, Y., Sinopoli, B.: False data injection attacks in electricity markets. In: Proceedings of 1st IEEE Smart Grid Communications Conference (SmartGridComm), pp. 226–231 (2010)

17. García-González, J., Muela, R.M.R., Santos, L.M., González, A.M.: Stochastic joint optimization of wind generation and pumped-storage units in an electricity market. IEEE Trans. Power Syst. **23**(2), 460–468 (2008)

18. García-González, J.: Hedging strategies for wind renewable generation in electricity markets. Proc. IEEE Power Energy Soc. Gen. Meet., Pittsburgh, USA (2008)

19. Angarita, J.M., Usaola, J.G.: Combining hydro-generation and wind energy biddings and operation on electricity spot markets. Electr. Power Syst. Res. **77**(5), 393–400 (2007)

20. Bathurst, G.N., Strbac, G.: Value of combining energy storage and wind in short-term energy and balancing markets. Electr. Power Syst. Res. **67**(1), 1–8 (2003)

21. Fabbri, A., San Roman, T.G., Abbad, J.R., Mendez Quezada, V.H.: Assessment of the cost associated with wind generation prediction errors in a liberalized electricity market. IEEE Trans. Power Syst. **20**(3), 1440–1446 (2005)

22. Bathurst, G., Weatherill, J., Strbac, G.: Trading wind generation in short term energy markets. IEEE Trans. Power Syst. **17**(3), 782–789 (2002)

23. Pinson, P., Chevallier, C., Kariniotakis, G.: Trading wind generation from short-term probabilistic forecasts of wind power. IEEE Trans. Power Syst. 22(3), 1148–1156 (2007)
24. Morales, J., Conejo, A., Pérez-Ruiz, J.: Short-term trading for a wind power producer. IEEE Trans. Power Syst. 25(1), 554–564 (2010)
25. http://www.esios.ree.es/web-publica/

GA-ANN Short-Term Electricity Load Forecasting

Joaquim L. Viegas[1], Susana M. Vieira[1], Rui Melício[1,2(✉)],
Victor M.F. Mendes[2,3], and João M.C. Sousa[1]

[1] IDMEC, LAETA, Instituto Superior Técnico, Universidade de Lisboa, Lisbon, Portugal
joaquim.viegas@tecnico.ulisboa.pt, ruimelicio@gmail.com
[2] Departament Física, Escola de Ciências e Tecnologia, Universidade de Évora, Évora, Portugal
[3] Instituto Superior de Engenharia de Lisboa, Lisbon, Portugal

Abstract. This paper presents a methodology for short-term load forecasting based on genetic algorithm feature selection and artificial neural network modeling. A feedforward artificial neural network is used to model the 24-h ahead load based on past consumption, weather and stock index data. A genetic algorithm is used in order to find the best subset of variables for modeling. Three datasets of different geographical locations, encompassing areas of different dimensions with distinct load profiles are used in order to evaluate the methodology. The developed approach was found to generate models achieving a minimum mean average percentage error under 2 %. The feature selection algorithm was able to significantly reduce the number of used features and increase the accuracy of the models.

Keywords: Load forecasting · Genetic algorithm · Feature selection · Artificial neural networks

1 Introduction

Energy is of great importance for the functioning and development of all activities in any nation. If the current pattern of energy consumption continues, at a global level, the total world demand will increase by more than 50 % before 2030 [1]. Meanwhile, the high demand for energy and its production causes degradation of the environment as most energy resources are non-renewable [2]. Allied to the fact that big changes are currently happening to the in the utility industry due to deregulation and an increase in competition, policy makers and utilities are continuously seeking to identify ways to increase energy efficiency and alternate energy sources [3].

The decision making in the energy sector has to be based in accurate forecasts of energy demand, making forecasts one of the most important tools for utilities and decision makers in the energy sector. Forecasts of different time horizons and different scopes are essential for operation of plants and of the whole power system, "system response follows closely the load requirement" [4, 5].

© IFIP International Federation for Information Processing 2016
Published by Springer International Publishing Switzerland 2016. All Rights Reserved
L. Camarinha-Matos et al. (Eds.): DoCEIS 2016, IFIP AICT 470, pp. 485–493, 2016.
DOI: 10.1007/978-3-319-31165-4_45

Short-term time horizon load forecasting is usually used for the one-day ahead forecasting and has a strong influence on the operation of electricity utilities. Many decisions depend on this type of forecast, namely scheduling of fuel purchases, scheduling of power generation, planning of energy transactions and assessment of system safety [6].

The load forecast can be related in complex and non-linear ways with various variables such as the past consumption pattern, the season of the year, climatic conditions and others. Several methods to model these relationships have been applied in the past such as regression, econometric, time series, decomposition, co-integration, ARIMA, artificial intelligence, fuzzy and support vector models [2, 7]. In the work of Lin et al. [8, 9], a stock exchange index (TAIEX) was used in order to better the performance of short-term load forecasting (STLF) at times of global economic downturn.

As intuition tells us that many different variables are correlated to the load patterns intended to be predicted, feature selection (FS) becomes increasingly important in order to improve the forecasting performance, provide faster, more scalable models and provide an understanding of the influence of the variables on the future load [10]. Genetic algorithm (GA) is a search metaheuristic inspired in the process of natural evolution [11] that can be used for the selection of the best feature subset to be used for forecasting [12].

This paper proposes the use of artificial neural networks (ANN) together with GA feature selection in order to model future load using the best possible subset of available variables. Stock index variables are used in the study, together with weather and past load data, with the intent of finding a relationship between financial markets behavior and electricity demand. The rest of this paper is organized as follows: Sect. 2 presents a relationship of the paper to cyber-physical systems. Section 3 deals with the applied load forecast methodology. Section 4 presents the case studies for experimental evaluation. Finally, concluding remarks are given in Sect. 5.

2   Relationship to Cyber-Physical Systems

The smart grid is a concept with the purpose of intelligently integrating the generation, transmission and consumption of electricity through technological means [13, 14]. Cyber-physical systems (CPS) are smart systems with physical and computational components, seamlessly integrated and interacting to sense the changing state of the world [15]. Meters, transformers, switchgear, lines, production plants are incorporating a growing number of automatic control components with the need to effectively sense the changes in the environment. Predicting the future state of the power grid, with the help of the proposed forecasting method, can potentially increase the effectiveness of CPS systems and reduce the complexity of the smart grid challenge by enabling prescriptive capabilities in order to give to the systems the ability of acting on change before it happens.

3   GA-ANN Load Forecasting Methodology

3.1   Feedforward Neural Network

ANN models were initially inspired by studies on brain modeling. This type of model generates a non-linear mapping from R^I to R^K; I and K are the dimension of the input and target space [11].

   In this work, a feedforward neural network (FFNN) is used, consisting of three layers: an input layer, a hidden layer and an output layer. The input layer is composed of a number of neurons equal to the number of inputs and each of the artificial neurons has a unique input. The hidden layer neurons are fully connected to both the input and output layer and the output layer is composed by a number of neurons equal to the number of outputs. The activation function used for the input and hidden layers is the *tansig* function and, the output layer makes use a linear or function. FFNN have been widely used for load forecasting with success [2, 16, 17] due their ease of application with inputs from different sources and good performance.

   In the proposed methodology the parameters of the model, such as the number neurons of the hidden layer, are optimized using grid search.

3.2   Feature Selection

The objective of FS is to find a subset of the available features by eliminating features with seemingly little or no information useful for prediction and also redundant features that are correlated [18]. FS techniques are usually divided into filter, embedded and wrapper methods. Wrapper and embedded are usually referred to as model-based methods and filter techniques as model-free methods.

   Filter techniques assess the features by looking only at the intrinsic properties of the data [12]. Wrapper methods include the classification model within the feature subset optimization. The selected set of features results from training and testing a specific model, rendering this approach tailored to a specific modeling method.

3.3   Genetic Algorithm

This works makes use of GA for wrapper feature selection. This evolution inspired optimization algorithm starts with a random population of individuals (solutions). The

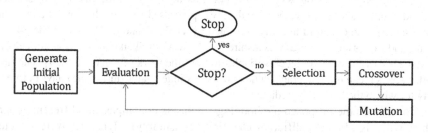

Fig. 1. GA optimization process.

main driving operators of a GA are selection (survival of the fittest) and recombination through crossover (reproduction), mutation is also used in order to escape local minima [11]. The typical flow of GA process is shown in Fig. 1.

The algorithm starts by generating a random population of individuals, binary encoded with dimension equal to the initial number of features, each chromosome representing if a feature is used in the subset represented by the individual [19]. The individuals are evaluated using the ANN modeling method and depending on the performance of the individuals, they are selected for the creation of the next generation. The selected individuals are then used to generate new solutions using the crossover operator and mutation is used in order to escape local minima. The selection, crossover and mutation operators used in this work are presented in Sect. 4.2.

4 Experimental Evaluation

4.1 Datasets

Three sets with different load patterns and available variables are used in order to better evaluate the proposed methodology. The first dataset is relative to Portugal, the second one to New York West and the third to the city of Rio de Janeiro. The data used is the following:

- **Portugal (PT)**: The historical load data is obtained from the European network of transmission system operators for electricity (ENTSOE) [20], the hourly load data is obtained between and including the days 2010-01-01 and 2013-12-31. The weather data is obtained from the Lisbon Airport Weather Station. The stock data used is from the Portuguese Stock Index 20 (PSI20) which tracks the prices of the twenty listings with the largest market capitalization and share turnover in Portugal [21].
- **New York West (NYW)**: The historical load data was obtained from NYISO [22], the hourly load data was obtained for the same period as the data from Portugal. The weather data was obtained from the Buffalo Weather Station. The stock data used in this case is from Standard & Poor's 500 (S&P 500), which is based on the capitalizations of the 500 largest companies having stock listed on the NYSE or NASDAQ [21].
- **Rio de Janeiro (RIO)**: The historical load and temperature data is the same used in the article of Hippert and Taylor [16] for the years of 1996 and 1997. No stock data was used with this dataset.

The pre-processing consisted in filling the missing values of the various time series by extending the last existing value. The data was processed for modeling and the input and output vector generated in the way presented by Tables 1 and 2. The use of the 24 prior hours and 7 days temperatures was inspired by the work of Sheikhan and Mohammadi [17] and the use of the mean of stock variables follows the work of Lin et al. [8, 9]. The stock index i days mean is the average value of the stock index value during the i days before the day for which the loads are predicted.

To validate the proposed methodology, datasets encompassing different geographies with significantly different electricity consumption dynamics were selected.

The datasets present the necessary characteristics to test the modelling approach and have not been extensively explored in the literature.

Table 1. List of inputs and outputs for the Portugal and New York West datasets

Outputs
$h + i$ hour's load ($i = 0, \ldots, 23$)
Inputs
$h - i$ hour's history load ($i = 1, \ldots, 24$)
$d - i$ day's maximum temperature ($i = 0, \ldots, 6$)
$d - i$ day's minimum temperature ($i = 0, \ldots, 6$)
Month
Season
Holiday Indicator
Weekday Indicator
Stock index i days mean ($i = 7, 15, 30, 60, 90$)
Day's precipitation

Table 2. List of inputs and outputs for the Rio de Janeiro dataset

Outputs
$h + i$ hour's load ($i = 0, \ldots, 23$)
Inputs
$h - i$ hour's history load ($i = 1, \ldots, 24$)
$d - i$ day's mean temperature ($i = 0, \ldots, 6$)
Month
Season
Holiday Indicator
Weekday Indicator

4.2 Training and Test Data

The data was divided into training and test sets in the following way: for the PT and NYW sets the first two years are used for training and the third for testing; for the RIO dataset, due to lack of data, one year is used for training and 266 days are used for testing.

4.3 Performance Evaluation

This work uses the mean absolute percentage error (MAPE) and normalized root-mean-square error (NRMSE) as performance evaluation metrics. The NRMSE is normalized using the maximum and minimum load values. The MAPE is the percentage based index of indicating forecasting accuracy level and is widely used in the field of electricity load forecasting and is calculated by:

$$MAPE\% = \frac{1}{n} \sum_{i=1}^{n} \frac{ol_i - pl_i}{ol_i} \times 100 \qquad (1)$$

where n is the total number of hours predicted by system, ol_i is the observed load and pl_i the predicted load.

The parameters used for the ANN and GA, determined through grid search, are presented in Tables 3 and 4.

The operators used by the applied genetic algorithm are the following:

- **Selection**: n three random individuals tournaments are run, where n is the size of the population;
- **Crossover**: Classic three-point crossover is used for selected individuals pairs with probability p_{co};
- **Mutation**: One-bit flip mutation is done on the selected individuals with probability p_m.

n_g represents the number of generations for which the genetic algorithm runs.

Table 3. ANN parameters

Parameter	PT	NYW	RIO
Hidden layer neurons	20		
Output layer neurons	24		
Training MSE goal	0.0035	0.006	0.00375
Relaxed FS training MSE goal	0.0045	0.007	0.00475

Table 4. GA parameters

n_{pop}	p_{co}	p_m	n_g
200	0.9	0.05	80

4.4 Results

Tables 5 and 6 present the final results for 24-h ahead load forecasting using the feature subsets selected by the genetic algorithm. 35 models were trained using the chosen feature subset for each one of the datasets. The minimum, mean and standard deviation of the MAPE is presented and compared with the performance of models trained without feature selection. The features used for the training of the models without feature selection are the ones used for the ANN parameter tuning with the economic variables removed. This was done in order to have a fairer comparison. For all the datasets the use of feature selection resulted in significantly better results, an improvement of, respectively, 10 %, 3.6 % and 11 % was obtained using the evolutionary approach for best performing models.

Table 5. Performance evaluation results – MAPE%

Dataset	Feature selection	Nr. of features	Min. MAPE%	μ (MAPE%)	σ (MAPE%)
PT	GA	16	**2.053**	2.148	0.059
	None	32	2.284	2.386	0.052
NW	GA	22	**2.549**	2.610	0.049
	None	33	2.645	2.752	0.052
RIO	GA	13	**1.935**	2.066	0.071
	None	30	2.178	2.353	0.076

Table 6. Performance evaluation results – NRMSE%

Dataset	Feature selection	Nr. of features	Min. NRMSE%	μ (NRMSE%)	σ (NRMSE%)
PT	GA	16	**3.698**	3.783	0.051
	None	32	3.795	3.932	0.064
NW	GA	22	**4.595**	4.666	0.033
	None	33	4.655	4.820	0.074
RIO	GA	13	**4.051**	4.252	0.104
	None	30	4.396	4.656	0.122

Regarding the features chosen by the selection algorithm, Table 7 presents the best feature subset selected by the algorithm for each one of the datasets. For the Portuguese dataset, 16 features are selected in total, consisting on 12 load values of the day before the one predicted, the maximum temperature 6 days before and the indicators for the season, holiday and weekday.

For the New York West dataset the number of loads used is equal to 16, the minimum temperature for the predicted day and 3 historic temperatures, prediction day's precipitation and weekday indicator are used. According to this, a stronger relationship between the load exists between the temperature and load for this dataset, in comparison to the Portuguese data. It is also interesting to note the lack of use of the season indicator. Maybe in this case the load follows closely the temperature in comparison to the Portuguese dataset, where the season is used for correction related to weather changes.

Regarding the Rio de Janeiro dataset, 13 features are selected. 10 historic loads, the prediction days and days before mean temperature and the weekday indicator are used.

It can be noted that the stock variables are not chosen for either one of the datasets. This way it is believed that these variables are not significant for the load forecasting in the Portuguese and New York West cases.

Table 7. Feature selection results

Dataset	Nr. of features	Features selected by feature selection
PT	16	$h - i$ hour's historic loads ($i = 1, 3, 4, 5, 6, 7, 10, 15, 17, 18, 21, 23$); $d - 6$ day's maximum temperature; Season; Holiday; Weekday
NW	22	$h - i$ hour's historic loads ($i = 1, 4, 5, 6, 7, 8, 9, 10, 13, 16, 17, 18, 20, 22, 23, 24$); $d - i$ day's minimum temperature ($i = 4, 3, 2, 0$); Weekday, Prediction day's precipitation
RIO	13	$h - i$ hour's historic loads ($i = 2, 3, 4, 6, 7, 9, 10, 12, 17, 22$); $d - i$ day's mean temperature ($i = 1, 0$); Weekday Indicator

5   Conclusions

According to the results presented in this report it is concluded that GA feature selection can provide a performance increase in load modeling, more specifically, short-term load forecasting using ANN. Three distinct datasets with very different load patterns were used in order to study the viability of the applied methods, the presented methodology achieved good performance for all of them.

The use of stock variables was found to not be suitable for this modeling approach. The genetic algorithm never selected these features and no improvements were found when forcing their use.

Acknowledgements. The work was supported by: PhD in Industry Scholarship SFRH/BDE/ 95414/2013, J.L. Viegas, from FCT and Novabase; FCT through IDMEC, under Project SusCity: Urban data driven models for creative and resourceful urban transitions, MITP-TB/CS/0026/2013. S.M. Vieira acknowledges support by Program Investigador FCT (IF/00833/2014) from FCT, co-funded by the European Social Fund (ESF) through the Operational Program Human Potential (POPH). Acknowledgement to FCT, through IDMEC, under LAETA, project UID/EMS/ 50022/2013.

References

1. CSIRO and The Natural Edge Project. Energy transformed: sustainable energy solutions for climate change mitigation (2007)
2. Suganthi, L., Samuel, A.A.: Energy models for demand forecasting: a review. Renew. Sustain. Energy Rev. **16**(2), 1223–1240 (2012)

3. Niu, D., Wang, Y., Wu, D.D.: Power load forecasting using support vector machine and ant colony optimization. Expert Syst. Appl. **37**(3), 2531–2539 (2010)
4. Kyriakides, E., Polycarpou, M.: Short term electric load forecasting: a tutorial. In: Chen, K., Wang, L. (eds.) Trends in Neural Computation, pp. 391–418. Springer, Berlin (2007)
5. Hahn, H., Meyer-Nieberg, S., Pickl, S.: Electric load forecasting methods: tools for decision making. Eur. J. Oper. Res. **199**(3), 902–907 (2009)
6. Ignizio, J.P.: Introduction to Expert Systems. McGraw-Hill, New York (1991)
7. Metaxiotis, K., Kagiannas, A., Askounis, D., Psarras, J.: Artificial intelligence in short term electric load forecasting: a state-of-the-art survey for the researcher. Energy Convers. Manage. **44**(9), 1525–1534 (2003)
8. Lin, C.-T., Chou, L.-D.: A novel economy reflecting short-term load forecasting approach. Energy Convers. Manage. **65**, 331–342 (2013)
9. Lin, C.-T., Chou, L.-D., Chen, Y.-M., Tseng, L.-M.: A hybrid economic indices based short-term load forecasting system. Int. J. Electr. Power Energy Syst. **54**, 293–305 (2014)
10. Guyon, I., Elisseeff, A.: An introduction to variable and feature selection. J. Mach. Learn. Res. **3**, 1157–1182 (2003)
11. Engelbrecht, A.P.: Computational Intelligence: An Introduction, 2nd edn. Wiley (2007)
12. Saeys, Y., Inza, I., Larrañaga, P.: A review of feature selection techniques in bioinformatics. Bioinformatics **23**(19), 2507–2517 (2007)
13. Welsch, M., Howells, M., Bazilian, M., DeCarolis, J.F., Hermann, S., Rogner, H.H.: Modelling elements of smart grids: enhancing the OSeMOSYS (open source energy modelling system) code. Energy **46**(1), 337–350 (2012)
14. International Energy Agency. Technology Roadmap: smart grids (2011)
15. Energetics Incorporated. Foundations for Innovation in Cyber-Physical Systems. In: Workshop Report. National Institute of Standards and Technology (2013)
16. Hippert, H.S., Taylor, J.W.: An evaluation of Bayesian techniques for controlling model complexity and selecting inputs in a neural network for short-term load forecasting. Neural Netw. **23**(3), 386–395 (2010)
17. Sheikhan, M., Mohammadi, N.: Neural-based electricity load forecasting using hybrid of GA and ACO for feature selection. Neural Comput. Appl. **21**(8), 1961–1970 (2011)
18. Vieira, S.M., Sousa, J.M.C., Runkler, T.A.: Two cooperative ant colonies for feature selection using fuzzy models. Expert Syst. Appl. **37**(4), 2714–2723 (2010)
19. Huang, C.-L., Wang, C.-J.: A GA-based feature selection and parameters optimization for support vector machines. Expert Syst. Appl. **31**(2), 231–240 (2006)
20. European network of transmission system operators for electricity (2015). https://www.entsoe.eu/
21. Google Finance (2015). http://www.google.com/finance
22. The New York Independent System Operator (2015). http://www.nyiso.com/

Optimal Bidding Strategies of Wind-Thermal Power Producers

R. Laia[1,2], H.M.I. Pousinho[1], R. Melício[1,2(✉)], and V.M.F. Mendes[2,3]

[1] IDMEC/LAETA, Instituto Superior Técnico, Universidade de Lisboa, Lisbon, Portugal
ruimelicio@gmail.com
[2] Department de Física, Escola de Ciências e Tecnologia, Universidade de Évora, Évora, Portugal
[3] Instituto Superior de Engenharia de Lisboa, Lisbon, Portugal

Abstract. This paper addresses a stochastic mixed-integer linear programming model for solving the self-scheduling problem of a thermal and wind power producer acting in an electricity market. Uncertainty on market prices and on wind power is modelled via a scenarios set. The mathematical formulation of thermal units takes into account variable and start-up costs and operational constraints like: ramp up/down limits and minimum up/down time limits. A mixed-integer linear formulation is used to obtain the offering strategies of the coordinated production of thermal and wind energy generation, aiming the profit maximization. Finally, a case study is presented and results are discussed.

Keywords: Mixed-integer linear programming · Stochastic optimization · Wind-thermal coordination · Offering strategies

1 Introduction

The adverse environmental impact of fossil fuel burning and the desire to reach energy supply sustainability promote exploitation of renewable sources. Mechanisms and policies provide subsidy and incentive for renewable energy conversion into electric energy [1], for instance, wind power conversion. But as the wind power technology matures and achieves breakeven costs, subsidy is due to be less significant and wind power conversion has to face the electricity markets for better profit [2]. Also, the incentives for wind power exploitation are feasible for low penetration levels but will become flawed as wind power integration rises [3]. EU in 2014 has of all new renewable installations a 43.7 % based on wind power and is the seventh year running that over 55 % of all additional power capacity is form renewable energy [4]. The growing worldwide usage of renewable energy is a fact, but electricity supply is still significantly dependent on fossil fuel burning, for instance, statistics for electricity supply in 2012 accounts that the usage of fossil fuel burning is more than 60 % [5].

Deregulation of electricity market imposes that a generation company (GENCO) has to face competition to obtain the economic revenue. Periodic nodal variations of electricity prices [6] have to be taken into consideration. The wind power producer (WiPP) has to address wind power and electricity price uncertainties to decide for realistic bids,

© IFIP International Federation for Information Processing 2016
Published by Springer International Publishing Switzerland 2016. All Rights Reserved
L. Camarinha-Matos et al. (Eds.): DoCEIS 2016, IFIP AICT 470, pp. 494–503, 2016.
DOI: 10.1007/978-3-319-31165-4_46

because cost is owed either in case of high bids due to the fact that other power producers must decrease or augment production to offset the deviation [7]. Thermal power producer has to address only electricity price uncertainty.

This paper focus on the coordinated trading of wind and thermal energy in order to achieve the optimal bidding strategies that provides the maximum profit. In the case study are matched the results from uncoordinated model with the results from the coordinated model.

2   Technological Innovation for Cyber-Physical Systems

Cyber-physical systems (CyPS) are systems whose operation is managed by a computing and communication core [8]. CyPS can be defined as smart systems that include computational and physical modules, effortlessly combined and strictly cooperating to sense the changing state of the real world [9].

On a first stage, cloud-based solutions can support the processing of models for helping trading in a pool-based electricity market so as to take more benefits of bids. Among these models the ones for the solution of the problems concerning with energy management and energy offers are specific vital for safeguarding a Wind-Thermal Power Producer (WTPP) business. The models for solving these problems are restricted by the computational resources, i.e., details about some reality are not considered in view of the extreme usage of computational requirements.

On a second stage, CyPS will make possible to connect the physical world, actuators and sensors, allowing the execution of the outputs of the systems decisions operating at a higher level. The strategy defined concerning with the commitment of thermal units (ThU) or the offers to the energy market can be implemented in real time with the CyPS.

3   State of the Art

Thermal energy conversion into electric energy has a significant state of art on optimization methods for solving the thermal scheduling problem (ScP), ranging from the old priorities list method to the traditional mathematical methods up to the more lately reported artificial intelligence methods [10]. The priority list method is easily implemented and requires a small processing time, but does not guarantee an appropriate solution near the global optimal one [11]. In the classical methods are considered dynamic programming (DyP) and Lagrangian relaxation-based (LR) methods [12]. DyP method is a flexible one but has a limitation known by the "curse of dimensionality". The LR can overwhelmed the aforementioned limitation, but does not necessarily lead to a feasible solution, implying further processing for satisfying the violated constraints so as find a feasible solution, which does not ensure optimal solution. The mixed integer linear programming (MILP) method is used with success for solving the thermal ScP [13]. MILP is a widely used method for ScPs due to the tractability and extensive modeling capability [14]. Although, artificial intelligence methods based on neural networks, evolutionary algorithms and simulating annealing have been used, the main drawback of the artificial intelligence methods concerning with the possibility to obtain a solution near the global optimum one is a disadvantage. So, classical methods are the main methods in use as long as the functions describing the mathematical model have conveniently smoothness.

Deregulated market and variability of the source of wind power impose uncertainties to WiPP. These uncertainties have to be conveniently considered, i.e., processed into the variables of the problems [15] to be treated by a WiPP in order to identify how much to produce and the price for bidding.

A WiPP in a deregulated market can benefit without depending on third-parties from: a coordination of wind power production with energy storage technology [16]; a financial options as a tool for WiPP to hedge against wind power Unc [17]; a stochastic model envisioned to determine optimal offer strategies for WiPP participating in a day-ahead (DaH) electricity market [18]. The stochastic model is a formulation explicitly taking into account the uncertainties tackled by the ScP of a WiPP [19], using multiple scenarios obtained by computer applications for wind power and market price forecasts [20].

4   Problem Formulation

4.1   Wind Power Producer

The uncertainties about the availability of wind power may imply differences between the energy traded with a WiPP and the actual quantity of energy supplied by the WiPP. The revenue H_x of the GENCO for period x is stated as:

$$H_x = \vartheta_x^D I_x^{offer} + R_x \tag{1}$$

In (1), ϑ_x^D is the energy price at period x, I_x^{offer} is the power at the close of the DaH electricity market accepted to be traded and R_x is the imbalance income derived from the balancing penalty of not acting in accordance with the accepted trade. The total deviation for period t is stated as:

$$\Theta_x = I_x^{act} - I_x^{offer} \tag{2}$$

Where I_x^{act} is the actual power for period x.

In (2), a positive deviation corresponds to the actual power traded higher than the traded in the DaH electricity market and a negative deviation corresponds to the power lower than the traded. Let ϑ_x^+ be the price paid for surplus of production and ϑ_t^- the price to be charged for scarcity of production. Consider the price ratios given by the equalities stated as:

$$h_x^+ = \frac{\vartheta_x^+}{\vartheta_x^D}, \; h_x^+ \leq 1 \text{ and } h_x^- = \frac{\vartheta_x^-}{\vartheta_x^D}, \; h_x^- \geq 1 \tag{3}$$

In (3), the inequalities at the right of the equalities mean, respectively, that the positive deviation never has a higher price of penalization and the negative one never has a lower price of penalization in comparison with the value of the closing price.

4.2   Thermal Power Producer

The operating cost, $A_{\phi rx}$, for a ThU is stated as:

$$A_{\phi rx} = F_r v_{\phi rx} + b_{\phi rx} + d_{\phi rx} + Z_r c_{\phi rx} \quad \forall \phi, \quad \forall r, \quad \forall x \tag{4}$$

In (4), $A_{\phi rx}$ is the operational cost for scenario ϕ of the ThU i at period x. $A_{\phi rx}$ is the sum of: the fixed production cost, F_r, a fixed associated with the unit state of operation; the added variable cost, $b_{\phi rx}$, part of this cost is associated with the quantity of fossil fuel used by the unit; and the start-up (SU) and shut-down (SD) costs, respectively, $d_{\phi rx}$ and Z_r, of the unit. The last three costs are in general described by nonlinear function and worse than that some of the functions are non- convex and non-differentiable functions, but some kind of smoothness is expected and required to use MILP, for instance, as being subdifferentiable functions.

The functions used to quantify the variable, the SU and SD costs of ThU in (4) are considered to be such that is possible to approximate those function by a piecewise linear or step functions. The variable cost, $b_{\phi rx}$, is stated as:

$$b_{\phi rx} = \sum_{q=1}^{Q} A_r^q \varsigma_{\phi rx}^q \quad \forall \phi, \quad \forall r, \quad \forall x \tag{5}$$

$$i_{\phi rx} = i_r^{\min} v_{\phi rx} + \sum_{q=1}^{Q} \varsigma_{\phi rx}^q \quad \forall \phi, \quad \forall r, \quad \forall x \tag{6}$$

$$(X_r^1 - i_r^{\min}) x_{\phi rx}^1 \leq \varsigma_{\phi rx}^1 \quad \forall \phi, \quad \forall r, \quad \forall x \tag{7}$$

$$\varsigma_{\phi rx}^1 \leq (X_r^1 - i_r^{\min}) v_{\phi rx} \quad \forall \phi, \quad \forall r, \quad \forall x \tag{8}$$

$$(X_r^q - X_r^{q-1}) x_{\phi rx}^q \leq \varsigma_{\phi rx}^q \quad \forall \phi, \quad \forall r, \quad \forall x, \quad \forall q = 2, \dots, Q - 1 \tag{9}$$

$$\varsigma_{\phi rx}^q \leq (X_r^q - X_r^{q-1}) x_{\phi rx}^{q-1} \quad \forall \phi, \quad \forall r, \quad \forall x, \quad \forall q = 2, \dots, Q - 1 \tag{10}$$

$$0 \leq \varsigma_{\phi rx}^Q \leq (i_r^{\max} - X_{\phi rx}^{Q-1}) x_{\phi rx}^{Q-1} \quad \forall \phi, \quad \forall r, \quad \forall x \tag{11}$$

In (5), the variable cost function is given by the sum of the product of the slope of each block, A_r^q, by the block power $\varsigma_{\phi rx}^q$. In (6), the power of the ThU is given by the minimum power production plus the summation of the block powers related with each block. The 0/1 variable $v_{\phi rx}$ guarantees that the power production is 0 if the ThU is in the state offline. In (7), if the 0/1 variable $x_{\phi rx}^q$ is equal to 0, then the block power $\varsigma_{\phi rx}^1$ can be lower than the block 1 maximum power; otherwise and in conjunction with (8), if the ThU is in the state on, then $\varsigma_{\phi rx}^1$ is equal to the block 1 maximum power. In (9), from the second bock to the second last one, if the 0/1 $x_{\phi rx}^q$ is 0, then the block power

$\varsigma_{\phi rx}^q$ can be lower than the block q maximum power; otherwise and in conjunction with (10), if the ThU is in the state on, then $\varsigma_{\phi rx}^q$ is equal to the block q maximum power. In (11), the block power must be between 0 and the last block maximum power.

The nonlinearities of the start-up costs, $d_{\phi rx}$, is normally considered to be described by an exponential function. This exponential function is estimated by a piecewise linear formulation as in [2] stated as:

$$d_{\phi rx} \geq K_r^\sigma \left(v_{\phi rx} - \sum_{h=1}^{\sigma} v_{\phi rx-h} \right) \quad \forall \phi, \quad \forall r, \quad \forall x \tag{12}$$

In (12), the second term models the lost of ThU, i.e., if the unit is a case of being in the state online at period x and has been in the state offline in the σ preceding periods, the term in parentheses is 1. So, in such a case a SU cost is incurred for the thermal energy that are not accountable for added value in a sense of that energy has not been converted into electric energy. The maximum number for σ is given by the number of periods need to cool down, i.e., completely lose all thermal energy. So, for every period at cooling and until total cooling one inequality like (12) is considered.

The units have to perform in accordance with technical constraints that limit the power between successive hours stated as:

$$i_r^{\min} v_{\phi rx} \leq i_{\phi rx} \leq i_{\phi rx}^{\max} \quad \forall \phi, \quad \forall r, \quad \forall x \tag{13}$$

$$i_{\phi rx}^{\max} \leq i_r^{\max} (v_{\phi rx} - c_{\phi rx+1}) + SD\, c_{\phi rx+1} \quad \forall \phi, \quad \forall r, \quad \forall x \tag{14}$$

$$i_{\phi rx}^{\max} \leq i_{\phi rx-1}^{\max} + RU v_{\phi rx-1} + SU y_{\phi rx} \quad \forall \phi, \quad \forall r, \quad \forall x \tag{15}$$

$$i_{\phi rx-1} - i_{\phi rx} \leq RD v_{\phi rx} + SD\, c_{\phi rx} \quad \forall \phi, \quad \forall r, \quad \forall x \tag{16}$$

In (13) and (14), the upper bound of $i_{\phi rx}^{\max}$ is defined as being the maximum available power of the ThU. This variable is used to consider: actual capacity of the ThU, SU/SD ramp rate and ramp-up limits. In (16), the ramps-down and SD ramp rate limits are defined. In (14), (15) and (16), the relation between the SU and SD variables of the ThU are provided, using 0/1 variables for describing the states and data parameters for ramp-down, SD and ramp-up rate limits.

The minimum down time (DT) constraint is stated as:

$$\sum_{x=1}^{J_r} v_{\phi rx} = 0 \quad \forall \phi, \quad \forall r \tag{17}$$

$$\sum_{x=k}^{k+DX_r-1} (1 - v_{\phi rx}) \geq DX_r c_{\phi rx} \quad \forall \phi, \quad \forall r, \quad \forall k = J_r + 1 \dots X - DX_r + 1 \tag{18}$$

$$\sum_{x=k}^{X} (1 - v_{\phi rx} - c_{\phi rx}) \geq 0 \quad \forall \phi, \quad \forall r, \quad \forall k = X - DX_r + 2 \dots X \tag{19}$$

$$J_r = \min\{X, (DX_r - s_{\phi r0})(1 - v_{\phi r0})\}$$

In (17), if the minimum DX is reached, then the unit will be offline at initial period. In (18), the minimum DX will be satisfied for all the sets of sequential periods of size DX_r. In (19), the minimum DX will be satisfied for the last $DX_r - 1$ periods.

The minimum up time (UX) constraint is also forced by constraints stated as:

$$\sum_{x=1}^{N_r} (1 - v_{\phi rx}) = 0 \quad \forall \phi, \quad \forall r \tag{20}$$

$$\sum_{x=k}^{k+UX_r-1} v_{\phi rx} \geq UX_r y_{\phi rx} \quad \forall \phi, \quad \forall r, \quad \forall k = N_r + 1 \dots X - UX_r + 1 \tag{21}$$

$$\sum_{x=k}^{X} (v_{\phi rx} - c_{\phi rx}) \geq 0 \quad \forall \phi, \quad \forall r, \quad \forall k = X - UX_r + 2 \dots X \tag{22}$$

$$N_r = \min\{X, (UX_r - V_{\phi r0})v_{\phi r0}\}$$

In (20), if the minimum UX is not reached, then the unit will be offline at initial period. In (21), the minimum UX will be satisfied for all the sets of sequential periods of size UX_r. In (22), the minimum UX will be satisfied for the last $UX_r - 1$ periods.

The operational status of the ThU is stated as:

$$y_{\phi rx} - c_{\phi rx} = v_{\phi rx} - v_{\phi rx-1} \quad \forall \phi, \quad \forall r, \quad \forall x \tag{23}$$

$$y_{\phi rx} + c_{\phi rx} \leq 1 \quad \forall \phi, \quad \forall r, \quad \forall x \tag{24}$$

The total power produced by the ThU is stated as:

$$i_{\phi x}^g = \sum_{r=1}^{R} i_{\phi rx} \quad \forall \phi, \quad \forall x \tag{25}$$

In (25), note by (13) to (16) and (23), (24) that if the unit is not in the state of online then the power of the unit is null.

4.3 Objective Function

The offer submitted by the GENCO, WTPP, is the summation of the power offered from the ThU and the power offered from the wind farm (WiF). The offer is stated as:

$$i_{\phi x}^{offer} = i_{\phi x}^{th} + i_{\phi x}^{D} \quad \forall \phi, \quad \forall x \tag{26}$$

The actual power produced by the GENCO is the summation of the power produced by ThU and the power produced by the WF. The actual power is stated as:

$$i_{\phi x}^{act} = i_{\phi x}^{g} + i_{\phi x}^{\phi d} \quad \forall \phi, \; \forall x \tag{27}$$

In (27), $i_{\phi x}^{g}$ is the actual power produced by ThU and $i_{\phi x}^{\phi d}$ is the actual power produced by the WF for scenario ϕ.

Consequently, the expected revenue of the GENCO is stated as:

$$\sum_{\varphi=1}^{N_{\Phi}} \sum_{x=1}^{N_X} \eta_{\phi} \left[\left(\vartheta_{\phi x}^{D} P_{\phi x}^{offer} + \vartheta_{\phi x}^{D} h_{\phi x}^{+} \Theta_{\phi x}^{+} - \vartheta_{\phi t}^{D} h_{\phi x}^{-} \Theta_{\phi x}^{-} \right) - \sum_{r=1}^{R} A_{\phi r x} \right] \quad \forall \phi, \; \forall x \tag{28}$$

Subject to:

$$0 \le i_{\phi x}^{offer} \le i_{\phi x}^{M} \quad \forall \phi, \; \forall x \tag{29}$$

$$\Theta_{x\phi} = \left(i_{\phi x}^{act} - i_{\phi x}^{offer} \right) \quad \forall \phi, \; \forall x \tag{30}$$

$$\Theta_{x\phi} = \Theta_{x\phi}^{+} - \Theta_{x\phi}^{-} \quad \forall \phi, \; \forall x \tag{31}$$

$$0 \le \Theta_{x\phi}^{+} \le I_{x\phi} d_x \quad \forall \phi, \; \forall x \tag{32}$$

In (29), $i_{\phi x}^{M}$ is maximum available power, limited by the sum of the installed capacity in the WF,i^{Emax}, with the maximum thermal production stated as:

$$i_{\phi x}^{M} = \sum_{r=1}^{R} i_{\phi r x}^{max} + i^{Emax} \quad \forall \phi, \; \forall x \tag{33}$$

Some system operators require non-decreasing offers to be submitted by the GENCO. Non-decreasing offers is considered by a constraint stated as:

$$(i_{\phi x}^{offer} - i_{\phi' x}^{offer})(\vartheta_{\phi x}^{D} - \vartheta_{\phi' x}^{D}) \ge 0 \quad \forall \; \phi, \; \phi', \quad \forall \; x \tag{34}$$

In (29), if the increment in price in two successive hours is not null, then the increment in offers in the two successive hours has two be of the same sign of the increment in price or a null value.

5   Case Study

The proposed SMILP model is applied to a case study of a GENCO with a WTPP, having 8 units with a total installed capacity of 1440 MW, the data is in [22]. Data from the Iberian electricity market for 10 days of June 2014 [21] are used for the energy prices and the energy produced from WF. This data is depicted in Fig. 1.

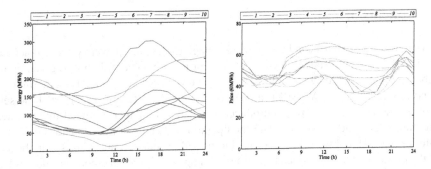

Fig. 1. Iberian electricity market June 2014 (ten days); left: prices, right: energy.

The non-decreasing offer is required. The energy produced is achieved through the total energy produced from wind scaled to the installed capacity in the WF, 360 MW. The expected results with and without coordination are depicted in Table 3.

Table 3. Results with and without coordination

Case	Expected profit (€)	Execution Time (s)
Wind uncoordinated	119 200	0.02
Thermal uncoordinated	516 848	0.13
Coordinated wind and thermal	642 326	0.13
Gain (%)	0,99	–

The non-decreasing energy offer for hours 5 and 20 is depicted in Fig. 2.

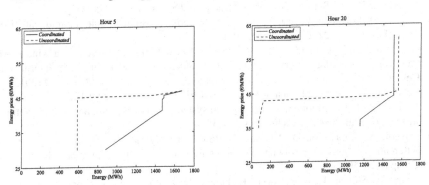

Fig. 2. Bidding energy offers.

In Fig. 2, the coordination allows for a minimum value of power offered higher than the one offered without coordination and allows for a lower price of the offering, which is a potential benefit to into operation.

6   Conclusion

Cyber-physical systems can be a great advantage for helping power systems to accommodate the changing state of the real world. Particularly, a contribution is given in this paper in what concerns the biding in the day-ahead electricity market for a thermal and wind power producer. A SMILP model for solving the offering strategy and the self-ScP of a thermal and wind power producer is settled in this paper. A mixed-integer linear program is considered to formulate the operational features of ThU. The coordinated offer of thermal and wind power proved to provide better revenue results than the sum of the isolated offers. The stochastic programming is a appropriate model to address Unc in modeling through scenarios. So, the SMILP model demonstrated to be accurate as well computationally acceptable. Since the bids are made in the DaH electricity market, this proposed SMILP model is a useful tool for the power producer.

Acknowledgments. The work presented in this paper is funded through Portuguese Funds by the Foundation for Science and Technology-FCT for project scope LAETA 2015-2020, UID/EMS/50022/2013.

References

1. Kongnam, C., Nuchprayoon, S.: Feed-in tariff scheme for promoting wind energy generation. In: IEEE Bucharest Power Tech Conference, pp. 1–6 (2009)
2. Morales, J.M., Conejo, A.J., Ruiz, J.P.: Short-term trading for a wind power producer. IEEE Trans. Power Syst. **25**(1), 554–564 (2010)
3. Bitar, E.Y., Rajagopal, R., Khargonekar, P.P., Poolla, K., Varaiya, P.: Bringing wind energy to market. IEEE Trans. Power Syst. **27**(3), 1225–1235 (2012)
4. Pineda, I., Wilkes, J.: Wind in power: 2014 European statistics (2015)
5. Key World Energy Statistics Report. International Energy Agency (2014)
6. Wu, A., Shahidehpour, L., Li, T.: Stochastic security-constrained unit commitment. IEEE Trans. Power Syst. **22**(2), 800–811 (2007)
7. Pousinho, H.M.I., Catalão, J.P.S., Mendes, V.M.F.: Offering strategies for a wind power producer considering uncertainty through a stochastic model. In: 12th International Conference on Probabilistic Methods Applied to Power Systems, pp. 1139–1144 (2012)
8. Rajkumar, R.: A cyber-physical future. Proc. IEEE **100**, 1309–1312 (2012)
9. Foundations for innovation in cyber-physical systems. Workshop summary report. National Institute of Standards and Technology (2013)
10. Trivedi, A., Srinivasan, D., Biswas, S., Reindl, T.: Hybridizing genetic algorithm with differential evolution for solving the unit commitment scheduling problem. Swarm Evol. Comput. **23**, 50–64 (2015)
11. Senjyu, T., Shimabukuro, K., Uezato, K., Funabashi, T.: A fast technique for unit commitment problem by extended priority list. IEEE Trans. Sustain. Energy **18**, 277–287 (2003)
12. Laia, R., Pousinho, H.M.I., Melício, R., Mendes, V.M.F., Reis, A.H.: Schedule of thermal units with emissions in a spot electricity market. In: Camarinha-Matos, L.M., Tomic, S., Graça, P. (eds.) DoCEIS 2013. IFIP AICT, vol. 394, pp. 361–370. Springer, Heidelberg (2013)
13. Ostrowski, J., Anjos, M.F., Vannelli, A.: Tight mixed integer linear programming formulations for the unit commitment problem. IEEE Trans. Power Syst. **27**(1), 39–46 (2012)

14. Floudas, C., Lin, X.: Mixed integer linear programming in process scheduling: modeling, algorithms, and applications. Ann. Oper. Res. **139**, 131–162 (2005)

15. El-Fouly, T.H.M., Zeineldin, H.H., El-Saadany, E.F., Salama, M.M.A.: Impact of wind generation control strategies, penetration level and installation location on electricity market prices. IET Renew. Power Gener. **2**, 162–169 (2008)

16. Angarita, J.L., Usaola, J., Crespo, J.M.: Combined hydro-wind generation bids in a pool-based electricity market. Electr. Power Syst. Res. **79**, 1038–1046 (2008)

17. Hedman, K., Sheble, G.: Comparing hedging methods for wind power: using pumped storage hydro units vs options purchasing. In: International Conference on Probabilistic Methods Applied to Power Systems, pp. 1–6 (2006)

18. Matevosyan, J., Soder, L.: Minimization of imbalance cost trading wind power on the short-term power market. IEEE Trans. Power Syst. **21**(3), 1396–1404 (2006)

19. Ruiz, P.A., Philbrick, C.R., Sauer, P.W.: Wind power day-ahead uncertainty management through stochastic unit commitment policies. In: IEEE/PES Power Systems Conference and Exposition, pp. 1–9 (2009)

20. Coelho, L.S., Santos, A.A.P.: A RBF neural network model with GARCH errors: application to electricity price forecasting. Electr. Power Syst. Res. **81**, 74–83 (2011)

21. Red Eléctrica de España. (October 2015) http://www.esios.ree.es/web-publica/

22. Laia, R., Pousinho, H.M.I., Melício, R., Mendes, V.M.F.: Self-scheduling and bidding strategies of thermal units with stochastic emission constraints. Energy Convers. Manage. **89**, 975–984 (2015)

Bio-energy

Wastewaters Reuse for Energy Crops Cultivation

Jorge Costa[1], Bruno Barbosa[1,2], and Ana Luísa Fernando[1(✉)]

[1] MEtRiCS, Departamento de Ciências e Tecnologia da Biomassa,
Faculdade de Ciências e Tecnologia, FCT, Universidade Nova de Lisboa, Campus de Caparica,
2829-516 Caparica, Portugal
jrgecosta@gmail.com, barbosabruno5@gmail.com, ala@fct.unl.pt
[2] Universidade Federal do Oeste da Bahia, Barreiras, Brazil

Abstract. This study evaluated wastewaters reuse in the production of perennial crops *Arundo donax* and *Miscanthus* x *giganteus*. The trials were conducted in pots under controlled conditions, with different water regimes (950, 475 and 238 mm) in two growing cycles. The results indicated that irrigation with wastewaters did not affect biomass productivity but the amount of irrigation did. Yet, biomass obtained from pots irrigated with wastewaters presented higher levels of ash and nitrogen content than biomass from control pots. The soil-plant system retained over 90 % of pollutant load resulting in wastewater depuration. Furthermore, the produced biomass can be economically valorized for energy or biomaterials, once irrigation with wastewater did not influence the contents in fiber and the calorific value. Still, the higher ash and nitrogen contents in the biomass can be detrimental especially when biomass is for combustion purposes.

Keywords: *Arundo donax* · *Miscanthus* x *giganteus* · Energy crops · Bioenergy · Phytoremediation · Wastewaters · Sustainability

1 Introduction

Most of the countries are too dependent on fossil fuels to meet their energy needs. The production of energy crops has been presented as a very promising alternative to partially replace the fossil fuels. Among the various species (preferably non-food) which can be grown to generate energy, *Arundo donax* L. (giant reed) and *Miscanthus* x *giganteus* Greef et Deu are presented as one of the most promising because of its high productivity, resistance to low water regimes, characteristic of the Mediterranean countries during summer, and pests [1, 2]. Giant reed is a woody rhizomatous grass that was spread throughout Asia, Southern Europe, North Africa and the Middle East as an agent to control erosion in drainage channels [2, 3]. *Miscanthus* is a woody rhizomatous grass originated in South-East Asia and was initially imported to Europe as an ornamental plant. It is a perennial plant, related to sugarcane, with an estimated productive lifetime of at least 10–15 years, and both the stems and leaves of the crop can be harvested annually [2, 4]. Both crops have the ability to combat desertification and prevent soil erosion, thanks to its high efficiencies in the use of resources and also to its robust root

© IFIP International Federation for Information Processing 2016
Published by Springer International Publishing Switzerland 2016. All Rights Reserved
L. Camarinha-Matos et al. (Eds.): DoCEIS 2016, IFIP AICT 470, pp. 507–514, 2016.
DOI: 10.1007/978-3-319-31165-4_47

system, high soil cover density and the fact that the stems keep alive during the winter [3, 5, 6]. Beyond the potential for use as energy crops, perennial grasses are associated with minimization of nutrient leaching as well as with the restoration of soil properties (fertility, structure, organic matter) due to the extensive radicular system [5–8]. Therefore it has been argued that these grasses can also be used to remove contaminants, such as nitrates or heavy metals, from soil and from wastewaters [7–12].

The irrigation of plants with treated wastewaters for contaminant removal is not new. This approach was developed as a final treatment stage in order to minimize the wastewater nutrient loads prior to its disposal in the environment. Because agriculture is a major consumer of water worldwide the same technique is also used to irrigate drought tolerant crops that are very efficient in the removal of nutrients and pollutants present in treated wastewaters. This can promote the creation of crop rotation systems more resilient, adding economic value and social benefits to water-scarce regions like the Mediterranean [7]. The use of wastewaters for energy crops irrigation, may allow not only counteracting the shortage or the precipitation seasonality but also to reduce the need of fertilizers, combining environmental and economic advantages. However, the presence of substances in wastewater, such as nitrates and heavy metals, may present environmental risks, besides representing a source of nutrients/toxics to the biomass. On this basis, this study was design to provide the answers to the following questions:

- Does irrigation with wastewaters affect the yields of *A. donax* and *Miscanthus,* in the Mediterranean region?
- Does irrigation with wastewaters affect the quality of the biomass as feedstock in energy and bioproducts uses?
- Does the amount of irrigation water added affect the yields and quality of biomass?
- Are these perennial grasses able to remediate the pollution load of wastewaters?

Some studies, but not many, have already evaluated the effect of giant reed and *Miscanthus* spp., irrigated with wastewaters [7]. However, most of them cover only the phytoremediation ability of the plants to the pollution load or the effects on yields, when no water stress is inflicted. Therefore, this study intends to provide more information on the adaptation of giant reed and *Miscanthus* spp. to the hydraulic loading. This will provide insights on the effects of the wastewater rate application on the yields and quality of biomass.

2   Relationship to Cyber-Physical Systems

The agricultural systems that generate food, feed, fiber, and fuels need to be more efficient and sustainable (economically, environmentally and socially). In order to accelerate this process, precision agriculture architecture based on cyber-physical systems (CPS) design technology is mandatory. CPS will help to increase efficiency throughout the value chain, improving our environmental footprint, and creating opportunities in the rural areas. For this, the scientific and technical challenges relay on the development of methods, tools, hardware and software components that can address this issue. Those need to be based upon transdisciplinary approaches, including the validation of the principles via pilot tests and field tests.

A precision agriculture architecture, that can provide autonomous irrigation management capabilities, based on CPS design technology, includes several layers: the physical layer, the network layer, the decision layer and the application layer.

The physical layer corresponds to the information acquisition process. Use of wireless sensors in field can monitor the moisture content of the soil in a continuous base, relaying the processed information to a central node. In the proposed architecture the information obtained from the wireless sensors will be transmitted by a tool that collects data from the field and provide the range of information required to the network nodes for the decision layer. In this decision layer, decisions will be obtained using artificial intelligence techniques. This system will create, store, analyse and process spatial information distributed through a computerized process regarding soil type, moisture content and correlate them with a certain plot of field. The application layer will provide solutions to incoming problems based on information processed and stored locally but also from knowledge bases. When the information processed indicates that the soil moisture content is below a defined value, the storage tanks with wastewaters can start pumping the water to the fields.

Another device installed in the storage tanks also monitors the existing volume of wastewaters and predicts the wastewaters incoming per hour based on the wastewater treatment plant design and process. By means of an intelligence mechanism (a developed software), the wastewater in the fields will be distributed according to both indicators, soil moisture and wastewater availability. The control mechanism will stop irrigation when a) the soil moisture returns to the optimal level for perennial grasses production, and/or b) the volume of wastewater is limited. Information on the growing cycle of the plants (needs of water per day, along the vegetation period) has to be uploaded in the system, in order to adjust the water distribution with minimal stresses for the plant. The main purpose of such an integrated system is to provide a solution for multispectral monitoring of perennial grasses irrigation with wastewaters based on mechatronic systems, in order to improve precision agricultural management. This work intends to provide new data that can be treated and uploaded in such a system, helping to design an optimized response.

3   Materials and Methods

The trials were established in April 2012 in pots. Both species, *Arundo donax* and *M. x giganteus* were tested. In each pot (0.06154 m^2, 12 kg of soil) 2 rhyzomes were established per pot and replicates were also established. After the establishment of the rhyzomes, pots were fertilized: 3 g N/m^2 (urea, 46 % N); 3 g N/m^2 (nitrolusal, mixture of NH$_4$NO$_3$ + CaCO$_3$, 27 % N); 17 g K$_2$O/m^2 (potassium sulphate, 51 % K$_2$O); 23 g P$_2$O$_5$/m^2 (superphosphate, 18 % P$_2$O$_5$). Three different wastewater irrigating regimes were applied in the pots: 950 mm, 475 mm and 238 mm. A piggery effluent was tested. In all experiments, control pots were also tested with the same water regimes but wastewater was replaced by tap water.

At the end of each growing season (January 2013 and January 2014), the plants were harvested and the productivity and biomass quality were monitored. To determinate the productivity of the biomass, the total aerial dry weight was determined at each harvest. The radicular productivity was determined also, but only on the 2nd harvest date (January 2014). The quality of the biomass harvested was analysed taking in consideration the following parameters: ash and nitrogen content, as also, fiber content and the calorific value. The analyses were performed according to the following procedures: (a) ash content: by calcination at 550 °C for two hours, in a muffler furnace; (b) nitrogen content: by the Kjeldahl method, after digestion of the sample; (c) Hemicellulose (H), cellulose (C) and lignin (L) were determined by the van Soest method [13]; (d) calorific value, by an adiabatic calorimeter. Tap water and wastewater were analyzed according to the Standard Methods for the examination of water and wastewater [14].

The statistical interpretation of the results was performed using analysis of variance (one-way ANOVA) (Statistica 6.0 program). LSD Fisher's test was applied to separate means when ANOVA revealed significant differences. The results were presented as the mean ± standard deviation.

4   Results and Discussion

Table 1 shows the physical-chemical characterization of the tap water (used in the control pots) and wastewater.

Table 1. Physicochemical characterization of tap water and piggery wastewater used in the trials

Parameter	Expression of results	Wastewater	Tap water
pH	Sorensen scale	6.9 ± 0.1	6.1 ± 0.2
Electrical conductivity	$mS\ cm^{-1}$	0.53 ± 0.04	0.42 ± 0.02
Oxidability	$mgO_2\ dm^{-3}$	12 ± 5	0.23 ± 0.08
Ammonia	$mg\ N\ dm^{-3}$	[3.3–27.7]	<0.14
Chlorides	$mg\ Cl^-\ dm^{-3}$	99 ± 6	81 ± 2
Nitrates	$mg\ N\ dm^{-3}$	5.4 ± 3.5	5.5 ± 0.08
Phosphates	$mg\ P\ dm^{-3}$	0.62 ± 0.25	0.042 ± 0.002
BOD$_5$	$mg\ O_2\ dm^{-3}$	6 ± 2	-

Table 1 show that the wastewater has higher organic matter content than tap water, as also, ammonia and chlorides. But, regarding nitrates and mineral composition, both waters presented similar results.

Figure 1 shows the yields after two growing cycles and allows highlight that the higher the water regime applied, the greater the yield. Similar findings were already presented and discussed in the literature [1, 15]. This indicates that these grasses prefer high water availability. Evaluating the plant as a whole, it is found that generally irrigation with wastewaters did not produce (in the tested water regimes) significant changes in biomass yield. Similar results were verified in literature by other authors [9], indicating that those grasses are tolerant to these type of wastewaters with this degree of salinity. Results obtained confirmed the results verified in the first growing season, but where the yields were lower [10, 12]. These plants give priority, in the first years, to the establishment of the root-rhizome system, compared to the other vegetative structures, a feature that had already been indicated by Angelini et al. [16]. *M.* x *giganteus* was significantly more productive than *A. donax*, a feature already observed by other authors [17]. Results also show that *M.* x *giganteus* accumulates 130 g m^{-2} biomass per additional 100 mm of irrigation water, and *A. donax* accumulates 80 g m^{-2} biomass per additional 100 mm of irrigation water, in the range studied [238–950 mm].

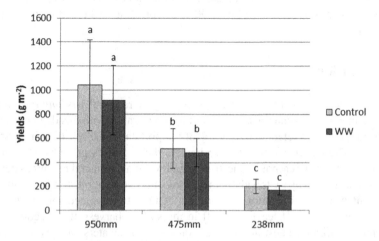

Fig. 1. Perennial grasses yields at different irrigation regimes at the end of the second growing cycle. Different lower-case letters indicate statistical significance ($p < 0.05$) between irrigation regimes. WW-wastewater.

Regarding the quality of the biomass for energy purposes, stems (the fraction of interest for combustion purposes and others) were the part of the plant that presented the lowest ash content, with average ash levels of 4.6 %, in the case of *A. donax* and 2.5 %, in the case of *M.* x *giganteus*. Due to the lower ash content, *M.* x *giganteus* present a better quality for energy purposes than *A. donax*. Both species presented a trend for a higher ash content when irrigated with wastewater. As wastewater is richer in salts (showed by the higher electrical conductivity, Table 1), those accumulate in higher amounts in the biomass, during the growing season. A trend for increasing ash content was observed, in the plants, by lowering the water regime. Nassi o Di Nasso et al. [18] highlighted the inverse variation between productivity and ash content, a fact that explains the observed trend.

Stems were the fraction of the plant with lower nitrogen content either. *A. donax* showed average nitrogen content of 0.5 % (dry basis) and *M.* x *giganteus* average nitrogen content of 0.11 % (dry basis). The lower nitrogen content presented by *M.* x *giganteus* stems corroborates the fact that this grass presents a better quality for energy purposes than *A. donax*. Both species showed a tendency to higher nitrogen accumulation in the biomass with wastewater irrigation. This increment reflects the wastewater richness in nitrogen (especially ammonia, Table 1). For both species, lower nitrogen content in the biomass was obtained with the application of higher water regimes, suggesting the dilution effect observed also by Nassi o Di Nasso et al. [18], with the increasing yields.

Furthermore, the produced biomass can be economically valorized for energy or biomaterials production, once irrigation with wastewater did not influence the contents in hemicellulose, cellulose and lignin, and the calorific value. *A. donax* biomass showed average fiber content of 87 % (H = 32 %, C = 31 % and L = 24 %) and a calorific value of 17.2 MJ/kg. *M.* x *giganteus* biomass showed average fiber content of 91 % (from this, H = 21 %, C = 38 % and L = 23 %) and a calorific value of 17.4 MJ/kg.

The soil-plant system retained over 90 % of pollutant load resulting in wastewater depuration.

5 Conclusions

Results of yields obtained for both energy crops (giant reed and *Miscanthus*) reflect that these perennial grasses can tolerate this sort of wastewaters with this degree of salinity. The soil-plant system retained over 90 % of the pollutant load resulting in wastewater depuration. The use of wastewaters to irrigate these grasses presents environmental advantages in terms of carbon sequestration and water and mineral resources depletion. Carbon sequestration by roots and rhizomes may improve also soil structure, soil organic matter content and soil aeration, factors that contribute to reduce soil erosion and to control desertification. If leaves are left in the soil, at harvest, this fraction can also contribute to carbon sequestration. Irrigation with wastewaters enables the recycle of nutrients (N, P and K) and minimizes the need for fertilizer application in soils. Consequently, the NPK production or extraction from mineral ore reserves is reduced, with economic and energetic revenues, along with mineral and fossil resources savings. Wastewater reuse represents also an approach to economize freshwater and may contribute to aquifer refilling. Furthermore, the produced biomass can be economically valorized for energy or biomaterials production, once irrigation with wastewater did not influence the contents in fiber and the calorific value. Still, wastewater irrigation induces in the biomass a higher accumulation of ash and nitrogen, which can be detrimental when processing biomass, especially for combustion.

Acknowledgments. This work was supported by the European Union (Project Optimization of perennial grasses for biomass production (OPTIMA), Grant Agreement No: 289642, Collaborative project, FP7-KBBE-2011.3.1-02.

References

1. Cosentino, S.L., Scordia, D., Sanzone, E., Testa, G., Copani, V.: Response of giant reed (*Arundo donax* L.) to nitrogen fertilization and soil water availability in semi-arid Mediterranean environment. Eur. J. Agron. **60**, 22–32 (2014)
2. El Bassam, N.: Handbook of Bioenergy Crops. A Complete Reference to Species, Development and Applications. Earthscan Ltd., London (2010)
3. Fernando, A.L., Barbosa, B., Costa, J., Papazoglou, E.G.: Giant reed (*Arundo donax* L.): a multipurpose crop bridging phytoremediation with sustainable bio-economy. In: Prasad, M.N.V. (ed.) Bioremediation and Bioeconomy, pp. 77–95. Elsevier Inc., UK (2016)
4. Fernando, A.L., Oliveira, J.F.S.: Caracterização do potencial da planta *Miscanthus* x *giganteus* em Portugal para fins energéticos e industriais. In: Robalo, M. (ed.) Biologia Vegetal e Agro-Industrial, vol. 2, pp. 195–204. Edições Sílabo, Lisboa (2005)
5. Oliveira, J.S, Duarte, M.P., Christian, D.G., Eppel-Hotz, A., Fernando, A.L.: Environmental aspects of Miscanthus production. In: Jones, M.B., Walsh, M. (eds.) *Miscanthus* for Energy and Fibre, pp. 172–178. James & James (Science Publishers) Ltd., London (2001)
6. Fernando, A.L., Duarte, M.P., Almeida, J., Boléo, S., Mendes, B.: Environmental impact assessment of energy crops cultivation in Europe. Biofuels Bioprod. Biorefin. **4**, 594–604 (2010)
7. Barbosa, B., Costa, J., Fernando, A.L., Papazoglou, E.G.: Wastewater reuse for fiber crops cultivation as a strategy to mitigate desertification. Ind. Crops Prod. **68**, 17–23 (2015)
8. Barbosa, B., Costa, J., Boléo, S., Duarte, M.P., Fernando, A.L.: Phytoremediation of inorganic compounds. In: Ribeiro, A.B., Mateus, E.P., Couto, N. (eds.) Electrokinetics Across Disciplines and Continents - New Strategies for Sustainable Development, pp. 373–400. Springer, Switzerland (2016)
9. Mavrogianopoulos, G., Vogli, V., Kyritsis, S.: Use of wastewaters as a nutrient solution in a closed gravel hydroponic cultures of giant reed (*Arundo donax*). Bioresour. Technol. **82**, 103–107 (2002)
10. Costa, J., Fernando, A.L., Coutinho, M., Barbosa, B., Sidella, S., Boléo, S., Bandarra, V., Duarte, M.P., Mendes, B.: Growth, productivity and biomass quality of *Arundo* irrigated with Zn and Cu contaminated wastewaters. In: Eldrup, A., Baxter, D., Grassi, A., Helm, P. (eds.) Proceedings of the 21st European Biomass Conference and Exhibition, Setting the Course for a Biobased Economy, pp. 308–310. ETA-Renewable Energies and WIP-Renewable Energies (2013)
11. Lino, J., Fernando, A.L., Barbosa, B., Boléo, S., Costa, J., Duarte, M.P., Mendes, B.: Phytoremediation of Cd and Ni contaminated wastewaters by Miscanthus. In: Hoffmann, C., Baxter, D., Maniatis, K., Grassi, A., Helm, P. (eds.) Proceedings of the 22nd European Biomass Conference and Exhibition, Setting the Course for a Biobased Economy, pp. 303–307. ETA-Renewable Energies (2014)
12. Bandarra, V., Fernando, A.L., Boléo, S., Barbosa, B., Costa, J., Sidella, S., Duarte, M.P., Mendes, B.: Growth, productivity and biomass quality of three miscanthus genotypes irrigated with Zn and Cu contaminated wastewaters. In: Eldrup, A., Baxter, D., Grassi, A., Helm, P. (eds.) Proceedings of the 21st European Biomass Conference and Exhibition, Setting the Course for a Biobased Economy, pp. 147–150. ETA-Renewable Energies and WIP-Renewable Energies (2013)
13. Van Soest, P.J., Robertson, J.B., Lewis, B.A.: Methods for dietary fiber, animal nutrition. J. Dairy Sci. **74**, 3583–3597 (1991)

14. APHA, AWWA e WPCF: Standard Methods for the examination of water and wastewater. 16th Ed. American Public Health Association, American Water Works Association e Water Pollution Control Federation, Washington D.C., USA (1985)

15. Cosentino, S.L., Patanè, C., Sanzone, E., Copani, V., Foti, S.: Effects of soil water content and nitrogen supply on the productivity of Miscanthus x giganteus Greef et Deu. in a Mediterranean environment. Ind. Crops Prod. **25**, 75–88 (2007)

16. Angelini, L.G., Ceccarini, L., Di Nasso, N.N., Bonari, E.: Comparison of *Arundo donax* L. and *Miscanthus* x *giganteus* in a long-term field experiment in Central Italy: analysis of productive characteristics and energy balance. Biomass Bioenergy **33**, 635–643 (2009)

17. Barbosa, B., Boléo, S., Sidella, S., Costa, J., Duarte, M.P., Mendes, B., Cosentino, S.L., Fernando, A.L.: Phytoremediation of heavy metal-contaminated soils using the perennial energy crops *Miscanthus* spp. and *Arundo donax* L. BioEnergy Res. **8**, 1500–1511 (2015)

18. Nassi o Di Nasso, N., Angelini, L.G., Bonan, E.: Influence of fertilisation and harvest time on fuel quality of giant reed (*Arundo donax* L.) in Central Italy. Eur. J. Agron. **32**, 219–227 (2010)

Removal of Chromium and Aluminum from Aqueous Solutions Using Refuse Derived Char

Catarina Nobre[1(✉)], Margarida Gonçalves[1], Dieimes Resende[2],
Cândida Vilarinho[3], and Benilde Mendes[1]

[1] MEtRICs, Departamento de Ciências e Tecnologia da Biomassa,
Faculdade de Ciências e Tecnologia, Universidade Nova de Lisboa, 2829-516 Caparica, Portugal
cp.nobre@campus.fct.unl.pt, {mmpg,bm}@fct.unl.pt
[2] Departamento de Ciências Florestais, Universidade Federal de Lavras,
Av. Doutor Sylvio Menicucci, 1001-Kennedy, Lavras-MG, 37200-000, Brazil
dieimes.rr@posgrad.ufla.br
[3] MEtRICs, Departamento de Engenharia Mecânica, Escola de Engenharia,
Universidade do Minho, 4804-533 Guimarães, Portugal
candida@dem.uminho.pt

Abstract. Refuse derived fuel (RDF) was subject to torrefaction in order to produce a char with higher homogeneity and lower moisture content than the RDF raw materials. The resulting product, RDF char, showed increased fixed carbon and ash contents, decreased moisture and volatile matter contents, and a very significant increase in density. The torrefaction of RDF may therefore contribute to reduce the landfill volume needed to accommodate these materials to one third of the presently used. This new char material was also tested for its adsorption capacities and the results show that it could be used for the removal of chromium and aluminum from aqueous solutions.

Keywords: RDF · Char · Adsorption · Chromium · Aluminum

1 Introduction

Refuse derived fuel (RDF) is an heterogeneous material comprehending polymeric and lignocellulosic waste that cannot be recycled and must be disposed of in landfills or used for energy purposes by co-combustion in cement kilns or power plants [1]. Landfill deposition is a source of environmental problems and it is not advised by current legislation that foresees a reduction of 59 % of the presently eliminated wastes to be achieved by 2020 [2].

On the other hand, the energetic valorization of waste derived fuels faces several constrains namely their high degree of heterogeneity, low density and high moisture, chlorine and ash contents. Moreover, the fuel applications of RDFs are often limited by their high content of polymeric wastes that causes agglomeration problems in the feeding systems of boilers or gasifiers and increases solid and gaseous emissions associated with incomplete combustion or gasification.

© IFIP International Federation for Information Processing 2016
Published by Springer International Publishing Switzerland 2016. All Rights Reserved
L. Camarinha-Matos et al. (Eds.): DoCEIS 2016, IFIP AICT 470, pp. 515–522, 2016.
DOI: 10.1007/978-3-319-31165-4_48

The research question addressed in this PhD program is the development sustainable solutions for the upgrading and valorization of RDFs either as better quality fuels or as refuse derived materials with industrial applications.

These technological solutions aim to modify the RDFs in order to increase density, homogeneity, hydrophobicity and calorific value by reduction of moisture, volatile matter, microbiological contamination and by rearrangement of the chemical bonds from the non-volatile fraction.

The techniques selected to achieve this goal are thermochemical processes, namely torrefaction, pyrolysis and gasification. These processes apply high temperatures, in the presence of oxidizing agents or inert gases, leading to an extensive reorganization of the molecular structure and contributing to a composition upgrading by eliminating oxygen, chlorine and other heteroatoms.

Torrefaction occurs at 200–300 °C under an inert atmosphere. It is the first step to obtain a stabilized RDF and has proven to be a useful tool to upgrade solid fuels such as raw biomass [3]. The torrefied RDF can then be directed to various final applications such as land reclamation, production of solid, liquid or gaseous fuels, incorporation into building materials or production of refuse derived materials such as refuse derived chars. The program of activities of this PhD program comprises the following tasks:

(a) Evaluate the influence of torrefaction conditions and initial RDF composition in the final physico-chemical properties of torrefied RDFs;
(b) Determine gaseous emissions associated with RDF torrefaction;
(c) Evaluate the combustion behavior of torrefied RDFs in what concerns efficiency and associated emissions;
(d) Study the conversion of torrefied RDFs in liquid fuels using pyrolysis or hydro-thermal liquefaction;
(e) Study the conversion of torrefied RDFs in gaseous fuels using thermal gasification;
(f) Evaluate the adsorbent characteristics of torrefied RDFs in different industrial applications (dyes, heavy metals and pesticides).

Modern industries use a significant amount of resources in the remediation of contaminated effluents that contain relevant concentrations of organic and inorganic hazardous compounds which are considered environmental priority pollutants [4].

Activated carbons are the most used adsorbents but they have high prices and are difficult to regenerate so they represent a constant investment as raw materials and a constant source of contaminated solid wastes.

In this context there is a growing interest in the production of low-cost adsorbents, derived from less expensive or residual materials such as agricultural wastes, industrial by-products, waste materials or natural substances [5].

In this work, a refused derived char (RDF char) was produced by torrefaction of RDF at 300 °C during 30 min. The RDF char was characterized for its proximate composition, heating value and density and its adsorption capacity towards chromium and aluminum was tested using different metal concentrations, adsorption times, initial pH and adsorbent-to-metal mass ratios.

2 Relationship to Cyber-Physical Systems

The torrefaction of RDF is a thermochemical process that requires the heating of the raw RDF at temperatures in the range of 250 °C to 350 °C during variable periods of time. The efficiency of the process depends strongly on the characteristics of the raw RDF namely its moisture content that influences the real temperature in the torrefaction unit. For the same amount of energy supplied, the maximum temperature measured in a given period of time is inversely proportional to the moisture content of the raw material. Thus, torrefaction is effective when the moisture and a substantial fraction of the volatile components are eliminated, therefore the automated monitoring of the torrefaction process requires the use of temperature sensors that enable the establishment of real temperature profiles during the process.

On the other hand, the thermal capacity of the raw RDF also influences the kinetics and extension of the torrefaction process, so the flow of gaseous products is also a key parameter to control this process.

The combination of those two variables (temperature and flow of the gaseous emissions) reflects the completion of the torrefaction process and may be used as a boundary condition to regulate the feed, for different RDF compositions.

The remediation of industrial effluents using adsorption processes is a technology that requires the implementation of automatic control and monitoring systems that comprehend sensors for the continuous evaluation of effluent composition and concentrations of critical components (for instance, electrochemical sensors), in association with intelligent systems for the control of mechanical operations such as pressure and flow control as well as diversion of the effluent flow to alternative adsorption columns or stopping the process when the adsorbent becomes saturated.

3 Methods and Materials

3.1 RDF Char Production and Characterization

Industrial refuse derived fuel was supplied by CITRI, S.A. This raw material was mainly composed of plastics, cigarette buds, rubber, fabric and lignocellulosic material. The torrefaction of RDF was performed in closed lid crucibles placed in a muffle furnace (Nabertherm) for 30 min at 300 °C.

Both the RDF and the RDF char were grinded and the char was sieved to a particle diameter of less than 500 μm and were characterized for proximate analysis (moisture, volatile matter, ash and fixed carbon contents) and density. All proximate analysis parameters were determined according to ASTM standards for RDF. The high heating values (HHV) were evaluated using an equation that relates this property to the proximate analysis of the materials [6].

3.2 Adsorption Experiments

The adsorption of Cr and Al was studied using solutions of $AlCl_3$ and $K_2Cr_2O_7$ prepared with ultra-pure water at a concentration of 100 mg/L. These solutions (100 mL) contacted with the RDF char (0.2 g), in stopped Schott glass flasks, at room temperature, for 2 h, at a pH of 8.0, under agitation (15 rpm) in an overhead shaker (Heidolph). The RDF char and metal solutions were separated by filtration and the concentrations of Cr and Al in the filtrate were determined by atomic absorption spectroscopy (Thermo Elemental Solaar). All the measurements were performed in triplicate and the reported values correspond to the average values.

The effect of pH on the adsorption of chromium and aluminum was studied in a range from 2.0 to 10.0 and the initial pH value was adjusted to the target values by using NaOH 1 M or HCl 1 M (Panreac). The influence of the contact time on the metal adsorption by the RDF char was evaluated for contact times from 0 to 600 min. The effect of initial metal concentration was determined varying the initial concentrations from 30 to 150 mg/L. The effect of adsorbent mass was tested in a range of 0.1 to 0.8 g, for metal solutions with an initial concentration of 100 mg/L. The adsorption capacity, q (mg/g), of the RDF char was determined according to Eq. 1:

$$q \ (mg/g) = \left(\left(C_0 - C_f \right) * V \right) / m .$$ (1)

The metal removal efficiency, R (%), was determined according to Eq. 2:

$$R \ (\%) = \left(\left(C_0 - C_f \right) / C_0 \right) .100.$$ (2)

where C_0 and C_f are the initial and final metal concentrations (mg/L), respectively, V is the volume of the solution (L), and m is the char mass (g).

4 Results and Discussion

4.1 RDF Char Characterization

The results for the characterization of standard RDF and its corresponding RDF char are presented in Table 1. When compared to the standard RDF, the RDF char shows a very significant increase in the fixed carbon content and density, which has very important implications in handling and storage. The ash content of the RDF char is very high (19.83 %), affecting negatively the estimated high heating value (HHV) and limiting the applications of these chars in energy conversion systems such as boilers, mainly due to their potential for slagging and fouling phenomena. Nevertheless the torrefaction process may be regarded as an advanced treatment to increase the density and reduce the moisture of RDFs, enabling important reductions in landfill land use and reducing the effluent emissions in storage sites.

Table 1. Fuel characteristics of standard RDF and corresponding RDF char.

	Standard RDF	RDF char
Proximate analysis		
Moisture (% w/w, wb)	6.02 ± 0.34	2.32 ± 0.18
Volatile matter (% w/w, db)	82.58 ± 1.03	59.69 ± 1.18
Ash (% w/w, db)	8.20 ± 1.21	19.83 ± 0.47
Fixed carbon (% w/w, db)	9.22 ± 0.26	20.48 ± 0.74
Estimated HHV (MJ/Kg)[1]	17.68	15.70
Density (g/cm^3)	0.12 ± 0.01	0.46 ± 0.04

Another approach regarding the valorization of this material is its use as a low cost adsorbent for industrial applications. Our goal is to eventually use this char to remediate the liquid effluents produced by the same companies that produce the RDF in an integrated perspective of waste management. Following this approach, the heavy metals chromium and aluminum (that are frequently present in these effluents) were chosen as adsorbates for a series of tests concerning the adsorption capacity of the RDF char.

4.2 Adsorption Experiments

Effect of pH. It is well documented that pH affects significantly the adsorption capacity of an adsorbent used for heavy metal removal from aqueous solutions, mostly because it influences the ionic forms of the metals in solution [7]. The effect of the initial pH of the metal solutions in the removal efficiency by the RDF char is shown in Fig. 1.

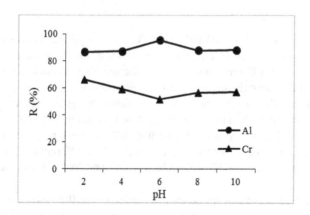

Fig. 1. Effect of solution pH on the removal efficiency (R) of Al and Cr.

Aluminum presents its higher removal efficiency value for pH 6 (95.5 %), which is in accordance with the determinations of other authors using different adsorbents [8]. On the other hand, chromium is better adsorbed at more acidic pH (pH 2), reaching a removal efficiency of 66.0 %. This high removal efficiency of chromium at low pH can be due to the increase in the protonation of functional groups from the char surface, producing a strong electrostatic interaction between the char and negatively charged oxyanion forms of Cr (VI) such as $HCrO_4^-$, $Cr_2O_7^{2-}$, and CrO_4^{2-} which are prevalent in the potassium dichromate solutions and may be significant in real effluents [9]. Regardless of the obtained removal efficiencies it was decided to perform the remaining tests at the natural pH of real landfill leachates samples (pH 8.0).

Effect of Adsorbent Mass. The adsorbent mass or adsorbent dose is a significant parameter considering removal efficiency because it is related to the adsorbent-adsorbate equilibrium [10]. The results obtained for this parameter are shown in Fig. 2.

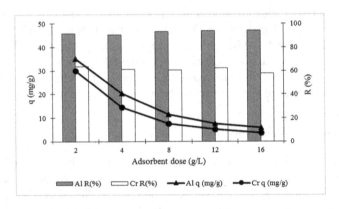

Fig. 2. Effect of adsorbent dose (g/L) on the adsorption capacity (q) and removal efficiency (R) of Al and Cr.

The removal efficiencies of Al and Cr didn't change significantly when the adsorbent dose changed from 2 g/L to 16 g/L, indicating that maximum extraction was achieved at a 2 g/L dose. The different ionic forms of the metals will bind to the char surface groups through electrostatic and hydrophobic interactions but as their concentration in solution decreases the concentration gradient favors a certain degree of desorption stabilizing in a concentration ratio that is constant for each metal. The average removal efficiency was of 93 % for Al while for Cr was of 61 %, what may result from a wider speciation of Cr, yielding a variety of ionic species with different net charges, limiting their removal by adsorption when the electrostatic and hydrophobic interactions are dominant.

On the other hand, the adsorption capacity decreases as the adsorbent dose increases indicating that the equilibrium between char and solution is achieved at 2 g/L. Increasing the adsorbent mass has a dilution effect in the adsorption capacity since the final concentration of metals in the char was inversely proportional to the adsorbent dose.

Effect of Metal Initial Concentration. The removal efficiency of the RDF char in both metal solutions showed a tendency to decrease with the initial metal concentration (Fig. 3).

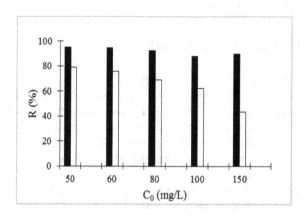

Fig. 3. Effect of initial metal concentration, C_0, (mg/L) on the removal efficiency R (%) of Al and Cr.

Aluminum presented its highest removal efficiency at 50 mg/L (95 %), and this property decreased slightly to values of 88–89 % for initial metal concentrations of 100 and 150 mg/L, suggesting that in this metal concentration range, binding sites present in the char surface were not saturated when the adsorbent dose was 2 g/L.

The removal efficiency for chromium was more sensitive to the initial metal concentration, varying from 79 %, at an initial concentration of 50 mg/L to 44 % at 150 mg/L.

Effect of Contact Time. The results for the adsorption capacities as a function of contact time are depicted in Fig. 4.

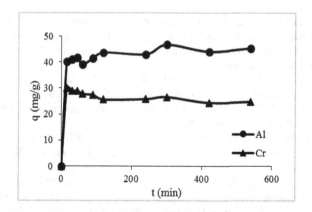

Fig. 4. Effect of contact time on adsorption capacities of Al and Cr.

RDF char showed a sharp increase in the adsorption capacity during the first 15 min of adsorption (for both metals) and remained almost constant for higher contact times, reinforcing the idea that the adsorption mechanisms are dominated by electrostatic and hydrophobic interactions with the external surface of the char.

5   Conclusions and Further Work

The torrefaction process can contribute to upgrade RDF materials yielding a RDF char that is more homogeneous, hydrophobic and dense, with a higher stability and less land use requirements if stored in landfills. This char presents high ash content and reduced HHV so it could be more efficiently used in material applications than for energy recovery purposes. The RDF char presented good adsorption characteristics for Al and Cr in aqueous solutions, reaching removal efficiencies above 90 % for aluminum and above 70 % for chromium.

These results highlight the potential use of RDF char in industrial effluent remediation and future work includes tests with other RDF chars to evaluate the influence of initial RDF composition, tests with other heavy metals and other organic and inorganic contaminants, both in model systems and in real industrial effluents.

Industrial implementation of these technologies will require the adoption of cyber physical approaches to monitoring and control in view of the dimension of the material flows to be converted and because the heterogeneity of the materials requires automated adjustment of process operational conditions.

References

1. Krüger, B., Mrotzek, A., Wirtz, S.: Separation of harmful impurities from refuse derived fuels (RDF) by fluidized bed. Waste Manage. **34**, 390–401 (2014)
2. Portuguese Diary of the Republic: Ministry Council Resolution no. 11-C/2015. 1st series, 52, 16 March 2015
3. Chen, W., Kuo, P.: A study on torrefaction of various biomass materials and its impact on lignocellulosic structure simulated by thermogravimetry. Energy **35**, 2580–2586 (2014)
4. Fu, F., Wang, Q.: Removal of heavy metal ions from wastewaters: a review. J. Environ. Manage. **92**, 407–418 (2011)
5. Salam, O.E.A., Reiad, N.A., ElShafei, M.M.: A study of the removal characteristics of heavy metals from wastewater by low cost-adsorbents. J. Adv. Res. **2**, 297–303 (2011)
6. Sheng, C., Azevedo, J.L.T.: Estimating the higher heating value of biomass fuels from basic analysis data. Biomass Bioenergy **28**, 499–507 (2005)
7. Pehlivan, E., Kahraman, H.: Sorption equilibrium of Cr (VI) ions on oak wood charcoal (Carbo Ligni) and charcoal ash as low-cost adsorbents. Fuel Process. Technol. **92**, 65–70 (2011)
8. Choksi, P.M., Joshi, V.Y.: Adsorption kinetic study for the removal of nickel (II) and aluminum (III) from an aqueous solution by natural adsorbents. Desalination **208**, 216–231 (2007)
9. Deveci, H., Kar, Y.: Adsorption of hexavalent chromium from aqueous solutions by bio-chars obtained during biomass pyrolysis. J. Ind. Eng. Chem. **19**, 190–196 (2013)
10. Kılıç, M., Kırbıyık, Ç., Çepelioğullar, O., Pütün, A.E.: Adsorption of heavy metal ions from aqueous solutions by bio-char, a by-product of pyrolysis. Appl. Surf. Sci. **283**, 856–862 (2013)

Bioremediation of Agro-industrial Effluents Using Chlorella Microalgae

Catarina Viegas[✉], Margarida Gonçalves, Liliana Soares, and Benilde Mendes

Department of Science and Technology of Biomass, Faculty of Sciences and Technology,
Mechanical Engineering and Resources Sustainability Centre,
New University of Lisbon, Caparica, Portugal
`cv.sousa@campus.fct.unl.pt`

Abstract. Two microalgae species (*Chlorella vulgaris* and *Chlorella protothecoides*) were tested at lab scale in order to select the optimal conditions for biomass production and the efficient remediation of effluents from poultry and pig industries. Both microalgae showed biomass productivities in the agro-industrial effluents that were comparable to the Chlorella synthetic medium used as control. *C. protothecoides* presented the higher productivities both for poultry effluents (46.13 and 41.75 mg.L^{-1}.day^{-1} for raw and flocculated effluents) and for pig manure (95.86 mg.L^{-1}.day^{-1}). The supplementation of pig effluents with biomass ash increased by 50 % the microalgae productivity with the highest results obtained for *C. protothecoides* and *C. vulgaris* at ash concentrations of 1.5 g/L and 3.0 g/L, respectively. The optical density of both effluents was efficiently reduced by both microalgae but particularly by *C. protothecoides* and in the presence of added ash, indicating that significant reductions of suspended solids and organic matter occurred. The results showed that poultry and pig effluents may be efficiently remediated with microalgae and the fortification with biomass ash benefits the process.

Keywords: Microalgae · Bioremediation · Poultry industry · Pig industry · *Chlorella vulgaris* · *Chlorella protothecoides*

1 Introduction

Current technologies for the treatment of urban wastewaters and agro-industrial effluents represent complex and costly procedures but are necessary because those effluents contain a high organic load and can cause eutrophication in freshwater and marine ecosystems if discharged without treatment [1, 2].

The animal production industry (poultry, pig and cattle industries) is a source of various wastes with significant environmental impact, including manure, effluents from cleaning activities and wastewaters from dead animal processing. All organic residues from the slaughterhouses are homogenized, treated thermally and sent to a solid-liquid separator. The decanted solid wastes are used for pet food production while the aqueous phase with a high fat content is subject to a series of unitary operations to reduce its

© IFIP International Federation for Information Processing 2016
Published by Springer International Publishing Switzerland 2016. All Rights Reserved
L. Camarinha-Matos et al. (Eds.): DoCEIS 2016, IFIP AICT 470, pp. 523–530, 2016.
DOI: 10.1007/978-3-319-31165-4_49

organic load and sent to the wastewater treatment plants. In traditional productions the mix of urine, faeces and wastewater were disposed on the ground as fertilizer, but with the increase in intensive farms this option is no longer usable [3].

Moreover the direct use of agro-industrial effluents as fertilizers has negative environmental impacts such as unpleasant odour emissions, contamination with pathogenic microorganisms and contamination of ground waters with components of those effluents.

Microalgae present very high biomass production rates (much higher than the vascular plants) and their low cultivation demands enable their production on degraded lands, wastelands, deserts and even on off-shore structures, thereby not competing with the food sector [4].

In the case of agro-industrial effluents with high levels of nutrients but not contaminated with hazardous elements the produced algal biomass can be incorporated at different stages of the industrial process, in a circular economy strategy. Microalgae applications include feed supplement [5], bio-fertilizers for feed crops [6] and/or feedstock for biofuel production [7].

Microalgae can be used in food and feed due to their chemical composition, with a high protein content, but also because they are a source of almost all the essential vitamins (eg A, B1, B2, B6, B12, C, E, biotin, folic acid, pantothenic acid, nicotinate) and they are rich in pigments such as chlorophylls, carotenoids and phycobilins [8–10].

This PhD program is focused in the optimisation of microalgae-based bioremediation processes for agro-industrial effluents and studying various applications of the produced microalgae biomass.

The different tasks to be implemented in order to achieve this main objective are:

(a) Optimization of the growth of different microalgae (*Chlorella vulgaris, Chlorella prototothecoides* and *Scenedesmus obliquus*) in effluents from poultry and piggery industries.

(b) Mixing of animal production effluents with other agro-industrial effluents with nutrient limitations in order to achieve the remediation of both effluents.

(c) Testing the use of biomass ashes as inorganic additives for microalgae growth as an alternative to the use of pure inorganic components of the culture medium.

(d) Characterisation of the microalgae biomass as a supplement for animal feed.

(e) Use of the produced microalgae biomass to increase the organic load of soils and as a fertilizer for fast growing vegetable species.

(f) Application of the optimized methods in the construction and test of a pilot plant for the integrated bioremediation of mixed agro-industrial effluents.

In the present work it was studied the bioremediation of effluents from the poultry and pig industries using the microalgae *Chlorella vulgaris* and *Chlorella prototothecoides*. Because these effluents have a strong organic load, the mixotrophic alga (*Chlorella prototothecoides*) was used to evaluate the contribution of the heterotrophic metabolism to the microalgae productivity. Biomass ashes (raw or pre-digested) were evaluated as mineral supplements for the pig effluents, in order to test growth limitations due to inorganic factors.

2 Relationship to Cyber-Physical Systems

The production of microalgae in an industrial scale and their use in the remediation of industrial effluents requires as other industrial processes, the implementation of automated systems for process monitoring and control. In microalgae production the automated control of liquid and gaseous flows and the continuous measurement of parameters such as pH, conductivity, temperature and optical density is critical to achieve stable operation and apply the necessary correction actions whenever these parameters deviate from optimum values.

Cyber-physical systems (CPS) are physical and engineered systems whose operations are monitored and controlled by a computing and communication core [11]. In the case of microalgae-base bioremediation systems there are two levels of parameter monitoring that should be implemented: process parameters such as temperature or pH that must be constant for ideal operation and culture parameters such as optical density or concentrations of specific elements that are being consumed during algae growth.

During microalgae growth automated systems should contemplate stirring of the culture medium and may include mechanical systems to adjust the position of the bioreactors in order to achieve a maximum exposure to solar light.

Mechanical systems for continuous cleaning of the internal surface of the bioreactor may also be implemented to ensure maximum light penetration in the culture medium. When the culture achieves a given value of optical density a fraction of the culture medium should be discharged to a decantation pond and a given volume of fresh medium should be added to the bioreactor in order to replenish the level of nutrients to sustain microalgae growth.

The association of continuous sensors, automated valves and computational control center are essential elements for the implementation of large scale microalgae bioremediation systems in which the culture achieves a steady state of biomass production.

3 Methods

Two microalgae (*Chlorella vulgaris* and *Chlorella protothecoides*) were grown in two different effluents from the poultry industry (raw effluent and flocculated effluent) and a pig manure effluent mainly composed by urine. For the control it was used a *Chlorella* culture synthetic medium [12].

Biomass ashes were added to the pig manure effluent at concentrations of 1.5 g/L and 3 g/L and in two different formulations: as received or pre-digested with concentrated sulfuric acid. The pig manure effluent was boiled and diluted (1:2) with tap water, before use.

The experiments were conducted at room temperature (22 °C ± 2 °C), under artificial lighting (10000 lux) with cycles of 12 h light/12 h dark, using 500 mL of effluent agitated by air bubbling. Algal growth was followed by measuring the culture optical density at 540 nm and dry weight. The pH was also measured and controlled in order to remain in the range of 7 to 8. The parameters evaluated in the effluents before and after microalgae growth were: total and suspended solids, total nitrogen, ammonia nitrogen, nitrate, nitrite, total phosphorus, chemical oxygen demand and biochemical oxygen demand. The growth experiments were performed in duplicate and the trials ended when the optical density started to decrease.

4   Results and Discussion

The characteristics of the tested effluents and the culture control medium were analyzed and presented in Table 1.

Table 1. Characterization of the effluents used in the trials.

Effluent	Total nitrogen (mg N/L)	Total phosphorus (mg P/L)	O.D. (540 nm)	Ash content (g/L)
Raw effluent (poultry)	160.78 ± 6.5	37.3 ± 0.5	0.778	0.46 ± 0.03
Floccul. effluent (poultry)	202.03 ± 2.6	105.4 ± 8.4	0.372	0.54 ± 0.04
Manure effluent 1:2	313.60 ± 15.8	83.7 ± 0.5	1.561	1.66 ± 0.01
Synthetic culture medium	173.1	283.9	0.087	4.64

It can be observed that nitrogen concentrations in the effluents are comparable or higher than in the culture medium therefore this is not a limiting nutrient in these agro-industrial waste streams. On the other hand, the presence of various organic components namely proteins and lipids, some with a limited solubility in water, explains why the agro-industrial effluents have an optical density considerably higher than the synthetic culture medium.

The flocculated poultry effluent had an optical density lower than the raw effluent because the flocculation/decantation process is mainly effective in the removal of fat and suspended solids but does not affect dissolved nitrogen and phosphorus species, whose concentration in both effluents depends on the mix of wastes being treated.

The raw effluent and the flocculated effluent used in this trial were supplied by the same poultry industry but correspond to different moments of collection so their mineral composition is not necessarily related. The high content of nitrogen of the pig effluent is expected since it is mainly composed of pig urine. On the other hand the poultry and pig effluents may have limitations of some inorganic components that are essential for the microalgae growth since total ash content is significantly lower than the synthetic culture media.

In a first set of experiments microalgae *C. vulgaris* and *C. protothecoides* were grown in the raw and flocculated poultry effluents and the productivities were comparable but inferior to those of the control medium (Fig. 1), although *C. protothecoides* had a better performance both in the control medium and in the effluents.

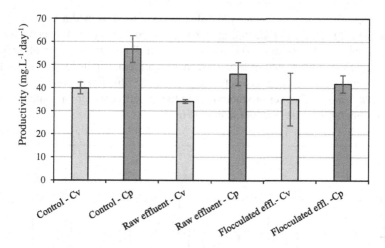

Fig. 1. Microalgae productivity in poultry effluents (mean values and standard deviation).

The lower microalgae productivity in both poultry effluents when compared with the control medium indicates that their lower mineral content and phosphorus concentrations are probably limiting microalgae growth.

The effect of algal growth in the optical density of the poultry effluents is presented in Fig. 2.

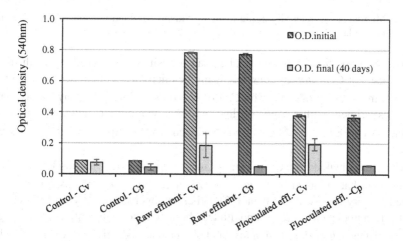

Fig. 2. Optical density of the poultry effluents measured before and after the microalgae growth experiments (mean values and standard deviation).

The microalgae growth caused an important reduction on the optical density of the effluents indicating that total solids and organic matter were used by the microalgae or aggregated to their cell wall. This effect was more pronounced for *C. prototecoides* that achieved final optical densities comparable to the control medium, probably due to its ability to metabolize organic compounds using heterotrophic pathways.

In a second set of experiments pig effluents were used as media to grow the same microalgae (Fig. 3). This effluent already has nitrogen and phosphorus contents higher than the poultry effluents so is expected to better fulfill the requirements of microalgae growth and the incorporation of biomass ashes as mineral supplements aims at testing the role of inorganic components as limiting nutrients. Effectively both *C. vulgaris* and *C. prototheoides* had better productivities in the pig effluent than in poultry achieving values comparable to the control medium (Fig. 3).

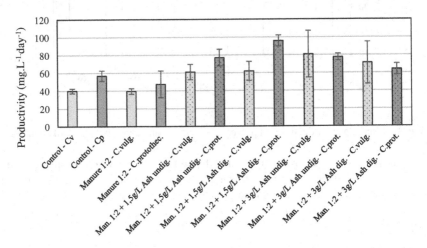

Fig. 3. Microalgae productivity in pig manure effluent, with and without ash addition (mean values and standard deviation).

The incorporation of biomass ashes had a positive effect in the growth of both microalgae but *C. prototheoides* was again the microalga with higher specific growth rate. The maximum productivity for *C. prototheoides* (95.9 mg.L^{-1}.day^{-1}) was attained in the experiment with 1.5 g/L of digested ashes while for *C. vulgaris* a maximum productivity of 80.9 mg.L^{-1}.day^{-1} was obtained with incorporation of 3 g/L of undi-gested ashes.

The yield obtained for *C. vulgaris* grown in pig manure (39,4 mg.L^{-1}.day^{-1}) was higher than that obtained by Reda *et al.* (23 mg.L^{-1}.day^{-1}) which used biologically-treated piggery wastewater effluent [13]. Other authors used piggery wastewater (COD of 11 g/L) autoclaved at 120 °C and diluted about 10 times with distilled water to grow the microalgae *Chlorella pyrenoidosa* and obtained a productivity of 6.3 mg.L^{-1}.day^{-1} [14]. The incorporation of biomass ash resulted in a 50 % increase in growth indicating that this mineral waste could be thus valorised as an inorganic supplement and no significant toxic effects were detected.

The effect of the microalgae growth on the optical density of the pig effluent is presented in Fig. 4.

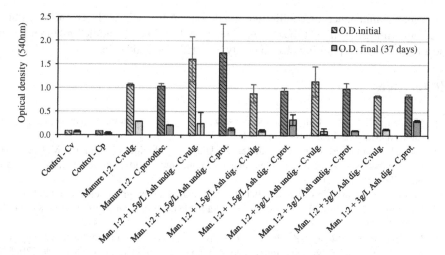

Fig. 4. Optical density of the pig effluents measured before and after the microalgae growth experiments (mean values and standard deviation).

The optical density of the pig effluent used for microalgae growth with or without ash supplementation suffered a significant reduction achieving in some cases values comparable to the control medium (Fig. 4).

The addition of biomass ashes contributed to an increase of the initial optical density of the pig effluent, especially for the 1.5 g/L dose in the undigested formulation. Nevertheless the final optical densities of the pig effluents used with added ash were generally lower than the pig effluents used without ash supplementation either because a higher microalgae productivity resulted in a better remediation of the effluent or because the biomass ashes induce the aggregation of the organic matter present in the effluent by altering their electrostatic status.

5 Conclusions

Microalgae can have an important role in the low cost remediation of complex effluents whose chemical and physical treatment methods are complex and expensive.

The results obtained in this series of experiments demonstrate the ability of *C. vulgaris* and *C. protothecoides* to bioremediate poultry and pig effluents, achieving a significant reduction in total solids and organic matter contents.

The incorporation of an inorganic waste (biomass ash) had a positive effect both in microalgae growth and effluent remediation. This approach constitutes an alternative use for these mineral wastes and reduces microalgae culture requirements of inorganic supplements, and therefore the production costs.

Future work comprehends the conception of a pilot scale microalgae remediation system applied to a small scale agroindustry to achieve an efficient and fast remediation of wastewaters and production of a biofertilizer. This system will comprise a cyber-physical monitoring system with automated data acquisition linked with alert systems to ensure maintenance of equilibrium conditions. Automatic replacement of culture medium with fresh effluent to achieve a steady state culture is another goal that could be attained by incorporating electronically actuated mechanical devices.

References

1. Mo, W., Zhang, Q.: Can municipal wastewater treatment systems be carbon neutral? J. Environ. Manage. **112**, 360–367 (2012)
2. González-García, S., Gomez-Fernández, Z., Dias, A.C., Feijoo, G., Moreira, M.T., Arroja, L.: Life cycle assessment of broiler chicken production: a portuguese case study. J. Clean. Prod. **74**, 125–134 (2014)
3. European Commission - Horizon 2020. http://ec.europa.eu/programmes/horizon2020
4. Demirbas, M.F.: Biofuels from algae for sustainable development. Appl. Energy **88**, 3473–3480 (2011)
5. Fredriksson, S., Elwinger, K., Pickova, J.: Fatty acid and carotenoid composition of egg yolk as an effect of microalgae addition to feed formula for laying hens. Food Chem. **99**, 530–537 (2006)
6. Wang, R., Peng, B., Huang, K.: The research progress of CO_2 sequestration by algal bio-fertilizer in China. J. CO_2 Util. **11**, 67–70 (2015)
7. Mata, T.M., Cardoso, N., Ornelas, M., Neves, S., Caetano, N.: Sustainable production of biodiesel from tallow, lard and poultry fat and its quality evaluation. Chem. Eng. Trans. **19**, 13–18 (2010)
8. Milledge, J.J.: Commercial application of microalgae other than as biofuels: a brief review. Rev. Environ. Sci. Biotechnol. **10**(1), 31–41 (2011)
9. Spolaore, P., Joannis-Cassan, C., Duran, E., Isambert, A.: Commercial applications of microalgae. J. Biosci. Bioeng. **101**(2), 87–96 (2006)
10. Dufossé, L., Galaup, P., Yaron, A., Arad, S., Blanc, P., Murthy, K., e Ravishankar, G.: Microorganisms and microalgae as sources of pigments for food use: a scientific oddity or an industrial reality? Trends Food Sci. Technol. **16**, 389–406 (2005)
11. Rajkumar, R., Lee, I., Sha, L., Stankovic, J.: Cyber-physical systems: the next computing revolution. In: 47th ACM/IEEE Design Automation Conference (DAC), pp. 731–736. IEEE (2010). ISBN: 978-1-4244-6677-1
12. Vonshak, A.: Laboratory techniques for the cultivation of microalgae. In: Richmond, A. (ed.) Handbook of Microalgal Mass Culture, pp. 117–143. CRC Press, Boca Raton (1986)
13. Reda, A., Min-Kyu, J., Hyun-Chul, K., Ki-Jung, P., Byong-Hun, J.: Microalgal species growing on piggery wastewater as a valuable candidate for nutrient removal and biodiesel production. J. Environ. Manage. **115**, 257–264 (2013)
14. Wang, H., Xiong, H., Hui, Z., Zeng, X.: Mixotrophic cultivation of *Chlorella pyrenoidosa* with diluted primary piggery wastewater to produce lipids. Bioresour. Technol. **104**, 215–220 (2012)

Flexible and Transparent Oxide Electronics

Oxide TFTs on Flexible Substrates for Designing and Fabricating Analog-to-Digital Converters

Ana Correia[1,2(✉)], João Goes[1], and Pedro Barquinha[2]

[1] CTS/UNINOVA, Departamento de Engenharia Electrotécnica,
Faculdade de Ciências e Tecnologia, Universidade NOVA de Lisboa, Campus da Caparica,
2829-516 Caparica, Portugal
a.correia@campus.fct.unl.pt, jg@uninova.pt
[2] I3N/CENIMAT, Departamento de Ciência dos Materiais, Faculdade de Ciências e Tecnologia,
Universidade NOVA de Lisboa and CEMOP/UNINOVA,
Campus da Caparica, 2829-516 Caparica, Portugal
pmcb@fct.unl.pt

Abstract. Thin-film transistors (TFTs) employing oxide semiconductors have recently emerged in electronics, offering excellent performance and stability, low processing temperature and large area processing, being indium-gallium-zinc oxide (IGZO) the most popular amorphous oxide semiconductor. In this work it is shown how IGZO TFTs can be integrated with multilayer high-κ dielectrics to obtain low operating voltages, both on glass and flexible PEN substrates. Then, the electrical properties extracted from these devices are used to design and simu-late a 2nd-order Sigma-Delta ($\Sigma\Delta$) analog-to-digital converter (ADC), showing superior performance (e.g. SNDR $\approx$ 57 dB, and DR $\approx$ 65 dB) over ADCs using competing thin-film technologies.

Keywords: Amorphous oxide semiconductors · IGZO TFTs · Circuit integration · ADCs · $\sum\Delta$ modulator

1 TFTs for Large-Area Electronics: A Materials Perspective

Displays are deeply incorporated in our daily lives in mobile phones, TVs, monitors, among others. The key electronic elements of these displays are thin-film transistors (TFTs) responsible for switching each pixel on and off. A TFT is a particular case of a field-effect transistor (FET) based on thin-film technologies, presenting a semiconductor located between drain and source electrodes and the dielectric layer placed between gate electrode and the semiconductor. Using these three terminals (drain, source and gate), the principle behind this device is to control the current between drain and source elec-trodes (I_{DS}) through the gate-source potential (V_{GS}) [1].

The first conceptual patents on TFTs were reported in 1930, however these devices only acquired a high relevance in the 70 s, with the advances on flat-panel displays, specially with the appearance of active matrix liquid-crystal displays (AMLCDs). Hydrogenated amorphous silicon (a-Si:H) soon started to be the

© IFIP International Federation for Information Processing 2016
Published by Springer International Publishing Switzerland 2016. All Rights Reserved
L. Camarinha-Matos et al. (Eds.): DoCEIS 2016, IFIP AICT 470, pp. 533–541, 2016.
DOI: 10.1007/978-3-319-31165-4_50

preferred choice for semiconductor owing to its amorphous structure perfectly suitable for large area fabrication. Still, its low field-effect mobility ($\mu_{FE} < 1$ cm^2/Vs) was not compatible either with the fabrication of the peripheral drivers of the displays or for other circuits operating in the MHz range. Poly-Si was introduced to deal with this limitation, enabling $\mu_{FE} > 100$ cm^2/Vs, but at the cost of larger processing temperatures and a polycrystalline structure, hindering low cost and uniform fabrication in large areas. On the other hand, organic semiconductor based TFTs were also introduced in the late 80 s, as the ultimate technology, when low-cost and mechanical flexibility are major requirements. However, their electrical performance was (and is) comparable to a-Si:H TFTs and degradation under normal environmental conditions is a major concern.

The new millennium set the stage for a new transistor technology, able to fulfil the needs of flexible, low cost but nonetheless fast electronics. Oxide semiconductor TFTs are such technology, offering excellent uniformity in large areas using temperatures below 150 °C and μ_{FE} exceeding 10 cm^2/Vs. In addition, most oxides have bandgaps (E_G) above 3 eV, turning them transparent materials in the visible spectrum, enabling exciting concepts such as fully transparent and flexible devices [2]. Even if some initial oxide TFTs were reported in the 60 s, the most relevant works showing the potential of such technology started in 2002 with polycrystalline ZnO TFTs [3–6]. Despite the encouraging results obtained using ZnO, rapidly amorphous multicomponent materials started to be used, specially indium gallium zinc oxide (IGZO). In fact, in these amorphous oxides a "continuous path" created by spherical isotropic ns orbitals of the metallic cations, when the radii of these orbitals is larger than the distance between cations, allows obtaining excellent μ_{FE} even in amorphous structures [7]. Although IGZO is the most widely used amorphous oxide semiconductor (AOS), some additional combinations have been considered, as indium-free alternatives such as zinc-tin oxide (ZTO).

Despite the semiconductor used in a TFT, the dielectric choice critically affects the device performance and stability. High-κ dielectrics have been used instead of traditional SiO$_2$ due to different reasons. When low temperature deposition processes are used, semiconductors and their interfaces are more prone to have large defect density and poor compactness. Hence, the large capacitance per unit area of high-κ dielectrics compensates this, by inducing a larger charge density per unit voltage. This also allows for low-voltage operation: as will be seen in this work, high-κ results in less than half of the operation voltage (seen by transconductance saturation) compared to thermal SiO$_2$ with similar thickness. Furthermore, the larger film thickness (without compromising a good overall capacitance) enabled by high-κ dielectrics compensates the more degraded dielectric film properties fabricated at low temperatures, avoiding for instance quantum tunnelling effects. Additionally, selecting high-κ materials with an amorphous structure typically results in smoother films and, consequently, in improved interfaces with amorphous semiconductors, while the lack of grain boundaries of amorphous structures decreases the gate leakage current (I_G). However, most of the high-κ dielectrics have lower E_G than lower-κ materials as SiO$_2$, which based on band alignment considerations can be problematic to obtain low I_G. Hence, multicomponent dielectrics based on combinations of materials with high-κ and high-E_G, such as Ta_2O_5 and SiO_2, respectively, has been studied in different configurations, taking advantage of structural and electrical properties of both materials [8].

2 Relationship to Cyber-Physical Systems: From Oxide TFTs to a Complete System-on-Foil

The current advances in oxide TFTs are changing and expanding the display market, enabling higher refresh rate combined with ultra-definition and stimulating new approaches such as curved or transparent displays. The big players in the multi-billion dollar display industry are naturally all aware of this TFT technology, with an increasing number of prototypes and even commercial products being available from companies as Samsung and LG. One of the most relevant examples is the line of curved OLED TVs by LG, making use of oxide TFT backplanes.

Despite all this activity in taking oxide TFTs to displays, efforts have also been done to take full advantage of oxide TFTs and integrate them in other electronic circuits and systems. In fact, good uniformity over large area, excellent electrical performance and stability, low production costs and the flexible or even recyclable concepts encourage their integration in full system-on-foil concepts, expanding their functionality and their applicability. To the author's best knowledge, the first circuits using oxide TFTs appeared in 2006, in which inverters and a five-stage ring oscillator (RO) have been reported using indium-gallium oxide (IGO) TFTs. Regarding inverters, a peak gain magnitude close to 1.5 was measured and a maximum oscillation frequency of 9.5 kHz has been experimentally demonstrated [9]. Several RO performance improvements have been achieved by decreasing the propagation delay and, therefore, allowing to increase the maximum oscillation frequency and the number of cascaded stages [10, 11]. Shifter registers, current mirrors and a digital-to-analog converter (DAC) were also produced using oxide TFTs. Regarding DAC, which is already a circuit realization with a considerable level of complexity, encouraging results were achieved using IGZO TFTs, showing a spurious-free dynamic range (SFDR) above 30 dB up to 300 kHz and a sampling rate of 1 MS/s [12]. It is also relevant to notice the relatively recent developments of using TFTs in applications spanning from near-field communications (NFC) to radio-frequency identification (RFID) tags. In 2011, the first RFID chip was produced using IGZO TFTs, showing a small power dissipation (20 μW) making it suitable for short-range wireless operation [13]. In 2015, a fully flexible NFC tag also using IGZO TFTs has been fabricated, exhibiting excellent performance, as a maximum data rate close to 72 kbits/s [14]. These last examples intensify the potential of using oxide TFTs for an entire system-on-foil concept where, besides performance and processability of electronic devices in low cost and flexible substrates, multifunctionality of the platform using only one semiconductor technology is a paramount advantage in terms of reliability and cost of the complete system.

The referred system-on-foil concept plays a relevant role nowadays in the global cyber-physical systems (CPS) approach, which combines computation and physical processes networking and it is transversal to multiple fields of knowledge, involving areas such as healthcare, energy or consumer appliances. In fact, the excellent performance and stability of oxide technologies are key enablers to develop innovative, more efficient and high frequency systems, improving strongly the communication in complex systems.

3 Oxide TFTs Produced at FCT-NOVA

Oxide TFTs have been studied using different approaches in order to optimize and enhance TFT performance, and reducing at the same time the involved fabrication costs. For that end different materials, structures and processes, such as the usage of organic insulators, nanowire structures and solution processes, respectively, have been studied in different research groups worldwide, including CENIMAT and CEMOP at FCT-NOVA. This section summarizes a small part of these developments, which were then used as the basis to design the ADC case study presented in Sect. 4. In this case, since complex circuit design demands for good performance and stable devices, sputtered IGZO was selected as the semiconductor, given that it is the most mature process/material available in-house. IGZO was then coupled with different sputtered high-κ dielectrics based on Ta_2O_5 and SiO_2, evaluating the effect of the dielectric on important device parameters such as μ_{FE}, turn-on voltage (V_{on}) and I_G.

3.1 Devices Fabrication and Characterization

IGZO TFTs were produced on Corning Eagle glass and on flexible Polyethylene Naphthalate (PEN), depositing all layers using RF magnetron sputtering. These devices present a staggered bottom gate, top contact structure and were fabricated as follows: Molybdenum (Mo) gate electrodes ($\approx$60 nm thick) were deposited in an Ar atmosphere; for the dielectric a multilayer structure with alternating SiO_2 and co-sputtered $Ta_2O_5+SiO_2$ (denoted TSiO) layers was fabricated, with a thickness of 100–150 nm. The semiconductor, IGZO ($\approx$40 nm thick and 2:1:1 In:Ga:Zn atomic ratio), was deposited in an Ar+O_2 atmosphere. At the end, Mo source and drain electrodes were fabricated using the same procedure as for gate electrode. Both the electrodes and the semiconductor were patterned by lift-off and, on the other hand, the dielectric layer by dry-etching, in a SF_6 atmosphere. To conclude the production, the devices fabricated on glass and on PEN were annealed at 180–200 and 150 °C, respectively, on a hotplate, in air.

Electrical characterization of devices was performed using a semiconductor parameter analyzer (Keithley 4200-SCS) and a probe station (Cascade Microtech M150) under dark room conditions.

3.2 IGZO TFTs on Glass Substrate

As discussed in [15] a multilayer dielectric comprising TSiO and thin SiO_2 (with the latter being in contact with IGZO) results in considerably improved device yield over single layer TSiO or Ta_2O_5. In fact, recent improvements in tuning the thickness of these multilayer dielectrics enabled breakdown field (E_B) > 7 MV/cm and $\kappa \approx 10$, resulting in IGZO TFTs with $I_G < 1$ pA and low operation voltage (yielding a transconductance saturation for $V_{GS} < 6$ V). Figure 1 shows transfer and output curves of one IGZO TFT with a 150 nm thick multilayer dielectric comprising 7 alternating SiO_2/TSiO layers, annealed at 180 °C, with a width-to-length ratio (W/L) of 40/20 (μm). Excellent performance is obtained: $\mu_{FE} \approx 14$ cm^2/Vs, On-Off ratio $\approx 10^9$, $V_{on} = -1.5$ V, subthreshold slope (S) of 0.16 V/dec, negligible hysteresis, hard saturation and no current crowding at low V_{DS}. Note that these results are quite similar to the ones obtained when the same IGZO layers are deposited on state-of-the-art thermal dry SiO_2. In fact, S and operating voltage are even improved with the present multilayer dielectric, owing to its larger κ (in thermal SiO_2 S = 0.20 V/dec and transconductance saturation occurs for $V_{GS} = 15$ V).

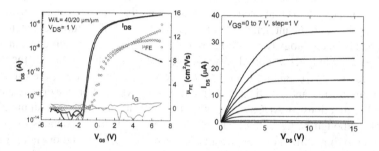

Fig. 1. Transfer (left) and output (right) curves of an IGZO TFT on glass with a sputtered 7 layer SiO_2/TSiO dielectric.

3.3 IGZO TFTs on Flexible PEN Substrate

The possibility of using flexible substrates brings significant advantages to a TFT technology. First, it enables the usage of roll-to-roll fabrication tools, highly desirable for large area electronics. Then, at a system level they enable weight and thickness reduction over conventional glass. Finally, bendable or even rollable products can be conceived. Given the low temperature fabrication required for IGZO TFTs with sputtered multilayer dielectrics, their performance was also analyzed on PEN substrates. Figure 2 shows the

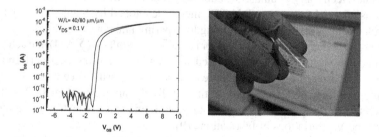

Fig. 2. Measured transfer curves (left) and a photo of IGZO TFTs on PEN substrate with W/L = 40/80 μm/μm and annealed at 150 °C (right) using a multilayer dielectric (7 layers).

transfer curves and a photo of such devices with a 7 layer dielectric, having a total thickness $\approx$100 nm, annealed at only 150 °C. It is noteworthy the excellent electrical performance obtained: in fact, besides the larger hysteresis, all the other parameters reveal to be even slightly improved over the ones achieved on glass substrates (Fig. 1), which is also a consequence of the thinner dielectric layer used on PEN. Electrical characterization under different bending radius is currently underway.

4   ADCs Using Oxide TFTs Produced at FCT-NOVA

As mentioned before, circuit integration is a crucial demonstration of the potential of a new transistor technology. ADCs are one of the most difficult circuits to implement and, given their importance in a variety of fields such as electronic, medical, telecommunications or other systems that need signal processing, their implementation using a new technology is a significant achievement. Despite the lack of maturity compared to the dominant TFT technology (i.e., a-Si:H), oxide TFTs have been implemented in systems with a reasonable level of complexity and integration but, to the best of the authors' knowledge and up to date, no kind of reference was found regarding the use of oxide TFTs in the design and practical implementation of ADCs. Some works related with practical realizations of ADCs using other thin-film technologies are summarized in [16–18]. Although being a small part of all developed work, it contains examples of ADCs employing a-Si:H, low-temperature poly-silicon (LTPS) and organic TFTs.

Excellent performance, good stability, large E_B and low I_G are some of the key characteristics of IGZO TFTS with multilayer dielectric previously optimized, encouraging to proceed to the design of a 2nd-order Sigma-Delta ($\sum\Delta$) ADC using the devices produced at FCT-NOVA. An ADC architecture based on oversampling and noise-shaping techniques has been adopted, since due its intrinsic feedback nature, it can easily deal with high process variations and mismatch errors in the physical/electrical parameters of the TFT devices during fabrication. It is important to notice that, due to the lack of a reproducible and stable p-type oxide TFTs, only n-type devices were used in the circuit design. Additionally, due to the expected high device mismatch and poor absolute accuracy when these passive elements are fabricated, SMD passive elements have been employed in order to minimize additional risks in the early phase of proof-of-concept.

Virtuoso[TM] Platform and Spectre[TM] Simulator from CADENCE have been used to design and electrically simulate all building blocks. Furthermore, a model for IGZO TFTs was adapted from an a-Si:H TFT model developed by Semiconductor Devices Research Group at RPI, with good fitting to experimental data.

Given the relevance of the comparator in the $\sum\Delta$ modulator ($\sum\Delta$M), the active block in circuit, it was designed and simulated in order to obtain the best performance. It employs a cascade of three pre-amplification stages, a positive-feedback analog latch stage and four logic inverters implementing a fully-dynamic digital latch. Results show a low offset, working at several tens of kHz, with an accuracy $\approx$10 mV. Moreover, it supports the V_{on} variations of fabrication [19].

Regarding the final design of continuous-time $\sum\Delta$M, it employs an analog loop filter (a cascade connection of two fully passive RC-type integrators), implementing the

2nd-order noise shaping transfer function. ADC simulations reveal: SNDR $\approx$ 57 dB, DR $\approx$ 65 dB, and power dissipation, approximately, of 22 mW (V_{DD} = 10 V). Comparing these simulation results, shown in Table 1, with the current state-of-the-art of $\sum\Delta$Ms, it can be observed that this work is clearly beyond prior art, for competing thin-film technologies such as organics or even LTPS, supporting the continuity of this work. However, the operation frequency and intrinsic gain of these devices can limit the performance of the $\sum\Delta$M. Nevertheless, results give an excellent stimulus to produce one of the first 2nd-order $\sum\Delta$ ADC using oxide TFTs.

Table 1. Simulated main performance parameters of the $\sum\Delta$M using IGZO TFTs.

Characteristic	Value
Sampling frequency (F_s)	128 kHz
Input signal amplitude (A_{in})	5 V_{diff}
Input signal frequency (F_{in})	101.5625 Hz
Oversampling ratio (OSR)	128
Bandwidth (BW)	500 Hz
Signal-to-noise distortion ratio (SNDR)	57 dB
Dynamic range (DR)	65 dB

Another paramount step in this kind of work is the layout, specially using new technologies for which a process-design-kit (PDK) is not available. Consequently, a parameterized cell (PCELL) needs to be constructed taking into account the structure of devices and all process/material constraints. Given the absence of PDK, automatic design rule check (DRC) and layout versus schematic (LVS) could not be performed using EDA (electronic design automation) tools. Hence, layout of the entire circuit with a die area close to 10 mm^2 still needs to be checked manually. More details regarding design, simulation results and layout can be found in [15].

5   Conclusions

Fully sputtered IGZO TFTs have been produced both on glass and PEN exhibiting comparable electrical performance, with μ_{FE} $\approx$ 14 cm^2/Vs, On-Off ratio $\approx$ 10^8–10^9, V_{on} = -1.0 to -1.5 V and subthreshold slope (S) of 0.16 V/dec. These properties are achieved with only 150–180 °C annealing temperature and are even superior to the ones obtained with high-temperature 100 nm thick dry-SiO$_2$ dielectric from conventional CMOS technology. A 2nd-order $\sum\Delta$M was designed and simulated based on the measured IGZO TFTs, showing a SNDR $\approx$ 57 dB, and a power dissipation of 22 mW. Circuit layout with a die area of 10 mm^2 is currently under final LVS and DRC verifications. Fabrication and experimental evaluation will be the next steps.

Acknowledgments. This work has been funded by FEDER funds through the COMPETE 2020 Programme and National Funds through FCT - Portuguese Foundation for Science and Technology, under the projects "Multifunctional nanoscale oxide materials" (EXCL/CTM-NAN/0201/2012), DISRUPTIVE (EXCL/EEI-ELC/0261/2012), INCENTIVO (EEI/UI0066/2014) and Strategic projects (UID/CTM/50025/2013) and (UID/EEA/00066/2013). The work has also received funding from the European Communities 7th Framework Programme under grant agreement i-FLEXIS project (ICT-2013-10-611070) and H2020 Programme under grant agreement ROLLOUT project (ICT-03-2014-644631).

References

1. Lee, J.-H., Wu, S.-T., Liu, D.N.: Introduction to Flat Panel Displays. Wiley, West Sussex (2008)
2. Grundmann, M., Frenzel, H., Lajn, A., Lorenz, M., Schein, F., von Wenckstern, H.: Transparent semiconducting oxides: materials and devices. Phys. Status Solidi. **207**, 1437–1449 (2010)
3. Masuda, S., Kitamura, K., Okumura, Y., Miyatake, S., Tabata, H., Kawai, T.: Transparent thin film transistors using ZnO as an active channel layer and their electrical properties. J. Appl. Phys. **93**, 1624–1630 (2003)
4. Carcia, P.F., McLean, R.S., Reilly, M.H., Nunes, G.: Transparent ZnO thin-film transistor fabricated by RF magnetron sputtering. Appl. Phys. Lett. **82**, 1117 (2003)
5. Hoffman, R.L.: ZnO-channel thin-film transistors: channel mobility. J. Appl. Phys. **95**, 5813–5819 (2004)
6. Fortunato, E.M.C., Barquinha, P.M.C., Pimentel, A.C.M.B.G., Gonçalves, A.M.F., Marques, A.J.S., Martins, R.F.P., Pereira, L.M.N.: Wide-bandgap high-mobility ZnO thin-film transistors produced at room temperature. Appl. Phys. Lett. **85**, 2541 (2004)
7. Nomura, K., Ohta, H., Takagi, A., Kamiya, T., Hirano, M., Hosono, H.: Room-temperature fabrication of transparent flexible thin-film transistors using amorphous oxide semiconductors. Nature **432**, 488–492 (2004)
8. Zhang, L., Li, J., Zhang, X.W., Jiang, X.Y., Zhang, Z.L.: High-performance ZnO thin film transistors with sputtering SiO$_2$/Ta$_2$O$_5$/SiO$_2$ multilayer gate dielectric. Thin Solid Films **518**, 6130–6133 (2010)
9. Presley, R.E., Hong, D., Chiang, H.Q., Hung, C.M., Hoffman, R.L., Wager, J.F.: Transparent ring oscillator based on indium gallium oxide thin-film transistors. Solid State Electron. **50**, 500–503 (2006)
10. Ofuji, M., Abe, K., Shimizu, H., Kaji, N., Hayashi, R., Sano, M., Kumomi, H., Nomura, K., Kamiya, T., Hosono, H.: Fast thin-film transistor circuits based on amorphous oxide semiconductor. IEEE Electron Device Lett. **28**, 273–275 (2007)
11. Mativenga, M., Choi, M.H., Choi, J.W., Jang, J.: Transparent flexible circuits based on amorphous-indium-gallium-zinc-oxide thin-film transistors. IEEE Electron Device Lett. **32**, 170–172 (2011)
12. Raiteri, D., Torricelli, F., Myny, K., Nag, M., Van der Putten, B., Smits, E., Steudel, S., Tempelaars, K., Tripathi, A., Gelinck, G., Van Roermund, A., Cantatore, E.: A 6b 10 MS/s current-steering DAC manufactured with amorphous gallium-indium-zinc-oxide TFTs achieving SFDR > 30 dB up to 300 kHz. In: 2012 IEEE International Solid-State Circuits Conference, pp. 314–316. IEEE (2012)

13. Ozaki, H., Kawamura, T., Wakana, H., Yamazoe, T., Uchiyama, H.: 20-μW Operation of an a-IGZO TFT-based RFID Chip Using Purely NMOS Active Load Logic Gates with Ultra-Low-Consumption Power (2011)

14. Myny, K., Cobb, B., van der Steen, J.-L., Tripathi, A.K., Genoe, J., Gelinck, G., Heremans, P.: 16.3 flexible thin-film NFC tags powered by commercial USB reader device at 13.56 MHz. In: 2015 IEEE International Solid-State Circuits Conference - (ISSCC) Digest of Technical Papers, pp. 1–3. IEEE (2015)

15. Correia, A.P.P., CândidoBarquinha, P.M., da PalmaGoes, J.C.: A Second-Order $\Sigma\Delta$ ADC Using Sputtered IGZO TFTs. Springer, Cham (2016)

16. Dey, A., Allee, D.R.: IEEE: Amorphous silicon 5 bit flash analog to digital converter. In: 2012 IEEE Custom Integrated Circuits Conference (2012)

17. Lin, W.-M., Lin, C.-F., Liu, S.-I.: A CBSC second-order sigma-delta modulator in 3 μm LTPS-TFT technology. In: 2009 IEEE Asian Solid-State Circuits Conference, pp. 133–136. IEEE (2009)

18. Marien, H., Steyaert, M.S.J., van Veenendaal, E., Heremans, P.: A fully integrated delta sigma ADC in organic thin-film transistor technology on flexible plastic foil. IEEE J. Solid-State Circ. **46**, 276–284 (2011)

19. Correia, A., Martins, R., Fortunato, E., Barquinha, P., Goes, J.: Design of a robust general-purpose low-offset comparator based on IGZO thin-film transistors. In: 2015 IEEE International Symposium on Circuits and Systems (ISCAS), pp. 261–264. IEEE (2015)

Electrochemical Transistor Based on Tungsten Oxide with Optoelectronic Properties

Paul Grey, Luís Pereira$^{(\boxtimes)}$, Sónia Pereira, Pedro Barquinha, Inês Cunha, Rodrigo Martins, and Elvira Fortunato$^{(\boxtimes)}$

CENIMAT/i3N, Departamento de Ciência dos Materiais, Faculdade de Ciências e Tecnologia, FCT, Universidade Nova de Lisboa and CEMOP-UNINOVA, Campus da Caparica, 2829-516 Caparica, Portugal
{lmnp,emf}@fct.unl.pt

Abstract. This paper reports the integration of an electrochromic inorganic oxide semiconductor (WO_3) into an electrolyte gated transistor device. The resulting electrochromic transistor (EC-T) is a novel optoelectronic device, exhibiting simultaneous optical and electrical modulation. These devices show an On-Off ratio of 5×10^6 and a transconductance (g_m) of 3.59 mS, for gate voltages (V_G) between -2 and 2 V, which, to the authors knowledge, are one of the best values ever reported for this type of electrochemical transistors. The simple and low-cost processing together with the electrical/optical performances, well supported into a comprehensive analysis of device physics, opens doors for a wide range of new applications in display technologies, biosensors, fuel cells or electrochemical logic circuits.

Keywords: Tungsten oxides (VI) · Electrochromism · Thin films · Electrolyte-gated transistors · Polymer electrolytes

1 Introduction

Electrolyte-gated transistors (EGTs) and electrochemical transistors (ECTs) have received special attention over the last decade due to their advantages, mainly deriving from the use of solid electrolytes as gate dielectrics. These include low operating voltages, printability and solution processability, low contact resistances, the possibility to fabricate new device architectures, and high driving currents. [1] The understanding of the involved phenomena in electrolyte-gated transistor (EGT) operation is an important aspect to enable their integration in applications like biosensors, [2] microelectronics [3] or even the emergent area of textronics [4].

EGTs using electrochromic (EC) semiconductors are particularly interesting as they combine electrical/electrochemical effects with a simultaneous optical modulation.

The optical absorption of an EC material can be modified through double insertion of electrons and charge compensating ions, [5] which can be easily achieved by the gate field in an EGT structure. Being a reversible process, this gives an EC

Published by Springer International Publishing Switzerland 2016. All Rights Reserved
L. Camarinha-Matos et al. (Eds.): DoCEIS 2016, IFIP AICT 470, pp. 542–550, 2016.
DOI: 10.1007/978-3-319-31165-4_51

material the capability of transiting between two distinct oxidation states: colored and bleached.

During the electrochemical reaction WO_3 also undergoes a change in its electrical properties, namely in its conductivity, making it a candidate for implementation as the semiconductor layer in transistors, as initially demonstrated by Natan *et al.* [6] Furthermore, inorganic oxide semiconductors as WO_3 possess higher chemical stability and higher conductance modulation than the organic semiconductors that compose most of the reported EGTs to date, which are great advantages for digital and analog circuit design with improved longevity.

The combination of both optoelectronic properties of EC-T opens possibilities in a wide range of applications and shows the value of the integration of smart materials into electronic circuitry.

2   Relationship to Cyber-Physical Systems

The use of electrolytes in active electronic components will be one of the fundamental building blocks of the interaction between common solid-state devices and ionic charge carriers. Figure 1 exemplifies the proposed interaction between an electrolyte and an ion permeable semiconductor.

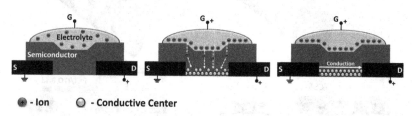

Fig. 1. Schematic representation of the interaction between an ion permeable semiconductor and an electrolyte in a transistor setup with source, drain and gate (S, D and G, respectively). In the present paper WO_3 not only generates an electric response (conductivity change), but also an optical one (transmittance change), turning the system double redundant.

Electrolytes are a big part of modern society and naturally occur throughout the whole spectrum of science. For instance solely in physiological systems there exist a plethora of electrolytes, responsible for intra- and extracellular interactions as well as nerve and muscle function control, blood pressure and pH. Consequently, reliable readout and double redundant values (electrical and optical) for electrolyte interpretation is crucial. Thus, creating a consistent interface between semiconductors and such electrolytes potentiates a main challenge for the development and study of EGTs. The technology of EGTs is therefore capable of adding gross value to a diverse range of cyber-physical systems (CPS), such as in chemical processes, healthcare and bio sensing schemes.

3 Results and Discussion

The complete device structure is depicted in Fig. 2. An interdigital source and drain titanium contact architecture was chosen in order to increase channel width (3000 μm) and consequently the reached drain current of the devices. Furthermore an encapsulated device was fabricated, where the gate electrode (in this case an ITO coated PET foil) was extended over the electrolyte, resulting, not only, in a vertical architecture, but also in a self-passivation effect (Fig. 2(b) and (d)).

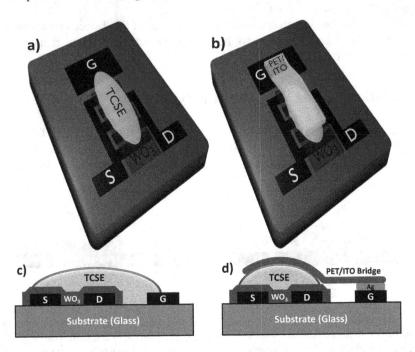

Fig. 2. Schematic device representation, where (a) represents the planar and (b) the vertical and encapsulated architecture and (c) and (d) show a longitudinal cross-section representation of both of the device architectures. The PET/ITO foil is kept in place by capillary forces of the electrolyte and a silver glue (Ag) on the gate electrode.

Subsequent characterization of the most crucial layers of the device was carried out. This includes electro-optical characterization of the sputtered WO_3 thin-films (100 nm), Electrochemical impedance spectroscopy (EIS) of the drop-casted electrolyte (a thermosetting composite solid electrolyte - TCSE) and lastly electrical characterization of the final devices.

3.1 Electro-Optical Measurements

Electrochromism for WO_3 is commonly introduced by reference to the following simplified redox reaction, also known as the double-injection model: [7]

$$WO_3 \ (transparent) + xe^- + xM^+ \leftrightarrow M_xWO_3(darkblue) \tag{1}$$

where M^+ is an ion of the alkali metal group (e.g. H^+, Li^+, Na^+ or K^+), provided by the electrolyte, e^- represents electrons and parameter x, designated as the insertion coefficient, represents the proportion of electro-reduced tungsten sites. Following Eq. 1, the pristine WO_3 is transparent and the resulting tungsten bronze (M_xWO_3), for low x, exhibits its characteristic blue coloration due to photo-intervalence charge transfer (CT) between adjacent W^V and W^{VI} sites [8].

It is widely accepted that the electro-optical centers are activated by the insertion of electrons and charge compensating ions, which distort and expand the lattice of the guest oxide, resulting in an overall modification of the optical properties and electronic structure of WO_3. Charge carriers are trapped by W^{6+} octahedra, reducing them to W^{5+}, while the alkali metals remain ionized in the interstitial space. [9] This results in a delocalization of electrons in the lattice and a charge carrier concentration increase of several orders of magnitude from the pristine oxide. This consequently increases the conductivity, which in terms of transistor operation is reflected in the transition between the Off- and On-states (oxidized and reduced states, respectively).

The WO_3 film was deposited on an ITO coated glass substrate, which was submersed in a Li^+ rich liquid electrolyte in a conventional three-electrode setup. By applying a potential of 1 and -1 V to the ITO film it is possible to influence the oxidation states (oxidized and reduced, respectively) and consequently monitor the change in transmittance at a

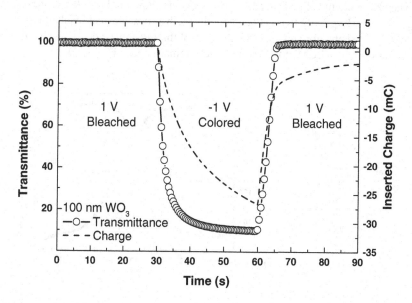

Fig. 3. Electro-optical measurements of a 100 nm thick WO^3 film on ITO/Glass Substrate in a Li^+ rich electrolyte. Transmittance was measured during a full bleaching/coloration/bleaching cycle (30 s each) for the characteristic wavelength of reduced WO_3 (633 nm). The inserted and extracted available charge-carriers were also measured as a function of the duration of the applied potential.

specific wavelength (λ = 633 nm), as well as the inserted charges into the WO_3 film (Fig. 3). The wavelength corresponds to the transmittance of the characteristic blue coloration of reduced WO_3 and it can be observed that a transmittance variation of over 90 % is achieved. The inserted charges during reduction (-1 V) remain as free available charge carriers in the film and are extracted upon oxidation (1 V). A maximum of 26 mC of charge induction was achieved, yielding (with a contact area of 0.833 cm^2) a charge density of about 2×10^{17} cm^{-2}. However the process showed to be not completely reversible as charges might remain in the film after the reduction reaction (charge does not reach 0 mC at 90 s, see Fig. 3). Nonetheless sputtered WO_3 films from previous works are capable of reaching very high lifetimes with more than 1000 cycles [10].

This characterization technique proves the inherent optoelectronic behavior of WO_3 and reveals its potential application in semiconducting devices, owing to reversible insulator-to-metal transitions, resulting from the electrochemical reaction.

3.2   Electrochemical Impedance Spectroscopy (EIS)

EIS is a well-known technique to determine the electrical properties of ionic materials, where an *ac* potential at different frequencies (f) is applied to an electrochemical cell, so the total capacitor impedance (Z) can be measured as a function of the frequency. This gives valuable information about ion migration, electric double layer (EDL) formation and its capacitance (C_{DL}).

The electrochemical cell prepared for EIS is, as shown in Fig. 4(a), composed of 3 distinct layers. Explicitly two ITO coated glass substrates and, sandwiched in between, the TCSE. The behavior and the obtained data of this cell can be described by an equivalent circuit model (ECM) for parameter determination (see Fig. 4(b)).

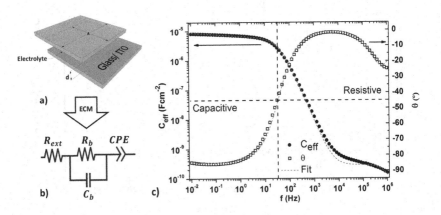

Fig. 4. Electrochemical impedance spectroscopy of the used thermosetting composite solid-electrolyte (TCSE). Where (a) shows the electrochemical cell (TCSE sandwiched in between two ITO (contact) coated glass substrates), (b) represents the equivalent circuit model (ECM) used for the fittings and (c) depicts the effective capacitance and the phase of the electrochemical cell in a frequency range between 10^{-2} and 10^6 Hz.

This ECM (describing an ideally polarizable electrochemical cell), is composed of a resistance (R_{ext}), associated to contact resistances (ITO), in series with a parallel R_bC_b circuit and a constant phase element (CPE). The R_bC_b circuit is associated to bulk properties of the electrolyte, namely the bulk resistance and capacitance, respectively. Whereas R_b gives information about σ_i, C_b accounts for the dipolar relaxation of electrolyte solvent molecules at higher frequencies. The CPE mimics a non-ideal capacitive behavior of the electric double layer, associated to interface inhomogeneities [11].

Upon polarization of the electrochemical cell, ions migrate from one electrode to the other, causing charge built-ups and EDL formation. As can be observed in Fig. 4(c) the ECM describes well the behavior of the cell throughout the whole frequency spectrum and serves for the determination of C_{DL} and the ionic conductivity (σ_i).

Here, as we are dealing with an electrochemical transistor (ECT), σ_i will be the parameter of interest. This parameter will determine how pronounced the effective net ion diffusion into the semiconductor will be and how fast the insulator-to-metal transition occurs. However, from an electrochemical point-of-view it is also important to determine the amount of accumulated charges at the interfaces (C_{DL}), which, in terms of transistor characterization, accounts for charge induction (field-effect) into the semiconductor. It was shown that both effects (electrochemical reaction and field-effect) can contribute to the drain current [12].

The ionic conductivity (for cell dimensions: $A = 5$ cm^2 and $d = 0.1$ mm) and $R_b = 87.34$ Ω from the fitting, σ_i of the electrolyte was determined as 2.29×10^{-5} Scm^{-1}, using the standard resistivity equation. For C_{DL} we considered the approach taken by Jović et al. [13] to empirically determine this value by using the fit for the CPE, yielding $C_{DL} = 5.10$ μFcm^{-2}.

A clear advantage for this type of dielectric becomes evident as the achieved dielectric capacitance reaches very high values. However, as both characteristic parameters (σ_i and C_{DL}) are highly frequency-dependent, transistors working with these materials are often restricted to lower frequencies, where complete ionic relaxation can take place (capacitive regime in Fig. 4(c)). Nonetheless, efforts are made towards an improvement of overall electrolyte behavior in order to augment the EGTs cut off frequency (f_{co} – frequency where $I_{On}/I_{Off} = 10$) [14].

3.3 Electrical Characterization

Electrical characterization of the final devices (as depicted in Fig. 5(a)) was conducted with a drain voltage (V_D) as low as 1 V and a voltage sweep at the gate (V_G) from -4 to 4 V and back at a scan-rate of 20 mVs^{-1}. Low-voltage operation of EGTs were found to be another strong factor of this type of devices, making their application in low power consuming applications feasible.

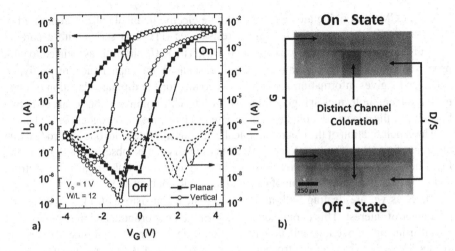

Fig. 5. (a) Transfer characteristics of the fabricated planar and vertical EC-Ts and (b) micrographs of channel region with distinct colorations between the On- and the Off-state. Curves obtained with $V_D = 1$ V and $V_G = (-4, 4, -4)$ V at a scan-rate of 20 mVs^{-1}. The arrow next to the main graphs indicates hysteresis direction.

The applied V_G drives ion migration in the electrolyte (as schematized in Fig. 1) and promotes reversible electrochemical doping (reduction and oxidation) of the channel region. As can be observed in Fig. 5(a), the fabricated EC-Ts start to switch close to $V_G = 0$ V to their respective On-states, remaining however with a normally On characteristic. As the channel reduces (and colors from transparent to blue, as seen in Fig. 5(b)) the drain current increases until reaching mA values around $V_G = 2$ V, with leakage currents (I_G) remaining in the order of µA (at most). The obtained On-Off ratios are considerable, reaching almost 7 orders of magnitude (vertical architecture). A hysteresis behavior is observed for both architectures planar and vertical, which is connected to the time-dependency of the electrochemical reaction, as well as ion migration processes. Nevertheless, it is possible to open and close both EC-Ts with symmetric VG (-4 and 4 V and -2 and 2 V for the planar and vertical architecture, respectively).

Table 1 resumes crucial device parameters, evidencing an advantage for the vertical and encapsulated EC-T, reaching higher values in each aspect. Furthermore, the vertical EC-T showed higher lifetimes than the planar one due to the self-passivation effect of the organic electrolyte, which is always subjected to environmental degradation factors.

Table 1. Crucial EC-T parameters, comparing the planar with the vertical structure. Values calculated from the curves of Fig. 5(a).

	I_{On}/I_{Off}	V_{On} (V)	S_s (Vdec^{-1})	g_m (mS)
Planar	6.55×10^5	0	0.28	2.24
Vertical	5.10×10^6	-1	0.22	3.59

Analyzing Fig. 5(a) and (b) together, the double functionality of the EC-Ts becomes visually clear, as it is possible to obtain a simultaneous electrical and optical response. The obtained information from these devices, regarding, for instance, bio-sensing applications, is on the one hand quantitative (I_D output) and on the other hand qualitative (color output). This leads to the above discussed double redundancy outputs, turning these devices highly reliable for sensing applications in healthcare or chemical and biological schemes. The results show how important but also how effective a controlled interaction between semiconductor/electrical components and ionic conductors is. A consistent integration of such devices into cyber-physical systems or as such will be feasible in the near future.

4   Conclusion

With this work we show an approach for a possible integration of ionic conductors (electrolytes) into electronic circuitry. The developed electrolyte-gated transistors (EGTs) based on the semiconductor WO_3 exhibit double functionality (electrical and optical), as the electrochemical reaction of WO_3 not only promotes a change in its coloration, but also a change of conductivity of several orders of magnitude. Characterization goes through electro-optical analysis of the WO_3 film, electrochemical impedance spectroscopy of the electrolyte (a thermosetting composite solid-electrolyte - TCSE) and electrical characterization of the final devices. Two transistor architectures with identical $W/L = 12$ were implemented (planar and vertical) in order to investigate structure influences and self-passivation effects. The best device (vertical) showed an On-Off ratio of almost 7 orders of magnitude (one of the highest ever reported for EGTs) for very low operation voltages ($V_D = 1$ V and V_G between -4 and 4 V) and additional hysteresis reduction. In future works the dynamical switching of such devices will be in focus, as well as testing and optimizing different types of electrolytes.

The presented research paves the way for sophisticated systems, where the analysis of electrolytes and the interaction of these with semiconductors is crucial. These devices could be integrated into applications as diverse as lab-on-chip, wearable electronics, healthcare monitoring and biosensors.

Acknowledgments. This work was supported by the FCT - Portuguese Foundation for Science and Technology, by the project EXCL/CTM-NAN/0201/2012 and by FEDER funds through the COMPETE 2020 Programme and National Funds through FCT under the project UID/CTM/ 50025/2013. The work was also supported by FP7 APPLE (grant number 262782-2) and SMART EC (grant number 258203) projects.

References

1. Kim, S.H., Hong, K., Xie, W., Lee, K.H., Zhang, S., Lodge, T.P., Frisbie, C.D.: Electrolyte-gated transistors for organic and printed electronics. Adv. Mater. **25**, 1822 (2013)
2. Berggren, M., Richter-Dahlfors, A.: Organic bioelectronics. Adv. Mater. **19**, 3201–3213 (2007)

3. Herlogsson, L., Crispin, X., Tierney, S., Berggren, M.: Polyelectrolyte-gated organic complementary circuits operating at low power and voltage. Adv. Mater. **23**, 4684–4689 (2011)

4. Hamedi, M., Herlogsson, L., Crispin, X., Marcilla, R., Berggren, M., Inganas, O.: Fiber-embedded electrolyte-gated field-effect transistors for e-textiles. Adv. Mater. **21**, 573–577 (2009)

5. Granqvist, C.G.: Electrochromic tungsten oxide films: review of progress 1993–1998. Solar Energy Mater. Solar Cells **60**, 201–262 (2000)

6. Natan, M.J., Mallouk, T.E., Wrighton, M.S.: PH-sensitive WO3-based microelectrochemical transistors. J. Phys. Chem. **91**, 648–654 (1987)

7. Granqvist, C.G.: Progress in electrochromics: tungsten oxide revisited. Electrochim. Acta **44**, 3005–3015 (1999)

8. Somani, P.R., Radhakrishnan, S.: Electrochromic materials and devices: present and future. Mater. Chem. Phys. **77**, 117–133 (2003)

9. Azens, A., Hjelm, A., LeBellac, D., Granqvist, C.G., Barczynskab, J., Pentjuss, E., Gabrusenoks, J., Wills, J.M.: Electrochromism of W-oxide-based thin films: recent advances. Solid State Ionics **86**, 943–948 (2007)

10. Barquinha, P., Pereira, S., Pereira, L., Wojcik, P.J., Grey, P., Martins, R., Fortunato, E.: Flexible and transparent WO3 transistor with electrical and optical modulation. Adv. Electron. Mater. **1** (2015)

11. Barsoukov, E., Macdonald, J.R.: Impedance Spectroscopy Theory, Experiment, and Applications. Wiley, New Jersey (2005)

12. Santos, L., Nunes, D., Calmeiro, T., Branquinho, R., Salgueiro, D., Barquinha, P., Pereira, L., Martins, R., Fortunato, E.: Solvothermal synthesis of gallium-indium-zinc-oxide nanoparticles for electrolyte-gated transistors. ACS Appl. Mater. Interfaces **7**, 638–646 (2015)

13. Jovic, V.D., Jovic, B.M.: EIS and differential capacitance measurements onto single crystal faces in different solutions - Part I: Ag(111) in 0.01 M NaCl. J. Electroanal. Chem. **541**, 1–11 (2003)

14. Liu, J., Herlogsson, L., Sawatdee, A., Favia, P., Sandberg, M., Crispin, X., Engquist, I., Berggren, M.: Vertical polyelectrolyte-gated organic field-effect transistors. Appl. Phys. Lett. **97**, 3 (2010)

TCAD Simulation of Amorphous Indium-Gallium-Zinc Oxide Thin-Film Transistors

Jorge Martins[1(✉)], Pedro Barquinha[1], and João Goes[2]

[1] i3N/CENIMAT, Department of Materials Science, Faculty of Science and Technology,
Universidade NOVA de Lisboa and CEMOP/UNINOVA,
Campus de Caparica, 2829-516 Caparica, Portugal
jorge.souto.martins@gmail.com
[2] CTS-UNINOVA, Departamento de Engenharia Electrotécnica,
Faculdade de Ciências e Tecnologia, Universidade Nova de Lisboa,
2825-149 Caparica, Portugal

Abstract. Indium-gallium-zinc oxide (IGZO) thin-film transistors (TFTs) are simulated using TCAD software. Nonlinearities observed in fabricated devices are obtained through simulation and corresponding physical characteristics are further investigated. For small channel length (below 1 μm) TFTs' simulations show short channel effects, namely drain-induced barrier lowering (DIBL), and effectively source-channel barrier is shown to decrease with drain bias. Simulations with increasing shallow donor-like states result in transfer characteristics presenting hump-like behavior as typically observed after gate bias stress. Additionally, dual-gate architecture is simulated, exhibiting threshold voltage modulation by the second gate biasing.

Keywords: IGZO · TCAD simulation · DOS · TFT · CPS

1 Introduction

Oxide TFTs are nowadays starting to play an important role in display industry, with leading companies as LG using them in the backplane of recent products, such as curved organic light emitting device (OLED) TV sets. Still, significant advances in the current state-of-the-art of oxide TFT technology are required before high performance fully transparent, flexible, low-cost and low-power dissipation electronics reach the maturity level required for commercialization. The implementation of Technology Computer-Aided Design (TCAD) [1, 2] tools for simulation of devices at a physical level can lead to understanding on how to further improve fabrication processes, material properties and device architectures. The ultimate goal is to enhance the performance of amorphous oxide TFTs, mimicking the routes followed for single-crystalline CMOS, a-Si and poly-Si technologies. Embedding physically-extracted parameters from the fabricated devices into the models for simulation, accurate models can be built and prediction on how changes in materials and processes affect physical response of the oxide TFTs can be achieved. This can allow viable process and device development and, simultaneously,

© IFIP International Federation for Information Processing 2016
Published by Springer International Publishing Switzerland 2016. All Rights Reserved
L. Camarinha-Matos et al. (Eds.): DoCEIS 2016, IFIP AICT 470, pp. 551–557, 2016.
DOI: 10.1007/978-3-319-31165-4_52

understanding, the otherwise difficult or even impossible to extract physical characteristics behind device operation. Some specific goals to be addressed are referred next.

As performance of amorphous oxide semiconductors (AOS) is controlled by the density of states (DOS) in a given material, and TCAD simulation is based in the DOS [3], it can provide understanding of the specific nature of states causing performance differences among different AOS. This can be used for instance to achieve a suitable and sustainable indium-free replacement for IGZO. Modeled IGZO TFTs based on DOS extracted from different techniques have shown to agree very well with measured characteristics, without employing any fitting parameter [4, 5]. Instabilities in oxide TFTs can be further investigated with TCAD, by implementing charge-trapping and defect creation in the simulations, allowing to understand the specific physical causes of degradation in TFTs, typically associated with Gate/Drain Bias Stress, illumination and temperature. Correlations between the defect-creation and the device performance degradation have been reported [6], with localization of defect creation being also investigated by means of simulations [7]. Some effects that limit device performance are related to device architecture, as parasitic capacitances that limit the maximum operation frequency, and short channel effects (SCE), which limit the scalability of the technology. As architectures can be arbitrarily defined in TCAD simulation optimization and understanding of physical phenomena in different geometries appears as a great advantage. Moreover, as recently more and more TFT architectures are proposed (e.g. dual-gate, floating-capping-metal, bulk-accumulation [8] and dual-active layer TFTs), TCAD can accelerate the optimization of those configurations, reducing fabrication runs and elucidating the physics behind the electrical behavior of each architecture [9].

In this work the short channel effect Drain Induced Barrier Lowering (DIBL) is shown with TCAD simulation in µm-scale IGZO TFTs. To the best of the authors' knowledge it's the first time that the barrier level for different drain bias is displayed for µm-scale oxide TFTs.

2   Integration in Cyber Physical Systems

Amorphous oxide electronics allow for production in a wide range of substrates, notably, inexpensive flexible and/or transparent substrates, as they can be processed at low temperatures (while still maintaining good performance) and have large area uniformity due to their amorphous nature. This way, they can be integrated into a wider range of systems, rather than their silicon technology counterparts. This technology thus shows to have great applicability in Cyber Physical Systems (CPS). In the CPS 5Cs architecture, amorphous oxides can be integrated in the Connection and the Conversion layers. In the Connection level they present the already discussed advantages of being applicable in wide range of substrates, as flexible and transparent ones. Several applications have already been reported as gas sensors [10], touch-panels [11] and even biological sensors [12]. Integrated with the connection level, the technology can also be in the conversion level as they have performance that allow for the processing of the information received at the Connection level. In fact, one of the current trends of oxide TFT technology is precisely focused on full integration and in increasingly more complex circuits, including signal processing, amplification and, ultimately, analog-to-digital conversion [13].

3 TCAD Simulation

In this section some preliminary simulations are presented, addressing specific scenarios in which IGZO TFT devices, as produced in our group, show the need for improvement. Sputtered staggered bottom-gate TFTs on glass have been produced. A 60 nm thick Molybdenum (Mo) gate electrode was sputtered. The dielectric layer is a sputtered 100 nm thick 7 layer multicomponent oxide (alternating layers of SiO_2 and Ta_2O_5). The active layer is a 40 nm thick amorphous IGZO (with 2:1:1 composition in In:Ga:Zn atomic ratio), also sputtered. Finally, 60 nm thick Mo source and drain electrodes have been sputtered. All layers were patterned by photolithography and lift-off with exception of the dielectric layer, in which, photolithography and dry-etching have been used. A schematic of the fabricated devices can be seen in Fig. 1. After all layer fabrication devices were annealed at $180°$ C for 10 min in air in a rapid thermal annealing system. This fabrication process is compatible with commercial polymeric films.

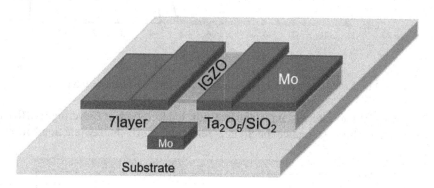

Fig. 1. Schematic of the sputtered IGZO TFTs.

For simulation purposes the TCAD Silvaco's 2D ATLAS^{TM} was used. A similar configuration than that of fabricated devices was employed. An electron mobility of 16 cm^2 (V s)$^{-1}$ and electron concentration of 10^{16} cm^{-3} in the a-IGZO was assumed. Both fabricated and simulated TFTs have a channel length of 20 μm unless otherwise is specified. Channel width of the characterized TFTs is 20 μm and for simulated devices the obtained drain current in 2D simulation is factored in order to correspond to a 20 μm width channel TFT.

3.1 Short Channel Effects

Sputtered IGZO TFTs with channel length (L) below 5 μm have exhibited SCEs, which are well known in silicon technology, where devices are of nm-scale. As low L (5 μm and below) IGZO TFTs have been reported with [14] and without [15] SCEs, TCAD simulation was used to investigate this physical limitation further, as the mechanisms for enabling or not the SCE are not clear yet. Preliminary simulations for similar TFTs showed that the short channel effects occur for L = 0.5 μm. As shown in Fig. 2, turn-on voltage (V_{on}) shifts negatively with increasing drain bias in the simulated transfer

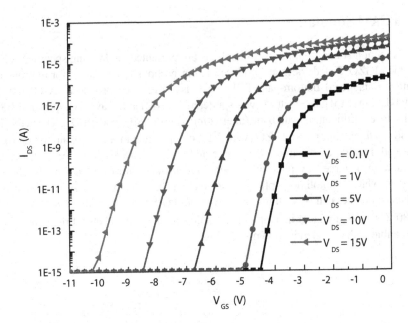

Fig. 2. V$_{On}$ shift with increasing drain bias for simulated L = 0.5 μm.

characteristics. This effect is explained as Drain Induced Barrier Lowering (DIBL), as for low L drain bias can reduce the Source-Semiconductor Schottky junction barrier.

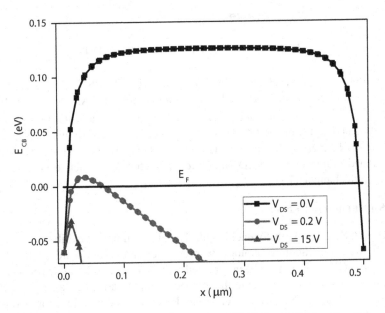

Fig. 3. Conduction band energy of IGZO in respect to the fermi level (E_F) for a L = 0.5 μm IGZO TFT, showing source-semiconductor barrier lowering (at $x = 0$ μm) for increasing drain bias.

By extracting the conduction band energy level across the channel the decreasing of the barrier between source and semiconductor with increasing drain bias is clear, as shown in Fig. 3.

3.2   Hump in Transfer Characteristics

Performance degradation by irradiation, temperature and/or bias stress is one the main challenges for the application of oxide electronics. Fabricated devices showed hump like behavior in transfer characteristics after gate bias stress. This hump has been reported on several scenarios, and many different explanations have been reported so far. Some authors report it as a result of an effect of shallow donor-like states creation in the semiconductor [6]. Figure 4 shows a simulation with increasing peak value of this shallow donor state, assuming 0.1 eV of peak variance and a state energy 0.2 eV below conduction band, since energy states at this level have been reported [16]. The concentration increase in these states leads to the creation of an increasingly more visible hump, suggesting that it might be the (or at least, one of the) creation mechanisms for the hump.

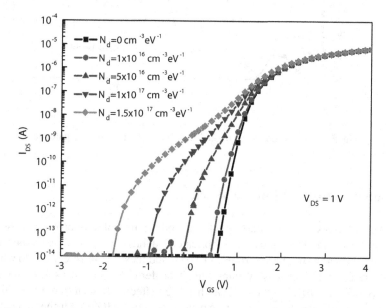

Fig. 4. Increasing of hump-like behavior in transfer characteristics of IGZO TFTs with increasing concentration of shallow donor-like states.

3.3   Dual-Gate TFT

Typically, fabricated IGZO TFT in our research group at CENIMAT/I3 N at FCT NOVA show a depletion mode operation, which complicates its application in circuits. Some architectures for threshold voltage modulation have been already reported, as applying a second-gate on the back-interface of the semiconductor. Figure 5 shows simulated transfer characteristics for dual-gated IGZO TFTs, with different fixed bias on the second gate. The added top gate and top gate oxide for dual-gate simulation are identical to the bottom gate and bottom gate oxide. As expected the V_{on} is modulated, which can be used to achieve enhancement mode devices.

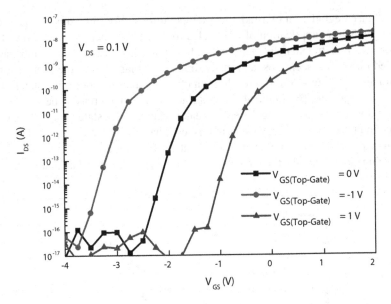

Fig. 5. Modulation of V_{on} in dual-gate IGZO TFT by second gate bias.

4   Conclusions and Future Work

TCAD simulations have demonstrated to be a powerful tool to improve understanding on amorphous oxide TFTs and reduce fabricated batches and costs. Short channel effects have been shown in simulations of IGZO TFTs, and these are in agreement to what was verified in fabricated devices by sputtering. Transfer characteristics shifted with drain bias, explained as DIBL, and the barrier has been effectively seen to reduce with drain bias observing the simulated conduction band energy in the IGZO. Further investigation on the SCEs, such as understanding the role of the Schottky barrier, might elucidate on how to suppress these effects. As for the hump behavior, simulation showed that it may be somehow related to shallow donor-like state creation. Further investigation on how specific DOS affect the device characteristics will lead to understanding the mechanisms of degradation in the oxide TFTs. Finally, simulation of a double-gate TFT was employed, and the V_{on} modulation by the second gate bias was observed. This modulation can be used to achieve enhancement mode TFTs as preferred for application in circuit.

Acknowledgements. This work is funded by FEDER funds through the COMPETE 2020 Programme and National Funds through FCT - Portuguese Foundation for Science and Technology under the Project Nos. UID/CTM/50025/2013 and EXCL/CTM-NAN/0201/2012. The work has also received funding from the European Communities 7th Framework Programme under grant agreement ICT-2013-10-611070 (i-FLEXIS project) and from H2020 program under ICT-03-2014-644631 (ROLL-OUT project).

References

1. DS Software: ATLAS User's Manual, pp. 567–1000 (2013)
2. CV Started G: Sentaurus Device User Guide. Synopsis (2009)
3. Hsieh, H.-H., Kamiya, T., Nomura, K., Hosono, H., Wu, C.-C.: Modeling of amorphous InGaZnO4 thin film transistors and their subgap density of states. Appl. Phys. Lett. **92**, 133503 (2008)
4. Jeon, K., Kim, C., Song, I., Park, J., Kim, S., Kim, S., Park, Y., Park, J.H., Lee, S., Kim, D.M., Kim, D.H.: Modeling of amorphous InGaZnO thin-film transistors based on the density of states extracted from the optical response of capacitance-voltage characteristics. Appl. Phys. Lett. **93**, 4–7 (2008)
5. Junfei Shi, Chengyuan Dong, Y.S.: Influence of Deep States in Active Layers on the Performance of Amorphous In-Ga-Zn-O Thin Film Transistors Junfei Shi, Chengyuan Dong and Yikai Su. Proc. China Display/Asia Disp. 2011. 356–359 (2011)
6. Im, H., Song, H., Jeong, J., Hong, Y., Hong, Y.: Effects of defect creation on bidirectional behavior with hump characteristics of InGaZnO TFTs under bias and thermal stress. Jpn. J. Appl. Phys. **54**, 153–156 (2015)
7. Lee, J., Choi, S., Kim, S.K., Choi, S., Kim, D.H., Park, J., Kim, D.M.: Modeling and characterization of the abnormal transistors after high positive bias stress. IEEE Electron Device Lett. **36**, 1047–1049 (2015)
8. Mativenga, M., An, S., Jang, J.: Bulk accumulation a-IGZO TFT for high current and turn-on voltage uniformity. IEEE Electron Device Lett. **34**, 1533–1535 (2013)
9. Lu, T., Chen, W., Zan, H., Ker, M.: Investigating electron depletion effect in amorphous indium–gallium– zinc-oxide thin-film transistor with a floating capping metal by technology computer-aided design simulation and leakage reduction. Jpn. J. Appl. Phys. **53**, 064302 (2014)
10. Chen, K.-L., Jiang, G.-J., Chang, K.-W., Chen, J.-H., Wu, C.-H.: Gas sensing properties of indium–gallium–zinc–oxide gas sensors in different light intensity. Anal. Chem. Res. **4**, 8–12 (2015)
11. Fu, R., Liao, C., Leng, C., Zhang, S.: An IGZO TFT based in-cell capacitance touch sensor. In: 2012 IEEE 11th International Conference on Solid-State and Integrated Circuit Technology (ICSICT). IEEE (2012)
12. Shen, Y.-C., Yang, C.-H., Chen, S.-W., Wu, S.-H., Yang, T.-L., Huang, J.-J.: IGZO thin film transistor biosensors functionalized with ZnO nanorods and antibodies. Biosens. Bioelectron. **54**, 306–310 (2014)
13. Correia, A.P.P., Barquinha, P., da Palma Goes, J.C.: A Second-Order ΣΔ ADC Using Sputtered IGZO TFTs. Springer International Publishing, Heidelberg (2015)
14. Chen, R., Zhou, W., Zhang, M., Wong, M., Kwok, H.S.: Self-aligned top-gate InGaZnO thin film transistors using SiO2/Al2O3 stack gate dielectric. Thin Solid Films **548**, 572–575 (2013)
15. Park, J.-S.: Characteristics of short-channel amorphous In-Ga-Zn-O thin film transistors and their circuit performance as a load inverter. J. Electroceramics. **28**, 74–79 (2012)
16. Oba, F., Nishitani, S.R., Isotani, S., Adachi, H., Tanaka, I.: Energetics of native defects in ZnO. J. Appl. Phys. **90**, 3665 (2001)

Electrochemically Gated Graphene Field-Effect Transistor for Extracellular Cell Signal Recording

Sanaz Asgarifar[1,2(✉)], Henrique L. Gomes[1,2], Ana Mestre[1,2], Pedro Inácio[1,2], J. Bragança[1], Jérôme Borme[3], George Machado Jr.[3,4], Fátima Cerqueira[3,4], and Pedro Alpuim[3,4]

[1] FCT, Universidade do Algarve, Faro, Portugal
sanaz.asgarifar@gmail.com
[2] IT-Instituto de Telecomunicações, Av. Rovisco, Pais, 1, Lisbon, Portugal
[3] INL – International Iberian Nanotechnology Laboratory,
Av. Mestre José Veiga, Braga, Portugal
[4] CFUM – Centre of Physics of the University of Minho,
Campus de Gualtar, 4710-057 Braga, Portugal

Abstract. This work presents an experimental characterization of electrochemically gated graphene field-effect transistors (EGFETs) to measure extracellular cell signals. The performance of the EGFETs was evaluated using cardiomyocytes cells. Extracellular signals with a peak value of 0.4 pico-amperes (pA) embedded in a noise level of 0.1 pA were recorded. Signals in current mode were compared with signals recorded as a voltage. Signals below 28 µV of magnitude can be detected in a noise floor of 7 µV with a signal-to-noise ratio of 4.

Keywords: Graphene · Field effect transistor · Extra-cellular cell signal recording

1 Introduction

Bioelectronic devices fabricated with emergent materials are currently being developed to establish electrical interfaces with living cells and tissues. The aim is to develop transducers that can record extracellular signals in complex biological environments [1–9]. In contrast with conventional semiconductors such as silicon some of these materials can be deposited into flexible, conformable and biocompatible substrates. Graphene deposited in thin layers is particular interesting for implantable biomedical devices because it offers a high electronic performance combined with biocompatibility and processing on a variety of flexible substrates [10–14].

The recent efforts in improving the fabrication of thin-layers of graphene have made possible to fabricate bioelectronic sensors know as electrochemically-gated field effect transistor (EGFETs). Although, these are field effect devices, unlike conventional transistors do not have a built-in dielectric layer. The gate dielectric is established when the device is immersed into the electrolyte solution. This occurs because when conductive or semiconductive materials are immersed into electrolytes a Helmholtz capacitive

L. Camarinha-Matos et al. (Eds.): DoCEIS 2016, IFIP AICT 470, pp. 558–564, 2016.
DOI: 10.1007/978-3-319-31165-4_53

double-layer is established at the material/electrolyte interface. This type of device has been used to record extracellular electrophysiological signals from cells [15, 16].

In this paper, we show that a field effect transistor based in electrochemically graphene gate can be used for extracellular signal recording. The aim is to demonstrate the detection limit of these sensing devices. The performance of the EGFETs was evaluated using contractile cells (cardiomyocytes).

This paper is organized as follows: first the measuring system is presented. Then the basic working principle of EGFETs is explained, including the fabrication and characterization. Finally, extracellular signals recorded in current as well as in voltage mode are presented and discussed.

2   Technological Innovation for Cyber-Physical Systems

Brain-related illnesses affect more than two billion people worldwide. Advances in treatments for brain disorders have to date relied largely upon a pharmaceutical approach, however the development of drugs, which do not have intolerable side effects, is becoming extremely complex and difficult. It is now believed that an electronic engineering-driven approach is needed, to develop solutions based on electrical signals. This is supported by a number of progresses in electronic transducers working as prosthetic and electroceutical devices. These are devices that aim to establish an electrical and chemical bidirectional communication interface with cells and tissues. These devices measure the signals so that researchers can develop a 'dictionary' of patterns associated with health and disease states. Once the signals are decoded, devices can also generate the correct signal patterns to modulate the neural impulses controlling the body, repair lost function and restore health. Field effect transistors based on graphene reported in this contribution are an interesting approach towards the development of these implantable brain-machine interfaces.

3   Methodology

3.1   CVD Growth and Graphene Field Effect Transistor Fabrication and Mouse Embryonic Stem Cells ESC Differentiation

Graphene has been grown with CVD on copper (Cu) catalyst with 99.99 % purity. The CVD graphene used in this experiment was grown on top of 25 μm copper foils (Alfa Aesar, 99.999 % purity). The copper foil was introduced into a thermal CVD system (EasyTube 3000EXT) and annealed at 1020°C for 20 min. in 0.5 Torr hydrogen atmosphere, in order to clean the surface from copper oxides and promote copper grain growth. After that, methane was introduced for 30 min. under a flow ratio $H_2:CH_4$ of 6:1 while keeping the same total pressure, to act as carbon source for the graphene growth. Graphene was then transferred to a final substrate using the copper dissolution method. First, oxygen plasma is used to remove graphene from the bottom side of the copper foil. The other side of the copper is then spin-coated with poly (methyl-2-methylpropenoate) (PMMA, AR-P 679.04) at 1000 rpm, followed by a bake at 80°C for 8 min., leading to a nominal PMMA thickness of 630 nm, sufficient to allow manipulation of the membrane without breaking the graphene layer. The copper foil was then dip into a $FeCl_3$ solution until copper was entirely dissolved. The remaining graphene/PMMA membrane, floating in the solution, was then cleaned in an HCl solution to eliminate iron precipitates and in DI water. At this step grapheme could be transferred onto a final substrate. This substrate was previously prepared by sputter depositing Cr 3 nm/Au 30 nm onto an oxidized silicon substrate, followed by optical lithography and ion milling. Finally, a 320 nm layer of Al_2O_3 was deposited on the substrate by lift-off to provide electrical insulation between the current lines and the electrolyte and a chemical protection. After the graphene/PMMA membrane was transferred onto these substrate from the solution, the substrates were dried for 12 h at 120°C in an oven to enhance graphene adhesion. The PMMA layer then was dissolved in acetone. In order to provide spatial resolution to the cell experiments, each array sample contains several transistors. The photograph in Fig. 1a shows one half of a 2 × 2 transistor array. The scheme in Fig. 1b shows a cross section view of the device. In our experiments, typical transistors have a channel length (L) of 25 μm and a width (W) of 75 μm.

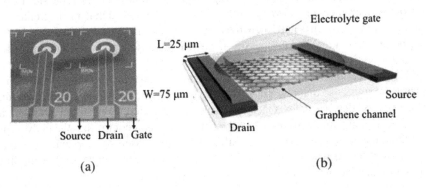

(a)   (b)

Fig. 1. (a) Photograph of two EGFETs on a silicon wafer, (b) schematic diagram showing the drian and source dimensions.

Mouse embryonic stem cells (ESC) differentiated into cardiomyocytes were used. ESC differentiation was performed using the hanging-drop method in medium containing 20 % of fetal bovine serum without Leukemia inhibitory factor (LIF) supplementation (differentiation medium). Briefly, 1000 cells were cultured in 20 μl hanging drops of differentiation medium for 48 h to initiate embryonic bodies (EB) formation. Next, EB were grown in differentiation medium in suspension, for 3 days, in a bacterial petri dish before being transferred to the devices (0.1 %) to allow cell attachment and further differentiation. Cells were kept alive for two days by changing half of the medium every 24 h. The sensing devices with cells were maintained at 37°C in an incubator (HERACell®150) with a humidified atmosphere with 5 % of CO_2. A photograph showing the cells on top of a recording device is shown in Fig. 2b.

All electrical measurements were performed with a Stanford low-noise current amplifier (SRS 570) or alternatively in voltage mode using a voltage amplifier (SRS 560) connected to a dynamic signal analyser (Agilent 35670A).

3.2   Interfacing of Cells with Graphene Based EGFETs

Figure 2a shows a schematic diagram of the EGFET and the electrical connections. A embryonic body of cardiomyocyte cells was placed on top of the gate terminal (Fig. 2b). Cell activity was measured without biasing the EGFET. No voltages were applied between the drain and source terminals. This measuring method avoids the risk of electrochemical reactions resulting from the use of DC voltages in a liquid environment. The extracellular signals can be detected as a current or as a voltage between the drain and source terminals.

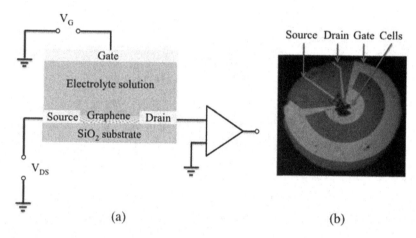

(a)   (b)

Fig. 2. (a) Schematic diagram showing the electrical connections to the EGFET. Low noise amplifiers were used to record the signals in current as well as in voltage, (b) Photograph showing a cluster of cardiomyocytes on top of the gate terminal.

4   Results and Discussion

A detail view of the signals in current and in voltage is show in Fig. 3. The spikes in current reach amplitude of 0.4 pA (peak-to-peak) and voltage signals reach values of 28 μV (peak-to-peak) in a noise level of 7 μV. In voltage amplification the signal-to-noise ratio (SNR) reaches 4, a value approximately near the value of 5.6 obtained using current amplification.

Figure 4 shows the thermal noise measured in current of the graphene based EGFET without cells. The noise is represented as a power spectral density (PSD) and it shows a flat noise behavior up to 10 Hz. This noise basically determines the device sensitivity. The origin of noise is not clear yet. Possible sources of noise include charge trapping at the interface to the substrate as well as the so-called charge noise [17, 18].

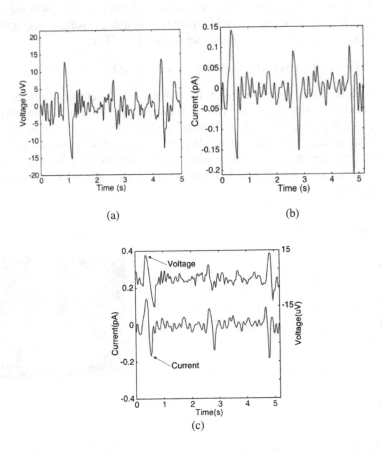

Fig. 3. Extracellular signals measured as a function of time in (a) voltage mode, (b) current mode, and (c) comparison between voltage and current signals. The voltage signal was shifted along the voltage axis.

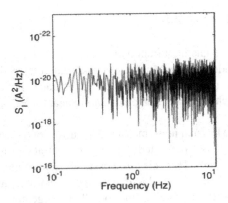

Fig. 4. The noise spectral density (PSD) of the graphene based EGFET without cells.

5 Conclusion

This paper reports on the use of a graphene based EGFET to record signals from contractile cells. During electrical measurements the transistor was not biased. In this operation mode the device is working as a simple microelectrode system. This strategy minimizes the electrical noise and the risk of electrochemical reactions. It allows us to explore the detection limits of this type of device. The noise level in current is below 1 pA and in current is below 10 μV. The signal-to noise ratio is approximately 5. This compares very well with the performance of current available microelectrode array technology to measure extracellular signals.

The measured signals are very weak. For contractile cells we would expect a much stronger signals. Among the possible reasons for these faint signals is a bad electrical coupling between the cells and the device. Clusters of cardiomyocytes do not adhere to the device surface. It is well possible that they float above the sensing surface.

References

1. Fromherz, P., Offenhausser, A., Vetter, J., Weis, T.: A neuron-silicon junction: a Retzius cell of the leech on an insulated-gate field-effect transistor. Science **252**, 1290 (1991)
2. Timko, B.P., Cohen-Karni, T., Qing, Q., Tian, B., Lieber, C.M.: Design and implementation of functional nanoelectronic interfaces with biomolecules, cells, and tissue using nanowire device arrays. IEEE Trans. Nanotechnol. **9**(3), 269–280 (2010)
3. Kotov, N.A., Winter, J.O., Clements, I.P., Jan, E., Timko, B.P., Campidelli, S., Pathak, S., Mazzatenta, A., Lieber, C.M., Prato, M., Bellamkonda, R.V., Silva, G.A., Kam, N.W.S., Patolsky, F., Ballerini, L.: Nanomaterials for neural interfaces. Adv. Mater. **21**(40), 3970–4004 (2009)
4. Patolsky, F., Timko, B.P., Zheng, G.F., Lieber, C.M.: Nanowire-based nanoelectronic devices in the life sciences. MRS Bull. **32**, 142–149 (2007)
5. Patolsky, F., Timko, B.P., Yu, G., Fang, Y., Greytak, A.B., Zheng, G., Lieber, C.M.: Detection, stimulation, and inhibition of neuronal signals with high-density nanowire transistor arrays. Science **313**, 1100–1104 (2006)
6. Timko, B.P., Cohen-Karni, T., Yu, G., Qing, Q., Tian, B., Lieber, C.M.: Electrical recording from hearts with flexible nanowire device arrays. Nano Lett. **9**, 914–918 (2009)
7. Pui, T.-S., Agarwal, A., Ye, F., Balasubramanian, N., Chen, P.S., Eschermann, J.F., Stockmann, R., Hueske, M., Vu, X.T., Ingebrandt, S., Offenhäusser, A.: CMOS-compatible nanowire sensor arrays for detection of cellular bioelectricity. Appl. Phys. Lett. **95**, 083703-1–083703-3 (2009)
8. Cohen-Karni, T., Timko, B.P., Weiss, L.E., Lieber, C.M.: Flexible electrical recordings from cells using nanowires transistor arrays. Proc. Natl. Acad. Sci. U.S.A. **106**, 7309–7313 (2009)
9. Qing, Q., Pal, S.K., Tian, B., Duan, X., Timko, B.P., Cohen-Karni, T., Murthy, V.N., Lieber, C.M.: Nanowire transistor arrays for mapping neural circuits in acute brain slices. Proc. Natl. Acad. Sci. U.S.A. **107**, 1882–1887 (2010). doi:10.1073/pnas.0914737107
10. Geim, A.K., Novoselov, K.: The rise of graphene. Nat. Mater. **6**, 183–191 (2007)

11. Anteroinen, J., Kim, W., Stadius, K., Riikonen, J., Lipsanen, H., Ryynanen, J.: Extraction of graphene-titanium contact resistances using transfer length measurement and a curve-fit method. World Acad. Sci. Eng. Technol. **6**, 08–25 (2012)

12. Stieglitz, T., Beutel, H., Schuettler, M., Meyer, J.U.: Biomed: micromachined, polyimide-based devices for flexible neural interfaces. Microdevices **2**, 283 (2000)

13. Hess, L.H., Jansen, M., Maybeck, V., Hauf, M.V., Seifert, M., Stutzmann, M., Sharp, I.D., Offenhñusser, A., Garrido, J.A.: Graphene transistor arrays for recording action potentials from electrogenic cells. Adv. Mater. **23**, 5045–5049 (2011)

14. Cohen-karni, T., Li, Q., Fang, Y., Lieber, C.M.: Graphene and nanowire transistors for cellular interfaces and electrical recording. Nano Lett. **10**, 1098 (2010)

15. Mohanty, N., Berry, V.: Graphene-based single-bacterium resolution biodevice and dna transistor: interfacing graphene derivatives with nanoscale and microscale biocomponents. Nano Lett. **8**, 4469–4476 (2008)

16. Nguyen, P., Berry, V.: Graphene interfaced with biological cells: opportunities and challenges. J. Phys. Chem. Lett. **3**, 1024–1029 (2012)

17. Hess, L.H., Seifert, M., Garrido, J.A.: Graphene transistors for bioelectronics. Proc. IEEE **101**(7), 1780–1792 (2011)

18. Heller, I., Chatoor, S., Männik, J., Zevenbergen, M.A.G., Oostinga, J.B., Morpurgo, A.F., Dekker, C., Lemay, S.G.: Charge noise in graphene transistors. Nano Lett. **10**(5), 1563–1567 (2010)

Author Index

Printed in the United States
By Bookmasters